INTRODUCTION TO ELECTRONICS
DC/AC CIRCUITS

INTRODUCTION TO ELECTRONICS
DC/AC CIRCUITS

Stephen C. Harsany
Mt. San Antonio College

Prentice Hall
Upper Saddle River, New Jersey   Columbus, Ohio

Library of Congress Cataloging-in-Publication Data

Harsany, Stephen C.
   Introduction to electronics: DC/AC circuits/Stephen C. Harsany.
      p.   cm.
   ISBN 0-13-359795-4
   1. Electronic circuits.   2. Electric circuit analysis.   I. Title.
TK7867.H366 2000
621.3815—dc21   99-12345
                                                 CIP

Cover art: Thomas Mack
Editor: Scott Sambucci
Production Editor: Stephen C. Robb
Design Coordinator: Karrie Converse-Jones
Text Designer: John Edeen
Cover Designer: Thomas Mack
Production Manager: Patricia A. Tonneman
Illustrations: York Graphic Services, Inc.
Marketing Manager: Ben Leonard

This book was set in Times Roman and Officina Sans by York
Graphic Services, Inc. and was printed and bound by World
Color Press, Inc. The cover was printed by Phoenix Color Corp.

Electronics Workbench® (EWB) is a registered trademark of
Interactive Image Technologies, Ltd., Toronto, ON, Canada.

The Sharp EL-506L Calculator is a product of Sharp Electronics
Corp., Mahwah, NJ.

The publisher does not endorse or warrant the products described
herein, nor does the publisher accept liability resulting from use
or misuse of those products.

Printed in the United States of America

10 9 8 7 6 5 4 3 2 1

ISBN: 0-13-359795-4

Prentice-Hall International (UK) Limited, *London*
Prentice-Hall of Australia Pty. Limited, *Sydney*
Prentice-Hall Canada, Inc., *Toronto*
Prentice-Hall Hispanoamericana, S. A., *Mexico*
Prentice-Hall of India Private Limited, *New Delhi*
Prentice-Hall of Japan, Inc., *Tokyo*
Prentice-Hall (Singapore) Pte. Ltd., *Singapore*
Editora Prentice-Hall do Brasil, Ltda., *Rio de Janeiro*

This text is dedicated to the three most important women in my life: my late mother, Mary, the guiding light of the family; my wife, Claudette, my partner in all things; and my daughter, Andrea, our inspiration.

PREFACE

Introduction to Electronics: DC/AC Circuits is intended for students who are taking their first course in electronics. The text is written for the beginning student and assumes no prior knowledge of electronics. The math prerequisites of the text are basic algebra with some knowledge of right-angle trigonometry. The content is suitable for community colleges, technical institutes, and beginning electronics-engineering technology programs.

The text provides thorough, comprehensive, and practical coverage of basic electrical and electronic concepts that include DC circuits, magnetism, and AC circuits. This includes circuit analysis problems (DC network theorems and AC complex number problems). Problem solving, applications, and troubleshooting receive special emphasis.

With 26 chapters in all, it can be used for either a one- or two-semester course. The pedagogy features a straightforward, matter-of-fact writing style that is easy to understand and student-friendly. Attention to detail is provided throughout. Topics receiving expansive coverage are new component-identification notations (for resistors and capacitors), additional components, rechargeable batteries, superconductivity, additional types of variable resistance, and AC test instruments (including both analog and digital scopes and scope probes).

Most basic electronic texts use the same circuit-analysis techniques. These techniques have been standard for years. This text covers these as well, but, in addition, it offers a fresh perspective through the optional use of a specific calculator readily available for less than \$20 from Sharp Electronics. Using this calculator, students find that both DC and AC circuit-analysis problems are easier to master. This is especially true in AC complex-number problems, where the Sharp calculator has a complex-number mode that allows either rectangular or polar numbers to be added, subtracted, multiplied, or divided directly. There is no need to convert from one notation system to the other when solving a problem.

Circuit-analysis problems (and, in particular, AC problems) have always been difficult for students. Using the Sharp calculator means that students can be taught to solve these problems using the same rules that were learned in DC analysis, thus reinforcing what they already know.

The Sharp calculator is one member of the EL-506 family of calculators that began with the EL-506A model. At the time of publication, the Sharp EL-506L version is the current model. The calculator keystrokes shown for problems in the text correspond to this model. As with all calculators, new models are periodically introduced. By the time this text is in the bookstores there will be available a new model, the Sharp EL-506R. This calculator uses keystrokes virtually identical to those referenced for the EL-506L model, although there is a minor difference in keystrokes for complex number problems.

Sharp has continuously demonstrated a commitment to education through their consistent efforts to improve the functions and features of the EL-506 calculator. All major features and functions have been retained with every model. This product family will doubtless evolve to the benefit of students.

The Sharp calculator *is not required* for use with the text. However, because of its built-in provisions for simultaneous equations and complex-number arithmetic, it naturally meets the requirements for easier DC/AC circuit analysis. Any good scientific calculator—and, in particular, a programmable calculator—will also meet the requirements of the text.

CHAPTER ORGANIZATION

Each chapter is organized with the following instructional elements:

- Performance-based objectives
- Introduction
- Numbered sections with accompanying self-checks (answers are given at the end of the chapter for immediate feedback)
- Numerous worked examples and illustrations with practice problems (again, answers appear at the end of the chapter)
- Summary
- List of important equations
- Review questions
- Critical thinking questions
- Problems
- Answers to practice problems
- Answers to self-checks

An innovative feature is that circuits from the text that are emphasized in the accompanying lab manual are referenced or keyed to one another. Electronics Workbench® (EWB) software has been used to check all lab assignments. All labs are provided on disk for use with EWB, and more than 100 circuits from the text are included on the enclosed CD-ROM. The CD-ROM also contains both a free demonstration version of EWB and a passcode-protected student version of EWB. The passcode to activate the full student version can be purchased by contacting Interactive Image Technologies at 1-800-263-5552, or by visiting IIT's web site, www.interactiv.com. The lab manual provides additional exercises for use with EWB. These labs may contain parts that are not readily available, are too costly, or contain circuits with component faults. Thus, these circuits can be simulated rather than actually breadboarded and tested, whereas other circuits can be built and used to reinforce basic skills or additional troubleshooting skills.

END-OF-TEXT MATERIAL

The following supplements are provided at the end of the text:

- Appendices
  Periodic Table of the Elements
  Selected Schematic Symbols for DC/AC Circuits
  Glossary of Selected DC/AC Terms
  Answers to Odd-Numbered Problems
- Index

ANCILLARIES PACKAGE

The following supplements are available to adopters of *Introduction to Electronics: DC/AC Circuits:*

- *Instructor's Resource Manual.* Written by the text author, this manual contains solutions to all even-numbered end-of-chapter problems, solutions to the lab exercises, a test item file, and sample syllabi and instructor's resources.
- *Laboratory Manual.* Cowritten by the text author and a colleague, the manual contains 50 exercises. The manual has been field-tested by students and simulated using Electronics Workbench software.
- *Prentice Hall Custom Test.* This computer-based testing software allows the instructor to prepare tests electronically from the Test Item File, as well as add custom material as needed.

- *PowerPoint® Slides.* More than 100 figures and graphs from the text are provided for multimedia presentations via PowerPoint.
- *Electronics Workbench Circuits.* More than 100 circuits from the text are built into EWB and packaged on the enclosed CD-ROM.
- *Companion Website.* Several forms of questions are given for each chapter that the student can answer and submit for checking. Real-time feedback of the correct answers is provided. Students can use the Website (www.prenhall.com/harsany) as an on-line study guide/tutorial.

A NOTE TO THE STUDENT

This text aims to provide you with the basics upon which to build your electronics understanding and expertise. Make sure that you answer the Self-Check and Practice Problems, because their answers are at the end of the chapter. This gives you immediate feedback so that you can judge how well you are absorbing the material. In addition, if you have access to the Internet, you should use the companion website (www.prenhall.com/harsany) as an on-line study guide/tutorial.

As you are well aware, the world has moved into an information age in which almost everything around us is controlled or affected by electronic devices or circuits. You made the right choice in wanting to expand your knowledge base into the electronics field. The field is growing, particularly in the areas of telecommunications, microwave, integrated-circuit (IC) fabrication, and related computer skills (repair, networking, etc.). Engineers are in great demand to design the circuits and systems needed for the future. The future certainly will be influenced by electronics. Study well!

ACKNOWLEDGMENTS

I gratefully acknowledge the following reviewers for their insightful suggestions: Tim Beecher, Wisconsin Indianhead Technical Institute; Bruce Bush, Albuquerque Technical Vocational Institute; Mauro Caputi, Hofstra University; Dr. William Croft, Indiana State University; Thomas Diskin, College of San Mateo; Doug Fuller, Humber College; Robert Martin, Northern Virginia Community College; Waldo Meeks, Georgia Southern University; Dr. Robert Powell, Oakland Community College; J. W. Roberts, West Georgia Technical Institute; Allen Sanderlin, York Technical College; Ken Simpson, Stark State College of Technology; Patrick Thomason, Patterson State Technical College; and Larry Wheeler, GE American Communications.

I also thank all the people at Prentice Hall who helped in the publication of this text. They include:

Dave Garza, vice president/publisher
Scott Sambucci, acquisitions editor
Steve Robb, production editor
Linda Thompson, copy editor

I extend a special thanks to Harry M. Smith, a colleague, for his help with the manuscript and with the lab manual, and to Thanut Lanlua, my student assistant, for help in checking the answers to text problems.

Finally, thanks once again to my wife, Claudette, for her patience and support while I spent so much time in the research and writing of the text.

Stephen C. Harsany
Mt. San Antonio College

CONTENTS

PART **III**

Magnetism and Magnetic Devices   239

PART **IV**

AC and Transformers   277

PART V

Capacitors, Inductors, and Transient Response 339

PART I

FUNDAMENTALS OF ELECTRICITY/ELECTRONICS

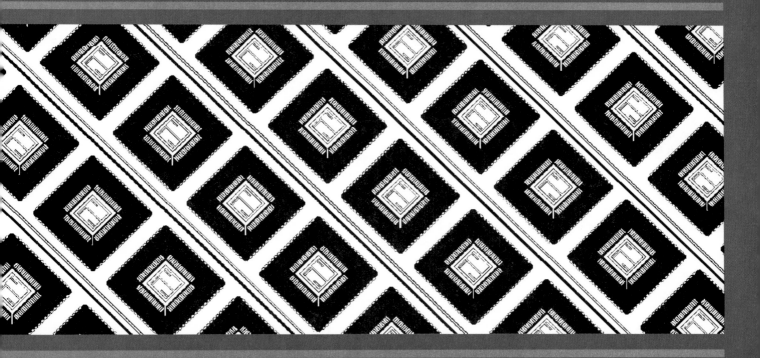

CHAPTER 1

SURVEY OF ELECTRONICS

OBJECTIVES

After completing this chapter, you will be able to:

1. State important historical events that led to the development of the field of electronics.
2. Define common electrical terms associated with the field of electronics.
3. Differentiate between the terms *electronics* and *electronic technology*.
4. Compare and contrast various electronic specialty fields.
5. Compare and contrast various electronic job descriptions.

INTRODUCTION

Electricity is a form of energy, a phenomenon that is a result of the existence of electrical charge. The theory of electricity and its inseparable effect, magnetism, is probably the most accurate and complete of all scientific theories. The understanding of electricity has led to the invention of motors, generators, telephones, radio and television, X-ray devices, computers, satellite communications, and nuclear energy systems, to name a few. Electricity is a necessity to modern civilization.

1.1   HISTORY OF ELECTRICITY/ELECTRONICS

The history of the electric charge can be traced to the Greeks around 600 B.C. The Greeks were aware of a peculiar property of amber, a yellowish, translucent mineral. When rubbed with a piece of fur, amber develops the ability to attract small pieces of material such as feathers. For centuries this strange, inexplicable property was thought to be unique to amber.

In 1600, William Gilbert (1544–1603), an English physician, proved that many other substances are electric (from the Greek word for amber, *elektron*) and that they have two electrical effects. When rubbed with fur, amber acquires resinous electricity; glass, however, when rubbed with silk, acquires vitreous electricity.

In 1747, Benjamin Franklin (1706–1790) in America and William Watson (1715–1787) in England independently reached the same conclusion; all materials possess a *single* kind of electrical "fluid" that can penetrate matter freely but can be neither created nor destroyed. The action of rubbing merely transfers the fluid from one body to another, electrifying both.

Franklin defined the fluid, which corresponds to vitreous electricity, as positive and the lack of fluid as negative. Therefore, according to Franklin, the direction of current flow is from positive to negative—the opposite of what is now known to be true in conductors. In conductors, current flow is accomplished via electron flow from negative to positive. (This text utilizes electron-flow theory. Many 2-year electronic programs use this type of current flow as well.) Franklin's effect is known as *conventional current flow* and is used in other texts, particularly engineering-level texts. Further details of conventional current flow are given in Section 4.4.

About 1766 the law that the force between electric charges varies inversely with the square of the distance between the charges was proved experimentally by British chemist Joseph Priestley (1733–1804). Priestley also demonstrated that an electric charge distributes itself uniformly over the surface of a hollow metal sphere and that no charge and no electric field exist within such a sphere. Charles Coulomb (1736–1806), a French physicist, perfected the measurement of the force exerted by electrical charges. The unit of electrical charge, the *coulomb,* is named in honor of him. He developed the laws of attraction and repulsion between charged bodies. This law (called Coulomb's law) is covered in Chapter 3.

The galvanometer, which is used to measure electrical current, was named after the Italian Luigi Galvani (1737–1798), who conducted many experiments with electrical current. Another Italian, Alessandro Volta (1745–1827), is famous for inventing the electric battery. The first battery was formed in 1800 by stacking plates made of zinc and copper (or silver) separated by felt soaked in brine. The earliest major market for these batteries was in telegraph systems in the 1830s. A battery separates electrical charge by chemical means. A battery can affect electrical charges by forcing them through an electric circuit. The ability to do work by electrical means is measured by the *volt,* named for Volta. The electrical ability of a battery to do work is called the electromotive force, or emf.

An electric charge in motion (electrodynamics) is called electric current. The unit of electrical current is the *ampere,* named in honor of the French physicist André-Marie Ampère (1775–1836).

In 1819, a Danish physicist, Hans C. Oersted (1777–1851), realized that electricity and magnetism were related. He found that a compass needle was affected by a current-carrying wire. The unit *oersted* was adopted for the unit of magnetic field strength in honor of him.

Ohm's law, the best known law for electrical circuits, was formulated by Georg S. Ohm (1787–1854), a German physicist, in 1826. He demonstrated the relationship between current, voltage, and wire area and length (the wire's resistance). The unit of resistance called the *ohm* (Greek letter Ω) was named in honor of his accomplishments.

In 1831, Michael Faraday (1791–1867), an English physicist, further explored Oersted's work. These findings are today referred to as Faraday's laws of electromagnetic induction. Faraday also investigated static electricity; in this area it is an acknowledgement of his work that the unit of capacitance is named the *farad*.

Heinrich Lenz (1804–1865), a German scientist, extended Faraday's findings and found that the current induced in a conductor is such that it opposes the change in the magnetic field producing it. This is known today as Lenz's law.

Joseph Henry (1797–1878), an American physicist, also conducted studies into electromagnetism. In recognition of his discovery of self-inductance in 1832, the unit of inductance is called the *henry*.

James C. Maxwell (1831–1879), a Scottish physicist, translated Faraday's experiments into mathematical notation. These became known as Maxwell's equations and are used to show the relationship between electricity and magnetism. He also proposed that electromagnetic waves and light are identical. His work paved the way for the German physicist Heinrich Hertz (1857–1894), who produced and detected electric waves in the atmosphere in 1886. In honor of his work in this field, the unit of frequency is called the *hertz*.

In 1896 the Italian engineer Guglielmo Marconi (1874–1937) harnessed these waves to produce the first practical radio signaling system. His work originated in "wireless" telegraphy. He is generally credited with being the inventor of radio.

The electron theory, which is the basis of modern electrical theory, was first advanced by Dutch physicist Hendrik Lorentz (1853–1928) in 1892. Sir Joseph Thomson (1856–1940), a British physicist, is considered to be the discoverer of the electron through his experiments on the stream of particles (electrons) emitted by cathode rays. The charge of the electron was first accurately measured by the American physicist Robert Millikan (1868–1953) in 1909.

Thomas Edison (1847–1931), the famous American inventor, developed a practical electric lightbulb, an electrical generating system, a sound-recording device, and a motion picture projector that had profound effects on the shaping of modern society. Edison promoted a direct current (DC) power-distribution system. His use of DC, however, lost out to the alternating current (AC) system developed by American inventors Nikola Tesla (1856–1943) and George Westinghouse (1846–1914). (DC refers to a constant level of current flow that travels in one direction only within a circuit. AC refers to current that changes magnitude and direction. Both DC and AC, among other things, are the subject of this text.)

In 1904 John Fleming (1849–1945), a British engineer, developed a two-electrode radio *rectifier,* the *Fleming valve,* patterned after the Edison effect. His invention could not boost or amplify the radio signal. That development was made by Lee De Forest (1873–1961), an American inventor. His "audion" was a type of vacuum tube—the three-electrode thermionic tube, or *triode.* This tube, devised in 1906, revolutionized the entire field of electronics. Modern electronics began with this invention, a device that could store and amplify electrical charges and signals respectively.

The father of television, Vladimir Zworykin (1889–1982), an American physicist and engineer, developed the television camera and picture tube in 1923.

During World War II there was a need for very high frequency (*microwave*) vacuum tubes. This was especially true for military radar. British physicists Henry Boot (1917–1983) and John Randall (1905–1984) invented an electron tube called the *magnetron* in 1939. At the same time, the Varian brothers, Russell (1898–1959) and Sigurd (1901–1961) invented the *klystron.* In 1943, the *traveling wave* tube was invented by Rudolf Komphner. These devices are still used in radar and communication circuits (particularly satellite communications).

In 1946 J. Presper Eckert and John Mauchly unveiled ENIAC, the world's first all-electronic computer. ENIAC (Electronic Numerical Integrator and Computer) contained 18,000 vacuum tubes and had a speed of several hundred multiplications per minute. During this time it was thought that seven ENIACs would be all the world would ever need to perform any required computations.

In 1947 Walter Brattain, William Shockley, and John Bardeen, at Bell Laboratories, invented the *transistor*. This device performed most of the functions of the vacuum tube and ushered in the era of *solid-state electronics*.

In 1959 Robert Noyce and others took part in the development of the *integrated circuit* (IC), which incorporated many transistors and other components on a small chip of semiconductor material.

Ted Hoff of Intel Corporation designed the *microprocessor* in 1971. The microprocessor is a central processor on a chip, and it enabled computer designers to replace dozens of ICs with a single chip, thereby further shrinking the size and cost of computers. This paved the way for the development of *personal computers*.

Mass-market personal computers emerged in 1977 with companies such as Apple, Commodore, Radio Shack, and Heath. IBM, who had up to this time dominated the big computer market, entered the personal computer market with an IBM PC in 1981.

Superconductivity was first discovered in 1911 by the Dutch physicist Heike Kamerlingh Onnes (1853–1926). *Superconductivity* is a low-temperature phenomenon in which a material (usually a metal) loses all electrical resistance below a certain temperature. Low-temperature superconductivity is called LTS. In 1986, at the IBM Zurich Research Laboratory, physicists J. George Bednorz and K. Alex Muller worked on high-temperature superconductivity (HTS) using ceramic oxide materials.

The first active communications *satellite* was launched in 1958. This was followed by Telstar I in 1962, which provided direct television transmission. Hundreds of active communication satellites are now in orbit. Hundreds more are proposed to be in varying earth orbits by the turn of the 21st century. The satellites have ushered in the era of personal communication services (PCS).

Further developments in electricity/electronics could be highlighted as well. There are enough developments to fill several chapters. What is important to note are the milestone events in the development and growth of electricity- and electronic-related devices. The events described here are among those milestones. All these developments have led to the exciting, dynamic field of electronics as we now know it.

SECTION 1.1 SELF-CHECK	Answers are given at the end of the chapter.

SECTION 1.1 SELF-CHECK

Answers are given at the end of the chapter.

1. What is the derivation of the word *electric*?
2. Who devised equations to show the relationship between electricity and magnetism?
3. Who is generally credited with being the inventor of radio?

1.2 ELECTRONICS TECHNOLOGY

As we saw in the previous section, electronics in a modern sense began with the pioneer days of radio communication and the invention of the triode vacuum tube. *Electronics* is a branch of physical science that deals with the behavior of electrons and other carriers of electric charge. A flow of *electric charge* is called an *electric current,* and a closed path that electric charges can follow is called an *electric circuit*. Most electric circuits have *transducers* (devices that convert energy from one form to another) at the input and/or output of the circuit. Between the input and output stages are devices that act on or control electric current. These devices are incorporated into circuits that are called *signal processors*.

FIGURE 1.1
A typical electronic circuit.

Input signal → Signal processor → Output signal

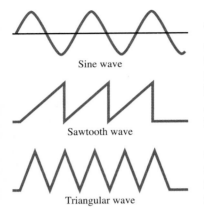

Sine wave

Sawtooth wave

Triangular wave

FIGURE 1.2
Common analog signals.

Electronic technology refers to how electrical components, circuits, and systems work. Typically there is some form of an electric signal applied to the input of a circuit. A typical electronic circuit is shown in Figure 1.1. If the amplitude of current and/or voltage of the signal (waveform) vary continuously over time (like the human voice), it is known as an *analog signal.* Some common analog signals are shown in Figure 1.2. The components between the input and output provide what is known as *analog signal processing* (ASP). Analog signal processing includes circuitry such as amplifiers, rectifiers, modulators, filters, waveshapers, multiplexers, and samplers.

Digital signals are those where the amplitude of current and/or voltage varies discretely or in steps with time. Binary signals used in computers are an example. Some common digital signals are shown in Figure 1.3. *Digital signal processing* (DSP) circuits are those that process digital signals. These circuits include gates, decoders, encoders, registers, counters, arithmetic logic units, microprocessors, and the like.

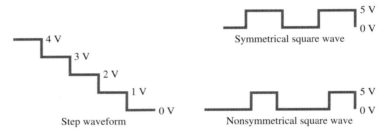

4 V
3 V
2 V
1 V
0 V

Step waveform

5 V
0 V

Symmetrical square wave

5 V
0 V

Nonsymmetrical square wave

FIGURE 1.3
Common digital signals.

There has been a marked effort in recent years to use digital signal-processing circuits whenever possible. This means that analog signals are converted to digital via analog-to-digital converters. They are then processed digitally and converted back to analog. This conversion is quite commonplace. The compact disc (CD) audio player is an example. The digital video disc (DVD) will eventually replace the videocassette recorder (VCR). The new standard for high-definition television (HDTV) is based on digital circuitry. Digital circuits have many advantages over analog, making it a more appealing technology. You will learn many of these attributes as you study electronics.

Signal processors contain components that can be further divided into two types. These are called *passive* and *active* components. Passive components are covered in this text. They comprise devices that alter or limit the flow of electric current. They do not put additional energy into a circuit. Resistors, inductors, and capacitors are the most common forms of passive devices. Active devices can rectify, boost, or amplify the signal being processed.

Signal-processing circuits and active devices are covered in depth in later electronics courses. The purpose of this text is to provide you with the necessary foundations (the basics) in order to understand and analyze electronic signal-processing circuits.

1.3   THE ELECTRONICS FIELD

There are logical groupings of electronic circuits that comprise a specialty field. The groupings are defined in this section. These groupings also define electronics fields in which the student can major. These groupings are not all-encompassing but nonetheless cover the major areas of electronics.

Communications   The communications field includes AM and FM radio, television, telephone, and satellite communications. Actually any transmitter/receiver combination is included here. Such combinations encompass aircraft (avionics), marine, and land-based radio systems. High-fidelity audio equipment is included in this area, as is data communications via networks. The personal communication services (PCS), a new and rapidly growing land-based and satellite communication system, is found within this group. Figure 1.4 shows a communications technician at work making power measurements.

FIGURE 1.4
A communications technician at work making power measurements. (Courtesy of Tektronix, Inc.)

Industrial Electronics   The industrial electronics field includes electronic controls of various machine operations. Examples include welding and heating processes, elevator control, automation, robotics, and CNC machines. The programmable logic controller (PLC) is a common device used in industrial electronics.

Digital or Microprocessor-Based Systems   The digital systems field includes digital or microprocessor control of various circuits. Computer operation is part of this grouping. Computer programming and microprocessor instruction sets are also covered here.

Consumer Home Electronics   The consumer home electronics field includes the installation, maintenance, and repair of consumer home electronics products, including audio/video components (TV, VCR, CD, DVD), computers, security systems, and the like. Courses within this field do not normally encompass the same level of circuit analysis or depth of those in the other three areas.

Courses within the consumer home electronics field are not offered at all schools, and in some cases they are the only offering. In addition, coursework within this field does not typically transfer to a 4-year school. The *Consumer Electronics Manufacturing Association* (CEMA), part of the *Electronics Industry Association* (EIA), sponsors and supports technical workshops in a variety of consumer electronics areas. These workshops allow field technicians an opportunity to keep abreast of the latest technology. Figure 1.5 shows an electronics service technician at work in a video studio.

FIGURE 1.5
An electronics service technician at work in a video studio. (Courtesy of Tektronix, Inc.)

1.3.1   Electronics Specialty Areas

There are several subgroupings of the first three areas that have more of a specific interest. Automotive electronics and medical electronics represent a couple of examples.

Automotive electronics has become more than the usual battery and electrical functions. It now encompasses microprocessor controls and sensors for the monitoring and control of engine performance. Many automotive dealerships have now added an electronic technician position to their service departments.

Medical electronics combines electronics with biology or physiology. Medical research, diagnosis, and treatment all use electronic equipment. A hospital contains a great deal of electronic monitoring equipment to assist in patient care. Technical personnel are needed to provide preventive maintenance and repairs on this equipment.

1.4 ELECTRONIC JOB TITLES AND CAREER DESCRIPTIONS

Various electronic job titles and classifications are provided in this section for technical personnel. Specific skills for each area are shown, along with the typical education required for each. Depending on the job, more or fewer units in specialty areas may be required.

Assembler   An assembler is involved in the production of electronic circuits and systems. Component placement and soldering to the circuit board are the prime skills needed to fabricate electronic circuits. Commercial electronic production utilizes through-hole and surface-mount technology (SMT). The latter area is growing rapidly. An assembler position does not typically require a degree, but there are levels of certification available that can increase salary.

Electromechanical Assembler/Fabricator   This job is very similar to that of an assembler, but it includes electromechanical fabrication techniques, wire harnessing, and wire-wrapping, among others. These jobs require more manual dexterity and usually are higher-paid than assemblers. A degree is not normally required.

Electronic Technician   Most 2-year electronic training programs provide the skills for this area. The math requirements are usually algebra- and trigonometry-based. A 2-year associate degree and some work experience is preferred by most employers that hire technicians. Duties include installation, calibration, maintenance, troubleshooting, problem-solving, final checkout, and the like. Typical job titles are R & D tech, calibration tech, electronic tech, field service tech, and bench tech.

Technicians working in the communications field might require FCC and/or NARTE certification. Computer service technicians might need A+ certification, whereas network technicians might need a Network+ certification. There are similar certifications available in the biomedical, consumer product servicing, and cable industry fields.

The Consumer Electronics Manufacturing Association sponsors a certification test that enables the 2-year student to acquire the Certified Technician Associate classification.

Electronics Engineer   Many 4-year schools provide for a bachelor's degree in Engineering Technology or Electrical Engineering. Engineers generally have the task of planning and designing electronic circuits and systems. Among degree requirements are physics and calculus courses.

The *Institute of Electrical and Electronic Engineers* (IEEE) is a professional society that provides an avenue for research, presenting papers, and continuing education for the engineering professional. Many engineers become affiliated with IEEE. Figure 1.6 shows an engineer at work designing digital circuits.

FIGURE 1.6
An engineer at work designing digital circuits. (Courtesy of Tektronix, Inc.)

**SECTION 1.4
SELF-CHECK**

Answers are given at the end of the chapter.

1. Name the job classification that most 2-year electronics program completers seek.
2. What certifications are available for those working in the field of electronic communications?
3. Many engineering professionals become affiliated with what organization?

■ SUMMARY

1. Electricity is a form of energy and is linked to its inseparable effect, magnetism.
2. The Greek word for amber is *elektron,* the basis for the modern word *electric.*
3. Coulomb perfected the measurement of the force exerted by electric charges.

4. Volta is famous for inventing the electric battery.
5. Ohm formulated the relationship between current, voltage, and resistance.
6. Maxwell's equations are used to show the relationship between electricity and magnetism.
7. Marconi originated wireless telegraphy and is generally credited with being the inventor of radio.

8. Fleming developed a two-electrode rectifier patterned after the *Edison effect*.

9. De Forest added a third electrode to the *Fleming valve* to produce a device that could store and amplify electrical signals. Modern electronics began with the invention of this device, called the *triode*.

10. The transistor was invented in 1947 by a team of scientists from Bell Laboratories. This ushered in the era of *solid-state electronics*.

11. Electronics is a branch of physical science that deals with the behavior of electrons and other carriers of electric charge.

12. Electronic technology refers to how electrical components, circuits, and systems work.

13. The signals to be processed by electronic circuits are either analog or digital. Further, processing circuits comprise passive and active components.

14. Three major fields of electronics are communications, industrial, and digital or microprocessor-based systems.

15. Most 2-year electronic training programs provide training for the electronic technician job classification.

■ REVIEW QUESTIONS

1. Define the term *electricity*.
2. Name some devices that have been invented that utilize electricity.
3. Where can the history of the electric charge be traced?
4. Where did the term *electric* come from?
5. Who is the father of conventional current flow?
6. What is conventional current flow?
7. What does Coulomb's law describe?
8. What was the first electric battery made from?
9. What affect does a battery have on electric charges?
10. Define the term *electrodynamics*.
11. What effect did Oersted discover about electricity and magnetism?
12. Describe Lenz's law.
13. What is *wireless telegraphy*?
14. What type of power-distribution system did Edison promote?
15. What vacuum-tube device revolutionized the field of electronics?
16. Name several microwave vacuum tubes used in the development of radar during WW II.
17. What differentiates the integrated circuit (IC) from the transistor?
18. What is the microprocessor and what did it accomplish?
19. What companies were first to market the personal computer?
20. Define the term *superconductivity*.
21. What is the difference between LTS and HTS superconductivity?
22. Define the term *electric current*.

23. Define the term *transducer*.
24. Define the term *electronics technology*.
25. What is the difference between an analog signal and a digital signal?
26. What is the difference between *passive* and *active* components?
27. What is included in the electronics communications field?
28. What is included in the industrial electronics field?
29. What is included in the microprocessor-based electronics field?
30. What is the difference between an assembler and an electro-mechanical assembler?
31. Describe the duties of an electronic technician.

■ CRITICAL THINKING

1. Correlate any differences between electronics and electronics technology.
2. Compare and contrast the differences between analog signal processing and digital signal processing.

■ ANSWERS TO SECTION SELF-CHECKS

SECTION 1.1 SELF-CHECK

1. From the Greek word for amber, *elektron*.
2. Maxwell
3. Marconi

SECTION 1.2 SELF-CHECK

1. Electronics is a branch of physical science that deals with the behavior of electrons and other carriers of charge. Electronics technology refers to how electrical components, circuits, and systems work.
2. The amplitude of an analog signal varies continuously with time (like the human voice). The amplitude of a digital signal varies discretely or in steps with time.
3. Resistors, capacitors, and inductors

SECTION 1.3 SELF-CHECK

1. Electronics communications, industrial electronics, microprocessor-based electronic systems, and consumer home electronics servicing
2. Automotive electronics and medical electronics

SECTION 1.4 SELF-CHECK

1. Electronic technician
2. FCC and NARTE
3. IEEE

CHAPTER 2

ELECTRICITY/ELECTRONIC UNITS AND NOTATIONS

OBJECTIVES

After completing this chapter, you will be able to:

1. Differentiate between fundamental units and derived units.
2. Demonstrate knowledge of the SI system of measurement.
3. Write numbers in scientific and engineering notation.
4. Recognize and properly use the symbols for SI prefixes.
5. Describe the advantage of engineering notation for use in electronics.
6. Convert SI prefixes from one form to another.
7. Utilize the calculator for entering numbers in engineering notation.

INTRODUCTION

Chapter 1 gave you some background information on electricity and electronics. You know some broad definitions of terms such as *electric charge* and *electric current*. Before you can develop more definitive characteristics of other terms used in the study of electricity and electronics, you need to learn how electronic quantities are symbolized. In addition, you must learn how electronic quantities are measured, a method for expressing the quantities, and how to convert from one notation method to another.

2.1 ELECTRICITY/ELECTRONICS UNITS

When anything can be quantified, it follows that some unit of measurement must accompany it. There are three quantities (length, mass, and time) that are known as *fundamental units.* Other quantities—and, therefore, units of measurement—can be obtained from the fundamental units. These are known as *derived units.* Three commonly used systems of measurement that have emerged are the MKS system, the CGS system, and the English, or customary, system of units. The International System of Units, or *SI,* represents a rationalized selection of the MKS system. It is commonly referred to as the *metric system.*

The most commonly used systems of measurement are the English and metric systems of units. The English system is based on the natural dimensions of physical objects. Units of length consist of the foot, rod, and chain, whereas mass units are the stone and slug. The metric (SI) system has been developed scientifically using precise measuring equipment. In the SI system, the unit of length is the *meter* (m), the unit of mass is the *kilogram* (kg), and the unit of time is the *second* (s).

In 1965, the Institute of Electrical and Electronic Engineers (IEEE) adopted the SI system as the standard for all engineering and scientific literature. The SI system uses prefixes to form multiples and submultiples of a unit. The multiple and submultiple units are different from each other by powers of 10. This aspect has led to specific notations known as scientific and engineering notations. These are covered in the next section.

The United States officially recognized the SI system in 1975 by passing the Metric Conversion Act. The SI system has seven base units: length, time, mass, electric current, temperature, luminous density, and molecular substance. There are many instances when the base units of measure are not suitable and derived units are required. These derived units are developed from the previously defined SI base units. As an example, frequency, velocity, and acceleration are measured with derived units that come from and include the base unit of time. Table 2.1 lists derived units considered essential to the study of electronics.

TABLE 2.1
SI derived units commonly used for electronic quantities

Quantity	Quantity Symbol	Unit	Unit Symbol
Capacitance	C	farad	F
Conductance	G	siemens	S
Electric charge	Q	coulomb	C
Electromotive force	E	volt	V
Energy, work	W	joule	J
Force	F	newton	N
Frequency	f	hertz	Hz
Inductance	L	henry	H
Magnetic flux	Φ	weber	Wb
Magnetic flux density	B	tesla	T
Power	P	watt	W
Resistance	R	ohm	Ω
Reactance	X	ohm	Ω
Impedance	Z	ohm	Ω

Answers are given at the end of the chapter.

1. What are the three fundamental units of measurement?
2. What are the three most commonly used systems of measurement?
3. Which system of measurement is employed in scientific endeavors?
4. What are the quantity and unit symbols for resistance?

2.2 SCIENTIFIC AND ENGINEERING NOTATION

2.2.1 Scientific Notation

As you will see as you study electronic quantities, there are some terms that are quite large and some that are extremely small. A method of expressing very large and very small numbers is to express them as a number multiplied by 10 or a power of 10. This method of expressing numbers is known as *scientific notation*. A number written in scientific notation is expressed as the product of a number greater than or equal to 1.0 and less than 10 and a power of 10. To express a number in scientific notation, the decimal point is moved (either left or right) until there is one significant digit to the left of the decimal point. The result is then multiplied by the appropriate power of 10 to return the quantity to its original value. On whole numbers where no decimal point appears, the decimal point is assumed to be to the right of the last digit in the number.

If the number in question is less than 1.0, the decimal point is moved to the right, and the appropriate power of 10 is a negative ($-$) exponent. Conversely, if the number in question is greater than 10, the decimal point is moved to the left and the appropriate power of 10 is a positive ($+$) exponent. The decimal point is not moved for numbers equal to or greater than 1.0 but less than 10. In these cases the appropriate power of 10 is 10^0. Note that 10^0 (or any value raised to the 0 power) is simply 1. Thus, the product of a number and 10^0 is the original number.

Most scientific calculators have four methods of assigning the decimal point. These include a *fixed (FIX)* number of decimal places, a *floating (no indicator)* decimal point, or a value in either *scientific (SCI)* or *engineering (ENG)* notation.

Optional Calculator Sequence

To put the Sharp calculator into the scientific notation mode, press the **2ndF** key (in yellow) and then the **FSE** key, labeled above the decimal point (.). Continue this process in turn until **SCI** appears across the top middle of the LCD display. To see a number displayed in scientific notation on the calculator, simply enter the number and press the = key. After the = key is pressed, several zeros will appear to the right of the decimal point. These zeros do not change the value of the number.

Note that in addition to displaying the number in scientific notation, the Sharp calculator displays a **×10** on the display to the right side and below the exponent. This reminds the student that **×10** is a constant already built in the calculator.

EXAMPLE 2.1

Express the following numbers in scientific notation using a calculator. Make sure your calculator is in the SCI decimal mode.

a. 82,400   **b.** 6.578   **c.** 0.0033   **d.** 6,250,000   **e.** 0.015

Solution

a. $82{,}400 = 8.24 \times 10^4$ **b.** $6.578 = 6.578 \times 10^0$

c. $0.0033 = 3.3 \times 10^{-3}$ **d.** $6{,}250{,}000 = 6.25 \times 10^6$

e. $0.015 = 1.5 \times 10^{-2}$

Practice Problems 2.1   Answers are given at the end of the chapter.

1. Express the number 5047 in scientific notation.

2. Express the number 0.00089 in scientific notation.

For convenience, various multiples and submultiples of the base units are assigned prefixes and symbols. These are shown in Table 2.2. The table encompasses those that are frequently used as electronic units. As your electronic training progresses, you will become very familiar with them.

TABLE 2.2
Prefixes for use with SI units

Prefix	Symbol	Powers of 10	Value
exa	E	10^{18}	1,000,000,000,000,000,000
peta	P	10^{15}	1,000,000,000,000,000
tera	T	10^{12}	1,000,000,000,000
giga	G	10^9	1,000,000,000
mega	M	10^6	1,000,000
kilo	k	10^3	1,000
milli	m	10^{-3}	0.001
micro	μ	10^{-6}	0.000001
nano	n	10^{-9}	0.000000001
pico	p	10^{-12}	0.000000000001
femto	f	10^{-15}	0.000000000000001
atto	a	10^{-18}	0.000000000000000001

2.2.2 Engineering Notation

You should note in Table 2.2 that all the prefixes listed involve powers of 10 (exponents) that are multiples of 3. Numbers expressed in powers of 10 notation having exponents that are multiples of 3 are expressed in what is known as *engineering notation*. This technique calls for the use of numbers between 1.0 and 999 times the appropriate third power of 10. Engineering notation is the *preferred* method of expressing numbers for electronics students. Most electronic quantities utilize the prefixes of Table 2.2.

Optional Calculator Sequence

To put your Sharp calculator into the **ENG** mode, press the **2ndF** and **FSE** keys as necessary until **ENG** appears near the top middle of the display. Then to see a number displayed in engineering notation, simply enter the number and press the = key. As with the **SCI** mode, several zeros will appear to the right of the decimal point when the = key is pressed.

EXAMPLE 2.2

Express the following numbers in engineering notation using a calculator. Make sure that your calculator is in the ENG decimal mode.
a. 545,000   **b.** 110   **c.** 6752   **d.** 0.000148   **e.** 0.0027

Solution
a. $545{,}000 = 545 \times 10^3$ **b** $110 = 110 \times 10^0$ **c.** $6752 = 6.752 \times 10^3$
d. $0.000148 = 148 \times 10^{-6}$ **e.** $0.0027 = 2.7 \times 10^{-3}$

Practice Problems 2.2   Answers are given at the end of the chapter.
1. Express the following numbers in engineering notation: 4700; 10,000; 0.001; 0.0000047; 7500.

Notice that in the solutions to Example 2.2, the decimal point was simply moved an appropriate number of places so that the power of 10 was always some multiple of 3. As with scientific notation, if the decimal point is moved to the left, then the resulting power of 10 is positive. When the decimal is moved to the right, the resulting power of 10 is negative.

Engineering notation provides a way of assigning prefixes to a number rather than displaying an entire number with a power of 10. The use of a prefix provides a convenient way to express electronic quantities. Remember that a base unit does not require a prefix and has a power of 10 of 0 (10^0).

EXAMPLE 2.3

Choose the appropriate prefix (with symbol) from those listed in Table 2.2 to replace the power of 10 in the following numbers in engineering notation.
a. 2.7×10^3 **b.** 150×10^{-6} **c.** 4.9×10^6
d. 5.0×10^{-3} **e.** 100×10^{-12}

Solution
a. $2.7 \times 10^3 = 2.7 \text{ k} \underline{\quad}$ **b.** $150 \times 10^{-6} = 150 \, \mu \underline{\quad}$
c. $4.9 \times 10^6 = 4.9 \text{ M} \underline{\quad}$ **d.** $5.0 \times 10^{-3} = 5.0 \text{ m} \underline{\quad}$
e. $100 \times 10^{-12} = 100 \text{ p} \underline{\quad}$
(The $\underline{\quad}$ is shown to indicate that further labeling for an electronic problem is necessary. This could be Ω, A, V, Hz, or a similar unit.)

Practice Problems 2.3   Answers are given at the end of the chapter.
1. Choose the appropriate prefix (with symbol) to replace the power of 10 in the following numbers in engineering notation.
2.5×10^{-9} 3.3×10^{-12}
12×10^6 8.6×10^3

SECTION 2.2
SELF-CHECK

Answers are given at the end of the chapter.
1. Describe the term *scientific notation.*
2. Express 0.000001 in scientific notation.
3. Describe the term *engineering notation.*
4. Express 3450 in engineering notation.
5. What is the prefix for 10^{-6}?

2.3   PREFIX CONVERSION

Another important skill for the electronics student to acquire is the ability to convert one prefix to another. For example, 2000 Ω is equal to 2.0 kΩ, and 1.5 mA is equal to 0.0015 A. To convert one prefix to another follow these rules:

1. Note that each prefix represents a power of 10 that is a multiple of 3.
2. If the prefix that you are converting to is smaller (for example, going from A to mA), then move the decimal point to the *right* a number of places equal to the difference in exponents of the two prefixes.
3. If the prefix that you are converting to is larger (for example, going from Ω to kΩ), then move the decimal point to the *left* a number of places equal to the difference in exponents of the two prefixes.
4. The decimal point is always moved in multiples of 3 regardless of the beginning and ending prefix value.
5. The base unit has a power of 10 equal to 10^0.

As an aid, the conversion rules can be put into a table format (Table 2.3). The left column lists the original prefix value, and the desired prefix value is listed across the top. The corresponding number indicates how many places to move the decimal point and the arrow indicates the direction.

To change 10 MHz to kHz, first find the original prefix value in the left column (mega) and then find the prefix value you are going to change to across the top (kilo). Find the intersection point. The chart indicates that the decimal point should be moved three places to the *right*. So, 10 MHz is equal to 10,000 kHz.

TABLE 2.3
Prefix conversion chart

Original Value	Desired Value							
	Giga (G)	Mega (M)	kilo (k)	Units	milli (m)	micro (μ)	nano (n)	pico (p)
Giga (G)		3 →	6 →	9 →	12 →	15 →	18 →	21 →
Mega (M)	← 3		3 →	6 →	9 →	12 →	15 →	18 →
kilo (k)	← 6	← 3		3 →	6 →	9 →	12 →	15 →
Units	← 9	← 6	← 3		3 →	6 →	9 →	12 →
milli (m)	← 12	← 9	← 6	← 3		3 →	6 →	9 →
micro (μ)	← 15	← 12	← 9	← 6	← 3		3 →	6 →
nano (n)	← 18	← 15	← 12	← 9	← 6	← 3		3 →
pico (p)	← 21	← 18	← 15	← 12	← 9	← 6	← 3	

EXAMPLE 2.4

Convert the following prefixes as shown.

a. 0.01 A = _____ mA   b. 1,500,000 Ω = _____ MΩ
c. 1200 Hz = _____ kHz   d. 5.1 mA = _____ μA
e. 40 μA = _____ mA   f. 1000 W = _____ kW
g. 0.0015 μF = _____ pF   h. 150 V = _____ kV
i. 250 kW = _____ MW   j. 3.7 MΩ = _____ kΩ

Solution
a. 0.01 A = 10 mA   **b.** 1,500,000 Ω = 1.5 MΩ
c. 1200 Hz = 1.2 kHz   **d.** 5.1 mA = 5100 μA
e. 40 μA = 0.04 mA   **f.** 1000 W = 1.0 kW
g. 0.0015 μF = 1500 pF   **h.** 150 V = 0.15 kV
i. 250 kW = 0.25 MW   **j.** 3.7 MΩ = 3700 kΩ

Practice Problems 2.4   Answers are given at the end of the chapter.
1. 4.2 mA = _____ A
2. 4.2 mA = _____ μA
3. 450 MΩ = _____ kΩ

**SECTION 2.3
SELF-CHECK**

Answers are given at the end of the chapter.

1. If the prefix you are converting to is a smaller unit, the decimal is moved to the _____.

2. What is the prefix for 10^{-9}?

2.4   ENTERING NUMBERS INTO THE CALCULATOR

Another skill needed when working with numbers expressed in SI form is simply being able to enter a number into a calculator. (This may seem elementary to some, but not everyone knows how to do it.) All scientific calculators accomplish this via the exponent key. The exponent key is labeled either **Exp** or **EE** on most calculators.

> ### Optional Calculator Sequence
>
> On the Sharp calculator, the exponent key is labeled **Exp.** The following example shows how to use this key.

EXAMPLE 2.5

Enter 180 kΩ on the Sharp calculator. Make sure that your calculator is still in the ENG decimal mode.

Solution
180 kΩ is equal to 180×10^3 Ω. The calculator sequence for this problem is as follows:

Step	*Keypad Entry*	*Top Display Response*
1	1	
2	8	
3	0	
4	Exp	
5	3	
6	=	180E03=
	Answer	180.0×10^{03}
		(bottom display)

(The **Exp** key raises 10 to some power. Thus, you do *not* need to enter **×10** and then the **Exp** key. Also note that after the **=** key is pressed, the calculator will display $180.0000000 \times 10^{03}$. These trailing zeros simply fill up the display.)

EXAMPLE 2.6

Enter 2.5 mA on the Sharp calculator. You should note that mA has a negative $(-)$ exponent when written in terms of amps.

Solution

2.5 mA is equal to 2.5×10^{-3} A. The calculator sequence for this problem is as follows:

Step	Keypad Entry	Top Display Response
1	2	
2	.	
3	5	
4	Exp	
5	+/−	
6	3	
7	=	2.5E−03=
	Answer	2.50×10^{-03}
		(bottom display)

(Note that the **+/−** key may be pressed either before or after the exponent 3.)

EXAMPLE 2.7

Enter the following numbers with prefixes on the Sharp calculator.
a. 220 kΩ
b. 25 mA
c. 18 kW
d. 0.01 μF
e. 15 pF

Solution
a. 220×10^{03}
b. 25×10^{-03}
c. 18×10^{03}
d. 0.01×10^{-06}
e. 15×10^{-12}

SECTION 2.4 SELF-CHECK

Answers are given at the end of the chapter.

1. Which key do you press to enter a number with an exponent?
2. How do you enter a number with a negative exponent?

■ SUMMARY

1. The three fundamental units of measurement are length, mass, and time.

2. Other quantities obtained from the fundamental units are known as derived units.

3. Three common measurement systems are the English, the CGS, and the MKS systems.

4. The SI system is the most common one used in scientific measurements and utilizes the MKS, or metric, system.

5. Frequency, velocity, and acceleration are derived units that are from and include the base unit of time.

6. Scientific notation is a method for expressing rather large or small numbers. It uses an exponential (power of 10) notation system.

7. 10^0 is equal to 1.

8. Most calculators utilize four methods of assigning the decimal point. They are fixed, floating, scientific, and engineering notation.

9. Prefixes are used for the various multiples and submultiples of the base units.

10. Engineering notation has numbers expressed in powers of 10 that have exponents that are multiples of 3.

11. Engineering notation allows for the use of prefixes. This simplifies the expression of electronic quantities.

12. Entering numbers into the calculator is done via the exponent key.

13. The exponent key on most calculators is labeled **Exp** or **EE.**

■ REVIEW QUESTIONS

1. What are three fundamental units for measurements?

2. What are derived units of measurements?

3. What are three commonly used systems of measurements?

4. The MKS system is referred to as the _____ system.

5. On what was the English system of measurement based?

6. How was the metric system developed?

7. What institution adopted the metric system as the standard for all engineering and scientific literature?

8. What are the quantity and unit symbols for resistance?

9. What are the quantity and unit symbols for current?

10. What are the quantity and unit symbols for electromotive force?

11. What are the quantity and unit symbols for power?

12. How is a number expressed in scientific notation?

13. Utilizing scientific notation, what power of 10 is used for the number 8.7?

14. What are the four methods of assigning a decimal point used by most scientific calculators?

15. What is the commonality in the prefixes listed in Table 2.2?

16. What is the prefix for 10^{-3}?

17. How are numbers expressed in engineering notation?

18. Using engineering notation, when the decimal is moved to the right, the resulting power of 10 is _____.

19. What is the advantage of using prefixes for number notations?

■ CRITICAL THINKING

1. Describe the merits of the English and metric systems of measurement.

2. What is the difference between base units and derived units? Give examples.

3. Why is engineering notation more commonly used in electronics than scientific notation?

■ PROBLEMS

Express the following numbers in scientific notation.

1. 3600
2. 0.0000015
3. 34.9
4. 10
5. 3.67
6. 278,900
7. 1,000,000
8. 0.002
9. 0.000057
10. 0.15

Express the following numbers in engineering notation.

11. 150
12. 5280
13. 12
14. 0.000005
15. 0.003
16. 1,000,000
17. 0.0033
18. 0.000047
19. 13,250
20. 8900

List the prefix for each of the following exponents.

21. 10^6
22. 10^{-3}
23. 10^9
24. 10^{-6}
25. 10^{12}
26. 10^{-9}
27. 10^{-12}
28. 10^3

Convert the following prefixes as shown.

29. 0.05 A = _____ mA
30. 0.05 A = _____ μA
31. 2700 Ω = _____ kΩ
32. 4500 W = _____ kW
33. 120 V = _____ mV
34. 120 V = _____ kV

35. 2400 Hz = _____ kHz
36. 0.00001 F = _____ μF
37. 1,500,000 Ω = _____ MΩ
38. 1,500,000 Ω = _____ kΩ
39. 0.00002 A = _____ mA
40. 0.00002 A = _____ μA

■ ANSWERS TO PRACTICE PROBLEMS

PRACTICE PROBLEMS 2.1

1. 5.047×10^3
2. 8.9×10^{-4}

PRACTICE PROBLEMS 2.2

1. 4.7×10^3; 10×10^3; 1.0×10^{-3}; 4.7×10^{-6}; 7.5×10^3

PRACTICE PROBLEMS 2.3

1. 2.5 n__; 3.3 p__
    12 M__; 8.6 k__

PRACTICE PROBLEMS 2.4

1. 0.0042 A
2. 4200 μA
3. 450,000 kΩ

■ ANSWERS TO SECTION SELF-CHECKS

SECTION 2.1

1. Length, mass, and time
2. MKS, CGS, and English systems
3. SI (metric)
4. R, Ω, respectively

SECTION 2.2

1. A method of expressing very large and very small numbers as a number multiplied by 10 or a multiple of 10
2. 1.0×10^{-6}
3. Numbers that are expressed in powers-of-10 notation having exponents that are multiples of 3
4. 3.45×10^3
5. μ

SECTION 2.3

1. Right
2. n

SECTION 2.4

1. Exp or EE
2. Use the $+/-$ key either before or after the number.

CHAPTER 3

ELECTRICAL PROPERTIES

OBJECTIVES

After completing this chapter, you will be able to:

1. Define common terms associated with atomic properties.
2. Differentiate between a molecule and a compound.
3. Describe the makeup of atoms via subatomic particles.
4. Compare and contrast the various energy levels within the atom and the basis for existence in the energy levels.
5. Calculate and label energy levels using Pauli's exclusion principle.
6. Identify the valence shell and state its importance to understanding electricity.
7. Differentiate ionization levels and their causes.
8. Compare and contrast the terms conductors, semiconductors, and insulators.
9. Define electrostatics and static electricity.
10. Evaluate various attributes of charged bodies and electric fields.

INTRODUCTION

If there are roots to western science, they no doubt lie in what was once ancient Greece. Except for the Greeks, ancient people had little interest in the structure of materials. They accepted a solid as just that—a continuous, uninterrupted substance. One Greek school of thought believed that if a piece of matter, such as copper, is subdivided, it could be done indefinitely and still only that material would be found. Others reasoned that there must be a limit to the number of subdivisions that can be done and have the material retain its original characteristics. They held fast to the idea that there must be a basic particle upon which all substances are built. Modern investigations have revealed that there are, indeed, several basic particles, or building blocks, within all substances. We examine these in this chapter.

3.1 ATOMIC PROPERTIES

3.1.1 Matter

Matter is defined as anything that occupies space and has mass; that is, the mass and dimensions of matter are measurable. Examples of matter are common and include air, water, automobiles, and even our bodies. We can say that matter may be found in one of three states: solid, liquid, or gaseous.

3.1.2 Elements and Compounds

An *element* is a substance that cannot be reduced to a simpler substance by chemical means. Examples of elements are iron, gold, silver, copper, and oxygen. There are more than 100 known elements. These are shown on the periodic chart of the elements found in Appendix B. Everything we know about consists of one or more of these elements.

When two or more elements chemically combine, the resulting substance is called a *compound*. A compound is a chemical combination of elements that can be separated by chemical but not by physical means. Examples of common compounds are water, which consists of hydrogen and oxygen, and table salt, which consists of sodium and chlorine.

3.1.3 Atoms

An *atom* is the smallest particle of an element that retains the characteristics of that element. The atom of one element, however, differs from the atoms of all other elements. Since there are more than 100 known elements, there must be more than 100 different atoms, or a different atom for each element. Just as thousands of words can be made by combining the proper letters of the alphabet, thousands of different materials can be made by chemically combining the proper atoms.

Any particle that is a chemical combination of two or more atoms is called a *molecule.* The oxygen molecule consists of two atoms of oxygen, and the hydrogen molecule consists of two atoms of hydrogen. Sugar, on the other hand, is a compound composed of atoms of carbon, hydrogen, and oxygen. These atoms combine into sugar molecules. Since the sugar molecules can be broken down by chemical means into smaller and simpler units, we cannot have sugar atoms.

The atoms of each element are made up of electrons, protons, and, in most cases, neutrons, which are collectively called *subatomic particles.* Furthermore, the electrons, protons, and neutrons of one element are identical to those of any other element. The reason that there are different kinds of elements is that the number and the arrangement of electrons and protons within the atom are different for each of the 100+ elements.

The *electron* is considered a small negative (−) charge of electricity. The *proton* has a positive (+) charge of electricity equal and opposite to the charge of the electron. The electron and proton have the same quantity of charge, although the mass of the proton is approximately 1837 times that of the electron. In some atoms there exists a neutral particle called a *neutron.* The neutron has a mass approximately equal to that of a proton, but it has no electrical charge.

One popular theory has the electrons, protons, and neutrons of the atom arranged in a manner similar to a miniature solar system. This was postulated by Sir Ernest Rutherford in 1911. The protons and neutrons form a heavy nucleus with a positive charge, around which the very light electrons revolve. Further elaboration of the electron orbits was done by Neils Bohr in 1913. Since the number of protons in the nucleus is equal to the electrons orbiting the nucleus, the atom is considered electrically neutral in charge.

Figure 3.1 shows one helium atom. It has a relatively simple structure. The helium atom has a nucleus made up of two protons and two neutrons, with two electrons rotating about the nucleus. Elements are classified numerically according to the complexity of their atoms. The *atomic number* of an atom is determined by the number of protons in its nucleus. Figure 3.2 shows a hydrogen atom and a carbon atom. Note the difference in the numbers of electrons, protons, and neutrons.

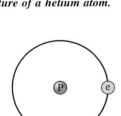

e = electron
P = proton
N = neutron

FIGURE 3.1
Structure of a helium atom.

Hydrogen atom
(one electron, one proton)

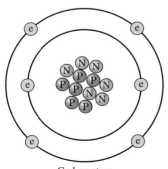

Carbon atom
(6 electrons, 6 protons, 6 neutrons)

FIGURE 3.2
Structures of hydrogen and carbon atoms.

Answers are given at the end of the chapter.

1. Define the term *matter.*
2. What is the difference between an element and a compound?
3. Define the term *atom.*
4. What are the subatomic particles?

3.2 ENERGY LEVELS

Because the electron in an atom has both mass and motion, it contains two types of energy. By virtue of its motion the electron contains *kinetic energy.* Due to its position it also contains *potential energy.* The total energy contained by an electron (kinetic and potential) is the factor that determines the radius of the electron orbit. In order for an electron to remain in its orbit, it must *neither* gain nor lose energy.

Light consists of tiny energy packets known as *photons.* Photons can contain various quantities of energy. The quantity depends upon the color of the light involved. Should a photon of sufficient energy collide with an orbital electron, the electron absorbs the photon's energy. The absorption of energy by the electron could also be done via heat or by friction. The electron, which now has a greater-than-normal amount of energy, jumps into a new orbit farther from the nucleus. These new orbits have radii many times as large as the radius of the original orbit. Thus, each orbit may be considered to represent one of several energy levels that the electron may attain. It must be emphasized that the electron cannot jump to just any orbit. The electron remains in its lowest orbit until a sufficient amount of energy is available, at which time the electron accepts the energy and jumps to one of a series of permissible orbits. An electron cannot exist in the space between energy levels. This shows that the electron does not accept the energy of heat or a photon unless it contains enough energy to elevate itself to one of the higher energy levels.

Once the electron has been elevated to an energy level higher than the lowest possible energy level, the atom is said to be in an *excited state.* The electron does not remain in this excited condition for more than a fraction of a second before it radiates the excess energy and returns to a lower energy orbit. The excess energy may be in the form of a photon or as heat.

Photon emission is the principle used in fluorescent light. Here ultraviolet light photons, which are not visible to the human eye, bombard a phosphor coating on the inside of a glass tube. The phosphor electrons, in returning to their normal orbits, emit photons of light that are visible. By using the proper chemicals for the phosphor coating, any color of light may be obtained, including white. This principle is also used in illuminating the screen of a television picture tube.

The basic principles just developed apply equally well to the atoms of more complex elements. In atoms containing two or more electrons, the electrons interact with each other, and the exact path of any one electron is very difficult to predict. However, each electron lies in a specific energy-level band, and the orbits are considered to be an average of the electron's position.

3.2.1 Shells and Subshells

The difference between the atoms, as far as their chemical activity and stability are concerned, depends on the number and position of the electrons included within the atom. How are these electrons positioned within the atom? In general, the electrons reside in groups of orbits called *shells.* These shells are elliptically shaped and are assumed to be at fixed intervals. Thus, the shells are arranged in steps that correspond to fixed energy levels. The shells and the number of electrons required to fill them may be predicted by the employment of Pauli's exclusion principle. Simply stated, this principle specifies that each shell contains a maximum of $2n^2$ electrons, where n corresponds to the shell number, starting with the one closest to the nucleus. By this principle, the second shell, for example, could contain 2(2^2), or 8, electrons when full.

Besides being numbered, the shells are also given letter designations, as illustrated in Figure 3.3. Starting with the shell closest to the nucleus and progressing outward, the shells are labeled K, L, M, N, O, P, and Q, respectively. The shells are considered full, or complete, when they contain the following quantities of electrons: 2 in the K shell, 8 in the L shell, 18 in the M shell, and so on, following the exclusion principle. Each of these shells is a major shell and can be divided into as many as four subshells, which are labeled *s*, *p*, *d*, and *f*. As with the major shells, the subshells are also limited in regard to the number of electrons that they can contain. Thus, the *s* subshell is complete when it contains 2 electrons, the *p* subshell when it contains 6, the *d* subshell when it contains 10, and the *f* subshell when it contains 14 electrons.

FIGURE 3.3
Shell designation.

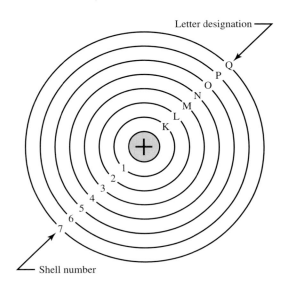

Inasmuch as the K shell can contain no more than two electrons, it must have only one subshell, the *s* subshell. The M shell is composed of three subshells: *s*, *p*, and *d*. If the electrons in the *s*, *p*, and *d* subshells are added, their total is 18, the exact number required to fill the M shell. Notice the electron configuration for copper illustrated in Figure 3.4. The copper atom contains 29 electrons, which completely fill the first three shells and subshells, leaving one electron in the *s* subshell of the N shell.

FIGURE 3.4
Copper atom.

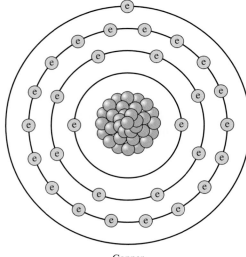

Copper
(29 electrons, 29 protons, 35 neutrons)

3.2.2 Valence

The number of electrons in the outermost shell determines the *valence* of an atom. Consequently, the outer shell of an atom is called the *valence shell,* and the electrons contained in this shell are called *valence electrons.* The valence of an atom determines its ability to gain or lose an electron, which in turn determines the chemical and electrical properties of the atom. An atom that is lacking only one or two electrons from its outer shell easily gains electrons to complete its shell, but a large amount of energy is required to free any of the electrons. An atom having a relatively small number of electrons in its outer shell in comparison to the number required to fill the shell easily loses its valence electrons. The maximum number of electrons that can exist in the valence shell is eight. The periodic chart of the elements (Appendix B) is shown with columns numbered I through VIII, representing numbers of electrons in the valence shell. Elements having the same number of valence electrons are placed in the appropriate column headed by I through VIII. Note that copper, atomic number 29, has one valence electron and can be found in the column headed by I.

SECTION 3.2 **SELF-CHECK**	Answers are given at the end of the chapter. **1.** What kind of energy does an electron possess? **2.** How are electrons distributed within an atom? **3.** Define the term *valence.*

3.3 CONDUCTORS, SEMICONDUCTORS, AND INSULATORS

The association of matter and electricity is important. Because every electronic device is constructed of parts made from ordinary matter, the effects of electricity on matter must be well understood. As a means of accomplishing this, all elements of which matter is made may be placed into one of three categories: *conductors, semiconductors,* and *insulators,* depending on their ability to conduct an electric current. Conductors are elements that conduct electricity very readily. Insulators have an extremely high resistance to the flow of electricity. All matter between these two extremes is called a semiconductor.

The electron theory states that all matter is composed of atoms and the atoms are composed of smaller particles called protons, electrons, and neutrons. The electrons orbit the nucleus that contains the protons and neutrons. It is the *valence electrons* that we are most concerned with in the study of electricity and electronics. These are the electrons that are the easiest to break free from their parent atom. Normally, conductors have three or fewer valence electrons; insulators have five or more valence electrons; and semiconductors usually have four valence electrons.

The electrical conductivity of matter depends on the atomic structure of the material from which the conductor is made. In any solid material, such as copper, the atoms that make up the molecular structure are bound firmly together. At room temperature, copper contains a considerable amount of heat energy. Since heat energy is one method of removing electrons from their orbits, copper contains many free electrons that can move from atom to atom. When not under the influence of an external force, these electrons move in a random manner within the conductor. This movement is equal in all directions so that electrons are not lost or gained by any part of the conductor. When controlled by an external force, the electrons move generally in the same (preferred) direction. The effect of this movement is felt almost instantly (at the speed of light) from one end of the conductor to the other, even though the electrons are traveling at a much slower speed. This electron movement is called an *electric current.*

Lab Reference: The conductivity of the conductors used within a protoboard is verified in Exercises 1, 2, and 3.

Some metals are better conductors of electricity than others. Silver, copper, gold and aluminum are materials with many free electrons (few valence electrons) that make good conductors. Silver is the best conductor, followed by copper, gold, and then aluminum. Copper is used more often than silver because of cost. Aluminum is used where weight is a major consideration, such as in high-tension power lines. Gold is used where oxidation or corrosion is a consideration and good conductivity is required. The ability of a conductor to handle current also depends upon its physical dimensions. Conductors are usually found in the form of wire but may be in the form of bars, tubes, or sheets.

Nonconductors have few free electrons. These materials are called *insulators*. Some examples of these materials are rubber, plastic, enamel, glass, and mica. Just as there is no perfect conductor, neither is there a perfect insulator.

Some materials are neither good conductors nor good insulators, because their electrical characteristics fall between those of conductors and insulators. These in-between materials are classified as *semiconductors*. Germanium and silicon are two common semiconductors used in solid-state devices.

The band-gap approach (Figure 3.5) represents another way to differentiate the differences in conductors, insulators, and semiconductors. The conduction band represents an energy level higher than the valence shell. Electrons in this level are completely mobile and free to drift. As can been seen in Figure 3.5(c), the valence and conduction band energy levels essentially overlap. Thus, it is very easy for conductor electrons to move around freely. As the other two figures show, there is a gap in the semiconductor and insulator. This gap is known as the *forbidden energy gap*. A valence electron cannot exist in this gap. The gap is so large in the insulator that for most practical considerations, the valence electrons cannot gain sufficient energy to jump to the conduction band.

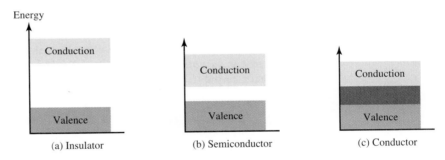

FIGURE 3.5
The band-gap approach to explain (a) insulator, (b) semiconductor, and (c) conductor.

SECTION 3.3 SELF-CHECK

Answers are given at the end of the chapter.

1. What are the three categories in which elements may be placed?
2. What differentiates these categories from each other?
3. What determines electrical conductivity?
4. Define an *electric current*.

3.4   ELECTROSTATICS

Electrostatics (electricity at rest) is a subject with which most people entering the field of electricity and electronics are somewhat familiar. For example, the way a person's hair stands on end after a vigorous rubbing is an example of electrostatics. The study of electrostatics provides you with the opportunity to gain important background knowledge and to develop concepts that are essential to the understanding of electricity and electronics.

As discussed in Chapter 1, interest in the subject of static electricity can be traced to the Greeks. Further study of this subject was done by Gilbert, an English scientist. In 1733, Charles Dufay, a French scientist, made an important discovery about electrification. He found that when a glass rod was rubbed with fur, both the glass rod and the fur became electrified. From experimentations he concluded that there must be two exactly opposite kinds of electricity.

Benjamin Franklin, American statesman, inventor, and philosopher, is credited with first using the terms positive and negative to describe the two opposite kinds of electricity. He labeled as positive the charge on a glass rod when it is rubbed with silk. He attached the term negative to the charge produced on the silk. He called those bodies neutral that were not electrified or charged.

3.4.1   Static Electricity

In a natural, or neutral, state each atom in a body of matter has the proper number of electrons in orbit around it. The number of electrons equals the number of protons in the nucleus. Consequently, the whole body of matter composed of neutral atoms is electrically neutral. In this state, it is said to have a *zero charge.* Electrons neither leave nor enter the neutrally charged body should they come in contact with other neutral bodies. If, however, any number of electrons are removed from the atoms of a body of matter, there remain more protons than electrons and the whole body of matter becomes electrically positive. Should the positively charged body come in contact with another body having a normal charge or having a negative (too many electrons) charge, an electric current flows between them. Electrons leave the more negative body and enter the positive body. This electron flow continues until both sides have equal charges. When two bodies of matter have unequal charges and are near one another, an electric force is exerted between them because of their unequal charges. However, because they are not in contact, their charges cannot equalize. The existence of such an electric force, where current cannot flow, is referred to as *static electricity.* ("Static" in this instance means "not moving.") This force is also referred to as an electrostatic force.

3.4.2   Ionization

When the atom loses electrons or gains electrons in the process of electron exchange, it is said to be *ionized.* For ionization to take place there must be a transfer of energy that results in a change in the internal energy of the atom. An atom having more than its normal amount of electrons acquires a negative charge and is called a *negative ion.* The atom that gives up some of its normal electrons is left with fewer negative charges than positive charges and is called a *positive ion.* Thus, *ionization* is the process by which an atom loses or gains electrons.

3.4.3   Nature of Charges

When in a natural, or neutral, state an atom has an equal number of electrons and protons. Because of this balance, the net negative charge of the electrons in orbit is exactly balanced by the net positive charge of the protons in the nucleus. The atom is therefore *neutral.*

Due to normal molecular activity, there are always ions present in any material. Further, the numbers of positive and negative ions present are equal, with a net neutral state in the material. If either ion numerically outnumbers its counterpart, the material has a net positive or negative charge.

Because ions are actually atoms without a normal number of electrons, it is the excess or lack of electrons in a substance that determines its charge. In most solids, the transfer of charges is by the movement of electrons rather than ions. The transfer of charges by ions becomes significant when one considers electrical activity in liquids or gases. In this text, the discussion of electrical behavior is in terms of *electron movement*.

3.4.4 Charged Bodies

One of the fundamental laws of electricity is that like charges repel each other, whereas unlike charges attract each other. A positive charge and negative charge, being unlike, tend to move toward each other. In the atom, electrons are drawn toward the proton as a result of their dissimilar charges. However, the attractive force is balanced by the electron's centrifugal force caused by its rotation about the nucleus. As a result, the electrons remain in orbit and are not drawn into the nucleus. Electrons repel each other because of their like negative charges, and protons repel each other because of their like positive charges.

The law of charged bodies may be demonstrated by a simple experiment. Two pith (paper pulp) balls are suspended near one another by threads. Each is charged as shown in Figure 3.6(a). Their movement is toward each other. As they touch, they remain in contact until the left-hand ball gains a portion of the negative charge from the other. At this time they swing apart. Figures 3.6(b) and 3.6(c) shows that when the charges are alike, the balls repel each other.

FIGURE 3.6
Reaction between charged bodies.

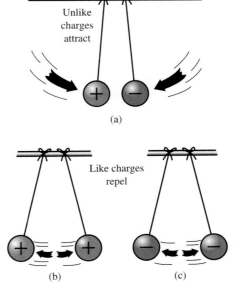

SECTION 3.4 SELF-CHECK

Answers are given at the end of the chapter.

1. Define the term *electrostatics*.
2. Define the term *static electricity*.
3. What differentiates a positive ion from a negative ion?
4. Like charges _____ each other, whereas unlike charges _____ each other.

3.5   COULOMB'S LAW OF CHARGES

The relationship between attracting or repelling charged bodies was first discovered and written about by the French scientist Charles A. Coulomb. His law is shown mathematically as

$$F = \frac{kq_1q_2}{d^2} \tag{3.1}$$

where F = the force in newtons (N)
q_1, q_2 = the charges in coulombs (C)
k = a constant (8.988×10^9) for converting the values to SI units
d = the distance in meters between the charges

A coulomb is a rather large quantity of charge. A coulomb is equal to 6.25×10^{18} electrons. Typical electrostatic values obtained in the laboratory seldom exceed a few millionths of a coulomb, so the microcoulomb (μC) is often used.

EXAMPLE 3.1

Determine the force of repulsion between two like charges, each having a charge of 0.25 μC and separated by a distance of 3 cm.

Solution $F = \dfrac{kq_1q_2}{d^2}$

$$= \frac{8.988 \times 10^9 \times 0.25 \times 10^{-6} \times 0.25 \times 10^{-6}}{(3 \times 10^{-2})^2}$$

$$= 0.624 \text{ N}$$

Practice Problems 3.1   Answers are given at the end of the chapter.
1. Determine the force of attraction between unlike charges where one has a charge of 0.3 μC and the other, 0.5 μC, separated by a distance of 5 cm.

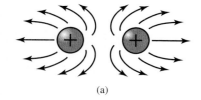

(a)

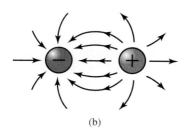

(b)

FIGURE 3.7
Electrostatic lines of force: (a) like charges repel; (b) opposite charges attract.

The importance of the equation is in the understanding that the force is directly proportional to the product of charges and inversely proportional to the square of the distance.

3.5.1   Electric Fields

The space between and around charged bodies in which their influence is felt is called an *electric field of force*. This field can exist in air, glass, paper, or a vacuum. Electrostatic fields and dielectric fields are other names used to refer to this region of force.

Fields of force spread out into space surrounding their point of origin and, in general, diminish in proportion to the square of the distance from their source. The field about a charged body is generally represented by lines that are referred to as electrostatic lines of force. These lines are imaginary and are used merely to represent the direction and strength of the field. To avoid confusion, the lines of force exerted by a positive charge are shown leaving the charge. For a negative charge, they are shown entering. This is illustrated in Figure 3.7. Figure 3.7(a) represents the repulsion of like-charged bodies and their associated fields. Figure 3.7(b) represents the attraction of unlike-charged bodies and their associated fields.

SECTION 3.5 SELF-CHECK

Answers are given at the end of the chapter.

1. The force of attraction or repulsion is _____ proportional to the product of charges and _____ proportional to the square of the distance.
2. What is an electric field of force?
3. Where can this field exist?

■ SUMMARY

1. Matter is defined as anything that occupies space and has mass.
2. Matter is found in one of three states, solid, liquid, or gaseous.
3. An element is a substance that cannot be reduced to a simpler substance by chemical means.
4. A compound is a chemical combination of two or more elements.
5. An atom is the smallest particle of an element that retains the characteristics of that element.
6. Atoms contain subatomic particles of electrons, protons, and, in most cases, neutrons.
7. Electrons are small negative charges, protons are larger positive charges, and neutrons have no electrical charge.
8. The number of electrons within an atom equals the number of protons.
9. Electrons contain both kinetic and potential energy.
10. Electrons orbit the nucleus of an atom in defined energy levels or shells.
11. The valence shell is the outermost shell surrounding the nucleus. It is the number of electrons in this shell that determines the electrical properties of the atom.
12. Conductors have three or fewer valence electrons, whereas semiconductors have four valence electrons and insulators have five or more valence electrons.
13. Electrons moving in a preferred direction are known as an electric current.
14. Good conductors are generally made from metals such as gold, silver, copper, or aluminum.
15. Static electricity is the force exerted between two bodies of matter of unequal charges and near one another.
16. When an atom loses a valence electron it becomes a positive ion; conversely, when an atom gains a valence electron it becomes a negative ion.
17. Like charges repel each other; unlike charges attract each other.
18. Coulomb's law shows that the force of attraction or repulsion between charged bodies varies directly as the product of their charges and is inversely proportional to the square of the distance between them.
19. The space between and around charged bodies in which their influence is felt is called an electric field of force. This field can exist in air, glass, paper, or a vacuum.

■ IMPORTANT EQUATIONS

$$F = \frac{kq_1q_2}{d^2} \tag{3.1}$$

■ REVIEW QUESTIONS

1. Define matter and give two examples.
2. What is an element?
3. Describe an atom.
4. Describe the atom's subatomic particles.
5. Describe the theory concerning atoms postulated by Rutherford.
6. Describe an electron's energy.
7. When can an electron move up to a higher energy level?
8. Describe the placement of the 29 electrons of the copper atom.
9. What determines the valence of an atom?
10. What is the importance of the valence of an atom?
11. What are the differences between conductors, semiconductors, and insulators?
12. What does heat energy do to valence electrons?
13. Describe the velocity of electron movement and its effect, electric current.
14. Compare and contrast the uses of gold, silver, aluminum, and copper as conductors.
15. Describe the band-gap approach to differentiate the differences in conductors, semiconductors, and insulators.
16. Describe the term *electrostatics*.
17. What is static electricity?
18. What does the process of ionization consist of?
19. How is the transfer of charge accomplished in most solids?
20. Why doesn't an orbiting electron fall into the nucleus of the atom?
21. The force of attraction between two dissimilar charged bodies _____ as the individual charge increases.
22. Describe an electric field of force.

■ CRITICAL THINKING

1. Compare and contrast the terms *atom, molecule, element,* and *compound.*

2. Describe the distribution of electrons around the nucleus utilizing Pauli's exclusion principle.
3. Describe the electrical conductivity levels of conductors, semiconductors, and insulators.

■ PROBLEMS

1. Calculate the force of attraction between two dissimilar charges where one charge is 0.50 μC and the other is 0.75 μC. They are separated by a distance of 5 cm.
2. Determine the force of repulsion between two charges each having 0.80 μC and separated by a distance of 40 mm.

■ ANSWERS TO PRACTICE PROBLEMS

PRACTICE PROBLEMS 3.1

1. 0.54 N

■ ANSWERS TO SELF-CHECKS

SECTION 3.1

1. Matter is defined as anything that occupies space and has mass.
2. An element is a substance that cannot be reduced to a simpler substance by chemical means. A compound is formed when two or more elements chemically combine.
3. An atom is the smallest particle of an element that retains the characteristics of that element.
4. The particles that make up an atom

SECTION 3.2

1. Both kinetic and potential energy
2. Electrons reside in groups of orbits called shells.
3. The number of electrons in the outermost shell determines the valence of an atom.

SECTION 3.3

1. Conductors, semiconductors, or insulators
2. The number of valence electrons the atom has
3. Electrical conductivity depends on the atomic structure of the material from which a conductor is made.
4. Electron movement within a conductor in a preferred direction

SECTION 3.4

1. Electrostatics is electricity at rest.
2. The existence of an electric force between charged bodies, where current cannot flow
3. A positive ion has lost a valence electron, whereas a negative charge has gained a valence electron.
4. Repel, attract

SECTION 3.5

1. Directly, inversely
2. The space between and around charged bodies in which their influence is felt
3. In air, glass, paper, or a vacuum

CHAPTER 4

ELECTRICAL QUANTITIES

OBJECTIVES

After completing this chapter, you will be able to:

1. Define *voltage* in terms of work and charge.
2. Compare and contrast potential difference and voltage.
3. Describe the conditions needed for electrons to flow.
4. Describe voltage references and ground.
5. Explain how voltage is produced.
6. Compare and contrast various methods for producing voltage.
7. Describe the term *voltage drop* and how it is labeled.
8. Define *electric current* in terms of charge and time.
9. Differentiate between electron flow and conventional current flow.
10. Describe how current is measured.
11. Define the term *electrical resistance*.
12. Describe the factors that affect resistance.
13. Compare and contrast conductance and resistance.
14. Describe the conditions necessary to make a practical electric circuit.

INTRODUCTION

We have seen how important the study of atomic properties and electrostatics is to our understanding of electricity. Electrostatics is the study of electricity (charges) at rest. To be useful, we must put the charges in motion (electrodynamics). In this way work can be accomplished. This chapter defines the criteria for charge to flow or to be put into motion, how the flow is quantified, and what affects the amount of flow. A practical circuit is also defined.

4.1 ELECTROMOTIVE FORCE—VOLTAGE

From the previous study of electrostatics you learned that a field of force exists in the space surrounding any electrical charge. The strength of the field depends directly on the force of the charge.

The charge of one electron might be used as a unit of electrical charge, since charges are created by the displacement of electrons, but the charge of one electron is so small that it is impractical to use. The charge was measured by Robert Millikan, American physicist, beginning in 1906 utilizing his famous oil-drop experiment. It was found to be 1.6×10^{-19} C. The coulomb is a practical unit adopted for measuring charges and named after Charles Coulomb. *One coulomb* (C) is equal to 6.25×10^{18} electrons (Chapter 3). Note that the charge of an electron and the coulomb are opposites (reciprocals); thus, $1/(1.6 \times 10^{-19}) = 6.25 \times 10^{18}$.

When a difference of charge of 1 C exists between two bodies, one unit of electrical potential energy exists, called the *potential difference* between the two bodies. This is also referred to as an *electromotive force (emf), electrical pressure,* or *voltage.* The unit of measure of emf, electrical pressure, or voltage is the *volt.*

A potential difference provides the mechanism for work to be done. Voltage is also related to the work that a source can do to move an electrical charge through a circuit. This gives us another definition for voltage:

$$E = \frac{W}{Q} \tag{4.1}$$

where E = the potential difference (V)
W = the amount of work or energy expended (J)
Q = the charge (C)

(Note that 1 J is the work done by a force of 1 N acting through a distance of 1 m.)

EXAMPLE 4.1

Calculate the potential difference (V) between two points if 24 J of work must be done to move 0.2 C of charge between the two points.

Solution $E = \dfrac{W}{Q}$

$\qquad = \dfrac{24 \text{ J}}{0.2 \text{ C}}$

$\qquad = 120 \text{ V}$

Practice Problems 4.1   Answers are given at the end of the chapter.
1. Calculate the potential difference (V) between two points if 32 J of work must be done to move 0.2 C of charge between the two points.
2. Calculate the charge moved between two points when the potential difference is 100 V and the work done is 40 J.

Bodies become electrically charged by the displacement of electrons, developing an excess of electrons at one point and a deficiency at another point. Consequently, a charge must always have either a negative or positive polarity. A body with an excess of electrons is considered to be *negative,* whereas a body with a deficiency of electrons is *positive.*

A potential difference can exist *between two points,* or bodies, *only* if they have a *difference of charges.* In other words, there is no difference in potential between two bodies if they both have a deficiency of electrons to the same degree. If, however, one body is deficient by 6 C (representing 6 V) and the other is deficient by 12 C (representing 12 V), there is a potential difference of 6 V. The body with the greater deficiency is positive with respect to the other.

When a potential difference exists between two charged bodies that are connected by a conductor, electrons flow in the conductor. This flow is from the negatively charged body to the positively charged body. The flow continues until the two charges are equalized and the potential difference no longer exists. The key point to remember is that *potential difference (voltage) does not flow; charges do.* Voltage supplies the "push" to put and keep the charges in motion.

An analogy for this action is shown in the two water tanks connected by a pipe and valve in Figure 4.1. At first the valve is closed, with all the water in tank A. Thus, the water pressure across the valve is at maximum. When the valve is opened, the water flows through the pipe from *A* to *B* until the water level becomes the same in both tanks. The water then stops flowing in the pipe, because there is no longer a difference in water pressure between the two tanks. This analogy shows how the term *electrical pressure* was coined to describe the potential difference in an electric circuit.

Electron movement through an electric circuit is directly proportional to the potential difference, or voltage, across the circuit, just as the flow of water through the pipe in Figure 4.1 is directly proportional to the difference in water level in the two tanks. If the voltage is increased, the electron flow is increased. If the voltage is decreased, the electron flow is decreased.

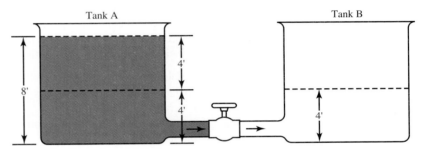

FIGURE 4.1
Water analogy for electric difference of potential.

4.1.1   Voltage Reference

In most electrical circuits, only the potential difference *between two points* is of importance. The absolute potentials of the points are of little concern. Very often it is convenient to use one standard reference for all the various potentials throughout an electric circuit. For this reason, the potentials at various points in a circuit are generally measured with respect to a metal chassis or some other common point (or *ground reference*). The chassis is considered to be at zero potential, and all other potentials are either positive or negative with respect to the chassis, or common point. When used as the reference point, the chassis is said to be at *ground potential.* When a ground is not indicated in a circuit, the negative terminal of the source is considered to be the reference. The voltage-measurement exercises in your lab course give you firsthand experience understanding terms such as voltage and ground. (The term *ground* is more fully explained in Section 7.6.1.)

4.1.2   Voltage Prefixes and Measuring Instruments

When large or small values of voltages are encountered, SI prefixes are used for convenience. The *kilovolt* (kV) is used for large values of voltage, whereas the *millivolt* (mV) and *microvolt* (μV) are used for small values of voltage.

The device used to measure voltage is the *voltmeter. Remember that potential difference, or voltage, exists between two points.* Thus, to measure voltage with a voltmeter, the voltmeter's two leads are connected across the component (from one side to the other). The black lead of the voltmeter is connected to the negative (−) side (ground side or lowest potential) of the component, and the red lead of the voltmeter is connected to the positive (+) side of the component. (The voltmeter is more fully described in Section 10.4.)

Lab Reference: The interpretation of scales on the voltmeter and other instruments is demonstrated in Exercise 5, and measuring voltage is demonstrated in Exercise 6.

**SECTION 4.1
SELF-CHECK**

Answers are given at the end of the chapter.

1. Define potential difference.
2. Under what conditions do electrons flow in a circuit?
3. What is a ground reference?

4.2   THE PRODUCTION OF VOLTAGE

To be a practical source of voltage, the potential difference must not be allowed to dissipate but must be maintained continuously. As one electron leaves the concentration of negative charge, another must be immediately provided to take its place, or the charge eventually diminishes to the point where no further work can be accomplished. A *voltage source,* therefore, is a device capable of supplying and maintaining voltage while some type of electrical device is connected to its terminals. The internal action of the source is such that it continuously removes electrons from one terminal, keeping it positive, and simultaneously supplies these electrons to the second terminal, which maintains a negative charge.

There are several methods for producing a voltage, or electromotive force. Some of these methods are more widely used than others, and some are used for specific applications. The following list gives several methods of producing a voltage.

1. Friction: Voltage is produced by rubbing certain materials together.
2. Pressure (piezoelectricity):   Voltage is produced by applying mechanical pressure to crystals of certain substances.
3. Heat (thermoelectricity):   Voltage is produced by heating the connecting junction of two dissimilar metals.
4. Light (photoelectricity):   Voltage is produced by light striking a photosensitive substance.
5. Chemical action: Voltage is produced by chemical reaction in a battery cell.
6. Magnetism: Voltage is produced in a conductor that moves through a magnetic field, or a magnetic field moves through a conductor in such a manner as to cut the magnetic lines of force of the field.

4.2.1   Voltage Produced by Friction

The first method discovered for creating voltage was that generated by friction. Rubbing a rod with fur is an example of the way in which a voltage is produced by friction. However, this method does not provide a practical source of voltage.

Friction is used when a voltage of a larger amplitude and of a more practical value is required. Machines have been developed in which charges are transferred from one terminal to another by means of rotating glass discs or moving belts. The most notable of these machines is the *Van de Graaff generator.* This generator is used to produce potentials on the order of millions of volts that are used for research.

4.2.2   Voltage Produced by Pressure

One specialized method of generating an emf utilizes the characteristics of certain ionic crystals such as quartz, Rochelle salts, or tourmaline. These crystals have the remarkable ability to generate a voltage whenever stresses are applied to their surfaces. If the force is reversed and the crystal is stretched, an emf of opposite polarity is produced. These crystals convert mechanical energy into electrical energy. This phenomenon, called the *piezoelectric effect,* is shown in Figure 4.2. Some common devices that make use of the piezoelectric effect are microphones, phonograph cartridges, and oscillators used in radio circuits.

FIGURE 4.2
Compression and decompression of a crystal causes voltage to be produced and then a reversal of polarity.

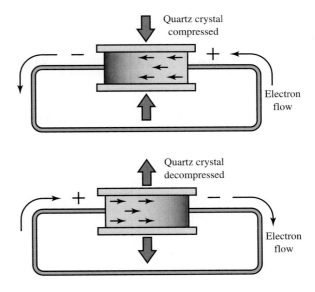

4.2.3 Voltage Produced by Heat

When a length of metal, such as copper, is heated at one end, electrons tend to move away from the hot end toward the colder end. This is true in most metals. However, in some metals, such as iron, the opposite takes place, and the electrons move toward the hot end. These characteristics are illustrated in Figure 4.3. The electrons are moving through the copper away from the heat and through the iron toward the heat. A difference of potential exists between the ends opposite the hot junction. This voltage causes an electron flow through the current meter, as shown. This device is generally referred to as a *thermocouple*.

Thermocouples provide somewhat greater power capacities than crystals, but their capacity is still small compared to other sources. They are widely used to measure temperature and as heat-sensitive devices in thermostats.

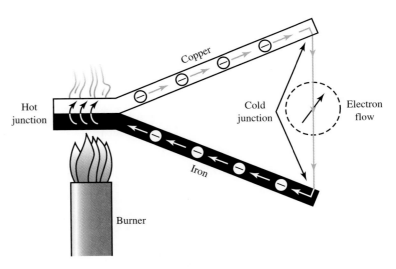

FIGURE 4.3
Thermocouple: voltage produced by heat.

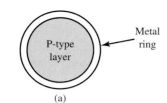

(a)

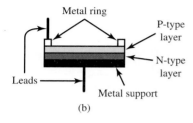

(b)

FIGURE 4.4
A basic photovoltaic cell: (a) top view; (b) side view.

4.2.4 Voltage Produced by Light

A *photoelectric cell* is a device that is activated by electromagnetic energy in the form of light waves (photons). Three basic kinds of photoelectric cells exist, corresponding to the three different forms of the photoelectric effect that they employ. These include the *photoconductive cell,* the *photoemissive cell,* and the *photovoltaic* or *solar cell.* The first two are passive devices, depending on an external current or voltage. The photoconductive cell is explained further in Section 5.3.2. Photoemissive cells such as photo diodes are described in detail in electronic devices texts.

The photovoltaic cell (Figure 4.4) is active, converting light energy directly into electricity. The photovoltaic cell employs a solid-state diode structure with a large area on a silicon wafer. The surface layer is very thin and transparent, so that light can reach the junction region of the silicon sandwich. In this region the photons are absorbed, releasing charges from the atomic bonds. These charges migrate to the terminals, raising the potential differences between the terminals. With no load attached, the terminal voltage approximates 0.6–1.0 V. With a short-circuit load, the current may be a few milliamperes. Cells may be connected in series to raise the voltage, and the output current can be raised by a parallel connection of cells. Efficiencies are under 30% except in space applications. Here the solar cell may be 75% efficient, because the sun's rays are not inhibited by an atmosphere.

Photovoltaic cells are used in calculators, wristwatches, space satellites, and remote terrestrial applications, such as buoys, oil-drilling platforms, and mountaintop microwave repeaters.

4.2.5 Voltage Produced by Chemical Action

An *electric cell,* or *battery,* is a device that converts chemical energy into electricity. Strictly speaking, a battery consists of two or more cells in series or parallel, but the term is also used for single cells. Any cell (Figure 4.5) consists of a liquid, paste, or solid electrolyte, a positive electrode, and a negative electrode. The electrolyte is an ionic conductor; one of the electrodes reacts, producing electrons, and the other accepts electrons. When the electrodes are connected to a device to be powered, called a *load,* an electrical current flows.

Batteries in which the chemicals cannot be reconstituted into their original form once the energy has been converted (that is, batteries that have been discharged) are called *primary cells,* or *voltaic cells.* Common examples include zinc-carbon, alkaline, and some lithium batteries. Batteries in which the chemicals can be reconstituted by passing an electric current through them in the direction opposite that of normal cell operation are called *secondary cells, rechargeable cells, storage cells,* or *accumulators.* Common examples include lead-acid car batteries and nickel-cadmium batteries for small electronics. (Detailed descriptions of battery types are found in Section 13.10.)

FIGURE 4.5
Simple electric cell.

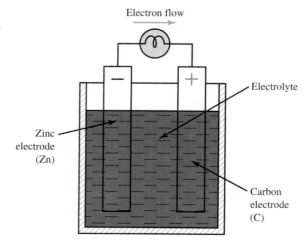

Combining several cells in series (Figure 4.6(a)) provides a higher voltage than an individual-cell voltage but yields the same current capacity. This is known as a series-aiding configuration. The common 9-V alkaline battery used for small electronics is an example. It is made up of six cells, each having a cell voltage of 1.5 V. To increase the current capacity, cells are connected in parallel (Figure 4.6(b)). However, this arrangement does not increase the voltage. To provide adequate power when both the voltage and current requirements are greater than the capacity of one cell, a series-parallel network of cells must be used. This is illustrated in Figure 4.6(c).

Lab Reference: The series-aiding and series-opposing characteristics of cells are demonstrated in Exercise 6.

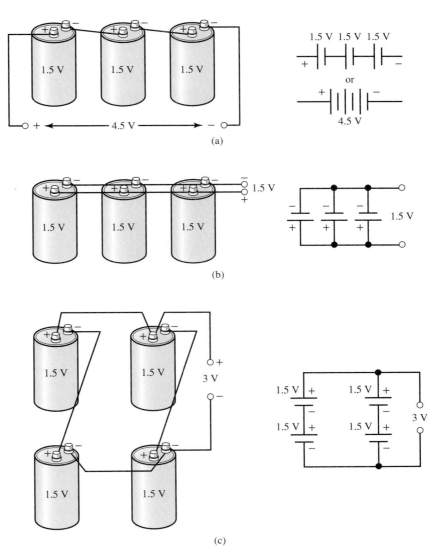

FIGURE 4.6
Increasing cell voltage and/or current: (a) series-connected cells, more voltage; (b) parallel cell connection, more current; (c) series-parallel cell connection, more voltage and current.

4.2.6   Voltage Produced by Magnetism

There are three fundamental conditions that must be met before a voltage can be produced by magnetism.

1. There must be a conductor in which the voltage will be produced.
2. There must be a magnetic field in the conductor's vicinity. The magnetic field can come from a permanent magnet or an electromagnet.

3. There must be relative motion between the field and conductor. The conductor must be moved to cut across the magnetic lines of force, or the field must be moved so that the lines of force are cut by the conductor.

In accordance with these conditions, when a conductor or conductors move across a magnetic field so as to cut the lines of force, electrons within the conductor are pushed in one direction or another (they are separated). Thus, an electric force, or voltage, is generated. The greatest voltage occurs when the angle of cutting is perpendicular (90°). This is illustrated in Figure 4.7. The direction of motion indicated by the large arrow produces a voltage across the conductor as shown, with electron flow in the closed loop of the conductor. More complex aspects of power generation, magnetism, and voltage are covered in later sections of this text.

FIGURE 4.7
Voltage produced by magnetism.

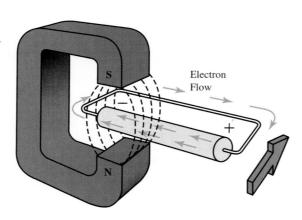

Electron
Flow

**SECTION 4.2
SELF-CHECK**

Answers are given at the end of the chapter.

1. What is the requirement for a voltage source?
2. Name four of the six methods used to produce voltage.
3. How do you get more voltage than that supplied by an individual cell?

4.3   VOLTAGE DROP

A voltage source represents a *voltage rise,* or the energy imparted to the free electrons for the formation and maintenance of electric current. A *voltage drop* represents the energy used by the free electrons while engaged in current flow. This energy is usually dissipated as heat in most electronic components. In this text, voltage sources will be identified as V_s, whereas a voltage drop is identified as V_1, V_A, or V_{R1}, for example. (The voltage drop uses the label of the component or the point with which it is associated.) Remember that voltage exists *between two points,* so if only one point is identified, the *negative terminal of the source or ground* is considered as the second point. (Further voltage identification and references are detailed in Section 7.6.)

**SECTION 4.3
SELF-CHECK**

Answers are given at the end of the chapter.

1. What does the voltage source represent?
2. What does a voltage drop represent?

4.4   ELECTRIC CURRENT

Electric current (*I*) is defined as the directed flow of free electrons and, as has been previously stated, is used throughout this text. The direction of electron movement is from a region of negative potential to a region of positive potential. Therefore, electric current can be said to flow from *negative* to *positive*. The direction of current flow in a material is determined by the polarity of the applied voltage. The end of a component where current enters is negative, whereas the end where current leaves is positive.

The motion of positive charges, in the opposite direction to electron flow, is considered *conventional current flow*. This direction is generally used for analysis in electrical engineering. The reason is based on some traditional definitions in the science of physics. An example of positive charge in motion for conventional current is the current or hole charges in *P*-type semiconductors. Also, a current of positive ions in liquids and gases moves in the opposite direction from electron flow.

It should be noted that with semiconductors, the arrow symbols for current flow are shown in the direction of conventional current. This method was standardized by the Electronic Industries Association (EIA) for all semiconductor devices.

Any circuit can be analyzed with either electron flow for current or by conventional current. *It is the work done by the moving charge that is important.*

4.4.1   Random Drift

All materials are composed of atoms, each of which is capable of being ionized. If some form of energy, such as heat, is applied to a material, some valence electrons acquire sufficient energy to move to a higher energy level. As a result, some electrons are freed from their parent atoms, the atoms becoming ions. Other forms of energy, particularly light or an electric field, can also cause ionization to occur.

The number of free electrons resulting from ionization depends upon the quantity of energy applied to a material as well as the atomic structure of the material. At room temperature, conductors have an abundance of free electrons.

In the study of electric current, conductors are of major concern. Conductors are made up of atoms that contain loosely bound electrons in their outer orbits. Due to the effects of increased energy, these outermost electrons frequently break away from their atoms and drift freely throughout the conductor. The free electrons, also called mobile electrons, take a path that is not predictable. Their movement is characterized as a *random drift*. Random drift occurs in all materials. The degree of random drift is greater in a conductor than in an insulator.

4.4.2   Directed Drift

Associated with every charged body is an electrostatic field. Bodies that are charged alike repel one another, and bodies with unlike charges attract each other. An electron is affected by an electrostatic field in exactly the same manner as any negatively charged body. It is repelled by negative charge and attracted by positive charge. If a conductor has a potential difference impressed across it, a direction of movement is imparted to the randomly drifting electrons. A general migration of electrons occurs from one end of the conductor to the other. The direction is from negative to positive.

The directed movement of electrons occurs at a relatively low velocity. The effect of this directed movement, however, is felt almost instantaneously. As a potential difference is impressed across the conductor, the positive terminal of the battery attracts electrons (Figure 4.8) beyond point *A*. Point *A* becomes positive because it now has an electron deficiency. As a result, electrons are attracted from point *B* to point *A*. Point *B* has now developed an electron deficiency; therefore, it attracts electrons. This same effect occurs throughout the conductor and repeats itself from points *D* to *C*. At the same instant,

FIGURE 4.8
Effect of directed drift.

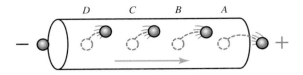

the positive battery terminal attracts electrons from point *A* and the negative terminal repels electrons toward point *D*. Point *D* attracts electrons as it gives up electrons to point *C*. This process continues for as long as a difference of potential exists across the conductor. Although the individual electron moves quite slowly, the overall action of the electron flow occurs at approximately the *speed of light.*

4.4.3 Magnitude of Current Flow

The magnitude of current flow is affected by the difference of potential in the following manner. Initially, the mobile electrons are given additional energy because of the repelling and attracting electrostatic field. If the potential difference is increased, the electric field is stronger, the amount of energy imparted to a mobile electron is greater, and the magnitude of current is increased. If the potential difference is decreased, the strength of the field is reduced, the energy supplied to the electron is diminished, and the magnitude of current is decreased.

4.4.4 Measurement of Current

The magnitude of current is measured in *amperes* (A). A current of 1 A is said to flow when 1 C of charge passes a point in 1 s. Remember, 1 C is equal to the charge of 6.25×10^{18} electrons. This may be expressed mathematically as

$$I = \frac{Q}{t} \qquad (4.2)$$

where
I = current (A)
Q = charge (C)
t = time (s)

EXAMPLE 4.2

If 2 C pass a point in 4 s, what is the magnitude of current flow in amperes?

Solution $I = \dfrac{2\,C}{4\,s}$

  $= 0.5\ A$

EXAMPLE 4.3

If 2 C pass a point in 0.75 s, what is the magnitude of current flow in amperes?

Solution $I = \dfrac{2\,C}{0.75\,s}$

  $= 2.67\ A$

Practice Problems 4.2   Answers are given at the end of the chapter.
1. If 0.5 μC pass a point in 0.5 ms, what is the magnitude of current flow in amperes?
2. What time elapses if 3 A of current flows when 4 C pass a point?

4.4.5 Amperage Prefixes

The ampere is frequently too large a unit for measuring current. Therefore, the *milliampere* (mA), one-thousandth of an ampere; the *microampere* (μA), one-millionth of an ampere; or the *picoampere* (pA), one-millionth-millionth of an ampere, are used. The device used to measure current is called an *ammeter* and is discussed in Section 10.3.

SECTION 4.4 SELF-CHECK

Answers are given at the end of the chapter.

1. Define electric current.
2. What is conventional current flow?
3. What is the velocity of current flow?

4.5   ELECTRICAL RESISTANCE

It has been stated that the directed movement of free electrons constitutes a current flow. Some materials offer little opposition to current flow, whereas others greatly oppose current flow. The opposition to current flow is known as *resistance (R);* the unit of measure is the *ohm* (Ω). A conductor has 1 Ω of resistance when an applied potential difference of 1 V produces a current of 1 A. A *resistor* is the name of a common component that offers resistance to a current flow. It is used extensively in electronic circuits and is detailed in Chapter 5.

4.5.1   Factors That Affect Resistance

The magnitude of resistance is determined in part by the number of free electrons available within the material. Because a decrease in the number of free electrons decreases the current flow, it can be said that the opposition to current flow (resistance) is greater in a material with fewer free electrons. Thus, the resistance of a material is determined by the number of free electrons available in the material.

Depending upon their atomic structures, different materials have different quantities of free electrons. Therefore, the various conductors used in electrical applications have different values of resistance.

Consider a simple metallic substance. Most metals are crystalline in structure and consist of atoms that are tightly bound in the lattice network. The atoms of such elements are so close together that the electrons in the outer shells of the parent atoms are associated with their neighbors as well (Figure 4.9(a)). As a result, the degree of attachment of an outer electron to an individual atom is *practically zero.* Depending on the metal, at least one electron, sometimes two, and, in a few cases, three electrons per atom exist in this state. In such a case, a relatively small amount of additional electron energy would free the outer electrons from the attraction of the nucleus. At normal room temperature, materials of this type have many free electrons and are good conductors. (Good conductors have a *low resistance.*)

If the atoms of a material are farther apart, as illustrated in Figure 4.9(b), the electrons in the outer shell are not equally attached to the associated atoms as they orbit the nucleus. They are attracted only by the nucleus of the parent atom. Therefore, a greater amount of energy is required to free any of these electrons. (Materials of this type are poor conductors and therefore have a *high resistance.*)

Silver, gold, copper, and aluminum are good conductors, and their atoms have a low resistance to current flow. Copper is the most widely used conductor in electrical applications. The effects on resistance value, as determined by a conductor's cross-sectional area, length, material, and temperature, are discussed in Section 13.2.

FIGURE 4.9
Atomic spacing in conductors: (a) atoms of good conducting metals; (b) atoms of poor conducting metals.

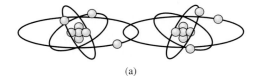

(a)

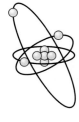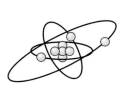

(b)

4.5.2 Resistance Prefixes

Resistance values for electrical and electronic components can range from several millions of ohms to values of less than 1 Ω. Typical values for commonly used components are in the *ohm* (Ω), the *kilohm* (kΩ), or thousands of ohms, and the *megohm* (MΩ), or millions of ohms, ranges. The device used to measure resistance is called the *ohmmeter* and is described in Section 10.5.

4.5.3 Conductance

Electricity is frequently explained in terms of opposites. The term that is the opposite of resistance is *conductance* (*G*). Conductance is the ability of a material to pass an electric current. The factors that affect the magnitude of resistance are exactly the same for conductors, but they affect conductance in the opposite manner.

The old unit of conductance was the mho, which is simply ohm spelled backwards. The SI unit, the siemens (S), is now more often used. It is named after Ernst von Siemens, a European inventor. The relationship that exists between resistance (*R*) and conductance (*G*) is a reciprocal one. In terms of resistance and conductance,

$$R = 1/G \qquad \qquad \textbf{(4.3)}$$

$$G = 1/R \qquad \qquad \textbf{(4.4)}$$

EXAMPLE 4.4

Determine the conductance in siemens when the resistance is 10 Ω.

Solution $G = 1/10 \ \Omega$
$\qquad\quad = 0.1 \ \text{S}$

EXAMPLE 4.5

Determine the resistance when the conductance is 0.4 mS.

Solution $R = 1/(0.4 \times 10^{-3} \text{s})$
$\qquad\quad = 2500 \ \Omega$
$\qquad\quad = 2.5 \ \text{k}\Omega$

Practice Problems 4.3   Answers are given at the end of the chapter.
1. Determine the conductance when the resistance is 5 kΩ.
2. Determine the resistance when the conductance is 0.1 mS.

Answers are given at the end of the chapter.

1. What is electrical resistance?
2. What affects the amount of resistance in a conductor?
3. Define *conductance*.

4.6   A PRACTICAL ELECTRIC CIRCUIT

An electric circuit has four important characteristics:

1. There must be a source of potential difference. Without an applied voltage, current cannot flow.
2. There must be a complete path (supplied by a conductor) for current flow, from one side of the applied voltage source, through the external circuit, and returning to the other side of the voltage source.
3. The current path normally has a resistance that may be in the form of several resistors. The resistance is in the circuit for the purpose of either generating heat or limiting the amount of current. The *practical circuit* contains a load. A *load* is any device through which current flows that changes the electrical energy into a more useful form.
4. The circuit is usually controlled by some device. This control device may be a switch, thermostat, or relay. The device controls the flow of current to the load.

Lab Reference: The attributes of a practical circuit are demonstrated in Example 4.

A flashlight is an example of a practical basic electric circuit. It contains a source of electrical energy (potential difference), a load (the bulb) that changes electrical energy into a more useful form of energy (light), and a switch to control the energy delivered to the load.

4.6.1   Schematic Representation

The technician's primary aid in troubleshooting a circuit is a schematic diagram. A *schematic diagram* is a picture of a circuit that uses symbols to represent the various circuit components. Figure 4.10 shows the symbols of components discussed in this chapter. These—and others like them—are used throughout the study of electricity and electronics.

FIGURE 4.10
Symbols commonly used in electricity.

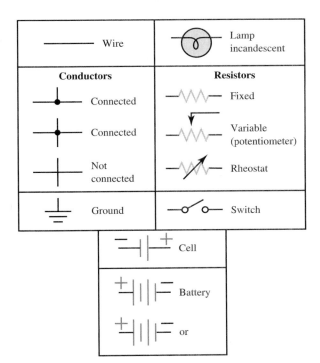

Note that the longer line used in the cell or battery symbol is the positive terminal. Thus, the short line is the negative terminal. A zigzag line is used for a resistor. Variable (adjustable) resistors are shown with an arrow and are in the form of a potentiometer or a rheostat.

The schematic diagram in Figure 4.11 represents a flashlight that is switched on. Current flows in the direction of the arrows from the negative terminal of the battery (BAT), through the switch (S1), through the lamp (DS1), and back to the positive terminal of the battery. With the switch closed, the path for current is complete. Note that the negative and positive sides of the lamp are indicated. A properly placed voltmeter to indicate the voltage drop across the lamp is also shown. Current continues to flow until the switch is opened or the battery is drained.

FIGURE 4.11
Basic flashlight schematic.

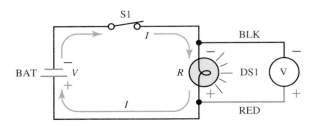

SECTION 4.6 SELF-CHECK

Answers are given at the end of the chapter.

1. What are the requirements for a practical circuit?
2. What is a schematic diagram?

■ SUMMARY

1. The charge of one electron is 1.6×10^{-19} C. This is the reciprocal of the number of electrons in 1 C (6.25×10^{18}).

2. A difference of charge is referred to as a potential difference, an electromotive force, or voltage. The unit of measure for these is the volt.

3. A potential difference can exist only between two points that have a difference of charge.

4. Electrons flow when a potential difference across two points is connected by a conductor.

5. Potential difference (voltage) does not flow; charges do.

6. Generally a reference point is needed when measuring voltage. The reference can be ground, a common point, or the negative terminal of the source.

7. A voltage source is a device capable of supplying and maintaining voltage (potential difference) while some type of electrical device is connected to its terminals.

8. Common methods for producing voltage include friction, pressure, heat, light, chemical action, and magnetism.

9. A voltage drop represents the energy used by the free electrons while engaged in current flow.

10. Electric current is defined as the directed flow of free electrons. The direction of flow is from a negative potential to a positive potential.

11. Conventional current flow is the motion of positive charges (from positive to negative).

12. A gain in energy causes free valence electrons to drift randomly.

13. A potential difference applied across a length of conductor causes the random drifting by electrons to become a directed movement of electrons from a negative to a positive potential.

14. Current is measured in amperes ($I = Q/t$).

15. The opposition to current flow is known as resistance (R), and the unit of measure is the ohm (Ω).

16. The relative resistance of a conductor is determined by its composition. The value is also affected by the conductor's cross-sectional area, length, and temperature.

17. Conductance is the ability of a material to pass an electric current. It is the opposite of resistance.

18. A practical electric circuit contains a voltage source, a complete path for current flow, a load, and some circuit-control device.

19. A schematic diagram is a picture of the circuit that uses symbols to represent various circuit components.

■ IMPORTANT EQUATIONS

$$E = \frac{W}{Q} \tag{4.1}$$

$$I = \frac{Q}{t} \tag{4.2}$$

$$R = \frac{1}{G} \tag{4.3}$$

$$G = \frac{1}{R} \tag{4.4}$$

■ REVIEW QUESTIONS

1. What is the difference between the charge of an electron and the coulomb?
2. Define the term *potential difference.*
3. How do bodies become electrically charged?
4. What is the difference between a negative charge and a positive charge?
5. What are the requirements for charge to flow?
6. What causes charges to stop flowing?
7. Describe the water tank analogy to illustrate how potential difference and electron movement are interrelated.
8. Where does potential difference exist?
9. What is a voltage reference?
10. What is a ground?
11. What instrument is used to measure voltage?
12. Describe how to use a voltmeter to measure voltage.
13. What requirements are necessary to produce a voltage source?
14. List four methods used for producing voltage, and describe each briefly.
15. What is the *piezoelectric effect*?
16. What is a thermocouple?
17. What is the difference between primary cells and secondary cells?
18. How does one increase the voltage using multiple cells?
19. Describe the composition of the common 9-V battery.
20. What conditions must be met in order for voltage to be produced by magnetism?
21. Describe a voltage drop.
22. Define the term *electric current.*
23. Compare and contrast random drift and directed drift of electrons.
24. Compare and contrast electron flow and conventional current flow.
25. What is the velocity of electrons in a conductor? How does this compare to the velocity of the current flow?
26. The amplitude of current is measured in _____.
27. If the same amount of charge passed a point in half the time, what would be the effect on the amount of current flowing?
28. Define the term *electrical resistance.*
29. What are the symbols for resistance and its unit of measurement?
30. What factors affect the amount of resistance that a conductor possesses?
31. Define the term *conductance.*
32. What are the requirements for making a practical circuit?

■ CRITICAL THINKING

1. What is voltage and where does it exist?
2. What are the differences between conventional flow and electron flow?
3. How is magnetism used to produce voltage?
4. What makes up a practical circuit?
5. Define the role of a load in circuit behavior.

■ PROBLEMS

1. Calculate the potential difference between two points if 16 J of work must be done to move 0.5 C of charge between the two points.
2. Calculate the potential difference between two points if 1 J of work must be done to move 500 μC of charge between the two points.
3. Calculate the amount of work done by a 12-V battery that causes 25 C of charge to be moved between two points.
4. Calculate the amount of charge that would be moved by a 9-V battery when 12 J of work is done.
5. What is the magnitude of current when 0.5 C of charge passes a point in 1 s?
6. What time would it take to produce 2 mA of current from the movement of 500 μC of charge?
7. If 1 mA of current is produced in 0.5 s, what amount of charge is involved?
8. Determine the conductance when the resistance is 1 kΩ.
9. Determine the resistance when the conductance is 50 μS.
10. Determine the conductance when the resistance is 470 Ω.

■ ANSWERS TO PRACTICE PROBLEMS

PRACTICE PROBLEMS 4.1

1. 64 V
2. 0.4 C

PRACTICE PROBLEMS 4.2

1. 0.001 A
2. 1.33 s

PRACTICE PROBLEMS 4.3

1. 0.2 mS
2. 10 kΩ

■ ANSWERS TO SELF-CHECKS

SECTION 4.1

1. A potential difference exists between two points when there is a difference of charge.
2. When a potential difference exists between two points that are connected by a conductor
3. The metal chassis or some other common point

SECTION 4.2

1. To provide a potential difference without dissipating the difference
2. Any four of the following: pressure, light, chemical action, magnetism, friction, or heat
3. Combining several cells in a series-aiding arrangement

SECTION 4.3

1. A voltage rise, or energy imparted to the free electrons for the formation and maintenance of electric current
2. The energy used by the free electrons while engaged in current flow

SECTION 4.4

1. The directed flow of free electrons
2. The motion of positive charges, in the opposite direction to electron flow
3. The effect of current flow is at the speed of light, whereas the individual electron travels at a much lower velocity.

SECTION 4.5

1. The opposition to current flow
2. The material of which it is made
3. The ability of a material to pass an electric current

SECTION 4.6

1. A potential difference, a complete path, a load, and (usually) a control device
2. A picture of the circuit using symbols to represent the various circuit components

CHAPTER 5

RESISTORS

OBJECTIVES

After completing this chapter, you will be able to:

1. Compare and contrast various resistor-packaging formats.
2. Describe the makeup of variable resistors.
3. Compare and contrast the potentiometer and the rheostat.
4. Differentiate between linear and nonlinear tapers.
5. Describe other (nonmechanical) types of variable resistors.
6. Describe a resistor's power rating.
7. Interpret resistor value codes, including color coding and alphanumeric.
8. Evaluate types and causes of resistor faults.

INTRODUCTION

Resistance is a property of every electrical component. At times, its effects are undesirable. However, resistance is used in many varied ways. Resistors are components manufactured to possess either a fixed value of resistance or a value in an adjustable format. They are manufactured in many types and sizes. The purpose of a resistor in a circuit is either to reduce or limit current (I) to a specific or safe value or, by using a series of resistors, to provide a desired voltage (V). The latter application is known as a voltage divider and is more fully described in Section 11.2.

5.1 TYPES OF RESISTORS

Fixed-value resistors are found primarily in seven different packaging formats. These include surface-mount, wirewound, carbon-composition, carbon-film, metal-film, and metal-oxide resistors and resistor networks. Several fixed-resistor packages are shown in Figure 5.1. Surface-mount resistors account for nearly half the market. Surface-mount technology (SMT) offers leadless components that do not require through-hole technology. That is, they do not have leads that fit into holes predrilled in printed circuit boards. This affords greater packaging density. The trend in electronics continues to be the reduction of component size, the size of printed circuit boards, and the overall size of the functioning circuit. Surface-mount devices (SMDs) are the logical keys to additional downsizing of electronic circuits, and their use continues to grow.

FIGURE 5.1
Common fixed-resistor packages.

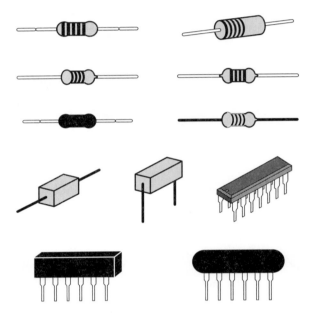

5.1.1 Carbon-Composition Resistors

One of the most common types of resistors is the molded-composition resistor, usually referred to as the carbon resistor. These resistors are manufactured in a variety of sizes and shapes. They have been used in designs for nearly half a century.

The chemical composition of the resistor determines its ohmic value and is accurately controlled by the manufacturer in the production process. They have a resistance range from 2.7 Ω to 100 MΩ. The physical size of the resistor is related to its power rating, which is the ability of the resistor to dissipate heat (thermal energy) caused by the flow of current (electrical energy) through the resistance.

Carbon resistors, as you might suspect, have as their principal ingredient the element carbon. In the manufacture of carbon resistors, fillers or binders are added to the carbon to obtain various resistor values. Examples of these fillers are clay, bakelite, rubber, and talc. These fillers are doping agents and cause the overall conduction characteristics to change. A carbon resistor is shown in Figure 5.2.

Carbon resistors have been popular because they are easy to manufacture, are inexpensive, and have a tolerance that is adequate for most electrical and electronic applications. Their disadvantages are that they have a tendency to change value as they age and they have limited power-handling capacity.

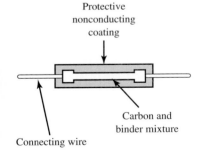

Protective
nonconducting
coating

Carbon and
binder mixture

Connecting wire

FIGURE 5.2
Construction of a carbon-composition resistor.

In order to get even better accuracies (closer tolerances), the carbon resistor comes in a carbon-film format. Carbon-film resistors have a thin coating of resistive material on a ceramic insulator. They have a resistance range from 10 Ω to 25 MΩ. The main advantage of using carbon-film resistors compared to carbon-composition resistors is that they can be manufactured to more exacting standards, allowing for tolerances of between 2% and 5%. They also generate less noise because of random electron motion. This is particularly helpful in audio and communication electronic circuits.

As manufacturing processes have improved, the carbon-composition resistor has lost favor with designers. Other types of resistors offer better tolerances and other attributes that are more desirable in today's electronic circuits. In May 1996, Allen-Bradley (a very large manufacturer of carbon-composition resistors) announced that they plan to discontinue manufacturing this type of resistor. Carbon-composition resistors will be replaced by carbon-film, metal-film, and wirewound resistors, according to application.

5.1.2 Metal-Film Resistors

Metal-film resistors are often used where a high degree of ohmic accuracy is required. These resistors are capable of providing very precise ohmic values and are often used as precision resistors with tolerances of less than 1%. Metal-film resistors are manufactured by spraying a relatively thin layer of metal on a ceramic cylinder. They have a resistance range from 0.1 Ω to 22 MΩ.

5.1.3 Metal-Oxide Film Resistors

Metal-oxide film resistors are made by oxidizing tin chloride on a heated-glass substrate. The ratio of the oxide insulator to the tin conductor determines the resistance of the device. These resistors have low noise and excellent temperature characteristics. Metal-oxide film resistors are popular where precision resistors of high ohmic value are required. They have a resistance range from 1 kΩ to 30 GΩ.

5.1.4 Resistor Networks

Low-power resistors are also available in packages that resemble integrated circuit chips; as a result, they are called *chip* resistors. They are also manufactured as single resistive elements. These resistors are small, high-quality components manufactured on small, rectangular ceramic chips. Chip packages are fabricated from thin-film and thick-film metals to exacting tolerances. They have a resistance range from 1.0 Ω to 100 MΩ. Chip resistors are designed for printed circuit board installations and are very popular in surface-mount technology.

5.1.5 Wirewound Resistors

Wirewound resistors (Figure 5.3) have very accurate values and possess a higher current-handling capability than carbon resistors. Several materials can be used to manufacture wirewound resistors. These include nickel-chromium, copper-nickel, and gold-platinum

FIGURE 5.3
Wirewound resistor formed by wrapping resistance wire around a ceramic core.

(with copper and silver). Nickel-chromium is the most common. The quantities and qualities of these elements present in the wire determine the resistivity of the wire. They have a resistance range of 0.01 Ω to 178 kΩ. One disadvantage of the wirewound resistor is that it takes a large amount of wire to manufacture a resistor of high ohmic value, thereby increasing the cost.

Unlike carbon-composition resistors, wirewound resistors do not use a color code to indicate the resistance value. The ohmic value of wirewound resistors is usually stamped on the resistor case.

SECTION 5.1 **SELF-CHECK**	Answers are given at the end of the chapter. **1.** What is the oldest resistor-packaging format? **2.** What is the most popular resistor-packaging format? **3.** What is the most common metal used for wirewound resistors?

5.2 VARIABLE RESISTORS

We have seen how a fixed resistor has one value and does not change (other than through temperature effects, aging, etc.). The other type of resistor has a variable (adjustable) value with a resistance range from 150 Ω to 5 MΩ.

There are two types of variable resistors, one called a *potentiometer* (*pot*) and the other a *rheostat*. An example of the pot is the volume control on your radio, and an example of the rheostat is the dimmer control for the dash lights in an automobile. There is a slight difference between them. Rheostats usually have two connections, one fixed and the other moveable. Any variable resistor can be properly called a rheostat. The pot always has three connections, two fixed and one moveable. The pot has a wide range of values, but it usually has a limited current-handling capability. Pots are always connected as voltage dividers. (Voltage dividers are discussed in Section 11.2.) Figure 5.4(a) shows the symbol for the pot. Figure 5.4(b) shows two symbols for the rheostat, a pot connected as a rheostat and the two-terminal rheostat.

FIGURE 5.4
Symbols for (a) potentiometer and (b) the potentiometer connected as a rheostat and the two-terminal rheostat.

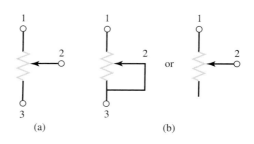

The pot and its internal construction are shown in Figure 5.5(a) and (b). Wirewound potentiometers are also common. Many have the same outward appearance as the pot shown in Figure 5.5(a). However, the internal construction is slightly different. As shown in Figure 5.5(c), resistance wire is wound around an insulating core. A contact arm moves along the bare wire, changing the resistance between the center and outside terminals.

Rheostats are used to control the circuit current by varying the amount of resistance in the resistive element. This is accomplished by "tapping" the resistive element. Usually the resistive element is on a circular track, and a contact, attached to a dial, sweeps from one end of the resistance to the other. The amount of resistance in the circuit is the section between the fixed end and the moveable contact (called the *wiper*).

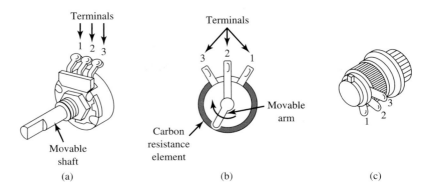

FIGURE 5.5
Variable resistors: (a) potentiometer; (b) internal construction of a pot; (c) wirewound pot.

Lab Reference: The applications of a potentiometer are demonstrated in Exercise 12.

The relationship between the angle of rotation and resistance, called the *taper,* is either linear or nonlinear (logarithmic). If the resistive element is deposited uniformly over the surface, the taper is linear. If it is deposited in a nonlinear fashion, the taper is logarithmic (also known as audio). Audio tapers are found in volume-control elements in electronic consumer products. This is because the human ear differentiates sound waves in a logarithmic fashion. Linear tapers, on the other hand, are commonly used to vary voltage in linear increments. They are used in test equipment, on printed circuit boards, and the like. An example is a voltage-calibration control on a piece of test equipment. The graphical relationship between the two tapers is shown in Figure 5.6.

FIGURE 5.6
Linear versus nonlinear resistive tapers.

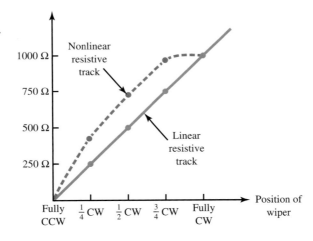

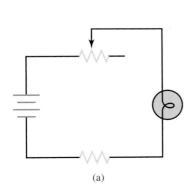

(a)

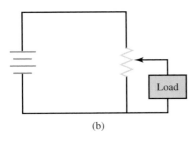

(b)

FIGURE 5.7
(a) Rheostat varies current in the lamp; (b) potentiometer varies voltage to the load.

Figure 5.7 summarizes the differences between the rheostat and the potentiometer. Figure 5.7(a) demonstrates how the rheostat varies the circuit current in the lamp, and Figure 5.7(b) shows how the potentiometer varies the load voltage.

5.2.1 Trimmer Resistors

Trimmer resistors are variable resistors that are used where small and infrequent adjustments of a resistance are necessary to maximize circuit performance. The adjustment of a trimmer resistor is usually made with a small screwdriver. These miniature potentiometers are also known as *trim pots.* They have a resistance range from 10 Ω to 100 kΩ. Figure 5.8 shows various examples of pots (including the trim pot) and demonstrates that they come in a wide variety of shapes and sizes.

FIGURE 5.8
Potentiometers are available in a variety of shapes and sizes.

5.3 ADDITIONAL TYPES OF VARIABLE RESISTORS

Resistance can be varied by other than a mechanical means (the rotation of the wiper contact). Heat and light, as well as other properties, can be used to vary the resistance of a component.

5.3.1 Bolometers

A *bolometer* is a device that changes its resistance when heat energy is applied. There are two types of bolometers, the barretter and the thermistor. The *barretter* is characterized by increasing resistance as the dissipated power increases (this is known as a *positive temperature coefficient,* or PTC). The *thermistor* decreases in resistance as the power increases (this is known as a *negative temperature coefficient,* or NTC). Both these components can be found in radio-frequency (RF) electronic test equipment used to measure heat energy (power) dissipated.

Barretter

The construction of a typical barretter is shown in Figure 5.9. The fine wire (usually tungsten) is extremely small in diameter. This thin diameter allows the RF current to penetrate to the center of the wire. The wire is supported in an insulating capsule between two metallic ends, which act as connectors. Because of these physical characteristics, the barretter resembles a cartridge-type fuse. The enclosure is a quartz capsule made in two parts. One part is an insert cemented in place after the tungsten wire has been mounted.

Thermistor

A high degree of precision is made possible by the thermistor; therefore, it is widely used. It is one of the more common heat-sensitive devices found in power meters. The negative temperature coefficient comes from the use of a semiconductor as the active material. Figure 5.10 shows the typical construction of a thermistor. Notice that the active mater-

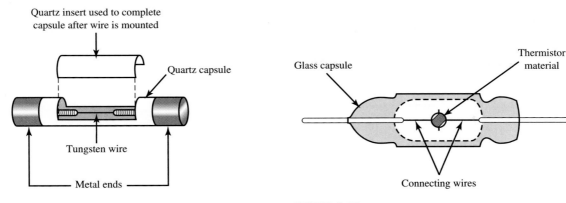

FIGURE 5.9
Typical barretter.

FIGURE 5.10
Bead-type thermistor.

ial is shaped in the form of a bead. It is supported between two pigtail leads by connecting wires. The pigtail ends are embedded in the ends of the surrounding glass capsule.

5.3.2 Light-Sensitive Devices

Light-sensitive (*optoelectronic*) devices respond to changes in light intensity by changing their internal resistance (or, as we saw in Section 4.2.4, by generating an output voltage).

Photoconductive Cells

The *photoconductive cell* is one of the oldest optoelectronic components. It is nothing more than a light-sensitive resistor (photoresistor) whose internal resistance varies with light intensity. The resistance decreases, but the decrease is not proportional to the increase in light.

Photoconductive cells are usually made from light-sensitive materials such as cadmium sulfide (CdS) or cadmium selenide (CdSe), although other materials are used. Figure 5.11 shows how a typical photoconductive cell is constructed. A thin layer of light-sensitive material is formed on an insulating substrate that is usually made from glass or ceramic materials. Two metal electrodes are deposited on the light-sensitive material as thin layers. The top view (Figure 5.11(a)) shows that the electrodes do not touch but leave an S-shaped portion of light-sensitive material exposed. This allows greater contact length but at the same time confines the light-sensitive material to a relatively small area between the electrodes. Two leads are inserted through the substrate and soldered to the electrodes, as shown in the side view in Figure 5.11(b). The photoconductive cell is often mounted in a metal or plastic case (not shown), which has a glass window that allows light to strike

FIGURE 5.11
A typical photoconductive cell: (a) top view; (b) side view.

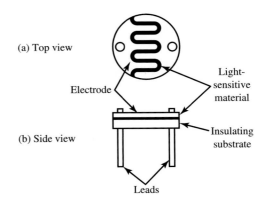

the light-sensitive material. The resistance of a typical cell might be as high as several hundred megohms under no light conditions to as low as several hundred ohms when the light illumination is good.

Photoconductive cells have many applications in electronics. They are often used in devices such as intrusion detectors and automatic door openers, where it is necessary to detect the presence or absence of light. They may also be used in precision test instruments that can measure light intensity.

SECTION 5.3 SELF-CHECK

Answers are given at the end of the chapter.

1. What is a bolometer?
2. What is NTC?
3. Describe how a photoresistor works.

5.4 RESISTOR POWER RATING

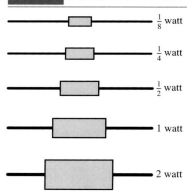

FIGURE 5.12
Carbon-resistor size comparison by wattage rating.

When a current is passed through a resistor, heat is developed within the resistor. The resistor must be capable of dissipating this heat into the surrounding air; otherwise, the temperature of the resistor rises, causing a change in resistance or possibly causing the resistor to burn out. *Hot-spot temperature* is the maximum temperature measured on the resistor due to both internal heating and ambient operating temperature. This value must be taken into account when selecting the proper resistor for an application.

The ability of the resistor to dissipate heat depends on the design of the resistor itself. Proper heat dissipation depends on the amount of surface area that is exposed to the air. A resistor designed to dissipate a large amount of heat must, therefore, have a large physical size. The heat-dissipating capability of a resistor is measured in watts (W; this unit is explained in Section 6.7). Some of the more common wattage ratings of carbon resistors are $\frac{1}{8}$ W, $\frac{1}{4}$ W, $\frac{1}{2}$ W, 1 W, and 2 W. The higher the wattage rating of the resistor, the larger the physical size. This is illustrated in Figure 5.12. Resistors that dissipate very large amounts of power (watts) are usually wirewound resistors. Wirewound resistors with wattage rating up to 250 W are available.

SECTION 5.4 SELF-CHECK

Answers are given at the end of the chapter.

1. What is the hot-spot temperature of a resistor?
2. What factors determine a resistor's ability to dissipate heat?
3. List some common wattage values for carbon resistors.

5.5 RESISTOR VALUE CODING

5.5.1 Alphanumeric Markings

When the physical size of a resistive component is large enough, such as with a potentiometer or a wirewound resistor, then value markings are simply stamped on the body of the component. Chip resistors also employ this type of value marking. This alphanumeric marking gives all the important information about the component: the dimensions

FIGURE 5.13
Example of alphanumeric markings on a chip resistor.

(wattage), temperature coefficient, value, and tolerance. Modern manufacturing techniques employ ink-jet or laser markings on the device. Figure 5.13 shows an individual chip resistor.

The alphanumeric markings for the chip resistor in Figure 5.13 are interpreted as follows:

RR 0510 P 102 D

where   RR represents the manufacturer parts code

0510 represents the type by dimension and wattage ($\frac{1}{16}$ W in this example)

P represents the temperature coefficient ($\pm$25 ppm/°C in this example)

102 represents the value in ohms (1000 Ω in this example)

D represents the tolerance (0.5% in this example)

The value code (102) for this example is always in ohms (Ω) and is interpreted as follows:

1 = 1st significant digit

0 = 2nd significant digit

2 = multiplier (multiply by 100, or add 2 zeros to the first two digits)

For resistive values less than 10 Ω, the letter *R* is placed between the first and second significant digits to indicate the decimal point; for example, 2R7 = 2.7 Ω.

EXAMPLE 5.1

Determine the value of a resistor marked numerically as 273.

Solution   2 = 1st significant digit

7 = 2nd significant digit

3 = multiplier (add 3 zeros to the first two digits)

resistor = 27,000 Ω, or 27 kΩ

EXAMPLE 5.2

Determine the value of a resistor marked numerically as 4R7.

Solution   4 = 1st significant digit

R = decimal point (between first and second digits)

7 = 2nd significant digit

resistor = 4.7 Ω

Practice Problems 5.1   Answers are given at the end of the chapter.
1. Determine the value of the resistors marked numerically as follows: 225; 103; 9R1; 824; 502; 151

5.5.2 Resistor Color Coding

General-Purpose Resistors

Color-coded bands are used on most general-purpose carbon-composition and carbon-film resistors. For the most part the sizes of these components do not lend themselves to alphanumeric markings. Color-coded bands have been in use for many years and provide a convenient (and inexpensive) method of indicating resistive values by the manufacturer. Likewise, this method offers an easy interpretation by the end user as long as you know the color code. Ten colors are used for indicating resistive value, and two colors are used for indicating the tolerance.

A four-band system for general-purpose resistors is very common and has been in use for many years. General-purpose resistors are classified as having tolerances between 5% and 20%. To determine the value of a resistor (Figure 5.14), it is held so that the value bands are on the left and the tolerance band is on the right. The tolerance band is always found in one of three forms: ±20% requires no color band, ±10% is a silver band, and ±5% is a gold band. The first and second value bands produce a two-digit number (known as the first and second significant digits), the third band indicates the multiplier (the two-digit number is multiplied by this value), and the fourth band designates the tolerance in percent.

FIGURE 5.14
Resistor color code.

Lab Reference: The use of the resistor color code is demonstrated in Exercise 7.

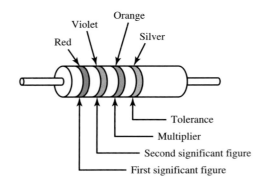

Table 5.1 contains the color code for the four-band, general-purpose, carbon-composition resistor. Three color bands near one end of the resistor indicate the resistance value, and an added gold or silver fourth band means the tolerance is ±5% or ±10%, respectively. If there are only three bands (no fourth band), the tolerance is taken as ±20%.

EXAMPLE 5.3	

Determine the value and tolerance of the resistor in Figure 5.14 that has the following four color bands (in order): red, violet, orange, silver.

Solution

$$red = 2 \quad \text{(1st significant digit)}$$
$$violet = 7 \quad \text{(2nd significant digit)}$$
$$orange = 10^3 \quad \text{(multiplier)}$$
$$silver = \pm 10\%$$
$$resistor = 27 \times 10^3 \pm 10\%$$
$$= 27 \times 1000 \pm 10\%$$
$$= 27{,}000 \; \Omega \pm 10\%$$
$$= 27 \, k\Omega \pm 10\%$$

TABLE 5.1
Four-band, general-purpose resistor color code

Color	First Band First Significant Digit	Second Band Second Significant Digit	Third Band Multiplier	Fourth Band Tolerance
Black	—	0	10^0	—
Brown	1	1	10^1	—
Red	2	2	10^2	—
Orange	3	3	10^3	—
Yellow	4	4	10^4	—
Green	5	5	10^5	—
Blue	6	6	10^6	—
Violet	7	7	10^7	—
Gray	8	8	10^8	—
White	9	9	10^9	—
Gold	—	—	10^{-1}	$\pm 5\%$
Silver	—	—	10^{-2}	$\pm 10\%$
None	—	—	—	$\pm 20\%$

EXAMPLE 5.4

Determine the value and tolerance of a resistor having the following four color bands (in order): brown, black, black, gold.

Solution   brown = 1   (1st significant digit)

black = 0   (2nd significant digit)

black = 10^0 (multiplier)

gold = $\pm 5\%$

resistor = $10 \times 10^0 \pm 5\%$ (Remember $10^0 = 1$.)

= $10 \times 1 \pm 5\%$

= $10\ \Omega \pm 5\%$

EXAMPLE 5.5

Determine the value and tolerance of a resistor having the following four color bands (in order): yellow, violet, gold, gold.

Solution   yellow = 4   (1st significant digit)

violet = 7   (2nd significant digit)

gold = 10^{-1} (multiplier)

gold = $\pm 5\%$

resistor = $47 \times 10^{-1} \pm 5\%$

= $47 \times 0.1 \pm 5\%$

= $4.7\ \Omega \pm 5\%$

Practice Problems 5.2 Answers are given at the end of the chapter.
1. Determine the value and tolerance of the following resistors having four color bands as indicated.
red–red–red–gold
yellow–violet–silver–gold
brown–black–orange–gold

There is a memory aid (mnemonic) that you can use to help remember the order of the colors. Each word of the mnemonic starts with the first letter of one of the 10 colors used for resistive values. There are several versions of the mnemonic; your instructor probably has a favorite one. Here is one:

Billy **B**ob **R**aises **O**ld **Y**ellow **G**rapes **B**ut **V**ery **G**ood **W**ine

or

0	**Black**	**Billy**
1	**Brown**	**Bob**
2	**Red**	**Raises**
3	**Orange**	**Old**
4	**Yellow**	**Yellow**
5	**Green**	**Grapes**
6	**Blue**	**But**
7	**Violet**	**Very**
8	**Gray**	**Good**
9	**White**	**Wine**

Precision Resistors

Precision resistors (usually a film-type) with tolerances of 2% or less use a five-band system, as shown in Table 5.2. The first three bands of a precision resistor indicate three significant digits. The fourth band is the multiplier, and the fifth band indicates tolerance.

EXAMPLE 5.6

What are the resistance and tolerance of a precision resistor having the following bands (in order): red, red, red, red, red.

Solution

red = 2   (1st significant digit)

red = 2   (2nd significant digit)

red = 2   (3rd significant digit)

red = 10^2 (multiplier)

red = ±2%

resistor = $222 \times 10^2 \pm 2\%$

= $222 \times 100 \pm 2\%$

= $22{,}200 \ \Omega \pm 2\%$

= $22.2 \ \text{k}\Omega \pm 2\%$

Practice Problems 5.3 Answers are given at the end of the chapter.
1. What is the resistance and tolerance of the following precision resistors having the following bands?
brown–brown–black–black–red
red–violet–red–brown–brown

TABLE 5.2
Five-band color code for precision resistors

Color	First Band First Significant Digit	Second Band Second Significant Digit	Third Band Third Significant Digit	Fourth Band Multiplier	Fifth Band Tolerance
Black	—	0	0	10^0	—
Brown	1	1	1	10^1	$\pm 1\%$
Red	2	2	2	10^2	$\pm 2\%$
Orange	3	3	3	10^3	—
Yellow	4	4	4	10^4	—
Green	5	5	5	10^5	$\pm 0.5\%$
Blue	6	6	6	10^6	$\pm 0.25\%$
Violet	7	7	7	10^7	$\pm 0.1\%$
Gray	8	8	8	10^8	—
White	9	9	9	10^9	—
Gold	—	—	—	10^{-1}	—
Silver	—	—	—	10^{-2}	—

**SECTION 5.5
SELF-CHECK**

Answers are given at the end of the chapter.

1. How are alphanumeric markings applied to resistors?
2. What does a gold third band indicate in the four-band system?
3. Precision resistors have a tolerance of _____ .

5.6 STANDARD RESISTOR VALUES

In order to minimize the problem of manufacturing different resistive values for an almost unlimited variety of circuits, specific values are made in large quantities so that they are less expensive and readily available.

Table 5.3 lists standard values for $\pm 10\%$, $\pm 5\%$, and $\pm 1\%$ tolerances. Note that for any value shown, the decimal multiple is also available. As examples, 27, 270, 2700, 27,000 and 270,000 are standard $\pm 10\%$ values and all are decimal multiples of each other. In this way, there is a standard value available within 10% of any resistor value needed. The 5% and 1% standard values provide for additional in-between resistances. The $\pm 5\%$ is the most common standard value, with the $\pm 1\%$ very popular because of 1% metal film types. There are standard values for resistors having a tolerance of less than 1% as well.

TABLE 5.3
Standard resistance values

±10% Tolerance	±5% Tolerance	±1% Tolerance	±10% Tolerance	±5% Tolerance	±1% Tolerance
10	10	10.0	33	33	33.2
	11	10.2		36	34.0
		10.5			34.8
		10.7			35.7
		11.0			36.5
		11.3			37.4
		11.5			38.3
		11.8	39	39	39.2
12	12	12.1		43	40.2
	13	12.4			41.2
		12.7			42.2
		13.0			43.2
		13.3			44.2
		13.7			45.3
		14.0			46.4
		14.3	47	47	47.5
		14.7		51	48.7
15	15	15.0			49.9
	16	15.4			51.1
		15.8			52.3
		16.2			53.6
		16.5			54.9
		16.9	56	56	56.2
		17.4		62	57.6
		17.8			59.0
18	18	18.2			60.4
	20	18.7			61.9
		19.1			63.4
		19.6			64.9
		20.0			66.5
		20.5	68	68	68.1
		21.0		75	69.8
		21.5			71.5
22	22	22.1			73.2
	24	22.6			75.0
		23.2			76.8
		23.7			78.7
		24.3			80.6
		24.9	82	82	82.5
		25.5		91	84.5
		26.1			86.6
		26.7			88.7
27	27	27.4			90.9
	30	28.0			93.1
		29.4			95.3
		30.1			97.6
		31.6			
		32.4			

Answers are given at the end of the chapter.

1. Why are standard resistor values used?
2. What is the most common standard value?
3. Where is the 1% standard used?

5.7 TROUBLESHOOTING RESISTOR FAULTS

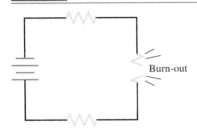

FIGURE 5.15
Open circuit path caused by resistor burnout.

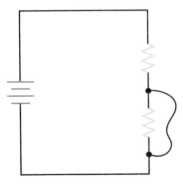

FIGURE 5.16
A short circuit across a resistor causes an increase in current.

The most common problems to occur with fixed resistors are a change in resistance or a complete failure. A complete failure occurs when the resistor overheats and burns out. When the resistor interior is burned out, it produces an *open circuit*. An open circuit causes a break in the path for current flow; thus the current ceases to flow. Figure 5.15 shows one resistor in the current path that has opened. As you can see, there is no path for current.

Open circuits caused by a resistor burnout are usually the result of some other component that failed. This, in turn, causes excessive current to flow and eventually results in a failure in the component and others in the same current path as well. The open circuit could also be caused by a *cold solder joint*. A cold solder joint is a poor solder connection of the component to the circuit board.

Another mode of failure is the *short circuit*. A short circuit occurs when a component's internal resistance reduces to zero. It is rare for a resistor to short-circuit, because the internal structure is made up of carbon and is quite ohmic. However, a short circuit of a resistor can occur because of excessive solder on the leads of a resistor, causing a bridge to form between them. The solder bridge effectively bypasses (shunts) the current around the resistor, causing an increase in current to flow. This condition can also occur when a bare wire touches the leads of the resistor. In any case, it is a situation that causes all kinds of havoc to a circuit as a result of an increased current flow. A short circuit is illustrated in Figure 5.16.

Variable resistors are susceptible to the same kinds of failures as fixed resistors. Their internal structure can burn out due to excessive current flow. A short can occur if a solder bridge or a wire crosses the terminals of a pot. A condition referred to as a *dirty,* or *noisy, pot* is very common with potentiometers. After aging the contact assembly between the wiper and the resistive track can become oxidized. This can cause noise to be generated in audio circuits via the volume or tone control. Oxidation can be reduced by chemical contact cleaners, but eventually it may be necessary to replace the pot.

The most common instrument used to check or troubleshoot resistors is the ohmmeter. This device is a resistance-measuring instrument that has an internal battery as its power source. Ohmmeters are usually found as part of a multimeter that is capable of measuring current and voltage. Figure 5.17 demonstrates an ohmmeter connected to an unknown resistance.

An ohmmeter should never be used in a "live" circuit — that is, one in which current is flowing. The voltage source should always be turned off or disconnected when measuring resistance (remember the ohmmeter has its own internal battery).

To get proper resistive measurements, the resistor being tested should be isolated from other components. This means removing it from the circuit or at least removing one lead of the resistor. Failure to do this may result in other devices connected to the resistor under test contributing to the resistance being read by the ohmmeter. (A detailed discussion of the ohmmeter is presented in Section 10.5.)

FIGURE 5.17
An ohmmeter being used to measure an unknown resistance.

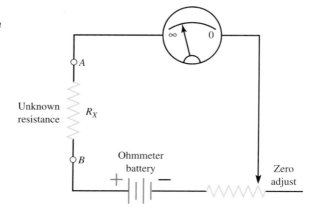

SECTION 5.7 SELF-CHECK	Answers are given at the end of the chapter.

1. What is the most common resistor failure?
2. How does a dirty pot occur?
3. What instrument is used to measure resistance?

■ SUMMARY

1. Fixed resistors come in seven different packaging formats. Surface-mount resistors account for nearly half the market.
2. Carbon-composition resistors have been in use for a very long time. They have lost favor with designers and are being replaced with carbon-film, metal-film and wirewound resistors.
3. Metal-film resistors are used where a high degree of ohmic accuracy is required.
4. Metal-oxide film resistors are precision resistors with high ohmic values.
5. Chip resistors are popular in surface-mount technology.
6. Wirewound resistors are most often made with nickel-chromium wire and have a high power rating.
7. Variable resistors come in one of two formats: the potentiometer and the rheostat.
8. A pot is used as a voltage divider, whereas a rheostat is used to limit current.
9. The moveable contact on a variable resistor is called a wiper.
10. A taper describes the relationship between an angle of rotation and an amount of resistance.
11. The two forms of a taper are linear and nonlinear.
12. Trim pots are miniature potentiometers.
13. Other types of variable resistances include the heat-varied barretter and thermistor.
14. Variable resistances also include light-sensitive devices such as the photoconductive cell.
15. A resistor's power rating is determined primarily by its physical size.

16. Resistor value coding includes a color-code system and an alphanumeric system.
17. Resistors are manufactured to have resistance values in certain ranges known as standard resistance values.
18. The most common failures in fixed resistors are a change in resistance or an open circuit.
19. Pots suffer from dirty or noisy contacts as they age.
20. The ohmmeter is the instrument used to measure resistance. It is usually found as a part of a multimeter.

■ REVIEW QUESTIONS

1. What packaging formats are available for resistors?
2. Which format accounts for nearly half the market?
3. Why are surface-mount devices increasingly used?
4. Describe how carbon composition resistors are made.
5. What are the advantages of carbon-film resistors?
6. What is the long-term prognosis for the use of carbon-composition resistors?
7. Describe the attributes of metal-film and metal-oxide resistors.
8. Describe how wirewound resistors are made.
9. Compare and contrast the pot and the rheostat.
10. Describe the differences between linear and nonlinear tapers.
11. Where are trim pots used?
12. Compare and contrast the barretter and the thermistor.
13. Describe how the photoconductive cell is manufactured.
14. What are the applications for photoconductive cells?
15. What contributes to a resistor's heat-dissipation ability?

16. What are common wattage ratings of carbon-composition resistors?

17. Compare and contrast the alphanumeric and color-code marking systems for resistors.

18. What are the values of chip resistors marked 223, 101, 7R5, and 504, respectively?

19. What is the difference between the four-color code and the five-color code for resistors?

20. What color is used to indicate a tolerance of ±2% using the five-color code for resistors?

21. What are standard resistor values?

22. Name the most common failures for fixed resistors.

23. What can a cold solder joint do to circuit performance?

24. How can a resistor short-circuit?

25. Describe the effect known as a dirty, or noisy, pot.

26. What instrument is used to measure resistance?

27. Describe the procedure for measuring resistance with an ohmmeter.

■ CRITICAL THINKING

1. Which resistor packaging type is the most popular? Why?

2. Compare and contrast the potentiometer and the rheostat.

3. What are the relative merits of the four-band versus the five-band resistor color code?

4. Describe the symptoms you would observe with resistors for an open, a short, and a dirty pot.

■ PROBLEMS

1. What are the values for chip resistors marked as shown?

220	473	104
2R2	102	153
105	391	272

2. Interpret the values for the following four-color-coded fixed resistors.

red–red–red–gold	brown–black–gold–silver
brown–black–orange–gold	green–red–yellow–silver
orange–orange–orange–gold	yellow–violet–red–gold
yellow–violet–silver–gold	white–brown–brown–gold

3. Interpret the values for the following five-color-coded fixed resistors.

red–red–red–red–red   brown–brown–black–brown–red
yellow–violet–green–gold–brown
                brown–violet–yellow–gold–brown
violet–green–black–red–red
                green–brown–brown–brown–brown
orange–orange–orange–red–red
                green–brown–black–brown–red

■ ANSWERS TO PRACTICE PROBLEMS

PRACTICE PROBLEMS 5.1

1. $225 = 2.2$ MΩ; $103 = 10$ kΩ; $9R1 = 9.1$ Ω; $824 = 820$ kΩ; $502 = 5$ kΩ; $151 = 150$ Ω

PRACTICE PROBLEMS 5.2

1. 2200 Ω ± 5%, or 2.2 kΩ ± 5%; 0.47 Ω ± 5%; 10,000 Ω ± 5%, or 10 kΩ ± 5%

PRACTICE PROBLEMS 5.3

1. 110 Ω ± 2%; 2720 Ω ± 1%, or 2.72 kΩ ± 1%

■ ANSWERS TO SELF-CHECKS

SECTION 5.1

1. Carbon composition

2. Surface-mount resistors

3. Nickel-chromium

SECTION 5.2

1. The potentiometer and the rheostat

2. As a voltage divider; as a current limiter

3. The resistive element is deposited uniformly over the surface.

SECTION 5.3

1. A device that changes its resistance when heat energy is applied

2. Negative temperature coefficient (Resistance decreases as heat energy increases.)

3. It varies resistance with light intensity.

SECTION 5.4

1. The maximum temperature measured on the resistor due to both internal heating and ambient operating temperature

2. Its design and physical size

3. $\frac{1}{8}, \frac{1}{4}, \frac{1}{2}$, 1, and 2 W

SECTION 5.5

1. Ink-jet or laser markings

2. Multiply by 0.1

3. 2% or less

SECTION 5.6

1. Standard resistive values are made in large quantities to reduce costs and to provide different values for an almost unlimited variety of circuits.

2. 5%

3. With metal-film resistors

SECTION 5.7

1. A change in resistance or a complete failure

2. From oxidation of the resistive track

3. An ohmmeter

CHAPTER 6

OHM'S LAW, WORK, ENERGY, AND POWER

OBJECTIVES

After completing this chapter, you will be able to:

1. Explain Ohm's law and its application.
2. Calculate various unknowns using Ohm's law.
3. Define the terms *work, energy, efficiency,* and *power.*
4. Describe why current is proportional to voltage and why it is inversely proportional to resistance.
5. Calculate power and efficiency.
6. Differentiate between independent and dependent variables.

INTRODUCTION

In the early part of the nineteenth century George Simon Ohm proved by experiment that a precise relationship exists between current, voltage, and resistance (as illustrated in Figure 6.1). His experiments involved an electromotive force (voltage source) impressed across conductors of various sizes. In 1826 he published what is known as Ohm's law. This law states that current flow in a circuit is directly proportional to the source voltage and inversely proportional to the resistance of the circuit.

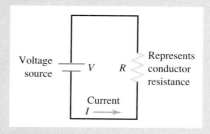

FIGURE 6.1
Typical setup for Ohm's law.

6.1   OHM'S LAW

Ohm's law may be stated mathematically as

$$I = \frac{E}{R}$$

where I = current in amperes

E = electromotive force (emf) in volts

R = resistance in ohms

During the period of Ohm and Franklin, what we now call *voltage* was rightly called electromotive force (emf). Thus, in the original form, Ohm's law shows voltage designated by an E in the equation. Many electronic texts and instructors still use this designation. Furthermore, E and V can be used interchangeably to represent voltage, electrical pressure, potential difference, or electromotive force. The letters V and E can easily be remembered for voltage, because they are the first and last letters of the word <u>voltage</u>. However, in order to maintain consistency for the student, this text uses V for voltage and simply states Ohm's law as follows:

$$I = \frac{V}{R} \tag{6.1}$$

where I = current in amperes (A)

V = voltage in volts (V)

R = resistance in ohms (Ω)

What this equation demonstrates is that the conductor resistance used by Ohm is constant regardless of the magnitude of the applied voltage. Assuming the temperature of the conductor does not change, then the direct proportionality between current and voltage is a linear relationship. This is illustrated by the graph in Figure 6.2. The quantity (voltage) that is intentionally varied is called the *independent* variable and is plotted on the horizontal axis. The second quantity (current), which varies as a result of changes in the first quantity, is called the *dependent* variable and is plotted on the vertical axis. The slope of the resulting curve represents voltage divided by current at any given instant. This V/I ratio represents the electrical effects of resistance; thus the slope symbolizes resistance and is linear.

FIGURE 6.2
Volt-ampere characteristic.

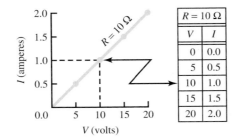

By algebraically manipulating Ohm's law, any one of the three variables can be determined, as long as two are known. Figure 6.3 is a circle diagram that is used as an aid to show all possible forms of Ohm's law. Note that V is above the line, with I and R below the line. To determine any variable, simply cover that quantity, and what is left is the mathematical calculation that yields the unknown you seek. The diagram should be used

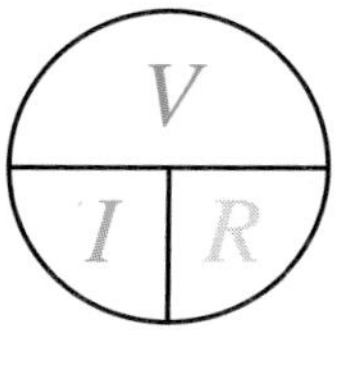

FIGURE 6.3
Ohm's law in diagram form.

Lab Reference: The use of Ohm's law is demonstrated in Exercise 8.

to supplement your basic knowledge of the algebraic method for transposing the formula to determine an unknown. Algebra is a basic tool in the solution of electrical and electronic problems and its understanding is essential to an electronics student.

From the diagram you can see that

$$I = \frac{V}{R}$$

$$V = IR$$

and

$$R = \frac{V}{I}$$

Ohm's law is so important to the analysis and understanding of electrical and electronic circuits that you will use it in circuits from the very simple to the most complex. In the sections that follow, various versions of Ohm's law are demonstrated.

SECTION 6.1 SELF-CHECK

Answers are given at the end of the chapter.

1. Describe Ohm's law in words.
2. Define the terms *independent* and *dependent variable.*
3. What is meant by *direct proportionality?*

6.2 CURRENT CALCULATIONS

Figure 6.4 shows a graph with the voltage held constant at 12 V. The independent variable is the resistance, which varies from 2 Ω to 12 Ω in 2-Ω increments. The current is the dependent variable. Values for current can be calculated as follows.

FIGURE 6.4
The relationship between current and resistance.

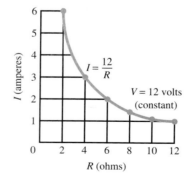

EXAMPLE 6.1

Given: $V = 12$ V

$R = 2\ \Omega$ to $12\ \Omega$,   in 2-Ω increments

$I = ?$

Solution $I = \dfrac{V}{R}$

$$I = \frac{12\ V}{2\ \Omega} = 6\ A$$

$$I = \frac{12\ V}{4\ \Omega} = 3\ A$$

$$I = \frac{12\ V}{6\ \Omega} = 2\ A \qquad\qquad \textbf{(6.1)}$$

$$I = \frac{12\ V}{8\ \Omega} = 1.5\ A$$

$$I = \frac{12\ V}{10\ \Omega} = 1.2\ A$$

$$I = \frac{12\ V}{12\ \Omega} = 1.0\ A$$

You can see that as the resistance is halved, the current is doubled; when the resistance is doubled, the current is halved. This graph also illustrates that the current varies *inversely* with the resistance when the applied voltage is held constant.

For most Ohm's law problems, your calculator should be put in **ENG** mode. In this way, the solution to the problem is already in a recognizable prefix form and can be readily converted to either a *basic unit* or a *unit* involving m, μ, k, M, and the like. Notice also that rounding the solution to *two decimal places* will suffice for most problems; we use this policy throughout this text.

EXAMPLE 6.2

Determine the current when the resistance is 4.7 kΩ and the voltage is 36 V.

Solution $I = \dfrac{V}{R}$

$$= \frac{36\ V}{4.7 \times 10^3\ \Omega} \qquad\qquad \textbf{(6.1)}$$

$= 7.66 \times 10^{-03}\ A$ (with the calculator in **ENG** mode and

$= 7.66\ mA$ rounding to 2 decimal places)

Optional Calculator Sequence

The Sharp calculator sequence for this problem is as follows:

Step	Keypad Entry	Top Display Response
1	36	
2	÷	36/
3	4.7	36/
4	Exp	36/
5	3	36/
6	=	36/4.7E03 =
	Answer	7.66×10^{-03}
		(bottom display)

EXAMPLE 6.3	Determine the current flow in a circuit when $R = 2 \text{ M}\Omega$ and the voltage is 24 V.

Solution $I = \dfrac{V}{R}$

$$= \dfrac{24 \, \text{V}}{2 \times 10^6 \, \Omega} \tag{6.1}$$

$$= 12 \times 10^{-06} \, \text{A} \qquad \text{(with calculator in } \textbf{ENG} \text{ mode)}$$

$$= 12 \, \mu\text{A}$$

Optional Calculator Sequence

The Sharp calculator sequence for this problem is as follows:

Step	Keypad Entry	Top Display Response
1	24	
2	÷	24/
3	2	24/
4	Exp	24/
5	6	24/
6	=	24/2E06=
	Answer	12.00×10^{-06}
		(bottom display)

You can see in both the preceding examples how convenient it is to express the solution in a prefix unit when the calculator is in the **ENG** mode.

Practice Problems 6.1   Answers are given at the end of the chapter.
1. Calculate I if $V = 24$ V and $R = 8 \, \Omega$.
2. Calculate I if $V = 24$ V and $R = 8 \, \text{k}\Omega$.

SECTION 6.2
SELF-CHECK

Answers are given at the end of the chapter.

1. Current varies _____ with resistance when V is held constant.
2. What prefix should be used if the current is calculated as 2.5×10^{-3} A?
3. Describe the term *inverse proportionality*.

6.3   VOLTAGE CALCULATIONS

Another version of Ohm's law relates the three factors V, I, and R using the equation

$$V = IR \tag{6.2}$$

Voltage is found by multiplying current (I) by resistance (R). Notice that in a simple circuit with only one resistance in the circuit, the voltage across the resistance is equal to the source voltage. This is because the resistance is directly connected across the battery. Thus, the potential difference across the battery and the resistance are the same. This is illustrated in Figure 6.5.

FIGURE 6.5
Source voltage equals resistor voltage.

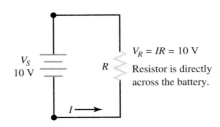

$V_R = IR = 10$ V

Resistor is directly
across the battery.

EXAMPLE 6.4

Determine V when $I = 4$ A and $R = 12\ \Omega$.

Solution $V = IR$
$\qquad = 4\text{ A} \times 12\ \Omega$ **(6.2)**
$\qquad = 48$ V

Notice that with I in amperes and R in ohms, the product, V, is in volts, because all units are basic units.

EXAMPLE 6.5

Determine V when $I = 4$ mA and $R = 15\ \Omega$.

Solution $V = IR$
$\qquad = 4 \times 10^{-03}\text{ A} \times 15\ \Omega$
$\qquad = 60 \times 10^{-03}\text{ V}$ **(6.2)**
$\qquad = 60$ mV

Optional Calculator Sequence

The Sharp calculator sequence for this problem is as follows:

Step	Keypad Entry	Top Display Response
1	4	
2	Exp	
3	+/−	
4	3	
5	×	4E−03*
6	15	4E−03*
7	=	4E−03*15=
	Answer	60×10^{-03}
		(bottom display)

Note that $+/-$ can be entered either before or after the numeral 3. Also note that with the calculator in the **ENG** mode, the exponent of the answer is recognizable as the prefix unit, milli, which is a multiple of the base unit. Thus, the answer is in mV but can be converted to the basic unit, volt, if desired. Prefix units are very common in many electronic applications.

EXAMPLE 6.6

Determine V when $I = 6.0$ mA and $R = 7.5$ kΩ.

Solution $V = IR$
$= 6 \times 10^{-03}$ A $\times 7.5 \times 10^{03}$ Ω
$= 45 \times 10^{00}$ V
$= 45$ V

(6.2)

Optional Calculator Sequence

The calculator sequence for this problem is as follows:

Step	Keypad Entry	Top Display Response
1	6	
2	Exp	
3	±	
4	3	
5	×	6E−03*
6	7.5	6E−03*
7	Exp	6E−03*
8	3	6E−03*
9	=	6E−03*7.5E03=
Answer		45.00×10^{00}
		(bottom display)

Notice here that the multiple units of mA and kΩ cancel, leaving the answer in the basic unit, volts.

EXAMPLE 6.7

Determine V when $I = 5.0$ μA and $R = 10$ kΩ.

Solution $V = IR$
$= 5 \times 10^{-06}$ A $\times 10 \times 10^{03}$ Ω
$= 50 \times 10^{-03}$ V
$= 50$ mV

(6.2)

EXAMPLE 6.8

Determine V when $I = 2.0$ μA and $R = 12$ MΩ.

Solution $V = IR$
$= 2 \times 10^{-06}$ A $\times 12 \times 10^{06}$ Ω
$= 24 \times 10^{00}$ V
$= 24$ V

(6.2)

Practice Problems 6.2   Answers are given at the end of the chapter.
1. Determine V when $I = 3.0$ mA and $R = 12$ kΩ.
2. Determine V when $I = 1.5$ μA and $R = 100$ kΩ.
3. Determine V when $I = 2$ A and $R = 25$ Ω.

Answers are given at the end of the chapter.

1. What is the proportionality of V to I with R constant?
2. What is the proportionality of V to R with I constant?
3. What prefix should be given to the value 1.5×10^{-3} V?

6.4 RESISTANCE CALCULATIONS

The third and final version of Ohm's law relates the three factors V, I, and R by the formula

$$R = \frac{V}{I} \tag{6.3}$$

When V and I are known, the resistance can be calculated as the voltage across R divided by the current through it. We do not need to know the physical construction of a resistance to analyze its effect in a circuit, so long as we know its V/I ratio. The V/I ratio represents the electrical effect of a resistance to a circuit.

EXAMPLE 6.9

Determine R when $V = 12$ V and $I = 4$ A.

Solution $R = \dfrac{V}{I}$

$$= \frac{12 \text{ V}}{4 \text{ A}} \tag{6.3}$$

$$= 3 \ \Omega$$

EXAMPLE 6.10

Determine R when $V = 10$ V and $I = 0.02$ A.

Solution $R = \dfrac{V}{I}$

$$= \frac{10 \text{ V}}{0.02 \text{ A}} \tag{6.3}$$

$$= 500 \ \Omega$$

EXAMPLE 6.11

Determine R when $V = 15$ V and $I = 2.0$ mA.

Solution $R = \dfrac{V}{I}$

$$= \frac{15 \text{ V}}{2 \times 10^{-03} \text{ A}} \tag{6.3}$$

$$= 7.5 \times 10^{03} \ \Omega$$

$$= 7.5 \text{ k}\Omega$$

EXAMPLE 6.12

Determine R when $V = 50$ V and $I = 10\ \mu\text{A}$.

Solution $R = \dfrac{V}{I}$

$$= \frac{50\ \text{V}}{10 \times 10^{-06}\ \text{A}}$$

(6.3)

$$= 5.0 \times 10^{06}\ \Omega$$

$$= 5.0\ \text{M}\Omega$$

Practice Problems 6.3   Answers are given at the end of the chapter.
1. Calculate R when $V = 12$ V and $I = 2$ mA.
2. Calculate R when $V = 10$ V and $I = 0.004$ A.
3. Calculate R when $V = 20$ V and $I = 5\ \mu\text{A}$.

**SECTION 6.4
SELF-CHECK**

Answers are given at the end of the chapter.

1. What does the V/I ratio represent?
2. True or false: You need to know the physical construction of a resistance in order to analyze its effect in a circuit.

6.5   WORK AND ENERGY

Whenever a force acts in such a way as to produce motion in a body, work is done. Motion must take place in a mechanical consideration of work, such as a rotating electric motor. Another way to look at work is to say that when energy is converted from mechanical energy to electrical energy, and vice versa, work is done. This implies that *energy* can be defined as the ability to do *work*. There are two kinds of energy, *potential* and *kinetic*. Potential energy is possessed by a system by virtue of its position, or condition. A compressed spring possesses potential energy. Kinetic energy is possessed by a body by virtue of its motion. A moving body of water possesses kinetic energy.

The SI unit for work is the joule (J). Work is performed when electrical energy is converted to mechanical energy, and vice versa. The joule is also used as a unit of measurement of electrical energy. Thus, the joule is used for both electrical energy and work. The definition of the joule can be stated in electrical terms and differs from the mechanical statement. Electrically speaking, 1 J of electrical energy is required to raise 1 C of electric charge through a potential difference of 1 V. Mathematically, this is expressed as

$$W = QV$$

(6.4)

where W = work or energy (J)
         Q = charge (C)
         V = potential difference (V)

(Note that this equation characterizes voltage as well; it was first defined in Chapter 4 as equation (4.1).)

Voltage, or potential difference, exists between any two charges that are not exactly equal to each other. Even an uncharged body has a potential difference with respect to a charged body. It is positive with respect to a negative charge and negative with respect to a positive charge. Voltage also exists between two unequal positive charges or between two unequal negative charges. Therefore, *voltage is purely relative* and is not used to express the actual amount of charge; instead, it is used to compare one charge to another. In doing this, voltage represents the electromotive force between the two charges being compared.

SECTION 6.5 SELF-CHECK

Answers are given at the end of the chapter.

1. Define energy.
2. Describe the two kinds of energy.
3. Where does voltage exist?

6.6   POWER

Power is an indication of how much work can be accomplished in a specified amount of time. From this statement, power can be defined as the *rate at which work is done*. In relation to electricity and electronics, power is the rate (t) at which electric energy (W) is transferred. Depending on the circuit, it may or may not be converted from one form to another. The electrical unit of measurement for power is the *watt* (W). The term watt is derived from the surname of James Watt, the developer of the steam engine. It originated in 1782 by Watt as a method of specifying how much power his steam engine developed in terms of the horses it was intended to replace. (One horsepower is equal to 746 W.)

One watt is the rate of doing work when 1 J of work is done in 1 s. Expressing this mathematically,

$$P = \frac{W}{t} \tag{6.5}$$

where P = power (W)
 W = work (J)
 t = time (s)

Since work (W) in equation (6.4) is also expressed in equation (6.5), we can substitute its value in equation (6.4) into equation (6.5) to yield

$$P = \frac{QV}{t} \tag{6.6}$$

From equation (4.2) we know that $I = Q/t$; thus, I can be substituted for Q/t in equation (6.6) to give us the more familiar form of the power equation:

$$P = IV \tag{6.7}$$

where P = power (W)
 I = current (A)
 V = potential difference (V)

A circle diagram similar to Ohm's law is used as an aid for determining the different forms of the power equation, as illustrated in Figure 6.6.

The amount of power changes when either voltage or current or both voltage and current change. In practice, the only factors that can be changed are voltage and resistance. In explaining the different forms that equations may take, current is sometimes presented as a quantity that is changed. Remember that if current changes, it is because either voltage or resistance has changed. *Current is a dependent quantity.*

FIGURE 6.6
The power equation in diagram form.

Because Ohm's law contains the same variables as the power equation of equation (6.7), they can be combined to form 12 basic equation variations with which you should be familiar. These are shown in the equation wheel of Figure 6.7.

FIGURE 6.7
Summary of basic electrical equations.

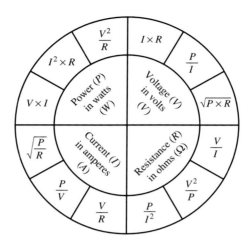

EXAMPLE 6.13

Determine P when $I = 5$ mA and $V = 20$ V.

Solution $P = IV$
$$= 5 \times 10^{-03} \text{ A} \times 20 \text{ V}$$
$$= 100 \times 10^{-03} \text{ W}$$
$$= 100 \text{ mW}$$

EXAMPLE 6.14

Determine P when $I = 6$ mA and $R = 5$ kΩ.

Solution $P = I^2R$ (from Figure 6.7)
$$= (6 \times 10^{-3} \text{ A})^2 \times 5 \times 10^3 \text{ Ω}$$
$$= 180 \times 10^{-3} \text{ W}$$
$$= 180 \text{ mW}$$

Optional Calculator Sequence

The calculator sequence for this problem is as follows:

Step	Keypad Entry	Top Display Response
1	6	
2	Exp	
3	$\pm$	
4	3	
5	X^2	$6E{-}03^2$
6	$\times$	$6E{-}03^2*$
7	5	
8	Exp	
9	3	
10	=	$6E{-}03^2*5E03=$
	Answer	180.00×10^{-03}
		(bottom display)

EXAMPLE 6.15

Determine P when $V = 20$ V and $R = 10$ kΩ.

Solution $P = \dfrac{V^2}{R}$ (from Figure 6.7)

$= \dfrac{(20 \text{ V})^2}{10 \times 10^3 \ \Omega}$

$= 40 \times 10^{-3}$ W

$= 40$ mW

Optional Calculator Sequence

The calculator sequence for this problem is as follows:

Step	Keypad Entry	Top Display Response
1	20	
2	X^2	20^2
3	$\div$	$20^2/$
4	10	
5	Exp	
6	3	
7	$=$	$20^2/10\text{E}03=$
	Answer	40.00×10^{-03}
		(bottom display)

EXAMPLE 6.16

Determine V when $P = 50$ mW and $R = 2$ kΩ.

Solution $V = \sqrt{PR}$ (from Figure 6.7)

$= \sqrt{50 \times 10^{-3} \text{ W} \times 2 \times 10^3 \ \Omega}$

$= 10$ V

Optional Calculator Sequence

The calculator sequence for this problem is as follows:

Step	Keypad Entry	Top Display Response
1	$\sqrt{}$	$\sqrt{}$
2	(	$\sqrt{}($
3	50	$\sqrt{}($
4	Exp	$\sqrt{}($
5	$\pm$	$\sqrt{}($
6	3	$\sqrt{}($
7	$\times$	$\sqrt{}(50\text{E}{-}03*$
8	2	$\sqrt{}(50\text{E}{-}03*$
9	Exp	$\sqrt{}(50\text{E}{-}03*$
10	3	$\sqrt{}(50\text{E}{-}03*$

(continued)

11	)	$\sqrt{(50E-03*2E03)}$
12	=	$\sqrt{(50E-03*2E03)}=$
	Answer	10.00×10^{00}
		(bottom display)

EXAMPLE 6.17

Determine I when $P = 100$ W and $R = 144\ \Omega$.

Solution $I = \sqrt{\dfrac{P}{R}}$ (from Figure 6.7)

$= \sqrt{\dfrac{100\text{ W}}{144\ \Omega}}$

$= 833.3$ mA

$= 0.83$ A

Practice Problems 6.4   Answers are given at the end of the chapter.
1. Find P when $I = 4$ mA and $R = 6$ kΩ.
2. Find P when $V = 25$ V and $R = 10$ kΩ.
3. Find R when $V = 10$ V and $P = 10$ mW.
4. Find V when $P = 100$ mW and $R = 10$ kΩ.

6.6.1   Wattage Prefixes

As with other electrical quantities, prefixes may be attached to the word *watt* when expressing very large or very small amounts of power. The previous examples showed this. Some of the more common of these are the kilowatt (kW), 1000 (10^3) W; the megawatt (MW), 1,000,000 (10^6) W; and the milliwatt (mW), 0.001 (10^{-3}) W.

SECTION 6.6 SELF-CHECK

Answers are given at the end of the chapter.

1. Define power.
2. In what unit is power measured?
3. What makes current vary in a circuit?

6.7   **POWER RATING**

Electrical components are often given a power rating. The power rating, in watts, indicates the rate at which the device converts electrical energy to another form of energy, such as light, heat, or motion. An example of such a rating is noted when comparing a 150-W lamp to a 100-W lamp. The higher wattage rating of the 150-W lamp indicates it is capable of converting more electrical energy into light energy than the lamp of the lower rating. Other common examples of devices with power ratings are soldering irons and small electric motors.

In some electrical devices the wattage rating indicates the maximum power the device is designed to use rather than the normal operating power. A 150-W lamp, for example, uses 150 W *when operated at the specified voltage printed on the bulb.* In con-

Lab Reference: The concepts of power and power rating are demonstrated in Exercise 8.

trast, a device such as a resistor is not normally given a voltage or a current rating. A resistor is given a power rating in watts and can be operated at any combination of *voltage* and *current* as long as the power rating is not exceeded. In most circuits, the actual power rating of the resistor is considerably less than the power rating of the resistor because a 100% *safety factor* is often used. For example, if a resistor normally consumes 2 W of power, a resistor with a power rating of 4 W is used. This ensures that in the event of a momentary surge in current through the device, its power rating is not exceeded.

**SECTION 6.7
SELF-CHECK**

Answers are given at the end of the chapter.

1. Define power rating.
2. What does the power rating of a resistor describe?
3. If a resistor uses 1.5 W of power, what wattage value should be used in an actual circuit?

6.8   POWER CONVERSION AND EFFICIENCY

The source supplies (delivers) the power the load uses. This value can be calculated by $P_S = V_S I$. The load uses (dissipates) this power and converts it from one form of energy to another. (Remember, most loads are some type of transducer.) The load power can be found by $P_L = V_L I$. (Note the subtle difference in the two equations.) An electric motor converts electrical energy to mechanical energy. An electric lightbulb converts electrical energy into light energy.

The energy used by electrical devices is measured in watt-hours. This practical unit of electrical energy is equal to 1 W of power used continuously for 1 h. The unit kilowatt-hour (kWh) is used more extensively on a daily basis and is equal to 1000 Wh. Every home and business has a kilowatt-hour meter attached to the power line to measure how much electrical energy is consumed so that the power utility can charge accordingly.

The *efficiency* of an electrical device is the ratio of power converted to useful energy (or useful output power) divided by the power consumed by the device (or useful input power). This number is always less than 1.00 because of the losses in any electrical device. If a device has an efficiency rating of 0.95 (95%), it transforms 95 W into useful energy for every 100 W of input power. The other 5 W are lost due to heat or other losses that cannot be used. The efficiency equation may be stated as

$$\text{efficiency} = \frac{\text{power converted}}{\text{power used}} \tag{6.8}$$

(Note that multiplying this quantity by 100 converts efficiency into a percentage.)

Calculating the amount of power converted by an electrical device is a simple matter. You need to know the length of time the device is operated and the input power or horsepower (hp) rating. Horsepower, a unit of work, is often found as a rating on electrical motors. (Remember that 1 hp equals 746 W.)

EXAMPLE 6.18

A $\frac{3}{4}$-hp motor operates 8 h a day. How much power (in kWh) is converted by the motor in a 30-day period? What is the efficiency of the motor if it actually uses 137 kWh in the same period?

Solution $P = \text{hp} \times \text{W}$
$\qquad = \frac{3}{4} \times 746 \text{ W}$
$\qquad = 559 \text{ W}$

Convert W to kWh:

$$P = \frac{\text{work} \times \text{time}}{1000}$$

$$= \frac{559 \times 8 \times 30}{1000}$$

$$= 134.16 \text{ kWh}$$

$$\text{efficiency} = \frac{\text{power converted}}{\text{power used}}$$

$$= \frac{134.16 \text{ kWh}}{137 \text{ kWh}}$$

$$= 0.98$$

$$= 98\%$$

Practice Problems 6.5   Answers are given at the end of the chapter.
1. A 200-W motor runs 10 h a day for 30 days. How much power is converted in this period? Determine the efficiency if it uses 75 kWh.

**SECTION 6.8
SELF-CHECK**

Answers are given at the end of the chapter.

1. How is power measured?
2. Define efficiency.
3. What is a kWh?

■ SUMMARY

1. Ohm's law $(I = V/R)$ states that current is directly proportional to voltage and inversely proportional to resistance.

2. A value that is intentionally varied in a circuit is called an independent variable. The second quantity that varies as a result of changes in the first quantity is called the dependent variable.

3. Voltage and resistance are independent variables, whereas current is a dependent variable.

4. The calculator should be put in the **ENG** decimal mode for most Ohm's law calculations.

5. The V/I ratio represents the electrical effects of resistance to a circuit.

6. When energy is converted from one form to another, work is done.

7. Energy can be defined as the ability to do work.

8. The two forms of energy are potential and kinetic.

9. Voltage exists between two unequal charges. It is purely relative.

10. Power is defined as the rate at which work is done. It is measured in watts.

11. In practice, the only factors that are changed in a circuit are voltage and resistance. Current changes are dependent on voltage and/or resistance changes.

12. The power rating of an electrical component indicates the rate at which the device converts electrical energy to another form of energy.

13. A resistor is given a power rating in watts and can be operated at any combination of voltage and current as long as the power rating is not exceeded.

14. The actual power rating of a resistor is less than the power rating of the device because a 100% safety factor is used.

15. The efficiency of an electrical device is the ratio of power converted to useful energy divided by the power consumed by the device (or, useful output power divided by useful input power). It is expressed as a numeric value less than 1 or as a percent.

■ IMPORTANT EQUATIONS

$$I = \frac{V}{R} \tag{6.1}$$

$$V = IR \tag{6.2}$$

$$R = \frac{V}{I} \tag{6.3}$$

$$W = QV \tag{6.4}$$

$$P = \frac{W}{t} \tag{6.5}$$

$$P = \frac{QV}{t} \qquad \textbf{(6.6)}$$

$$P = IV \qquad \textbf{(6.7)}$$

$$\text{efficiency} = \frac{\text{power converted}}{\text{power used}} \qquad \textbf{(6.8)}$$

■ REVIEW QUESTIONS

1. State Ohm's law in words.
2. What other words may be used interchangeably with voltage?
3. Differentiate the terms *independent* and *dependent* variables.
4. When graphing, which variable is plotted on the horizontal axis?
5. What does the slope of a *VI* curve represent?
6. What advantage is gained by having your calculator in the **ENG** mode?
7. State three forms of Ohm's law.
8. Which metric prefixes are commonly used for current?
9. The electrical effect of a resistance to a circuit is represented by what ratio?
10. If voltage is held constant and circuit resistance is decreased, current will _____.
11. If applied voltage is increased, the total circuit resistance will _____.
12. The unit of measurement for electrical power is the _____.
13. Define energy.
14. Describe the differences in potential and kinetic energy.
15. Define the joule in electrical terms.
16. Define the term horsepower.
17. Of the factors *V*, *I*, *R*, and *P*, which are actually physically changeable in a circuit?
18. Show how $P = I^2R$ can be derived from Ohm's law.
19. Define the term *power rating*.
20. What is meant by a 100% *safety factor,* as applied to a resistor?
21. Define the term *efficiency*.
22. Why is efficiency always less than 100%?
23. What do you need to know in order to calculate the amount of power converted by an electrical device?
24. Prove that 1 kWh is equal to 3.6×10^6 J.
25. How much will it cost to run a 5-W device for 30 days at a cost of $0.10 per kWh?

■ CRITICAL THINKING

1. Describe the relationship between a *dependent* and an *independent* variable.
2. Define energy and its relationship to work.
3. Describe the power rating as it applies to a resistor.

■ PROBLEMS

1. A load operates at 48 V and draws 30 mA. What is its resistance?
2. What is the resistance of a device if it draws 45 mA from a 9-V source?
3. A 1.5-MΩ resistor is connected across a 100-V source. What is the current through the resistor?
4. What voltage is dropped across a 1-MΩ resistor when 6 μA of current is flowing through it?
5. A 50-Ω load has 2 A of current. Calculate its voltage drop.
6. An insulator has 15 μA of current flowing through it with 4.5 kV across it. What is its resistance?
7. What is the resistance of a car radio that draws 120 mA from a 12-V battery?
8. A device draws 2.5 A from a 100-V source. What current will it draw from a 200-V source?
9. A load is connected to a 100-V source with two leads. The resistance of each lead is 0.2 Ω. What is the voltage across the load?
10. What current will a 2-Ω resistor draw from a 24-V source?

Solve for the unknowns in Problems 11–20.

	Resistance	Voltage	Current
11.	3.3 kΩ	18 V	_____
12.	_____	80 mV	2 mA
13.	270 kΩ	_____	150 μA
14.	_____	30 V	10 mA
15.	_____	100 V	2.5 μA
16.	10 kΩ	_____	40 μA
17.	1 kΩ	12 V	_____
18.	470 Ω	_____	5 mA
19.	12 Ω	12 V	_____
20.	_____	90 mV	30 μA

21. If a DC power supply can deliver 2 A of current at 12 V, what is the maximum available power?
22. The voltage drop of a 22-kΩ resistor is 18 V. How much power is the resistor dissipating?
23. A 1-kΩ resistor has 15 V across it. Calculate its power dissipation.
24. Calculate the maximum current that can safely flow through a 150-Ω, $\frac{1}{2}$-W resistor.
25. A 4-Ω load dissipates 6 W of power. How much voltage is across the load?
26. A 50-Ω load dissipates 100 W of power. How much voltage is across the load?
27. An 8-Ω load dissipates 20 W of power. How much current flows through the load?
28. How much current does a 75-W lightbulb draw from a 120-V source?
29. Calculate the maximum voltage drop a 1-kΩ, $\frac{1}{4}$-W resistor can handle safely.
30. Calculate the maximum voltage drop an 8-Ω, 200-W load can handle safely.

Solve for the unknowns in Problems 31–40.

	Power	Resistance	Current	Voltage
31.	_____	_____	8 A	440 V
32.	800 mW	_____	40 mA	_____
33.	15 kW	_____	20 A	_____
34.	120 μW	_____		120 V
35.	120 mW	1.2 kΩ	_____	
36.	250 mW	_____	_____	16 V
37.	_____	1.8 kΩ	_____	12 V
38.	_____	220 Ω	220 mA	_____
39.	2 W	_____	_____	50 V
40.	_____	39 Ω	3 A	_____

41. A 4-hp motor draws 20 A from the 220-V power line. What is the efficiency of the motor?

42. A 5-hp motor is 80% efficient. Calculate the current it draws from a 220-V power line.

43. A $\frac{1}{4}$-hp motor draws 2 A of current from a 120-V power line. Calculate its efficiency.

44. A television set draws 1.5 A from the 120-V power line. It is on 50% of the time. How much will it cost the user to operate it over a period of 30 days at a cost of $0.07 per kWh?

45. How much will it cost to light a 100-W lamp for 30 days at a cost of $0.12 per kWh?

■ ANSWERS TO PRACTICE PROBLEMS

PRACTICE PROBLEMS 6.1

1. $I = 3$ A
2. $I = 3$ mA

PRACTICE PROBLEMS 6.2

1. $V = 36$ V
2. $V = 0.15$ V, or 150 mV
3. $V = 50$ V

PRACTICE PROBLEMS 6.3

1. $R = 6$ kΩ
2. $R = 2.5$ kΩ
3. $R = 4$ MΩ

PRACTICE PROBLEMS 6.4

1. $P = 96$ mW
2. $P = 62.5$ mW
3. $R = 10$ kΩ
4. $V = 31.6$ V

PRACTICE PROBLEMS 6.5

1. 60 kW; 0.80 (80%)

■ ANSWERS TO SECTION SELF-CHECKS

SECTION 6.1

1. The current is directly proportional to the voltage and inversely proportional to the resistance.

2. An independent variable is the one that physically varies. The dependent variable depends on the value of the independent variable.

3. Direct proportionality means that both quantities vary directly; that is, if one increases, so does the other. Example: voltage and current.

SECTION 6.2

1. Inversely

2. mA

3. Inverse proportionality means that both quantities vary inversely; that is, if one increases, the other decreases. Example: current and resistance.

SECTION 6.3

1. Direct

2. Direct

3. mV

SECTION 6.4

1. Effective circuit resistance

2. F

SECTION 6.5

1. Energy is the ability to do work.

2. Potential energy is possessed by a device because of its position or condition. Example: a compressed spring. Kinetic energy is possessed by a device because of its motion. Example: running water.

3. Voltage exists only between two points that differ in charge or potential.

SECTION 6.6

1. Power is the rate of doing work measured in watts.

2. Watt (W)

3. A change in either V or R or both

SECTION 6.7

1. Power rating is the rate at which a device converts electrical energy into another form.

2. A resistor's power rating is any combination of voltage or current that it handles as long as its rating is not exceeded.

3. 3 W at a 100% safety factor

SECTION 6.8

1. Power is measured in watts.

2. Efficiency is the ratio of power converted to useful energy divided by the power consumed by the device.

3. 1000 W/h

PART II

DC CIRCUITS

CHAPTER 7

SERIES DC CIRCUITS

OBJECTIVES

After completing this chapter, you will be able to:

1. State the rules for a series-connected circuit.
2. Solve for circuit resistance, voltage, current, and power.
3. Solve for component resistance, voltage, current, and power.
4. Analyze series circuits to determine unknown quantities.
5. Determine voltage polarity using a variety of notations.
6. Determine the appropriate reference to use for voltage drops.
7. Compare and contrast the terms ground, common, and reference.
8. Troubleshoot series circuits and determine faults.
9. Differentiate between opens and shorts in series circuits.

INTRODUCTION

A *series circuit* can be defined as a circuit in which there is only *one current path* and all components are connected end to end along the path from the negative terminal of the source to the positive terminal of the source. If any component in this path opens, current flow ceases. All components connected in series depend on each other for a continuous flow of current through the circuit. Common examples of this dependency are series-strung holiday lights. In any string, all the lamps go out if *any* lamp burns out. The filament in a lamp opens when it burns out, preventing current from flowing through the rest of the circuit.

Series circuits are among the easiest of all circuits to understand and analyze. They are also among the most common types of connections used in electrical and electronic circuitry.

7.1   RESISTANCE IN SERIES CIRCUITS

Lab Reference: The concepts of resistances in series are demonstrated in Exercise 9.

In a series circuit, the total resistance is the sum of the individual resistances. Because there is only one path for current, it must flow through all components. Thus, the total resistance through which the current must flow is cumulative. Mathematically this may be stated as:

$$R_T = R_1 + R_2 + R_3 + \cdots + R_n \qquad (7.1)$$

where

R_T = total resistance (Ω)

R_1, R_2, R_3 = individual resistance (Ω)

R_n = any number of additional resistances (Ω)

Total resistance (R_T) is the only opposition to current the source can sense; it does not "see" the individual, separate resistors. It sees one equivalent resistance. Thus, a series circuit can be reduced to a single resistance that offers the same opposition as the sum of the individual resistances.

EXAMPLE 7.1

Determine the total resistance (R_T) of the circuit in Figure 7.1.

Solution $R_T = R_1 + R_2 + R_3 + R_4$
  $= 2\,k\Omega + 5\,k\Omega + 10\,k\Omega + 8\,k\Omega$
  $= 25\,k\Omega$

FIGURE 7.1
Determining total resistance.

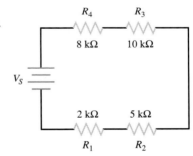

EXAMPLE 7.2

Connect the resistors in Figure 7.2 to produce a series circuit.

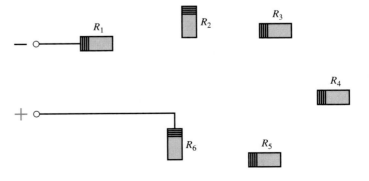

FIGURE 7.2
Assorted resistors to be connected in series.

Solution
The resistors should be connected together end to end, as shown in Figure 7.3.

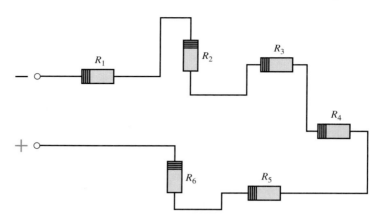

FIGURE 7.3
Solution to Example 7.2.

Practice Problems 7.1 Answers are given at the end of the chapter.
1. Find the total resistance for a series circuit consisting of five resistors having the following values: 1 kΩ, 2.7 kΩ, 3.3 kΩ, 6.8 kΩ, and 12 kΩ.
2. Find the total resistance for the series circuit shown in Figure 7.4.

FIGURE 7.4
Circuit for Practice Problems 7.1,
Problem 2.

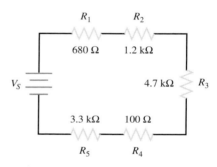

SECTION 7.1 SELF-CHECK

Answers are given at the end of the chapter.

1. Describe a series circuit.
2. How is the total resistance of a series circuit determined?

7.2 CURRENT IN SERIES CIRCUITS

Because there is only one path for current in a series circuit, the same current must flow through each component of the circuit. When measured by an ammeter, the current is the same at any point in the circuit. Current is often measured as part of a lab exercise.

To determine the current in a series circuit, only the current through one of the components must be known. Mathematically this is expressed as

Lab Reference: The concept of current in a series circuit is demonstrated in Exercise 9.

$$I_T = I_1 = I_2 = I_3 = \cdots = I_n \tag{7.2}$$

I_T is equal to any I in a series circuit. I_T, I_1, I_2, I_3, etc., could be used interchangeably in a series circuit. However, the distinction in the equation between $I_T, I_1, I_2,$ and I_3 should be observed. The reason for this is that future circuits may have several currents, and it will be necessary to differentiate between them and I_T. In order to solve for I_T, one must use Ohm's law and know the total resistance and source voltage or an individual resistance and individual voltage drop.

An important fact to keep in mind when applying Ohm's law to a series circuit is to consider whether the values used are component values or total values. When the information available enables the use of Ohm's law to find total resistance, total voltage, and total current, then total values must be inserted into the appropriate equation.

To compute any quantity ($V, I, R,$ or P) associated with a *single* given resistor, the values used in the equation must be obtained from that particular resistor. For example, to find the value of an unknown resistance, the voltage across and the current through that particular resistor are used.

EXAMPLE 7.3

Determine I_T for the circuit in Figure 7.5.

FIGURE 7.5
Circuit for Example 7.3.

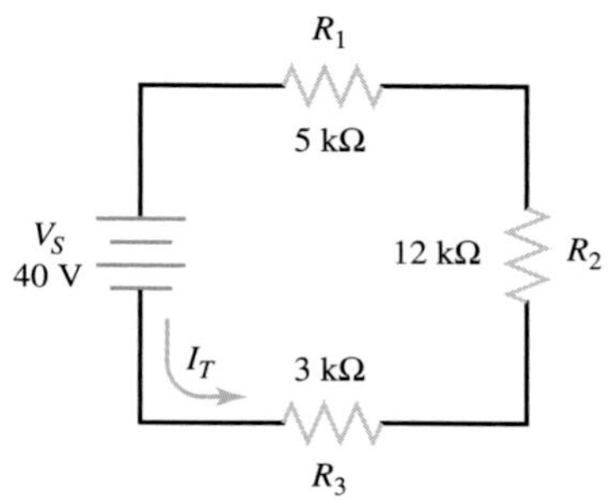

R_1
5 kΩ

V_S
40 V

12 kΩ R_2

I_T 3 kΩ

R_3

Solution $R_T = R_1 + R_2 + R_3$

$\qquad = 5\text{ k}\Omega + 12\text{ k}\Omega + 3\text{ k}\Omega$

$\qquad = 20\text{ k}\Omega$

$I_T = \dfrac{V_S}{R_T}$

$\qquad = \dfrac{40\text{ V}}{20 \times 10^3\ \Omega}$

$\qquad = 2 \times 10^{-3}\text{ A}$

$\qquad = 2\text{ mA}$

EXAMPLE 7.4

Determine I_T when the voltage drop across R_2 in Figure 7.5 is 36 V.

Solution $I_T = I_2 = \dfrac{V_2}{R_2}$

$\qquad = \dfrac{36\text{ V}}{12 \times 10^3\ \Omega}$

$\qquad = 3.0 \times 10^{-03}\text{ A}$

$\qquad = 3.0\text{ mA}$

Practice Problems 7.2   Answers are given at the end of the chapter.
1. If $V_S = 12$ V and $R_T = 6$ kΩ, determine I_T.
2. If $V_1 = 6$ V and $R_1 = 4$ kΩ, determine I_1.

**SECTION 7.2
SELF-CHECK**

Answers are given at the end of the chapter.

1. How many paths for current exist in a series circuit?
2. Where do you measure for current in a series circuit?

7.3   VOLTAGE IN SERIES CIRCUITS

Lab Reference: The concept of voltage in a series circuit is demonstrated in Exercise 9.

As was shown in Section 4.3, a battery is a voltage source and is also called a *voltage rise*, because it imparts energy to the free electrons to constitute current flow. Whenever current flows in a circuit, a *voltage drop* occurs across a resistive device. The voltage drop occurs because the resistive device offers opposition to the flow of current, and energy is expended overcoming this opposition. The amount of voltage drop depends on both resistor and current values. The voltage drops that occur in a series circuit are directly proportional to the resistance values. This is the result of having the same current flow through each resistor—the larger the ohmic value of the resistor, the larger the voltage drop across it.

Because of the direct proportionality that a resistor voltage has to resistance, we are provided with a way other than Ohm's law to calculate a voltage drop. This technique is called the *voltage-divider rule* (more fully described in Chapter 11). We simply take the ratio of individual resistance to total resistance times the source voltage to get the voltage drop across a resistor. This gives us

$$V_X = \frac{R_X}{R_T} \times V_S \qquad (7.3)$$

where V_X = voltage drop of resistor in question (V)
R_X = resistance of resistor in question (Ω)
R_T = total resistance of series circuit (Ω)
V_S = source voltage (V)

EXAMPLE 7.5

Find the voltage drop across R_1 in Figure 7.5.

Solution $V_X = \dfrac{R_X}{R_T} \times V_S$

$= \dfrac{5 \text{ k}\Omega}{20 \text{ k}\Omega} \times 40 \text{ V}$

$= 0.25 \times 40 \text{ V}$

$= 10 \text{ V}$

The voltage dropped across a resistor in a circuit consisting of a *single* resistor and a voltage source is the total voltage across the circuit and is equal to the applied voltage. This was shown in Figure 6.5. The total voltage across a series circuit consisting of more than one resistor is also equal to the applied voltage, but it consists of the sum of the individual voltage drops. In any series circuit then, the sum of the resistor voltage drops must equal the source voltage. For a series circuit,

$$V_S = V_1 + V_2 + V_3 + \cdots + V_n \tag{7.4}$$

where $\qquad V_S$ = source voltage (V)
V_1, V_2, V_3 = voltage drops (V)
V_n = any additional number of voltage drops (V)

V_S may be found by adding the individual voltage drops of a series circuit or by using Ohm's law when I_T and R_T are known. Equation 7.3 is a form of Kirchhoff's voltage law (KVL), which is a powerful tool used in circuit analysis. It is explored more fully in Chapter 11.

Voltage drops are often called *IR drops* because of the relationship between current and resistance in the Ohm's law equation. Thus, equation (7.4) may be restated as:

$$V_S = I_1R_1 + I_2R_2 + I_3R_3 + \cdots + I_nR_n \tag{7.5}$$

EXAMPLE 7.6

Find V_S in the circuit in Figure 7.6. There are two possible solutions to this problem.

FIGURE 7.6
Circuit for Example 7.6.

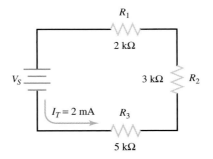

Solution 1 $\quad I_T = I_1 = I_2 = I_3 = 2$ mA
$V_1 = I_1R_1$
$\quad = 2 \times 10^{-3}$ A $\times 2 \times 10^3 \ \Omega$
$\quad = 4$ V
$V_2 = I_2R_2$
$\quad = 2 \times 10^{-3}$ A $\times 3 \times 10^3 \ \Omega$
$\quad = 6$ V
$V_3 = I_3R_3$
$\quad = 2 \times 10^{-3}$ A $\times 5 \times 10^3 \ \Omega$
$\quad = 10$ V
$V_S = V_1 + V_2 + V_3$
$\quad = 4$ V $+ 6$ V $+ 10$ V
$\quad = 20$ V

Solution 2 $\quad R_T = R_1 + R_2 + R_3$
$\quad = 2$ kΩ $+ 3$ kΩ $+ 5$ kΩ
$\quad = 10$ kΩ
$V_S = I_TR_T$
$\quad = 2 \times 10^{-3}$ A $\times 10 \times 10^3 \ \Omega$
$\quad = 20$ V

Practice Problems 7.3 Answers are given at the end of the chapter.
1. $R_1 = 10$ kΩ and $I_1 = 3.3$ mA. Determine V_1.
2. Find V_S in a series circuit when $I_2 = 2$ mA and $R_T = 10$ kΩ.

SECTION 7.3 SELF-CHECK

Answers are given at the end of the chapter.

1. To find the source voltage in a series circuit, the individual voltage drops are _____.

2. The voltage drop of a resistor is _____ _____ to the ohmic value of its resistance.

3. A voltage source is also known as a voltage _____.

| 7.4 | **POWER IN SERIES CIRCUITS** |

Power is the rate at which work is done. Each of the resistors in a series circuit consumes power that is dissipated in the form of heat. Because this power must come from the source, the total power must be equal to the power consumed by the individual circuit resistances. The power is measured in watts. Mathematically this is stated as:

$$P_T = P_1 + P_2 + P_3 + \cdots + P_n \qquad (7.6)$$

where
$$P_T = \text{total power (W)}$$
$$P_1, P_2, P_3 = \text{individual resistor power (W)}$$
$$P_n = \text{any additional resistor power (W)}$$

| EXAMPLE 7.7 | Find P_T in the circuit in Figure 7.7. |

FIGURE 7.7
Circuit for Example 7.7.

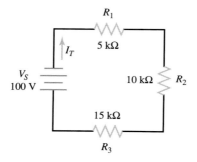

Solution $R_T = R_1 + R_2 + R_3$

$$= 5\,\text{k}\Omega + 10\,\text{k}\Omega + 15\,\text{k}\Omega$$

$$= 30\,\text{k}\Omega$$

$$I_T = \frac{V_S}{R_T}$$

$$= \frac{100\,\text{V}}{30 \times 10^3\,\Omega}$$

$$= 3.33\,\text{mA}$$

$$P_1 = I_1^2 R_1$$

$$= (3.33 \times 10^{-3}\,\text{A})^2 \times 5 \times 10^3\,\Omega$$

$$= 55.4\,\text{mW}$$

$$P_2 = I_2^2 R_2$$

$$= (3.33 \times 10^{-3}\,\text{A})^2 \times 10 \times 10^3\,\Omega$$

$$= 110.9\,\text{mW}$$

$$P_3 = I_3^2 R_3$$
$$= (3.33 \times 10^{-3}\text{ A})^2 \times 15 \times 10^3\ \Omega$$
$$= 166.3\text{ mW}$$
$$P_T = P_1 + P_2 + P_3$$
$$= 55.4\text{ mW} + 110.9\text{ mW} + 166.3\text{ mW}$$
$$= 332.6\text{ mW}$$

As a check, the total power delivered by the source can be calculated as

$$P_T = I_T V_S$$
$$= 3.33 \times 10^{-3}\text{A} \times 100\text{ V}$$
$$= 333\text{ mW}$$

(Note: There can be rounding errors between the two answers.)

Practice Problems 7.4   Answers are given at the end of the chapter.
1. $R_3 = 12\text{ k}\Omega$ and $I_3 = 2$ mA. Determine P_3.
2. Find P_T when $P_1 = 15$ mW, $P_2 = 30$ mW, and $P_3 = 50$ mW.

7.4.1   Rules for Series DC Circuits

1. The same current flows through each part of a series circuit.
2. The total resistance of a series circuit is equal to the sum of the individual resistances.
3. The total voltage across a series circuit is equal to the sum of the individual voltage drops.
4. The voltage drop across a resistor in a series circuit is proportional to the ohmic value of the resistor.
5. The total power in a series circuit is equal to the sum of the individual powers used by each circuit component.

**SECTION 7.4
SELF-CHECK**

Answers are given at the end of the chapter.

1. Define the term *power.*
2. How is power dissipated in a resistor?

7.5   SERIES CIRCUIT ANALYSIS

Problems involving the determination of resistance, voltage, current, and power in a series circuit require a procedure. The procedure is as follows:

1. Observe the circuit diagram carefully, or draw one if necessary.
2. Note the given values and the values to be found.
3. Select the appropriate equations to be used in solving for the unknown quantities based on the known quantities.
4. Substitute the known values in the equation you have selected and solve for the unknown value.

Some sample problems are shown in order to practice the procedure for solving series circuits with several unknowns.

EXAMPLE 7.8

Use the circuit in Figure 7.8.

FIGURE 7.8
Circuit for Example 7.8.

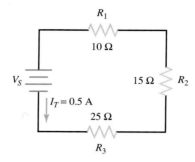

(a) Find the source voltage.
(b) Find the voltage across each resistor.
(c) Find the power dissipation of each resistor.
(d) Find the total power.

Solution

(a) $R_T = R_1 + R_2 + R_3$
$\quad\quad = 10\ \Omega + 15\ \Omega + 25\ \Omega$
$\quad\quad = 50\ \Omega$
$\quad V_S = I_T R_T$
$\quad\quad = 0.5\ \text{A} \times 50\ \Omega$
$\quad\quad = 25\ \text{V}$

(b) $V_1 = I_1 R_1$
$\quad\quad = 0.5\ \text{A} \times 10\ \Omega$
$\quad\quad = 5\ \text{V}$
$\quad V_2 = I_2 R_2$
$\quad\quad = 0.5\ \text{A} \times 15\ \Omega$
$\quad\quad = 7.5\ \text{V}$
$\quad V_3 = I_3 R_3$
$\quad\quad = 0.5\ \text{A} \times 25\ \Omega$
$\quad\quad = 12.5\ \text{V}$

(c) $P_1 = I_1 V_1$
$\quad\quad = 0.5\ \text{A} \times 5\ \text{V}$
$\quad\quad = 2.5\ \text{W}$
$\quad P_2 = I_2 V_2$
$\quad\quad = 0.5\ \text{A} \times 7.5\ \text{V}$
$\quad\quad = 3.75\ \text{W}$
$\quad P_3 = I_3 V_3$
$\quad\quad = 0.5\ \text{A} \times 12.5\ \text{V}$
$\quad\quad = 6.25\ \text{W}$

(d) $P_T = P_1 + P_2 + P_3$
$\quad\quad = 2.5\ \text{W} + 3.75\ \text{W} + 6.25\ \text{W}$
$\quad\quad = 12.5\ \text{W}$

or

$\quad P_T = I_T V_S$
$\quad\quad = 0.5\ \text{A} \times 25\ \text{V}$
$\quad\quad = 12.5\ \text{W}$

or

$\quad P_T = I_T^2 R_T$
$\quad\quad = (0.5\ \text{A})^2 \times 50\ \Omega$
$\quad\quad = 12.5\ \text{W}$

or

$$P_T = \frac{V_S^2}{R_T}$$

$$= \frac{(25\ \text{V})^2}{50\ \Omega}$$

$$= 12.5\ \text{W}$$

EXAMPLE 7.9

Find the missing information in Table 7.1 using the values in Figure 7.9.

FIGURE 7.9
Circuit for Example 7.9.

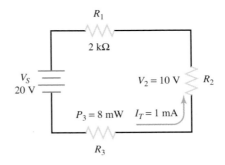

TABLE 7.1
Series circuit analysis

V_S	V_1	V_2	V_3	I_T	R_T	R_1	R_2	R_3	P_T	P_1	P_2	P_3
20 V		10 V		1 mA		2 kΩ						8 mW

Solution $V_1 = I_1 R_1$

$$= 1 \times 10^{-3}\ \text{A} \times 2 \times 10^3\ \Omega$$

$$= 2\ \text{V}$$

$$V_3 = \frac{P_3}{I_3}$$

$$= \frac{8 \times 10^{-3}\ \text{W}}{1 \times 10^{-3}\ \text{A}}$$

$$= 8\ \text{V}$$

$$R_T = \frac{V_S}{I_T}$$

$$= \frac{20\ \text{V}}{1 \times 10^{-3}\ \text{A}}$$

$$= 20\ \text{k}\Omega$$

$$R_2 = \frac{V_2}{I_2}$$

$$= \frac{10\ \text{V}}{1 \times 10^{-3}\ \text{A}}$$

$$= 10\ \text{k}\Omega$$

$$R_3 = \frac{V_3}{I_3}$$
$$= \frac{8 \text{ V}}{1 \times 10^{-3} \text{ A}}$$
$$= 8 \text{ k}\Omega$$
$$P_T = I_T V_S$$
$$= 1 \times 10^{-3} \text{ A} \times 20 \text{ V}$$
$$= 20 \text{ mW}$$
$$P_1 = I_1 V_1$$
$$= 1 \times 10^{-3} \text{ A} \times 2 \text{ V}$$
$$= 2 \text{ mW}$$
$$P_2 = I_2 V_2$$
$$= 1 \times 10^{-3} \text{ A} \times 10 \text{ V}$$
$$= 10 \text{ mW}$$

Notice in this solution that the unknowns were solved in order across the table. This may not always be the case. The key to circuit analysis is to note all given values, to determine what can logically be found by calculation, and then to proceed based upon those calculations.

Practice Problems 7.5   Answers are given at the end of the chapter.
1. $R_2 = 3.3$ kΩ and $I_1 = 2$ mA. Determine V_2.
2. Resistors R_1 and R_2 are connected in series across a 20-V source. $I_2 = 4$ mA and $V_1 = 4$ V. Determine I_T, R_1, R_2, V_2, P_1, P_2, and P_T.

**SECTION 7.5
SELF-CHECK**

Answers are given at the end of the chapter.

1. What is the key to series circuit analysis?

7.6   VOLTAGE POLARITY AND REFERENCES

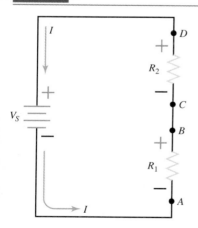

FIGURE 7.10
Voltage polarities.

The meaning of voltage polarity must be understood to more fully analyze a circuit and to be able to measure voltages correctly in lab. In the circuit shown in Figure 7.10, the current is shown flowing in a counterclockwise direction. Notice that the end of resistor R_1, into which the current flows, is marked negative ($-$). The end of R_1 where the current leaves is marked positive ($+$). These polarity markings are used to show that the end of R_1 into which the current flows is at a higher negative potential than the end of the resistor at which the current leaves. Point A is more negative than point B.

Point C, which is at the same potential as point B, is labeled negative. This is to indicate that point C is more negative than point D. To say a point is positive (or negative) without stating where the polarity is referenced has no meaning. Voltage polarities are relative, just as potential differences or voltage drops.

A more fundamental way to consider the polarity of a voltage drop is to consider the fact that between any two points, the one nearer to the positive terminal of the voltage source is more positive; also, the point nearer to the negative terminal of the applied voltage is more negative. (A point being nearer means there is less resistance in its path.)

In the final analysis however, in a DC circuit the direction of current flow determines the polarity of a device and is governed by the orientation of the voltage source.

7.6.1 Ground

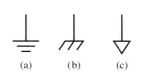

FIGURE 7.11
Ground symbols: (a) Earth ground; (b) chassis ground; (c) common connection.

Lab Reference: The attributes of ground and references are demonstrated in Exercise 11.

The polarity of a voltage source or drop must be taken with respect to a reference. That reference is often *ground*. An electrical ground utilizes the earth or an equipment chassis as a reservoir of charge. Because the earth is said to be at a zero potential (0 V), the term ground is used to denote a common electrical point of zero potential. Originally, the term ground meant a point in a circuit directly connected to the earth. Although this meaning is still true, ground is now frequently applied to a circuit point connected to a chassis, frame, or large metal object. The terms *ground, common,* and *reference* all apply to chassis ground systems and are represented schematically by the symbols shown in Figure 7.11. The terms are used interchangeably.

Ground provides a source for electrons and can also act as a conductor for the supply of charges to the system. Although the symbol in Figure 7.11(c) is a ground connection, it is also used to show a common connection (tie point) in a more complicated circuit—that is, a circuit with a large number of connections to a particular device or conductor.

In the automotive electrical system, the negative terminal of the battery is connected to the chassis of the automobile. This is referred to as a *negative ground*. In this type of system, the chassis acts as a conductor between the battery and all electrical and electronic devices in the system. The majority of the electronic circuits you will come across also utilize a negative ground system.

What makes the term *ground* confusing is that it is used both as a reference and a common tie point. To help alleviate ground referencing problems, subscripts are used to indicate voltage drops and their polarities. There are three common methods to indicate a voltage drop. All use subscripts for labeling purposes.

1. The simplest of these uses the component label in the notation system. Thus, the voltage drop for R_1 in a circuit is labeled V_1 or V_{R_1}.
2. Another method is to label a point in the circuit with a letter. This single-letter subscript *must* be referenced to an indicated ground symbol in the circuit. (Remember that a potential difference exists between two points; thus, a single-letter subscript must assume ground as the second point (reference). If no ground is indicated in the circuit, then the negative ($-$) terminal of the battery is considered the reference.)
3. A double-letter subscript notation is also used and is a better method. In this notation, the second subscript indicates the reference point.

The use of double subscripts identifies not only which points are being measured in a circuit but also the relative polarity of the voltage between these two points. Double-letter subscripts are commonly used to indicate voltage values in circuits of solid-state devices.

Figure 7.12 shows two resistors connected in series. Note the various types of labeling employed in the schematic diagram. To measure the voltage drop across R_1 (V_{R_1}), you need to trace the current flow to know the polarity at each end of the resistor. Next, the voltmeter must be correctly connected in order to *observe polarity*. This is done by connecting the black lead of the voltmeter to the negative side of the component. The red lead is connected to the other side (the positive side of the component). If V_A is measured, then the ground symbol is used as the second point (reference). (The black lead of the voltmeter is connected to ground and the red lead is connected to point A.) If V_{AB} is measured, the reference is the second letter. The voltage drop exists between points A and B, or across R_1. (The black lead of the voltmeter is connected to point B and the red is connected to point A.) V_B represents the voltage from point B to ground (across R_2). Note that $V_{AB} = -V_{BA}$.

FIGURE 7.12
Voltage drops using subscripts.

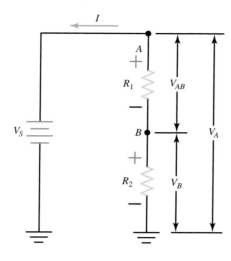

EXAMPLE 7.10

In the circuit in Figure 7.13, determine the potential difference and polarity for each of the following: V_1, V_2, V_3, V_B, V_C, V_{AB}, V_{AD}, V_{CA}, and V_{DC}.

FIGURE 7.13
Determining potential difference and polarity using voltage subscripts.

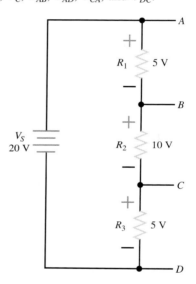

Solution
$$V_1 = 5 \text{ V}$$
$$V_2 = 10 \text{ V}$$
$$V_3 = 5 \text{ V}$$
$$V_B = 15 \text{ V}$$
$$V_C = 5 \text{ V}$$
$$V_{AB} = 5 \text{ V}$$
$$V_{AD} = 20 \text{ V}$$
$$V_{CA} = -15 \text{ V}$$
$$V_{DC} = -5 \text{ V}$$

SECTION 7.6
SELF-CHECK

Answers are given at the end of the chapter.

1. What is a voltage polarity in relation to a resistor's voltage drop?
2. What determines the voltage polarity in a series circuit?
3. Define the term *ground*.
4. What makes the term *ground* confusing?

7.7   TROUBLESHOOTING SERIES CIRCUITS

The ability to locate faulty components in electronic circuits is a skill that must be learned by all electronic technicians. The process is known as *troubleshooting*. This ability is gained by combining proper troubleshooting procedures with a knowledge of the theory of the circuit. Essentially, troubleshooting is a practical application of theory. It is part art and part science. The science part is learned in class. The art part is learned on the job.

Troubleshooting is a logical, systematic process of proceeding from *effect* to *cause*. It is also a process of elimination. For the most part, troubleshooting procedures are based on narrowing the problem by identifying what parts of the circuit work and what parts do not work. To properly troubleshoot an electronic circuit, the technician must know what should be there. A service manual and/or any technical literature available for the circuit greatly assist in assessing the problem and arriving at a solution.

The following steps represent a practical approach to troubleshooting electronic circuits:

Step 1.   *Analyze the symptom:* Determine what is not working in the circuit, and understand what the circuit should be doing when it is functioning properly.

Step 2.   *Locate obvious problems:* Check the source voltage, fuses, and circuit breakers to ensure that the circuit has applied voltage. If troubleshooting a printed circuit board, look for signs of excessive heat, such as burned components and traces, damaged solder traces, loose connections, and so on.

Step 3.   *Localize the problem:* Isolate the trouble to as small an area as possible in the circuit. When the problem has been isolated, measure voltage and resistance and compare the results with what should be there. Knowing how to use test equipment properly is a high priority when trying to isolate the problem in terms of a defective component.

Step 4.   *Replace the defective component:* Once the problem has been resolved and the defective component found, install a new component. Simply replacing a burned-out component with another may not solve the problem. Make sure that you have determined what caused the problem.

7.7.1   Troubleshooting Open Circuits

A circuit is said to be *open* when a break exists in a conducting pathway. Although an open occurs when a switch is used to deenergize a circuit, an open may also develop accidentally. To restore a circuit to proper operation, the open must be located, its cause determined, and repairs made.

Sometimes an open can be located visually by a close inspection of the circuit components. Defective components, such as burned-out resistors, can usually be discovered by this method. Other problems, such as a break in a wire covered by insulation or the melted element of an enclosed fuse, are not visible to the eye. Under such conditions, the understanding of the effect an open has on circuit conditions enables a technician to make use of test equipment to locate the open component.

In Figure 7.14, the series circuit consists of two resistors and a fuse. Notice the effects on circuit conditions when the fuse opens. Current ceases to flow; therefore, there is no longer a voltage drop across the resistors. Each end of the open conducting path becomes an *extension* of the battery terminals and the *voltage felt across the open is equal to the source voltage.*

An open circuit has *infinite* resistance. Infinity represents a quantity so large it cannot be measured. The symbol for infinity is ∞. In an open circuit, $R_T = ∞$.

7.7.2   Troubleshooting Short Circuits

A *short* circuit is an accidental path of low resistance that passes an abnormally high amount of current. A short circuit exists whenever the resistance of a circuit or the

FIGURE 7.14
Normal and open circuit conditions: (a) normal current;
(b) zero current.

Lab Reference: Opens and shorts
in series circuits are demonstrated
in Exercise 10.

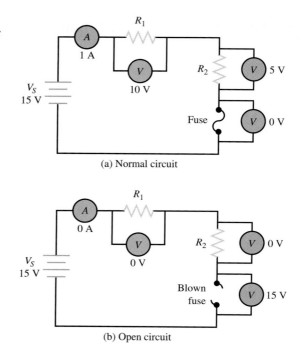

(a) Normal circuit

(b) Open circuit

resistance of a component in a circuit drops in value to almost 0 Ω. A short often occurs as a result of improper wiring, poor soldering, or broken insulation.

In Figure 7.15, a short is caused by improper wiring. Note the effect on current flow. Since the resistor has, in effect, been replaced with a piece of wire, practically all the current flows through the short and very little current flows through the resistor. Electrons flow through the short (the path of least resistance) and the remainder of the circuit by passing through the 10-Ω resistor and the battery. The amount of current flow increases greatly because its resistive path has decreased. Due to excessive current flow, the 10-Ω resistor becomes heated. As it attempts to dissipate this heat, the resistor will probably be destroyed.

FIGURE 7.15
Normal and short-circuit conditions: (a) normal current;
(b) excessive current.

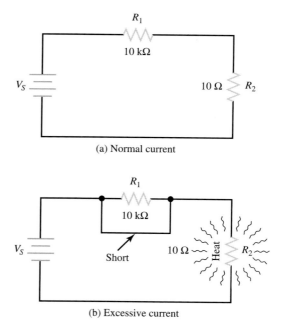

(a) Normal current

(b) Excessive current

SECTION 7.7 SELF-CHECK

Answers are given at the end of the chapter.

1. Define the term *troubleshooting.*
2. Troubleshooting is a logical, systematic process of proceeding from _____ to _____.
3. Define the term *open.*
4. Define the term *short.*

■ SUMMARY

1. A series circuit is one where there is only one current path.
2. If any component opens in a series circuit, current ceases to flow.
3. The total resistance in a series circuit is cumulative.
4. A voltage drop occurs across a resistor because the resistance offered by the resistance expends energy from the source.
5. The voltage drop across a resistor is directly proportional to its resistance.
6. The voltage drops in a series circuit are cumulative. The sum of the voltage drops equals the source voltage.
7. Power is the rate at which work is done.
8. Resistors consume power and dissipate energy in the form of heat.
9. The key to circuit analysis is to note all given values and then determine what can logically be found by calculations.
10. A voltage polarity is relative.
11. A negative potential labeled at a point means that this point is more negative than another point or reference.
12. Voltage polarity is determined by the direction of current flow, which in turn is determined by the source orientation.
13. Ground is a circuit's reference point and has a zero potential.
14. The terms ground, common, and reference are used interchangeably.
15. Ground can be used both as a reference and a common tie point.
16. Subscript notation is used to alleviate ground referencing problems.
17. When ground is not indicated in a circuit, the negative terminal of the battery is assumed to be the reference.
18. Troubleshooting skills help to locate faulty components in electronic circuits. It is a logical, systematic process of proceeding from effect to cause.
19. Troubleshooting is a practical application of theory that is part science and part art.
20. An open occurs when a break exists in a conducting pathway. Current ceases to flow when an open occurs in a series circuit.
21. A short is an accidental path of low resistance that passes an abnormally high amount of current.

■ IMPORTANT EQUATIONS

$$R_T = R_1 + R_2 + R_3 + \cdots + R_n \qquad (7.1)$$

$$I_T = I_1 = I_2 = I_3 = \cdots = I_n \qquad (7.2)$$

$$V_X = \frac{R_X}{R_T} \times V_S \qquad (7.3)$$

$$V_S = V_1 + V_2 + V_3 + \cdots + V_n \qquad (7.4)$$

$$V_S = I_1 R_1 + I_2 R_2 + I_3 R_3 + \cdots + I_n R_n \qquad (7.5)$$

$$P_T = P_1 + P_2 + P_3 + \cdots + P_n \qquad (7.6)$$

■ REVIEW QUESTIONS

1. What constitutes a series circuit?
2. What does component dependency mean in series-strung lights?
3. Total resistance in a series circuit is the _____ of the individual resistances.
4. Where should an ammeter be connected to measure current in a series circuit?
5. How does a voltage drop differ from a voltage rise?
6. Why does a resistor drop voltage?
7. Voltage drops are _____ proportional to the resistance in a series circuit.
8. Voltage drops are _____ in a series circuit.
9. Equation (7.4) is another form of what law?
10. Voltage drops are often called _____ drops.
11. Define the term *power.*
12. How is power dissipated by resistors?
13. In what units is power measured?
14. The individual power consumed is _____ in a series circuit.
15. True or False: Power values are additive in any form of circuit.
16. Summarize the rules for series DC circuits.
17. What is the key to circuit analysis problems?
18. What does the term *voltage polarity* refer to?
19. Why must voltage polarity be observed when measuring voltage?
20. Describe two ways in which voltage polarity can be determined.
21. What is a ground?
22. How is an automobile electrical system grounded?
23. What makes the term *ground* confusing?
24. What are the merits of a double-subscript notation for indicating voltages?
25. Where are double-subscript notations used?

26. Describe how to measure voltage with a voltmeter.

27. How would you describe the term *troubleshooting*?

28. The key to troubleshooting is knowing what _____.

29. Describe an open circuit. What voltage exists across an open? Why?

30. Describe a short circuit. Why do the majority of electrons always flow through a short?

■ CRITICAL THINKING

1. What causes a voltage drop?

2. Describe the term *voltage polarity* and define what it constitutes.

3. Describe the concept of ground and indicate why it is confusing.

4. Compare the merits of each subscript notation as it applies to voltage drops and polarities.

5. Describe the symptoms you would observe if a series circuit had an open. A short. Why does source voltage appear across an open in a series circuit?

■ PROBLEMS

1. For the circuit in Figure 7.16, determine R_T, I_T, V_1, V_2, V_3, P_T, P_1, P_2, and P_3.

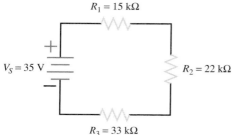

$R_1 = 15 \ k\Omega$

$V_S = 35 \ V$

$R_2 = 22 \ k\Omega$

$R_3 = 33 \ k\Omega$

FIGURE 7.16
Circuit for Problem 1.

2. For the circuit in Figure 7.17, determine R_T, I_T, V_1, V_2, V_3, P_T, P_1, P_2, and P_3.

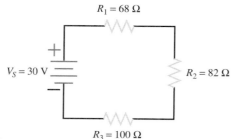

$R_1 = 68 \ \Omega$

$V_S = 30 \ V$

$R_2 = 82 \ \Omega$

$R_3 = 100 \ \Omega$

FIGURE 7.17
Circuit for Problem 2.

3. For the circuit in Figure 7.18, determine R_T, I_T, V_1, V_2, V_3, P_T, P_1, P_2, and P_3.

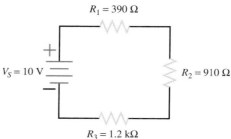

$R_1 = 390 \ \Omega$

$V_S = 10 \ V$

$R_2 = 910 \ \Omega$

$R_3 = 1.2 \ k\Omega$

FIGURE 7.18
Circuit for Problem 3.

4. For the circuit in Figure 7.19, determine R_T, I_T, V_1, V_2, V_3, P_T, P_1, P_2, and P_3.

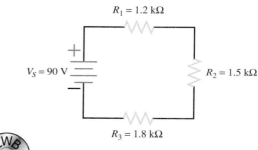

$R_1 = 1.2 \ k\Omega$

$V_S = 90 \ V$

$R_2 = 1.5 \ k\Omega$

$R_3 = 1.8 \ k\Omega$

FIGURE 7.19
Circuit for Problem 4.

5. For the circuit in Figure 7.20, determine R_T, I_T, V_1, V_2, V_3, P_T, P_1, P_2, and P_3.

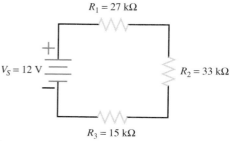

$R_1 = 27 \ k\Omega$

$V_S = 12 \ V$

$R_2 = 33 \ k\Omega$

$R_3 = 15 \ k\Omega$

FIGURE 7.20
Circuit for Problem 5.

6. For the circuit in Figure 7.21, determine R_T, I_T, V_1, V_3, R_1, P_T, P_1, P_2, and P_3.

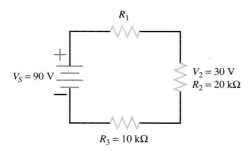

FIGURE 7.21
Circuit for Problem 6.

7. For the circuit in Figure 7.22, determine R_T, I_T, V_1, V_2, R_2, P_T, P_1, P_2, and P_3.

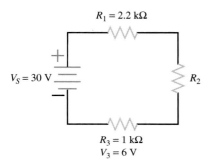

FIGURE 7.22
Circuit for Problem 7.

8. For the circuit in Figure 7.23, determine R_T, I_T, V_S, V_2, R_3, P_T, P_1, P_2, and P_3.

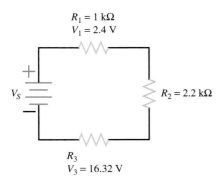

FIGURE 7.23
Circuit for Problem 8.

9. For the circuit in Figure 7.24, determine R_T, I_T, R_2, V_1, V_2, P_T, P_1, P_2, and P_3.

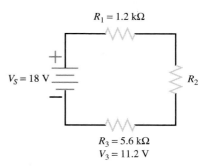

FIGURE 7.24
Circuit for Problem 9.

10. For the circuit in Figure 7.25, determine R_T, I_T, V_1, V_2, V_3, V_{AG}, V_{BG}, V_{CG}, P_T, P_1, P_2, and P_3.

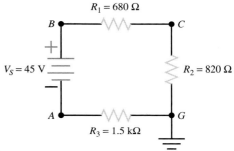

FIGURE 7.25
Circuit for Problem 10.

11. For the circuit in Figure 7.26, determine R_T, V_1, V_2, V_4, V_S, R_3, V_{AG}, V_{BG}, V_{CG}, V_{DG}, P_T, P_1, P_2, P_3, and P_4.

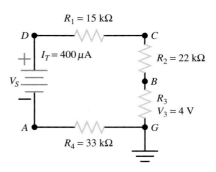

FIGURE 7.26
Circuit for Problem 11.

12. For the circuit in Figure 7.27, determine V_S, I_T, V_1, V_2, V_3, V_4, R_3, V_{AG}, V_{BG}, V_{CG}, V_{DG}, P_T, P_1, P_3, and P_4.

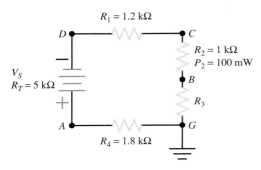

FIGURE 7.27
Circuit for Problem 12.

13. For the circuit in Figure 7.28, determine V_S, R_T, V_1, V_2, V_3, V_4, R_4, R_2, V_{AG}, V_{BG}, V_{CG}, V_{AB}, V_{DC}, P_1, P_3, and P_4.

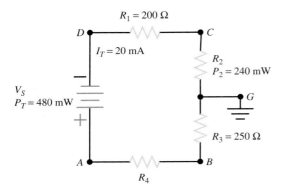

FIGURE 7.28
Circuit for Problem 13.

Figure 7.29 represents a functional series circuit. Use it as a reference for troubleshooting each circuit in Problems 14–18. Based on the data given for each set of measurements shown in Problems 14–18, identify the defective component. Describe the defect (open, short, or a change in value).

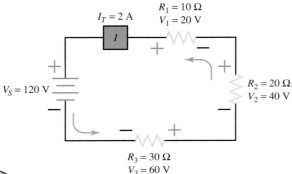

FIGURE 7.29
Reference circuit for Problems 14–18. The voltage drops shown exist when the circuit is functioning normally.

14. $V_1 = 0$ V, $V_2 = 48$ V, $V_3 = 72$ V, $I_T = 2.4$ A
15. $V_1 = 40$ V, $V_2 = 60$ V, $V_3 = 0$ V, $I_T = 4$ A
16. $V_1 = 0$ V, $V_2 = 120$ V, $V_3 = 0$ V, $I_T = 0$ A
17. $V_1 = 30$ V, $V_2 = 0$ V, $V_3 = 90$ V, $I_T = 3$ A
18. $V_1 = 24$ V, $V_2 = 24$ V, $V_3 = 72$ V, $I_T = 2.4$ A
19. Three resistors of 1 kΩ, 2 kΩ, and 3 kΩ are in series across a 20-V source. If the 2-kΩ resistor shorted, what is the voltage across the 3-kΩ resistor?
20. Three resistors connected in series have a total resistance of 3.6 kΩ. The resistors are labeled R_1, R_2, and R_3. R_2 is twice the value of R_1, whereas R_3 is three times the value of R_1. Determine the value of each resistor.

■ ANSWERS TO PRACTICE PROBLEMS

PRACTICE PROBLEMS 7.1

1. $R_T = 25.8$ kΩ
2. $R_T = 9.98$ kΩ

PRACTICE PROBLEMS 7.2

1. $I_T = 2$ mA
2. $I_1 = 1.5$ mA

PRACTICE PROBLEMS 7.3

1. $V_1 = 33$ V
2. $V_S = 20$ V

PRACTICE PROBLEMS 7.4

1. $P_3 = 48$ mW
2. $P_T = 95$ mW

PRACTICE PROBLEMS 7.5

1. $V_2 = 6.6$ V
2. $I_T = 4$ mA, $R_1 = 1$ kΩ, $R_2 = 4$ kΩ, $V_2 = 16$ V, $P_1 = 16$ mW, $P_2 = 64$ mW, $P_T = 80$ mW

■ ANSWERS TO SELF-CHECKS

SECTION 7.1

1. A circuit where there is only one current path
2. As the sum of the individual resistances

SECTION 7.2

1. One
2. At any point in the circuit

SECTION 7.3

1. Added
2. Directly proportional
3. Rise

SECTION 7.4

1. Power is the rate at which work is done.
2. As heat

SECTION 7.5

1. To note all given values and then to determine what can logically be found by calculation

SECTION 7.6

1. The voltage drop has two points. The one nearer to the positive terminal of the voltage source is more positive, whereas the one nearer to the negative terminal of the voltage source is more negative.

2. The direction of current flow
3. A common electrical point of zero potential
4. It is used both as a reference and a common tie point.

SECTION 7.7

1. The ability to locate faulty components in electronic circuits
2. Effect to cause
3. An open exists when there is a break in a conducting pathway.
4. A short is an accidental path of low resistance that passes an abnormally high amount of current.

CHAPTER 8

PARALLEL DC CIRCUITS

OBJECTIVES

After completing this chapter, you will be able to:

1. State the rules for a parallel connected circuit.
2. Solve for circuit resistance, voltage, current, and power.
3. Solve for component resistance, voltage, current, and power.
4. Compare and contrast various equations for solving for total resistance.
5. Analyze parallel circuits to determine unknown quantities.
6. Troubleshoot parallel circuits and determine faults.

INTRODUCTION

A *parallel circuit* is defined as one having more than one current path connected to a common voltage source. Parallel circuits must, therefore, contain two or more resistances that are not connected in series. A parallel connection is when resistances or other circuit components are connected so that they have the same pair of terminal points, or nodes. A *node* is any point in a circuit where two or more circuit paths intersect.

An example of a basic parallel circuit is shown in Figure 8.1. By starting at the voltage source (V_S) and tracing counterclockwise around the circuit, note that there are two complete and separate paths that can be identified in which current can flow. One path is traced from the source, through resistance R_1, and back to the source. The other path is from the source, through resistance R_2, and back to the source.

In text format, to indicate that two resistors are connected in parallel, two vertical lines (∥) are used between the resistor notations. Thus, parallel resistors R_1 and R_2 are shown as

$$R_1 \| R_2$$

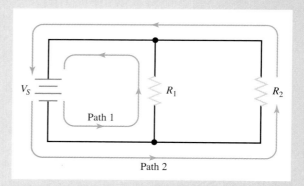

FIGURE 8.1
Example of a basic parallel circuit.

Parallel circuits are very common in electrical and electronic systems. The lights, receptacles, and appliances in a home are connected in parallel to the voltage source so that they can be operated independently. The headlights, taillights, radio, and all other accessories in an automobile are connected in parallel to the voltage source. The preamplifier, voltage amplifier, and power amplifier stages of a stereo system are also connected in parallel to the voltage source. Whenever a device is required to be switched off without affecting other devices, it is wired in parallel.

8.1 VOLTAGE IN PARALLEL CIRCUITS

Lab Reference: Voltage distribution in parallel circuits is demonstrated in Exercise 13.

You will notice in Figure 8.1 that each resistor is connected directly across the terminals of the source voltage. Thus, the parallel circuit may be described as a circuit having a common voltage across its components. The voltage across any branch of a parallel combination is equal to the voltage across each of the other branches in parallel. (A *branch* is a section of a circuit that has a complete path for current.) The voltage in a parallel circuit may be expressed mathematically as

$$V_S = V_1 = V_2 = V_3 = \cdots = V_n \qquad (8.1)$$

When a load is connected in parallel to a voltage source, the voltage measured across the load does not change if the load varies, or opens. As shown in Figure 8.2, even if the load opens, a voltmeter still measures the source voltage across the load. Each end of a parallel connected component is just an *extension* of the voltage source terminals.

FIGURE 8.2
Even with an open-load resistor, the voltmeter will still measure V$_S$.

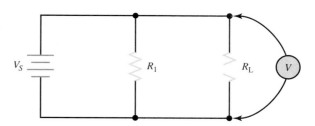

SECTION 8.1 SELF-CHECK

Answers are given at the end of the chapter.

1. Describe a parallel circuit.
2. How is the source voltage distributed in a parallel circuit?

8.2 CURRENT IN PARALLEL CIRCUITS

Ohm's law states that the current in a circuit is inversely proportional to the circuit resistance. This fact is true in both series and parallel circuits. There is a single path for current in a series circuit. The amount of current is determined by the total resistance of the circuit and the source voltage. In a parallel circuit the current divides among available paths.

The behavior of current in parallel circuits is shown by a series of illustrations. Figure 8.3(a) shows a basic series circuit. Here, the total current must pass through a single resistor. The amount of current is determined using methods previously shown in series circuit analysis.

Lab Reference: Current flow in a parallel circuit is demonstrated in Exercise 13.

Given $V_S = 50 \text{ V}$
$R_1 = 10 \ \Omega$

Solution $I_T = \dfrac{V_S}{R_1}$

$= \dfrac{50 \text{ V}}{10 \ \Omega}$

$= 5 \text{ A}$

Figure 8.3(b) shows the same R_1 but with a second resistor (R_2) of equal value connected in parallel across the voltage source. When Ohm's law is applied, the current flow through each resistor is found to be the same as the current through the single resistor in Figure 8.3(a).

Given $V_S = 50$ V
$R_1 = 10$ Ω
$R_2 = 10$ Ω

Solution $I = \dfrac{V}{R}$

$V_S = V_1 = V_2$

$I_1 = \dfrac{V_1}{R_1}$

$\quad = \dfrac{50 \text{ V}}{10 \text{ Ω}}$

$\quad = 5$ A

$I_2 = \dfrac{V_2}{R_2}$

$\quad = \dfrac{50 \text{ V}}{10 \text{ Ω}}$

$\quad = 5$ A

FIGURE 8.3
Analysis of current in a parallel circuit.

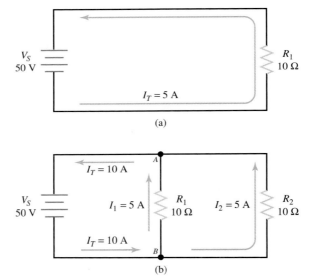

It is apparent that if there is 5 A of current through each of the two resistors, there must be a total current of 10 A drawn from the source. The total current of 10 A, as illustrated in Figure 8.3(b), leaves the negative terminal of the battery and flows to point A. Point A is a connecting point for two resistors, so it is called a *node*. At node A, the total current (I_T) divides into two currents of 5 A each. These two currents flow through their respective resistors and rejoin at node B. The total current then flows from node B back to the positive terminal of the source. The source supplies a total current of 10 A, and each of the two equal resistors carries one-half the total current.

Each individual current path in the circuit of Figure 8.3(b) is referred to as a *branch*. Each branch carries a current that is a portion of the total current. Two or more branches form a *network*. Circuits containing networks are covered in Chapter 12.

From the previous explanation, the characteristics of current in a parallel circuit can be expressed in terms of the following general equation:

$$I_T = I_1 + I_2 + I_3 + \cdots + I_n \tag{8.2}$$

Compare Figure 8.4(a) with the circuit in Figure 8.3(b). Notice that doubling the value of the second branch resistor (R_2) has no effect on the current in the first branch (I_1), but it does reduce the second branch current (I_2) to one-half its original value. The total circuit current decreases to a value equal to the sum of the branch currents.

These facts are verified by the following computations.

Given $V_S = 50$ V
$R_1 = 10\ \Omega$
$R_2 = 20\ \Omega$

Solution $I = \dfrac{V}{R}$

$V_S = V_1 = V_2$

$I_1 = \dfrac{V_1}{R_1}$

$= \dfrac{50\ \text{V}}{10\ \Omega}$

$= 5$ A

$I_2 = \dfrac{V_2}{R_2}$

$= \dfrac{50\ \text{V}}{20\ \Omega}$

$= 2.5$ A

$I_T = I_1 + I_2$

$= 5\ \text{A} + 2.5\ \text{A}$

$= 7.5$ A

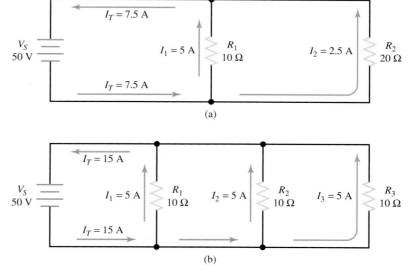

(a)

(b)

FIGURE 8.4
Current behavior in parallel circuits.

Look at the circuit in Figure 8.4(b). The amount of current flow in the branch circuits and the total current in the circuit are determined by the following computations.

Given $V_S = 50$ V
$$R_1 = 10 \ \Omega$$
$$R_2 = 10 \ \Omega$$
$$R_3 = 10 \ \Omega$$

Solution $I = \dfrac{V}{R}$

$$V_S = V_1 = V_2 = V_3$$

$$I_1 = \dfrac{V_1}{R_1}$$

$$= \dfrac{50 \text{ V}}{10 \ \Omega}$$

$$= 5 \text{ A}$$

$$I_2 = \dfrac{V_2}{R_2}$$

$$= \dfrac{50 \text{ V}}{10 \ \Omega}$$

$$= 5 \text{ A}$$

$$I_3 = \dfrac{V_3}{R_3}$$

$$= \dfrac{50 \text{ V}}{10 \ \Omega}$$

$$= 5 \text{ A}$$

$$I_T = I_1 + I_2 + I_3$$

$$= 5 \text{ A} + 5 \text{ A} + 5 \text{ A}$$

$$= 15 \text{ A}$$

Notice that the *sums* of the ohmic values of the resistors in both circuits shown in Figure 8.4 are equal (30 Ω) and that the source voltage is the same (50 V). However, the total current in Figure 8.4(b) is *twice* the amount of Figure 8.4(a). It is apparent, therefore, that the manner in which resistors are connected in a circuit, as well as their number and ohmic values, affects the total current and, therefore, the total resistance.

The reason that the current in Figure 8.4(b) is larger than in Figure 8.4(a) is because there is an extra current path in the circuit in Figure 8.4(b). Resistor R_3 provides this path. Every additional current path lowers the overall opposition to the current flow. At the same time, every additional path requires more current to be drawn from the same voltage source. A common voltage source cannot provide more current *unless the resistance it sees has decreased.* Thus, the key to analyzing the current flow in parallel circuits is in the number of current paths the circuit provides.

**SECTION 8.2
SELF-CHECK**

Answers are given at the end of the chapter.

1. How is current distributed in a parallel circuit?
2. What affects the value of total current in a parallel circuit?
3. Providing more current paths _____ the resistance the source sees.

8.3 RESISTANCE IN PARALLEL CIRCUITS

There are two resistors connected in parallel across a 5-V battery in Figure 8.5. Each has a resistance value of 10 Ω. A complete circuit consisting of two parallel paths is formed and the current flows as shown.

Computing the individual currents shows that there is 0.5 A of current through each resistance. The total current flowing from the battery to the junction of the resistors and returning from the resistors to the battery is equal to 1.0 A. The total resistance of the circuit can be calculated using Ohm's law and the values of source voltage (V_S) and total current (I_T).

Lab Reference: The concept of resistance in parallel is demonstrated in Exercise 13.

Given $V_S = 5\ \text{V}$
$I_T = 1.0\ \text{A}$

Solution $R_T = \dfrac{V_S}{I_T}$

$= \dfrac{5\ \text{V}}{1\ \text{A}}$

$= 5\ \Omega$

FIGURE 8.5

Determining total resistance for two equal resistors connected in parallel.

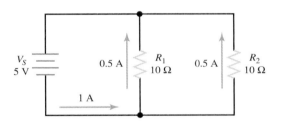

This computation shows the total resistance to be 5 Ω, one-half the value of either of the two resistors. Because the total resistance of a parallel circuit is *smaller* than any of the individual resistors, the total resistance of a parallel circuit *is not the sum* of the individual resistor values, as was the case in a series circuit. For any parallel resistive circuit, the total resistance *is always less than the smallest value of any individual resistance.*

The total resistance in parallel is also referred to as an *equivalent resistance* (R_{EQ}). The terms total resistance and equivalent resistance can be used interchangeably.

There are several methods used to determine the equivalent resistance of parallel circuits. The best method for a given circuit depends on the number and value of the resistors.

8.3.1 Equal-Value Method

For the circuit in Figure 8.5, the resistors have the same value. For any parallel circuit where the resistors are the same, the following simple equation is used to determine the equivalent resistance:

$$R_{EQ} = \frac{R}{N} \tag{8.3}$$

where R_{EQ} = equivalent parallel resistance
R = ohmic value of one resistor
N = number of resistors

This equation is valid for any number of parallel resistors of equal value.

EXAMPLE 8.1

Four 40-Ω resistors are connected in parallel. What is their equivalent resistance?

Given $R_1 = 40\ \Omega$
$R_1 = R_2 = R_3 = R_4$

Solution $R_{EQ} = \dfrac{R}{N}$

$$= \dfrac{40\ \Omega}{4}$$

$$= 10\ \Omega$$

Practice Problems 8.1   Answers are given at the end of the chapter.
1. Four equal 10-Ω resistors have an $R_{EQ} = $ _____.
2. Three equal 3.3-kΩ resistors have an $R_{EQ} = $ _____.

We know that in a parallel circuit the total current is the sum of the individual currents. This was previously shown as

$$I_T = I_1 + I_2 + I_3 + \cdots + I_n \tag{8.2}$$

Since current can be found by Ohm's law as the ratio of V/R, equation (8.2) can be restated as

$$\frac{V_S}{R_{EQ}} = \frac{V_S}{R_1} + \frac{V_S}{R_2} + \frac{V_S}{R_3} + \cdots + \frac{V_S}{R_n} \tag{8.4}$$

Dividing both sides of equation (8.4) by V_S gives

$$\frac{1}{R_{EQ}} = \frac{1}{R_1} + \frac{1}{R_2} + \frac{1}{R_3} + \cdots + \frac{1}{R_n} \tag{8.5}$$

Note that equation (8.5) does not solve for R_{EQ} directly but rather the reciprocal of R_{EQ}. Therefore, the reciprocal of both sides of the equation must be taken in order to solve for R_{EQ}.

8.3.2 Reciprocal Method

Taking the reciprocal of both sides of Equation 8.5 gives

$$R_{EQ} = \frac{1}{1/R_1 + 1/R_2 + 1/R_3 + \cdots + 1/R_n} \tag{8.6}$$

This equation is used to solve for the equivalent resistance of a number of unequal parallel resistors. It can be used as a *general equation* for *any* parallel resistive circuit (having equal or unequal values of resistance).

EXAMPLE 8.2

What is the equivalent resistance of the circuit in Figure 8.6?

Given $R_1 = 20\ \Omega$
$R_2 = 30\ \Omega$
$R_3 = 40\ \Omega$

Solution $R_{EQ} = \dfrac{1}{1/R_1 + 1/R_2 + 1/R_3}$

$= \dfrac{1}{1/20\ \Omega + 1/30\ \Omega + 1/40\ \Omega}$

$= 9.23\ \Omega$

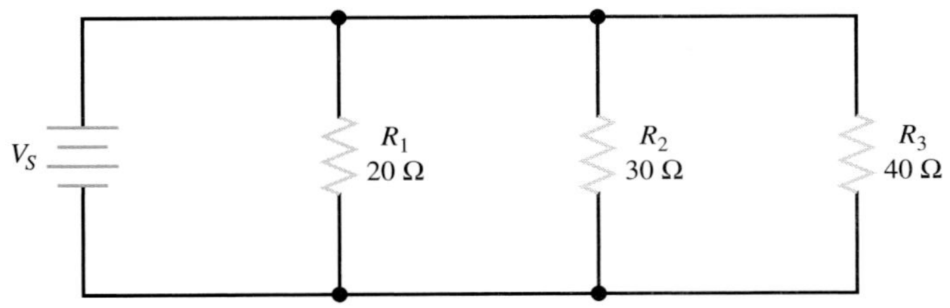

FIGURE 8.6
Circuit for Example 8.2.

Optional Calculator Sequence

There is a reciprocal function on most scientific calculators. This is usually labeled as $1/x$ or x^{-1}. On the Sharp calculator it is labeled x^{-1} (as a second function of x^2). Solving for the equivalent resistance of the circuit in Example 8.2 can be done in either of two ways. Since the Sharp calculator is a direct algebraic logic (DAL) calculator, Equation 8.6 can be keyed in directly as shown. The calculator sequence for this is as follows.

Step	Keypad Entry	Top Display Response
1	1	
2	÷	1/
3	(	1/(
4	1	1/(
5	÷	1/(1/
6	20	1/(1/
7	+	1/(1/20+
8	1	1/(1/20+
9	÷	1/(1/20+1/
10	30	1/(1/20+1/
11	+	1/(1/20+1/30+
12	1	1/(1/20+1/30+
13	÷	1/(1/20+1/30+1/
14	40	1/(1/20+1/30+1/
15	)	1/(1/20+1/30+1/40)
16	=	1/(1/20+1/30+1/40)=
	Answer	9.23 (bottom display)

If you use the reciprocal key (x^{-1}), the calculator sequence for the same problem takes fewer steps and is still keyed in directly as before. (Remember on the Sharp calculator that x^{-1} is entered by using the second function of x^2.)

Step	Keypad Entry	Top Display Response
1	1	
2	÷	1/

(continued)

3	(	1/(
4	20	1/(
5	x^{-1}	$1/(20^{-1}$
6	+	$1/(20^{-1}+$
7	30	$1/(20^{-1}+$
8	x^{-1}	$1/(20^{-1}+30^{-1}$
9	+	$1/(20^{-1}+30^{-1}+$
10	40	$1/(20^{-1}+30^{-1}+$
11	x^{-1}	$1/(20^{-1}+30^{-1}+40^{-1}$
12	)	$1/(20^{-1}+30^{-1}+40^{-1})$
13	=	$1/(20^{-1}+30^{-1}+40^{-1})=$
	Answer	9.23 (bottom display)

Practice Problems 8.2   Answers are given at the end of the chapter.
1. Find the equivalent resistance of three parallel-connected resistances having 3 kΩ, 4 kΩ, and 5 kΩ, respectively.
2. Find R_T for 10-kΩ, 10-kΩ, and 40-kΩ parallel-connected resistors.

8.3.3   Product-over-the-Sum Method

A convenient method for finding the equivalent, or total, resistance for two (*and only two*) parallel resistors is by using the following formula:

$$R_{EQ} = \frac{R_1 \times R_2}{R_1 + R_2} \tag{8.7}$$

This equation, called the product-over-the-sum formula, is used so frequently that it should be committed to memory.

| **EXAMPLE 8.3** | What is the equivalent resistance of the two resistors in Figure 8.7? |

Given $R_1 = 20\ \Omega$
$R_2 = 30\ \Omega$

Solution $R_{EQ} = \dfrac{R_1 \times R_2}{R_1 + R_2}$

$$= \frac{20\ \Omega \times 30\ \Omega}{20\ \Omega + 30\ \Omega}$$

$$= 12\ \Omega$$

FIGURE 8.7
Circuit for Examples 8.3, 8.4, and 8.5.

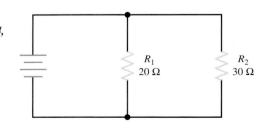

8.3.4   The 10-to-1 Approximation Rule

One method of approximation used in circuits with two parallel resistances is the *10-to-1* rule. If two resistors are connected in parallel and one resistor is 10 or more times greater than the other resistor, then the greater-value resistor may be ignored. The equivalent resistance is approximately equal to (≈) the smaller value. The following example illustrates this rule.

EXAMPLE 8.4

If R_2 in Figure 8.7 is replaced by a 470-Ω resistor (which is more than 10 times the 20-Ω resistor), determine the new equivalent resistance.

Solution $R_{EQ} \approx 20 \, \Omega$ (by ignoring the 470 Ω)

$$R_{EQ} = \frac{R_1 \times R_2}{R_1 + R_2}$$

$$= \frac{20 \, \Omega \times 470 \, \Omega}{20 \, \Omega + 470 \, \Omega}$$

$$= 19.18 \, \Omega$$

The difference between the true equivalent value and the approximate value is 0.82 Ω, which represents a percentage of error of less than 5%. This rule should always be kept in mind, because there are numerous situations where two parallel resistors exhibit this resistance characteristic. Therefore, a quick approximation of the equivalent resistance is easy to perform.

You should note that when the two resistors do not exhibit resistance the characteristic shown in Example 8.4, the equivalent resistance decreases. This is known as resistive *loading* and causes an increase in current flow. The technician should be aware of loading, because it is a common electrical occurrence.

**SECTION 8.3
SELF-CHECK**

Answers are given at the end of the chapter.

1. The total resistance in parallel is _____ than the smallest value of the individual resistances.
2. Define the term *resistive loading*.
3. If two resistances are in parallel, under what conditions is loading no longer a consideration?

8.4   CONDUCTANCE

Since conductance, G, is equal to $1/R$, then the reciprocal equation (equation (8.5)) can be stated for conductance as

$$G_T = G_1 + G_2 + G_3 + \cdots + G_n \tag{8.8}$$

As discussed in Chapter 3, conductance is a term used to describe a circuit or component's ability to pass current. The SI unit of conductance is the siemens (S). The ease with which current flows through a circuit represents the circuit's conductance. As the number of resistors connected in parallel increases, the ease with which current flows also increases. Because resistance is the inverse of conductance, equations (8.5) and (8.8) are combined to form

$$G_T = \frac{1}{R_1} + \frac{1}{R_2} + \frac{1}{R_3} + \cdots + \frac{1}{R_n} \tag{8.9}$$

The conductance of a parallel circuit and the resistance of a series circuit are often called the *complements* of each other because it is possible to develop some of the relationships for the parallel circuit directly from those of a series circuit by simply interchanging R and G. Another example of this is that the current is the same in a series circuit, whereas in a parallel circuit the voltage is the same.

EXAMPLE 8.5

Determine the total conductance of the circuit in Figure 8.7 (used for Example 8.3).

Solution　$G_T = \dfrac{1}{R_1} + \dfrac{1}{R_2}$

$$= \frac{1}{20\ \Omega} + \frac{1}{30\ \Omega}$$

$$= 0.83\ \text{S}$$

Practice Problems 8.4　Answers are given at the end of the chapter.
1. If $R_{EQ} = 10\ \text{k}\Omega$, find total conductance.
2. What is the total conductance of two parallel resistors with values of 5 kΩ and 10 kΩ, respectively?

**SECTION 8.4
SELF-CHECK**

Answers are given at the end of the chapter.

1. Define the term *conductance*.
2. Resistance in a series circuit and conductance in a parallel circuit are called
　_____.

8.5 POWER IN PARALLEL CIRCUITS

Power computations in a parallel circuit are essentially the same as those used for the series circuit. Because power dissipation in resistors consists of heat loss, power dissipations are additive, regardless of how the resistors are connected in the circuit. The total power is equal to the sum of the power dissipated by the individual resistors. Any of the power equations used in series circuits can be used in a parallel circuit.

EXAMPLE 8.6

Find the total power consumed by the circuit in Figure 8.8.

Solution　$P_1 = I_1^2 R_1$
$$= (5 \times 10^{-3}\text{A})^2 \times 10 \times 10^3\ \Omega$$
$$= 250\ \text{mW}$$

$$P_2 = I_2^2 R_2$$
$$= (2 \times 10^{-3} \text{A})^2 \times 25 \times 10^3 \ \Omega$$
$$= 100 \text{ mW}$$
$$P_3 = I_3^2 R_3$$
$$= (1 \times 10^{-3} \text{A})^2 \times 50 \times 10^3 \ \Omega$$
$$= 50 \text{ mW}$$
$$P_T = P_1 + P_2 + P_3$$
$$= 250 \text{ mW} + 100 \text{ mW} + 50 \text{ mW}$$
$$= 400 \text{ mW}$$

Because the total current and voltage source are known, the total power can also be computed by

$$I_T = I_1 + I_2 + I_3$$
$$= 5 \text{ mA} + 2 \text{ mA} + 1 \text{ mA}$$
$$= 8 \text{ mA}$$
$$P_T = I_T V_S$$
$$= 8 \times 10^{-3} \text{A} \times 50 \text{ V}$$
$$= 400 \text{ mW}$$

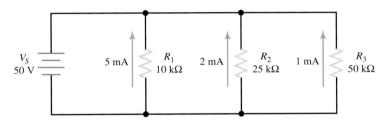

FIGURE 8.8
Circuit for Example 8.6.

Practice Problems 8.5   Answers are given at the end of the chapter.
1. $I = 2$ mA, $R = 12$ kΩ. What is the power dissipation?
2. Determine P_1, P_2, P_3, and P_T for the circuit in Figure 8.9.

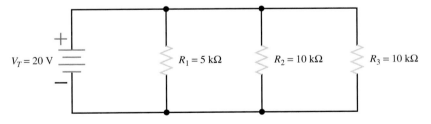

FIGURE 8.9
Circuit for Practice Problems 8.5, Problem 2.

8.5.1   Rules for Parallel DC Circuits

1. The same voltage exists across each branch of a parallel circuit and is equal to the source voltage.
2. The current through a parallel branch is inversely proportional to the amount of resistance of the branch.

3. The total current of a parallel circuit is equal to the sum of the individual branch currents of the circuit.

4. The equivalent resistance of a parallel circuit is found by the general equation

$$R_{EQ} = \frac{1}{1/R_1 + 1/R_2 + 1/R_3 + \cdots + 1/R_n} \tag{8.6}$$

5. The total power consumed in a parallel circuit is equal to the sum of the power consumed by the individual resistors.

SECTION 8.5
SELF-CHECK

Answers are given at the end of the chapter.

1. Power dissipation in any circuit is _____.
2. Branch currents in a parallel circuit are _____.
3. The current in a parallel branch is _____ proportional to the branch resistance.

8.6 PARALLEL CIRCUIT ANALYSIS

Problems involving the determination of resistance, voltage, current, and power in a parallel circuit are solved as simply as in a series circuit. The procedure is the same as stated in Chapter 7 for series circuits:

1. Observe the circuit diagram or draw one as necessary.
2. Note the given values and the values to be found.
3. Select the appropriate equations to be used in solving for the unknown quantities based on the known quantities.
4. Substitute the known values in the equation you have selected and solve for the unknown value.

EXAMPLE 8.7

Find all possible unknowns in the circuit in Figure 8.10.

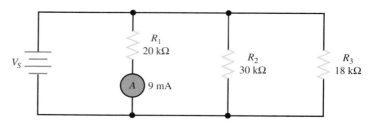

FIGURE 8.10
Circuit for Example 8.7.

Solution
This may appear to be a large amount of mathematical manipulation. However, if you use the step-by-step approach, the circuit falls into place. There are several ways to approach this problem. With the values given, you could solve for R_{EQ}, the power used by R_1 (P_1), or the voltage across R_1 (V_1). The voltage across R_1 is equal to the source voltage (V_S) and the voltage across each of the other resistors.

Solving for R_{EQ} or P_1 does not help you to calculate other unknowns; therefore, as the first step, the logical unknown to solve for is V_1.

$$V_S = V_1 = V_2 = V_3$$

$$V_1 = I_1 R_1$$
$$= 9 \times 10^{-3}\,\text{A} \times 2 \times 10^3\,\Omega$$
$$= 18\,\text{V}$$

$$I_2 = \frac{V_2}{R_2}$$
$$= \frac{18\,\text{V}}{3 \times 10^3\,\Omega}$$
$$= 6 \times 10^{-3}\,\text{A}$$
$$= 6\,\text{mA}$$

$$I_3 = \frac{V_3}{R_3}$$
$$= \frac{18\,\text{V}}{1.8 \times 10^3\,\Omega}$$
$$= 10 \times 10^{-3}\,\text{A}$$
$$= 10\,\text{mA}$$

$$I_T = I_1 + I_2 + I_3$$
$$= 9\,\text{mA} + 6\,\text{mA} + 10\,\text{mA}$$
$$= 25\,\text{mA}$$

$$P_1 = I_1 V_1$$
$$= 9 \times 10^{-3}\,\text{A} \times 18\,\text{V}$$
$$= 162\,\text{mW}$$

$$P_2 = I_2 V_2$$
$$= 6 \times 10^{-3}\,\text{A} \times 18\,\text{V}$$
$$= 108\,\text{mW}$$

$$P_3 = I_3 V_3$$
$$= 10 \times 10^{-3}\,\text{A} \times 18\,\text{V}$$
$$= 180\,\text{mW}$$

$$P_T = P_1 + P_2 + P_3$$
$$= 162\,\text{mW} + 108\,\text{mW} + 180\,\text{mW}$$
$$= 450\,\text{mW}$$

$$R_{EQ} = \frac{1}{1/R_1 + 1/R_2 + 1/R_3}$$
$$= \frac{1}{1/(2 \times 10^3\,\Omega) + 1/(3 \times 10^3\,\Omega) + 1/(1.8 \times 10^3\,\Omega)}$$
$$= 720\,\Omega$$

As a check, P_T and R_{EQ} can be found by

$$P_T = I_T V_S$$
$$= 25 \times 10^{-3}\,\text{A} \times 18\,\text{V}$$
$$= 450\,\text{mW}$$

$$R_{EQ} = \frac{V_s}{I_T}$$
$$= \frac{18\,\text{V}}{25 \times 10^{-3}\,\text{A}}$$
$$= 720\,\Omega$$

Practice Problems 8.6   Answers are given at the end of the chapter.
1. For the circuit in Figure 8.11, determine all unknown quantities.

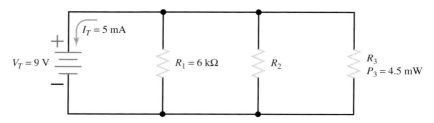

FIGURE 8.11
Circuit for Practice Problems 8.6, Problem 1.

**SECTION 8.6
SELF-CHECK**

Answers are given at the end of the chapter.

1. Describe the step-by-step approach to parallel circuit analysis.

8.7   TROUBLESHOOTING PARALLEL CIRCUITS

The same principles of troubleshooting that were applied to series circuits can also be applied to parallel circuits. The same two faults can occur in parallel circuits that occurred in series, the open and the short circuit. Unlike series circuits, an open circuit does not necessarily interrupt or stop all current from flowing. The circuit in Figure 8.12 shows an open in the line directly connected to the source; this type of open is the only one that would make all current cease to flow.

When an open occurs in a branch of a parallel network, the resistance of the branch increases and the total resistance of the circuit also increases. This causes a decrease in total current. All normal branches still have normal current flow. Measuring the voltage across an open branch is pointless, the voltmeter provides the same reading across a good resistor as it would a bad (open) resistor. To determine which branch has a faulty resistor, either the branch currents have to be determined or an ohmmeter has to be used. Inserting ammeters requires a break inside the branch, whereas the use of the ohmmeter requires isolation to test the suspected faulty component. Both these procedures are time consuming and somewhat involved. You need to have your thinking cap on when troubleshooting these circuits.

FIGURE 8.12
Open in a parallel circuit; all current ceases to flow.

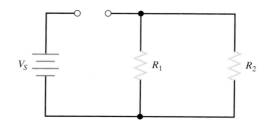

In a parallel circuit, a short in one branch always results in no current flowing in any other branch. The shorted branch has the least resistance, and current always takes the path of least resistance. Shorts in parallel are particularly dangerous, because the source is essentially shorted out, allowing abnormally high current to flow through the loop. This is shown in Figure 8.13. The voltage across the shorted component is 0 V because it exhibits zero resistance ($I \times R = 0$). The source voltage sees no load resistance; thus, current is limited only by the internal resistance of the source (which is very small). You can see what havoc shorts cause in parallel circuits.

Lab Reference: Opens and shorts in a parallel circuit are demonstrated in Exercise 14.

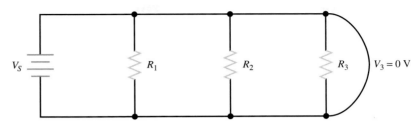

FIGURE 8.13
Short-circuit voltage equals 0 V.

Voltmeters and ammeters are harder to use in troubleshooting shorts, because the excessive current flowing in an energized circuit could cause damage to other components due to extra heat. The ohmmeter is the best bet to troubleshoot with, because it requires the source voltage to be removed or turned off. Finding the shorted branch should be obvious with the ohmmeter, as long as you remember to isolate the branch under test.

SECTION 8.7 SELF-CHECK

Answers are given at the end of the chapter.

1. What is the voltage across an open resistance of a three-branch parallel circuit?
2. What is the voltage across a normal resistance of a three-branch parallel circuit?
3. An open branch in a parallel circuit causes a _____ in circuit resistance and a _____ in circuit current.

■ SUMMARY

1. A parallel circuit is defined as one having more than one current path connected to a common voltage source.
2. A node is any point in a circuit where two or more circuit paths connect.
3. The voltage across each branch in a parallel circuit is equal to the source voltage.
4. Current is additive in parallel circuits. The current in any branch is inversely proportional to the resistance of the branch.
5. A common voltage source cannot provide more current unless the resistance it sees has decreased.
6. In a parallel circuit, the total resistance is always less than the smallest value of the individual resistances.
7. Equation (8.6) is a general equation for any parallel resistive circuit.
8. To speed up total resistance calculations in a parallel circuit, use the reciprocal function on the calculator.
9. The 10-to-1 rule is useful for a quick approximation of the equivalent resistance of two parallel resistors.
10. Conductance is defined as the ease with which current flows through a circuit.

11. The conductance of a parallel circuit and the resistance of a series circuit are often called the complements of each other.
12. Power computations for a parallel circuit are done in the same way as for a series circuit.
13. Power is additive in resistive circuits regardless of how the resistors are connected.
14. Circuit analysis for parallel circuits is done the same way as for series circuits.
15. Troubleshooting an open parallel branch with a voltmeter is unproductive as the voltage across the branch is the same regardless of whether the resistance is normal or opened.
16. To determine which branch has an open, either branch currents have to be determined or an ohmmeter has to be used.
17. A shorted branch causes excessive circuit current to flow limited only by the internal resistance of the source.

■ IMPORTANT EQUATIONS

$$V_S = V_1 = V_2 = V_3 = \cdots = V_n \qquad (8.1)$$

$$I_T = I_1 + I_2 + I_3 + \cdots + I_n \qquad (8.2)$$

$$R_{EQ} = \frac{R}{N} \tag{8.3}$$

$$V_S / R_{EQ} = \frac{V_S}{R_1} + \frac{V_S}{R_2} + \frac{V_S}{R_3} + \cdots + \frac{V_S}{R_n} \tag{8.4}$$

$$1/R_{EQ} = \frac{1}{R_1} + \frac{1}{R_2} + \frac{1}{R_3} + \cdots + \frac{1}{R_n} \tag{8.5}$$

$$R_{EQ} = \frac{1}{1/R_1 + 1/R_2 + 1/R_3 + \cdots + 1/R_n} \tag{8.6}$$

$$R_{EQ} = \frac{R_1 \times R_2}{R_1 + R_2} \tag{8.7}$$

$$G_T = G_1 + G_2 + G_3 + \cdots + G_n \tag{8.8}$$

$$G_T = \frac{1}{R_1} + \frac{1}{R_2} + \frac{1}{R_3} + \cdots + \frac{1}{R_n} \tag{8.9}$$

■ REVIEW QUESTIONS

1. Describe a parallel circuit.
2. What is a circuit *node*?
3. Describe a common parallel-connected circuit with which you are familiar.
4. Define the term *branch*.
5. Why is the branch voltage of a parallel circuit equal to the source voltage?
6. Two or more branches form a _____.
7. Each additional current path added to a parallel circuit _____ the circuit resistance.
8. Branch currents in a parallel circuit are _____.
9. What effect on the current in the first branch of a parallel circuit occurs by doubling the resistance of a second branch resistance?
10. The total resistance of a series circuit is found by adding the individual resistances. Why doesn't this same operation work for a parallel circuit?
11. A common voltage source cannot provide more current unless the resistance it sees has _____.

12. What is the key to analyzing the current flow in parallel circuits?
13. Compare the value of total resistance for a parallel circuit and the value of the individual resistances.
14. What determines the method for solving for total resistance in a parallel circuit one should use?
15. What is the disadvantage of the product-over-sum method for solving for total resistance of a parallel circuit compared to the two other methods that can also be used?
16. Describe the 10-to-1 approximation rule for resistances in a parallel circuit.
17. Describe the term *resistive loading*. What does it cause?
18. Define the term *conductance*.
19. Why are the conductance of a parallel circuit and the resistance of a series circuit called *complements* of each other?
20. How is power calculated in a parallel circuit?
21. Describe the rules for parallel DC circuits.
22. Describe the procedure used in solving parallel circuit analysis problems.
23. Describe the symptoms of an open branch resistance in a parallel circuit.
24. Compare and contrast the use of a voltmeter, ammeter, and ohmmeter for troubleshooting an open branch resistance.
25. Describe the procedure for troubleshooting a shorted branch in a parallel circuit.

■ CRITICAL THINKING

1. Compare the applications for parallel circuits versus series circuits.
2. Explain how the total resistance in parallel is less than any branch resistance.
3. What is loading and how does it apply to the 10-to-1 approximation rule?
4. What are the symptoms associated with opens and shorts in parallel circuits? Compare and contrast these to those associated with series circuits.

■ PROBLEMS

1. For the circuit in Figure 8.14, determine R_T, I_T, I_1, I_2, P_1, P_2, P_3, and P_T.

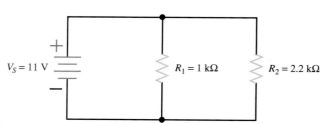

FIGURE 8.14
Circuit for Problem 1.

2. For the circuit in Figure 8.15, determine R_T, I_T, I_1, I_2, I_3, P_1, P_2, P_3, and P_T.

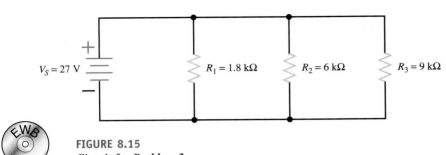

FIGURE 8.15
Circuit for Problem 2.

3. For the circuit in Figure 8.16, determine R_T, I_T, I_1, I_2, I_3, P_1, P_2, P_3, and P_T.

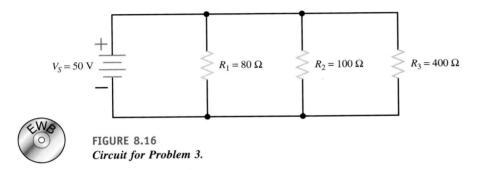

FIGURE 8.16
Circuit for Problem 3.

4. For the circuit in Figure 8.17, determine R_T, I_T, I_1, I_2, I_3, P_1, P_2, P_3, and P_T.

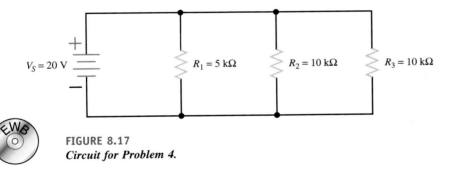

FIGURE 8.17
Circuit for Problem 4.

5. For the circuit in Figure 8.18, determine R_T, I_T, I_1, I_2, I_3, P_1, P_2, P_3, and P_T.

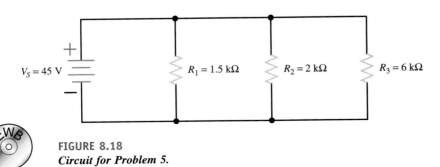

FIGURE 8.18
Circuit for Problem 5.

6. How much resistance must be connected in parallel with a 12-kΩ resistor to obtain an equivalent resistance of 8 kΩ?

7. A 12-kΩ resistor is in parallel with a 4-kΩ resistor. How much resistance must be connected in parallel with this combination to obtain an equivalent resistance of 2 kΩ?

8. Three 1-kΩ resistors are connected in parallel with a 24-V DC source. If the source is doubled, what effect does this have on R_{EQ}? Why?

9. Three parallel resistances of $R_1 = 20\ \Omega$, $R_2 = 25\ \Omega$, and $R_3 = 100\ \Omega$ have 2.5 A of total current. What are the individual branch currents?

10. For the circuit in Figure 8.19, determine R_{EQ}, I_1, I_2, R_2, R_3, P_T, P_1, P_2, and P_3.

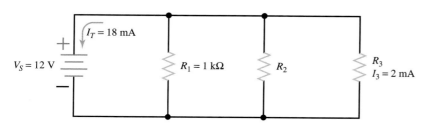

FIGURE 8.19
Circuit for Problem 10.

11. For the circuit in Figure 8.20, determine R_T, I_T, I_1, I_2, I_3, P_1, P_2, and P_3.

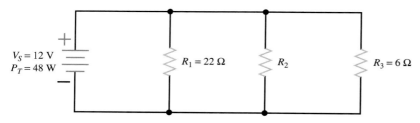

FIGURE 8.20
Circuit for Problem 11.

12. For the circuit in Figure 8.21, determine V_S, I_T, I_1, I_3, R_2, R_3, P_T, P_1, P_2, and P_3.

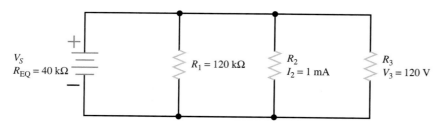

FIGURE 8.21
Circuit for Problem 12.

13. For the circuit in Figure 8.22, determine I_T, I_1, I_2, R_1, R_2, R_3, P_T, P_2, and P_3.

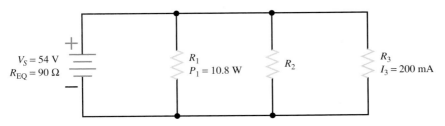

FIGURE 8.22
Circuit for Problem 13.

14. For the circuit in Figure 8.23, determine I_T, I_1, I_2, I_4, R_3, R_4, P_T, P_1, P_2, P_3, and P_4.

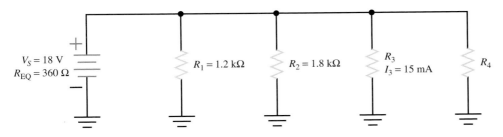

FIGURE 8.23
Circuit for Problem 14.

15. For the circuit in Figure 8.24, determine R_{EQ}, I_T, I_2, I_3, I_4, R_1, R_3, P_T, P_1, P_2, P_3, and P_4.

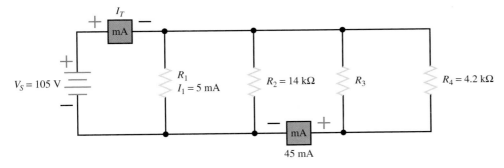

FIGURE 8.24
Circuit for Problem 15.

16. For the circuit in Figure 8.25, determine R_{EQ}, I_T, I_1, I_2, I_3, R_2, R_4, P_T, P_1, P_2, P_3, and P_4.

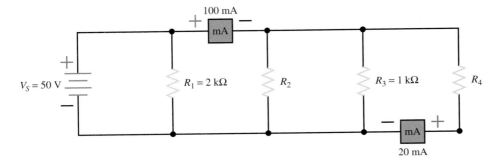

FIGURE 8.25
Circuit for Problem 16.

Figure 8.26 represents a functional parallel circuit. Use it as a reference for troubleshooting each circuit in Problems 17–20. Based on the data given for each set of measurements shown in Problems 17–20, identify the defective component. Describe the defect (open or short).

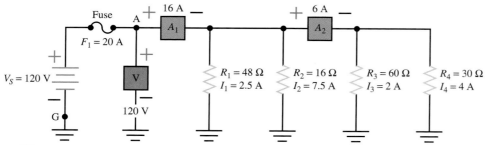

FIGURE 8.26
Reference circuit for troubleshooting parallel circuits in Problems 17–20. Values shown exist when the circuit is functioning normally.

17. $V_{AG} = 120$ V, $A_1 = 8.5$ A, $A_2 = 6$ A
18. $V_{AG} = 120$ V, $A_1 = 14$ A, $A_2 = 4$ A
19. $V_{AG} = 120$ V, $A_1 = 12$ A, $A_2 = 2$ A
20. $V_{AG} = 0$ V, $A_1 = 0$ A, $A_2 = 0$ A

■ ANSWERS TO PRACTICE PROBLEMS

PRACTICE PROBLEMS 8.1

1. $2.5\ \Omega$
2. $1.1\ k\Omega$

PRACTICE PROBLEMS 8.2

1. $1.28\ k\Omega$
2. $4.44\ k\Omega$

PRACTICE PROBLEMS 8.3

1. $8.57\ k\Omega$
2. $2\ \Omega$

PRACTICE PROBLEMS 8.4

1. 0.1 mS
2. 0.3 mS

PRACTICE PROBLEMS 8.5

1. 48 mW
2. $P_1 = 80$ mW, $P_2 = 40$ mW, $P_3 = 40$ mW, $P_T = 160$ mW

PRACTICE PROBLEMS 8.6

1. $V_1 = V_2 = V_3 = 9$ V, $I_1 = 1.5$ mA, $I_2 = 3$ mA, $I_3 = 0.5$ mA, $P_1 = 13.5$ mW, $P_2 = 27$ mW, $P_T = 45$ mW, $R_2 = 3\ k\Omega$, $R_3 = 18\ k\Omega$, $R_T = 1.8\ k\Omega$

■ ANSWERS TO SELF-CHECKS

SECTION 8.1

1. A parallel circuit is defined as one having more than one current path connected to a common voltage source.

2. It is directly connected across all parallel branches.

SECTION 8.2

1. In a parallel circuit, the current divides among available paths.
2. The manner in which resistors are connected as well as their number and ohmic values affect the total current and, therefore, the total resistance.
3. Decreases

SECTION 8.3

1. Less
2. When a parallel-connected resistance lowers the equivalent resistance and increases current flow
3. When one resistance has a value of 10 times or greater than the other

SECTION 8.4

1. The ease with which current flows through a circuit represents the circuit's conductance.
2. Complements

SECTION 8.5

1. Additive
2. Additive
3. Inversely

SECTION 8.6

1. (1) Observe the circuit diagram or draw one as necessary. (2) Note the given values and the values to be found. (3) Select the appropriate equations to be used in solving for the unknown quantities based on the known quantities. (4) Substitute the known value in the equation you have selected and solve for the unknown value.

SECTION 8.7

1. Source voltage
2. Source voltage
3. Increase, decrease

CHAPTER 9

SERIES-PARALLEL DC CIRCUITS

OBJECTIVES

After completing this chapter, you will be able to:

1. Find the equivalent resistance of a series-parallel circuit.
2. Redraw a series-parallel circuit for clarity.
3. Analyze series-parallel circuits to determine unknown quantities.
4. Describe applications for the Wheatstone bridge.
5. Analyze the balanced bridge to determine unknown quantities.
6. Troubleshoot series-parallel circuits and determine faults.

INTRODUCTION

In the preceding chapters, series and parallel DC circuits have been considered separately. However, a technician will encounter circuits consisting of both series and parallel elements. A circuit of this type is called a *series-parallel,* or *combination,* circuit. In many cases, a complete series-parallel combination can be replaced by a single resistor that is electrically equivalent to the entire network. When calculating values of voltage, current, and resistance, it is often useful to simplify the circuit as these values are solved for. For example, three resistors in parallel can be redrawn as one resistor with an ohmic value equal to the equivalent value of the three resistors. The same procedure can be used for other combinations of resistors. Series circuit rules and equations are applied to components of a circuit that are connected in series, and parallel circuit rules and equations are applied to the components of a circuit that are connected in parallel.

9.1 **RESISTANCE IN SERIES-PARALLEL DC CIRCUITS**

The basic technique used for solving DC series-parallel circuit problems is the use of equivalent circuits. To simplify a combination circuit to a simple circuit containing only one load, equivalent circuits are substituted (on paper) for the combination circuit they represent.

To demonstrate the method used to solve problems involving combination circuits, the network shown in Figure 9.1(a) is used to find an equivalent resistance. A close inspection of the circuit shows that the only quantity that can be computed with the given information is the equivalent resistance of parallel resistors, R_2 and R_3.

Given $R_2 = 20\ \Omega$
$R_3 = 30\ \Omega$

Solution $R_{EQ_1} = \dfrac{R_2 \times R_3}{R_2 + R_3}$

$= (20\ \Omega \times 30\ \Omega)/20\ \Omega + 30\ \Omega$

$= 12\ \Omega$

Now that the equivalent resistance for R_2 and R_3 has been calculated, the circuit can be redrawn as a series circuit, as shown in Figure 9.1(b). Using series circuit rules and equations, the equivalent resistance of this circuit can now be calculated.

FIGURE 9.1
Example series-parallel circuit.

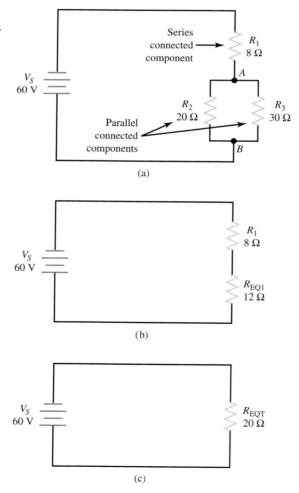

Given $R_1 = 8\ \Omega$
              $R_{EQ_1} = 12\ \Omega$

Solution $R_{EQ_T} = R_1 + R_{EQ_1}$
              $= 8\ \Omega + 12\ \Omega$
              $= 20\ \Omega$

The original circuit can be redrawn with a single resistor that represents the equivalent resistance of the entire circuit, as shown in Figure 9.1(c).

Figure 9.2 represents another typical combination circuit. Note that R_1 and R_2 are connected in series. To solve for the equivalent resistance (total resistance), use the following calculations.

Given $R_1 = 300\ \Omega$
              $R_2 = 100\ \Omega$

Solution $R_{EQ_1} = R_1 + R_2$
              $= 300\ \Omega + 100\ \Omega$
              $= 400\ \Omega$

Redraw the circuit as shown in Figure 9.2(b). Note that R_{EQ_1} and R_3 are in parallel. Using the rules and equations of parallel circuits gives the following.

Given $R_{EQ_1} \| R_3$
              $R_{EQ_1} = 400\ \Omega$
              $R_3 = 400\ \Omega$

Solution $R_{EQ_2} = \dfrac{R}{N}$

              $= \dfrac{400\ \Omega}{2}$

              $= 200\ \Omega$

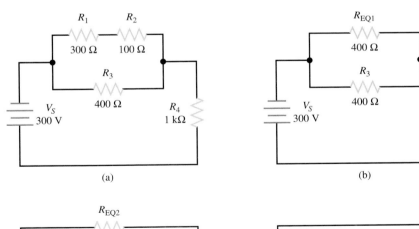

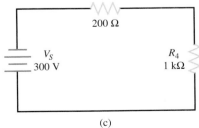

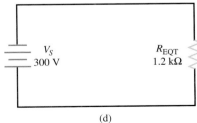

FIGURE 9.2
Series-parallel practice circuit.

Redraw the circuit as shown in Figure 9.2(c). R_{EQ_2} and R_4 are in series. Use the rules and equations of series circuits.

Given $\quad R_{EQ_2} = 200\ \Omega$
$\qquad\qquad R_4 = 1\ \text{k}\Omega$

Solution $\quad R_{EQ_T} = R_{EQ_2} + R_4$
$\qquad\qquad\quad = 200\ \Omega + 1\ \text{k}\Omega$
$\qquad\qquad\quad = 1.2\ \text{k}\Omega$

Practice Problems 9.1 Answers are given at the end of the chapter.
1. Find the equivalent circuit resistance for the circuit in Figure 9.3(a).
2. Find the equivalent circuit resistance for the circuit in Figure 9.3(b).

FIGURE 9.3
Series-parallel circuits: (a) circuit for Practice Problems 9.1, Problem 1; (b) circuit for Practice Problems 9.1, Problem 2.

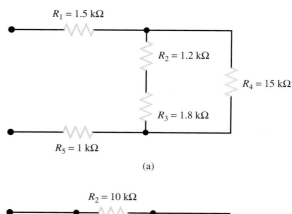

(a)

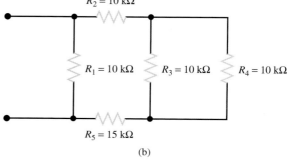

(b)

9.2 REDRAWING CIRCUITS FOR CLARITY

The schematic diagrams with which you have been working have shown parallel circuits drawn as neat square figures, with each branch easily identified. In actual practice wired circuits and more complex schematics are rarely laid out in this simple form. For this reason, it is important for you to recognize that circuits can be drawn in a variety of ways and to learn some of the techniques for redrawing them into their simplified, more recognizable form.

9.2.1 Steps for Redrawing the Circuit for Clarity

When a circuit is redrawn for clarity or to its simplest form the following steps are used.

1. Trace the current paths in the circuit.
2. Label the nodes (junctions) in the circuit.
3. Recognize points that are at the same potential.
4. Visualize a rearrangement, "stretching" or "shrinking" connecting wires.
5. Redraw the circuit into a simpler form (through stages if necessary).

To redraw any circuit, start at the source and trace the path of current flow through the circuit. At points where the current divides, called *nodes,* parallel branches begin. These nodes are key points of reference in any circuit and should be labeled as you find them. The wires in circuit schematics are assumed to have *no resistance* and there is *no voltage drop* along any wire. This means that any unbroken wire is at the same voltage (potential) all along its length, until it is interrupted by a resistor, battery, or some other circuit component. In redrawing a circuit, a wire can be *stretched* or *shrunk* as much as you like without changing any electrical characteristic of the circuit.

Figure 9.4(a) is a schematic of a circuit that is not drawn in the boxlike fashion used in previous illustrations. To redraw this circuit, start at the voltage source and trace the path for current to the node marked (*a*). At this node the current divides into three paths. If you were to stretch the wire to show the three current paths, the circuit would appear as shown in Figure 9.4(b).

Although these circuits may appear to be different, the two drawings actually represent the same circuit. The drawing in Figure 9.4(b) is the familiar boxlike structure and may be easier to work with.

FIGURE 9.4
Redrawing a simple parallel circuit.

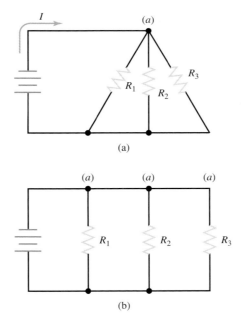

(a)

(b)

Figure 9.5(a) is a schematic of a circuit shown in a boxlike structure, but it may be misleading. This circuit is in reality a series-parallel circuit that may be redrawn as shown in Figure 9.5(b). The drawing in part (b) of the figure is a simpler representation of the original circuit and could be reduced to just two resistors in parallel.

FIGURE 9.5
Redrawing a simple series-parallel circuit.

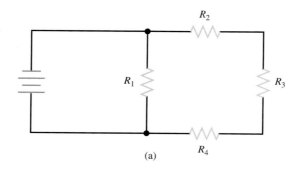

(a)

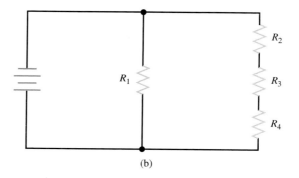

(b)

9.3 ■ REDRAWING A SERIES-PARALLEL CIRCUIT

Figure 9.6(a) shows a series-parallel circuit that may be redrawn for clarification in the following steps. As you redraw the circuit, draw it in simple, boxlike form. Each time you reach a node, create a new branch by stretching or shrinking the wires.

To begin, start at the negative terminal of the voltage source and trace the current (Figure 9.6(b)). Current flows through R_1 to a node and divides into three paths; label this junction (*a*). Follow one of the paths of current through R_2 and R_3 to a node where the current divides into two more paths. This node is labeled (*b*).

The current through one branch of this node goes through R_5 and back to the source (the most direct path). Now that you have completed a path for current to the source, return to the last node, (*b*). Follow current through the other branch from this node. Current flows from node (*b*) through R_4 to the source. All the paths from node (*b*) have been traced. Only one path from node (*a*) has been completed. You must now return to node (*a*) to complete the other two paths. From node (*a*) the current flows through R_7 back to the source. (There are no additional nodes on this path.) Return to node (*a*) to trace the third path from this node. Current flows through R_6 and R_8 and comes to another node. Label this node (*c*). From node (*c*) one path for current is through R_9 to the source. The other path for current from node (*c*) is through R_{10} to the source. All the nodes in this circuit have now been labeled. The circuit and the nodes can be redrawn as shown in Figure 9.6(c). It is much easier to recognize the series and parallel paths in this figure.

FIGURE 9.6
Redrawing a complex circuit.

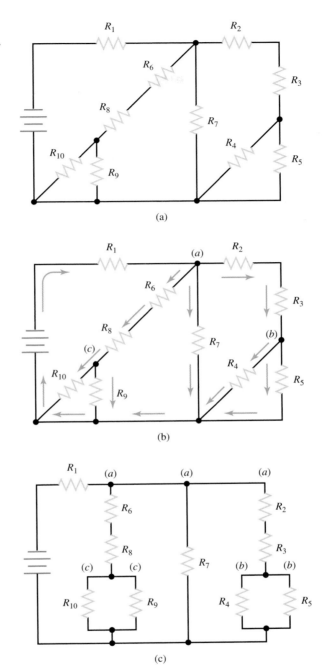

(a)

(b)

(c)

EXAMPLE 9.1	Assume all resistors in Figure 9.6(a) are 4 Ω in value. Find the equivalent resistance of the whole circuit.

Solution $R_{EQ_1} = R_{10} \| R_9$

$$= \frac{R}{N}$$

$$= \frac{4\ \Omega}{2}$$

$$= 2\ \Omega$$

$$R_{EQ_2} = R_6 + R_8 + R_{EQ_1}$$
$$= 4\,\Omega + 4\,\Omega + 2\,\Omega$$
$$= 10\,\Omega$$
$$R_{EQ_3} = R_4 \| R_5$$
$$= \frac{R}{N}$$
$$= \frac{4\,\Omega}{2}$$
$$= 2\,\Omega$$
$$R_{EQ_4} = R_{EQ_2} \| R_7 \| R_{EQ_3}$$
$$= \frac{1}{1/R_{EQ_2} + 1/R_7 + 1/R_{EQ_3}}$$
$$= \frac{1}{1/10\,\Omega + 1/4\,\Omega + 1/2\,\Omega}$$
$$= 1.18\,\Omega$$
$$R_{EQ_T} = R_1 + R_{EQ_4}$$
$$= 4\,\Omega + 1.18\,\Omega$$
$$= 5.18\,\Omega$$

Practice Problems 9.2   Answers are given at the end of the chapter.
1. Find R_{EQ_T} for the circuit in Figure 9.7.

FIGURE 9.7
Circuit for Practice Problems 9.2, Problem 1.

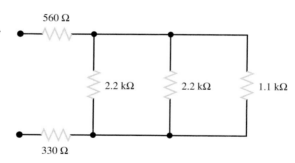

560 Ω

2.2 kΩ   2.2 kΩ   1.1 kΩ

330 Ω

**SECTION 9.3
SELF-CHECK**

Answers are given at the end of the chapter.

1. What shape should the redrawn circuit look like?
2. Where do you start in the redrawing of a series-parallel circuit?

9.4   SERIES-PARALLEL CIRCUIT ANALYSIS

When solving Ohm's law problems relating to series-parallel circuits, it is advisable to begin on those components where two of the three Ohm's law factors are known. A series-parallel circuit can be analyzed using the steps for redrawing for clarity previously shown and the following steps:

Lab Reference: Series-parallel circuit analysis results are demonstrated in Exercise 15.

1. Reduce all parallel circuits to series-equivalent resistances.
2. Combine all the branches containing more than one resistance in series into a single resistance.

3. Redraw the resulting equivalent circuit and determine the equivalent resistance of the whole circuit.
4. When the series-parallel circuit has been simplified, solve for the total current, voltage, or power, as required.
5. To obtain a complete solution for the series-parallel circuit, find the individual component values by using the values obtained in the equivalent circuit and applying them to the original circuit. Solve for individual voltage, current, or power, as required.

EXAMPLE 9.2

Find all unknown values for the circuit in Figure 9.1 (Section 9.1).

Solution

$$I_T = V_S/R_{EQ_T}$$
$$= 60 \text{ V}/20 \ \Omega$$
$$= 3 \text{ A}$$

$$P_T = V_S I_T$$
$$= 60 \text{ V} \times 3 \text{ A}$$
$$= 180 \text{ W}$$

$$V_1 = I_1 R_1$$
$$= 3 \text{ A} \times 8 \ \Omega$$
$$= 24 \text{ V}$$

$$V_2 = V_3 = V_{REQ_1}$$
$$V_{REQ_1} = I_T R_{EQ_1}$$
$$= 3 \text{ A} \times 12 \ \Omega$$
$$= 36 \text{ V}$$

$$P_1 = I_1 V_1$$
$$= 3 \text{ A} \times 24 \text{ V}$$
$$= 72 \text{ W}$$

$$P_2 = \frac{V_2^2}{R_2}$$
$$= \frac{(36 \text{ V})^2}{20 \ \Omega}$$
$$= 64.8 \text{ W}$$

$$P_3 = \frac{V_3^2}{R_3}$$
$$= \frac{(36 \text{ V})^2}{30 \ \Omega}$$
$$= 43.2 \text{ W}$$

Note that P_2 and P_3 were found directly from the known values at the time, which are V_2, V_3, R_2, and R_3. These powers could also be found by first determining I_2 and I_3 and combining these with V_2 and V_3. As a check, we have

$$I_2 = \frac{V_2}{R_2}$$
$$= \frac{36 \text{ V}}{20 \ \Omega}$$
$$= 1.8 \text{ A}$$

$$I_3 = \frac{V_3}{R_3}$$

$$= \frac{36 \text{ V}}{30 \text{ }\Omega}$$

$$= 1.2 \text{ A}$$

$$P_2 = I_2 V_2$$

$$= 1.8 \text{ A} \times 36 \text{ V}$$

$$= 64.8 \text{ W}$$

$$P_3 = I_3 V_3$$

$$= 1.2 \text{ A} \times 36 \text{ V}$$

$$= 43.2 \text{ W}$$

Practice Problems 9.3   Answers are given at the end of the chapter.

1. What is R_{EQ} for the circuit shown in Figure 9.8?

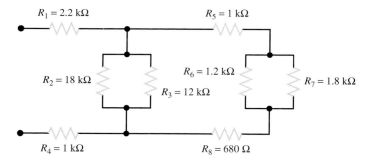

$R_1 = 2.2 \text{ k}\Omega$ $R_5 = 1 \text{ k}\Omega$

$R_2 = 18 \text{ k}\Omega$ $R_6 = 1.2 \text{ k}\Omega$ $R_7 = 1.8 \text{ k}\Omega$

$R_3 = 12 \text{ k}\Omega$

$R_4 = 1 \text{ k}\Omega$ $R_8 = 680 \text{ }\Omega$

FIGURE 9.8
Circuit for Practice Problems 9.3, Problem 1.

2. Find all unknowns for the circuit in Figure 9.2 (Section 9.1).

Now that you have solved for the unknown quantities in this circuit you can apply what you have learned to any series, parallel, or combination circuit. It is important to remember to first look at the circuit and from observation make a determination of the type of circuit, what is known, and what you are looking for. A minute spent in this manner may save you many unnecessary calculations.

**SECTION 9.4
SELF-CHECK**

Answers are given at the end of the chapter.

1. In simplifying a series-parallel circuit, start by reducing any _____ resistances to a _____ resistance.

9.5   WHEATSTONE BRIDGE

One of the most useful applications for series-parallel circuits is the *bridge*. The bridge circuit is used in a wide variety of electronic circuits, including rectifiers and measuring instruments. One such measuring instrument is the Wheatstone bridge, invented in 1850 by Charles Wheatstone. The Wheatstone bridge is an instrument that employs a special type of ammeter in conjunction with several resistors. It can be used to measure resistance with much greater accuracy than the conventional ohmmeter. Typical accuracies for an ohmmeter are within 5% to 10%, but utilizing a Wheatstone bridge, accuracies to within 1% or less are possible. Thus, the Wheatstone bridge can be used as a *precision ohmmeter.*

A Wheatstone bridge is shown in Figure 9.9. Four resistances are connected to form a diamond pattern. One of the resistances is an unknown (R_X), one is variable (R_1), and two resistors are fixed (R_2 and R_3). A battery is connected to two opposite corners of the diamond. A galvanometer is connected across the other two corners. A *galvanometer* is a special ammeter that measures small currents in either direction and has zero at its center scale.

FIGURE 9.9
Wheatstone bridge circuit. When balanced, the potentials at points A and B are equal.

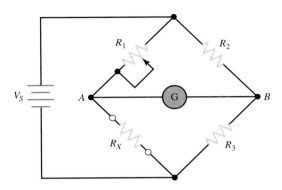

Lab Reference: The balanced wheatstone bridge is demonstrated in Exercise 20.

9.5.1 Balancing the Bridge

R_1 is adjusted until the galvanometer reads zero current. This operation is known as *balancing the bridge*. When balanced, the potentials at points A and B (in Figure 9.9) must be the same. Note that the potential at point A is developed by the series voltage divider combination of R_1 and R_X. The potential at point B is developed by the series voltage divider combination of R_2 and R_3. Note that there is no reference indicated in the circuit; thus, the negative terminal of the source becomes the reference for the voltages at A and B. Each voltage divider branch is connected in parallel across the battery. At balance, the equal voltage ratios in the two branches can be stated as

$$I_A R_1 / I_A R_X = I_B R_2 / I_B R_3 \quad \text{or} \quad R_1 / R_X = R_2 / R_3$$

Solving for R_X yields

$$R_X = \frac{R_1 R_3}{R_2} \tag{9.1}$$

EXAMPLE 9.6

Determine the unknown resistance of a Wheatstone bridge (like the one in Figure 9.9) when $R_1 = 1\ k\Omega$, $R_2 = 2\ k\Omega$, and $R_3 = 4\ k\Omega$ and the bridge is balanced.

Solution
$$R_X = \frac{1\ k\Omega \times 4\ k\Omega}{2\ k\Omega}$$
$$= 2\ k\Omega$$

Practice Problems 9.4 Answers are given at the end of the chapter.
1. Determine the unknown resistance of a Wheatstone bridge (like the one in Figure 9.9) when $R_1 = 400\ \Omega$, $R_2 = 200\ \Omega$, and $R_3 = 2\ k\Omega$ and the bridge is balanced.
2. If $R_1 = 2\ k\Omega$, $R_2 = 2\ k\Omega$, and $R_3 = 500\ \Omega$, determine the value of R_X.

When the bridge is *not* balanced, there is a current flow through the galvanometer. Unbalanced bridges must be analyzed by other methods, as described in Chapter 12.

In addition to measuring resistance, the Wheatstone bridge can be configured to measure quantities such as power, temperature, pressure, and light. This is done utilizing transducers that convert the form of energy being monitored into a resistance that the bridge can accommodate.

9.6 TROUBLESHOOTING SERIES-PARALLEL CIRCUITS

The major difference between an open in a parallel circuit and an open in a series circuit is that in the parallel circuit the open does not necessarily disable the circuit. If the open condition occurs in a series portion of the circuit, there is no current because there is no complete path for current flow. If, on the other hand, the open occurs in a parallel path, some current still flows in the circuit (in other branches). The parallel branch where the open occurs is effectively disabled, total resistance of the circuit increases, and total current decreases.

To clarify these points consider the series-parallel circuit of Figure 9.10. First, the effect of an open in the series portion of this circuit is examined. Figure 9.10(a) shows the normal circuit; $R_T = 40 \ \Omega$ and $I_T = 3$ A. In Figure 9.10(b) an open is shown in the series portion of the circuit. There is no complete path for current, and the resistance of the circuit is considered to be infinite.

In Figure 9.10(c) an open is shown in the parallel branch of R_3. There is no path for current through R_3. In the circuit, current flows through R_1 and R_2 only. Because there is only one path for current flow, R_1 and R_2 are effectively in series.

FIGURE 9.10
Series-parallel circuit with opens.

Lab Reference: Opens and shorts in series-parallel circuits are demonstrated in Exercise 16.

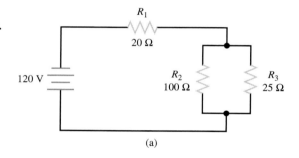

(a)

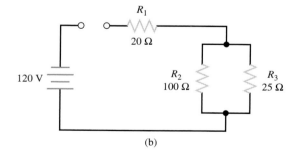

(b)

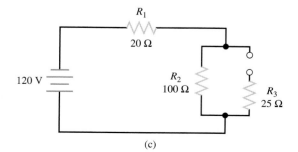

(c)

Under these conditions $R_T = 120 \ \Omega$ and $I_T = 1$ A. As you can see, total circuit resistance increased and total circuit current decreased.

A short circuit in a parallel network has an effect similar to a short in a series circuit. In general, the short causes an increase in current and the possibility of component damage, regardless of the type of circuit involved. To illustrate this point, Figure 9.11(a) shows a series-parallel network in which shorts are developed. In Figure 9.11(a) the normal circuit is shown; $R_T = 40 \ \Omega$ and $I_T = 3$ A.

In Figure 9.11(b), R_1 has shorted. R_1 now has zero resistance. The total resistance of the circuit is now equal to the resistance of the parallel network of R_2 and R_3, or 20 Ω. Circuit current increases to 6 A. All this current goes through the parallel network (R_2, R_3), and this increase in current will most likely damage the components.

In Figure 9.11(c), R_3 has shorted. With R_3 shorted, there is a short circuit in parallel with R_2. The short circuit routes the current around R_2, effectively removing R_2 from the circuit. Total circuit resistance is now equal to the resistance of R_1, or 20 Ω. The total circuit current with R_3 shorted is 6 A. All this current flows through R_1 and will most likely damage it. Notice that even though only one portion of the parallel network is shorted, the *entire* parallel network is disabled.

FIGURE 9.11
Series-parallel circuit with shorts.

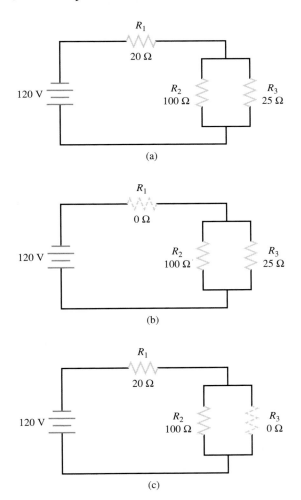

Opens and shorts alike, if occurring in a circuit, result in an overall change in the equivalent resistance. This can cause undesirable effects in other parts of the circuit due to the corresponding change in the total current flow. A short usually causes components to fail in a circuit that is not properly fused or otherwise protected. The failure may take the form of a burned-out resistor, damaged source, or a fire in the circuit components and wiring.

Fuses and other circuit-protection devices are installed in electronic circuits to prevent damage caused by increases in current. These circuit-protection devices are designed to open if current increases beyond a predetermined value. Circuit-protection devices are connected *in series* with the circuit or portion of the circuit that the device is protecting. When the circuit-protection device opens, current flow ceases in the circuit. A more thorough explanation of fuses and other circuit-protection devices is presented in Chapter 13.

SECTION 9.6 SELF-CHECK

Answers are given at the end of the chapter.

1. What does an open in the series portion of a series-parallel circuit do to circuit operation?
2. Circuit protection devices are connected in _____ with the circuit that the device is protecting.

■ SUMMARY

1. The basic technique used for solving DC series-parallel circuits involves equivalent circuits.
2. Points where current divides in a circuit are known as nodes.
3. The wires in schematic diagrams are assumed to have no resistance, and there is no voltage drop along the wire.
4. In redrawing a circuit for clarity, a wire can be shrunk or stretched as necessary without changing any electrical characteristic of the circuit.
5. The redrawn circuit should be drawn in a simple boxlike form.
6. To redraw the circuit start at the negative terminal of the voltage source.
7. The steps in circuit analysis of series-parallel circuits allow that all parallel circuits be reduced to series equivalents. All series resistances are combined into one value. Solve for total resistance, current, voltage, or power as required. Find the individual component values by using the values obtained in the equivalent circuit and applying them to the original circuit.
8. The Wheatstone bridge employs four resistances connected in a diamond pattern. A galvanometer is connected between two corners, with the source connected between the other two corners.
9. The Wheatstone bridge was first used as a precision ohmmeter.
10. Other uses of the Wheatstone bridge include rectifiers and measuring instruments (power, temperature, pressure, or light).
11. A galvanometer is a special ammeter that measures small currents in either direction and has zero as its center scale.
12. When the galvanometer reads zero current, the bridge is said to be balanced.
13. Analyzing unbalanced bridges requires methods described in Chapter 12.
14. If an open occurs in the series portion of a series-parallel circuit, there is no path for current. If an open occurs in a parallel portion of a series-parallel circuit, there are still paths for current through other parallel branches.
15. A short in a parallel network has an effect similar to a short in a series circuit.

16. A short usually causes components to fail in a circuit that is not properly fused or otherwise protected.
17. Circuit-protection devices are connected in series with the circuit or portion of the circuit that the device is protecting.

■ IMPORTANT EQUATIONS

$$R_X = \frac{R_1 R_3}{R_2} \tag{9.1}$$

■ QUESTIONS

1. What is another name for a series-parallel circuit?
2. What is the basic technique used to solve DC series-parallel circuits?
3. What steps are employed to redraw a circuit for clarity?
4. What is a node?
5. A wire is considered to have _____ resistance and _____ voltage drop.
6. Where do you start in redrawing a circuit?
7. Why can wires be stretched or shrunk without affecting circuit behavior?
8. What are the steps in circuit analysis for a series-parallel circuit?
9. Describe the makeup of the Wheatstone bridge.
10. Describe the electrical characteristics of the galvanometer.
11. Describe the applications of the Wheatstone bridge.
12. What is meant by *balancing a bridge*?
13. If the bridge is not balanced, what techniques are employed to analyze it?
14. Compare and contrast an open in a series portion of a series-parallel circuit and one in a parallel branch of the same circuit.
15. Compare and contrast a short in a series portion of a series-parallel circuit and one in a parallel branch of the same circuit.

16. What are circuit-protection devices used for?
17. How are circuit-protection devices connected in a circuit?
18. What can happen if a short occurs and the circuit is not protected by a fuse or other form of protection device?

■ CRITICAL THINKING

1. Describe the process used for redrawing a circuit for clarity.
2. What is the equivalent resistance of the circuit in Figure 9.12?

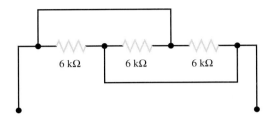

FIGURE 9.12
Circuit for Critical Thinking, Question 2. Find R_{EQ}.

3. Compare and contrast the effects of opens and shorts in the series portion of a combination circuit versus the parallel portion of the same circuit.

■ PROBLEMS

1. For the circuit in Figure 9.13, determine R_{EQ}.

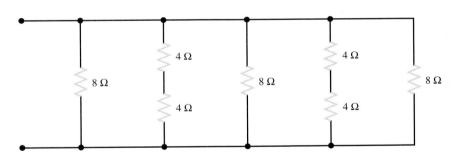

FIGURE 9.13
Circuit for Problem 1.

2. For the circuit in Figure 9.14, determine R_{EQ}.

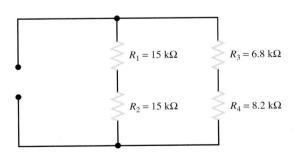

FIGURE 9.14
Circuit for Problem 2.

3. For the circuit in Figure 9.15, determine R_{EQ}.

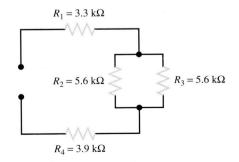

FIGURE 9.15
Circuit for Problem 3.

4. For the circuit in Figure 9.16, determine R_{EQ}.

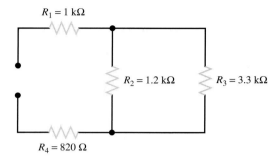

$R_1 = 1$ kΩ

$R_2 = 1.2$ kΩ $R_3 = 3.3$ kΩ

$R_4 = 820$ Ω

FIGURE 9.16
Circuit for Problem 4.

5. For the circuit in Figure 9.17, determine R_{EQ}.

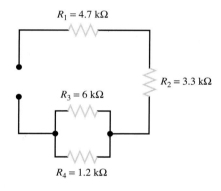

$R_1 = 4.7$ kΩ

$R_2 = 3.3$ kΩ

$R_3 = 6$ kΩ

$R_4 = 1.2$ kΩ

FIGURE 9.17
Circuit for Problem 5.

6. For the circuit in Figure 9.18, determine R_{EQ}.

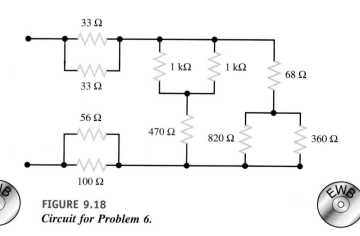

33 Ω 1 kΩ 1 kΩ 68 Ω
33 Ω
56 Ω 470 Ω 820 Ω 360 Ω
100 Ω

FIGURE 9.18
Circuit for Problem 6.

7. For the circuit in Figure 9.19, determine R_{EQ}.

$R_1 = 270$ Ω $R_2 = 330$ Ω

$R_3 = 1.2$ kΩ $R_4 = 1.2$ kΩ

FIGURE 9.19
Circuit for Problem 7.

8. For the circuit in Figure 9.20, determine R_{EQ}.

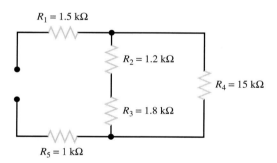

$R_1 = 1.5$ kΩ

$R_2 = 1.2$ kΩ

$R_4 = 15$ kΩ

$R_3 = 1.8$ kΩ

$R_5 = 1$ kΩ

FIGURE 9.20
Circuit for Problem 8.

9. For the circuit in Figure 9.21, determine R_{EQ}.

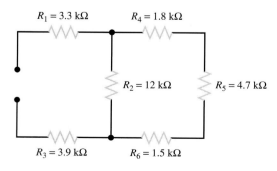

$R_1 = 3.3$ kΩ $R_4 = 1.8$ kΩ

$R_2 = 12$ kΩ $R_5 = 4.7$ kΩ

$R_3 = 3.9$ kΩ $R_6 = 1.5$ kΩ

FIGURE 9.21
Circuit for Problem 9.

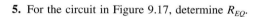

10. For the circuit in Figure 9.22, determine R_{EQ}, I_T, I_1, I_2, I_3, I_4, V_1, V_2, V_3, and V_4.

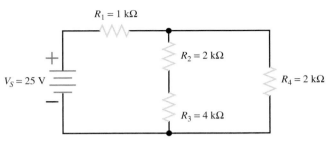

FIGURE 9.22
Circuit for Problem 10.

11. For the circuit in Figure 9.23, determine R_{EQ}, I_T, I_1, I_2, I_3, I_4, I_5, V_1, V_2, V_3, V_4, and V_5.

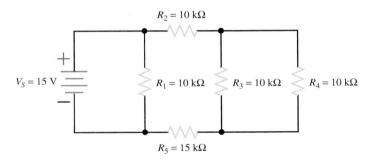

FIGURE 9.23
Circuit for Problem 11.

12. For the circuit in Figure 9.24, determine R_{EQ}, I_T, I_1, I_2, I_3, I_4, I_5, I_6, V_1, V_2, V_3, V_4, V_5, and V_6.

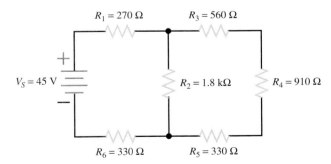

FIGURE 9.24
Circuit for Problem 12.

13. For the circuit in Figure 9.25, determine R_{EQ}, I_T, I_1, I_2, I_3, I_4, I_5, I_6, I_7, V_1, V_2, V_3, V_4, V_5, V_6, and V_7.

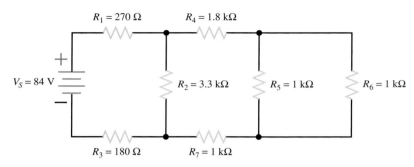

FIGURE 9.25
Circuit for Problem 13.

14. For the circuit in Figure 9.26, determine R_{EQ}, I_T, I_1, I_2, I_3, I_4, I_5, I_6, I_7, I_8, I_9, V_1, V_2, V_3, V_4, V_5, V_6, V_7, V_8, and V_9.

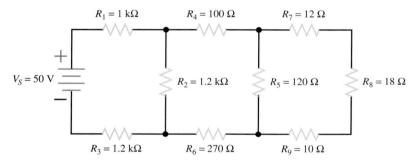

FIGURE 9.26
Circuit for Problem 14.

15. For the circuit in Figure 9.27, determine R_{EQ}, I_T, I_1, I_2, I_3, I_4, I_5, I_6, I_7, I_8, V_1, V_2, V_3, V_4, V_5, V_6, V_7, and V_8.

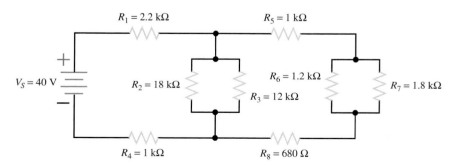

FIGURE 9.27
Circuit for Problem 15.

16. For the circuit in Figure 9.28, determine R_T, V_S, I_T, V_1, V_2, V_3, V_4, V_6, and I_2.

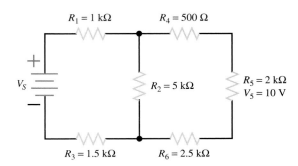

FIGURE 9.28
Circuit for Problem 16.

17. For the circuit in Figure 9.29, determine R_T, I_T, V_S, I_2, I_5, V_1, V_3, V_4, V_5, and V_6.

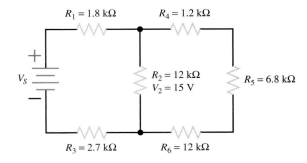

FIGURE 9.29
Circuit for Problem 17.

18. For the circuit in Figure 9.30, determine R_T, I_T, V_S, I_2, I_5, I_6, I_7, V_1, V_2, V_3, V_5, V_6, and V_7.

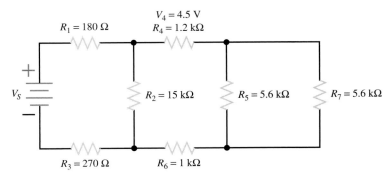

FIGURE 9.30
Circuit for Problem 18.

19. For the circuit in Figure 9.31, determine R_T, I_T, V_S, I_1, I_2, I_3, I_4, I_5, V_1, V_2, V_3, V_4, V_6, and V_7.

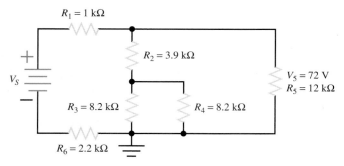

FIGURE 9.31
Circuit for Problem 19.

20. To balance a Wheatstone bridge when $R_1 = 4.7\ \text{k}\Omega$, $R_2 = 10\ \text{k}\Omega$, and $R_3 = 6.8\ \text{k}\Omega$, determine the value of R_X.

■ ANSWERS TO PRACTICE PROBLEMS

PRACTICE PROBLEMS 9.1

1. $5\ \text{k}\Omega$
2. $7.5\ \text{k}\Omega$

PRACTICE PROBLEMS 9.2

1. $1.44\ \text{k}\Omega$

PRACTICE PROBLEMS 9.3

1. $5\ \text{k}\Omega$
2. $I_T = 0.25$ A, $V_1 = 37.5$ V, $V_2 = 12.5$ V, $V_3 = 50$ V, $V_4 = 250$ V, $I_1 = 0.125$ A, $I_2 = 0.125$ A, $I_3 = 0.125$ A, $I_4 = 0.25$ A, $P_1 = 4.69$ W, $P_2 = 1.56$ W, $P_3 = 6.25$ W, $P_4 = 62.5$ W, $P_T = 75$ W

PRACTICE PROBLEMS 9.4

1. $4\ \text{k}\Omega$
2. $500\ \Omega$

■ ANSWERS TO SELF-CHECKS

SECTION 9.1

1. To simplify the circuit to one containing a single resistance

SECTION 9.2

1. A node is a point where current divides.
2. Zero
3. Stretched, shrunk

SECTION 9.3

1. A simple boxlike form
2. Start at the negative terminal of the source and trace the current flow.

SECTION 9.4

1. Parallel, series

SECTION 9.5

1. A galvanometer is a special ammeter that measures small currents in either direction and has zero at its center scale.
2. A precision ohmmeter or to measure power, temperature, pressure, and light
3. Transducer

SECTION 9.6

1. There is no current flow in that portion of the circuit.
2. Series

CHAPTER 10

CIRCUIT MEASUREMENT

OBJECTIVES

After completing this chapter, you will be able to:

1. Describe the basic analog meter movement.
2. Compare and contrast other meter movements to the basic meter movement.
3. Define the terms *meter sensitivity* and *meter loading* as they apply to ammeters and voltmeters.
4. Calculate the resistances required to extend the range of the ammeter and voltmeter.
5. Describe the proper procedures to follow when using an ammeter, a voltmeter, and an ohmmeter.
6. Compare and contrast the digital multimeter and the analog multimeter.

INTRODUCTION

Circuit measurement is used to monitor the operation of an electrical or electronic device or to determine the reason a device is not operating properly. Because electricity is invisible, you must use some sort of device to determine what is happening in an electrical circuit. Various devices, called ammeters, voltmeters, and ohmmeters in single-function form and test instruments in multiple-function form, are used to measure electrical quantities. The most common types of test equipment use some kind of metering device.

In this chapter we limit our discussion of instruments to those that measure current, voltage, or resistance. There are other quantities involved in electrical circuits that can be measured; some of these include power, capacitance, inductance, impedance, and frequency. It is possible to measure any circuit quantity once you are able to select and use the proper circuit-measuring device. There are many measuring devices available to measure any quantity, but basic measurements and troubleshooting techniques begin with current, voltage, or resistance values.

The measurement of the electrical quantities in a circuit is an integral part of working on electrical and electronic circuits. It is one that by now you have begun to master as you gain experience by doing the exercises in lab. A competent electronic technician must master essential test-equipment skills.

10.1 THE BASIC METER MOVEMENT

The meter movement is, as the name implies, the part of a meter that moves (giving an indication of some value). Meter movements convert electrical energy into mechanical energy. There are many types of meter movements. The earliest type and one that is still in use is the *analog meter movement*. Analog is a word that, in this context, means "similar to" and comes from the word *analogous*. With an analog readout meter, the amount of pointer deflection across a scale is an analog, or similar, to the magnitude of the electrical property being measured. This type of indicator is called an analog display, as opposed to a digital display, which makes use of decimal digits to display the magnitude of the electrical property. Digital-display meters are explained later in this chapter.

The basic meter movement encompasses a permanent-magnet moving-coil movement based on a fixed permanent magnet and a coil of wire that is able to move. When current passes through the coil, it establishes a magnetic field that reacts to the magnetic field of the permanent magnet. The magnetic interaction of both fields causes the coil to rotate or move.

The coil of wire is wound on an aluminum frame, or bobbin, that is supported by jeweled bearings, which allows it to move freely. This is illustrated in Figure 10.1.

To use the moving-coil movement as a meter, two problems must be solved. First, a way must be found to return the coil to its original position when there is no current through the coil. Second, a method is needed to indicate the amount of coil movement.

The first problem is solved by the use of hairsprings attached to each end of the coil, as shown in Figure 10.2. These hairsprings are also used to make the electrical connections to the coil. With the use of hairsprings, the coil returns to its initial position when there is no current flow through the meter.

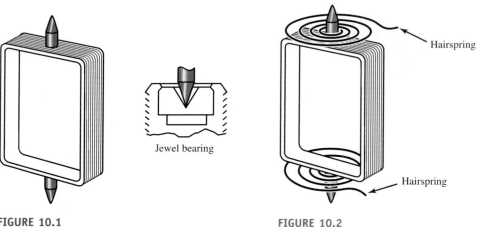

FIGURE 10.1
A basic coil arrangement.

FIGURE 10.2
Coil and hairsprings.

For indicating purposes, a pointer is attached to the coil and extended out to a scale. As the coil moves, the pointer moves across the face of the scale. The coil (and thus the pointer) moves in relation to the *average value* of current through it. The scale is calibrated to indicate the amount of current through the coil. This is shown in Figure 10.3. The scale can also be used to indicate voltage or resistance values. Sound recording and mixing equipment use a scale calibrated in volume units (VU).

Two other features are used to increase the accuracy and efficiency of this meter movement. First, an iron core is placed inside the coil to concentrate the magnetic fields. Second, curved pole pieces are attached to the magnet to ensure that the turning force on the coil increases steadily as the current increases.

FIGURE 10.3
The pointer assembly.

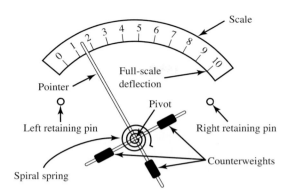

The meter movement as it appears when fully assembled is shown in Figure 10.4. This permanent-magnet moving-coil meter movement is the basic movement in most analog measuring instruments. The earliest instrument that used the moving-coil movement was the *galvanometer*. It measures very small amounts of current and was first used with the Wheatstone bridge. The moving-coil arrangement is also called a D'Arsonval movement because it was first employed by the Frenchman D'Arsonval in 1881 while making electrical measurements. The Weston-type instrument, another common meter movement, adapted this principle as well. Edward Weston (1850–1936), an American electrical engineer, formed the Weston Electrical Instrument Co. in 1888. Weston instruments are still in use today.

FIGURE 10.4
A permanent-magnet moving-coil meter movement.

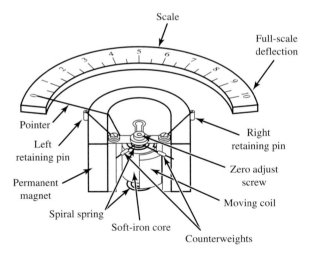

10.1.1 Rectifier for AC Measurement

The D'Arsonval meter movement can also be used to indicate values of alternating current (AC). This is done by passing AC through a rectifier. A rectifier is a device that changes alternating current to a form of direct current known as pulsating DC. This is shown in Figure 10.5.

FIGURE 10.5
Rectifier action for measuring AC.

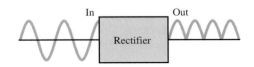

10.1.2 Damping

By connecting a rectifier to a D'Arsonval meter movement, an alternating current–measuring device is created. When AC is converted to pulsating DC, the D'Arsonval movement reacts to the average value of the pulsating DC. Because the current is pulsating, the pointer of the movement tends to vibrate (oscillate) around the average value indication. This oscillation makes the meter difficult to read and needs to be smoothed out.

The process of smoothing out the oscillation of the pointer is known as *damping*. Damping is accomplished in two ways. The current through the coil causes the coil to rotate. This rotation within a magnetic field causes a current to be induced into the coil opposite to the current that caused the movement of the coil. This induced current acts to dampen oscillations. A second method of damping employs the use of a vane attached to the coil. As the vane moves, friction is created against air molecules, which tends to reduce oscillations.

SECTION 10.1 SELF-CHECK	Answers are given at the end of the chapter.

Answers are given at the end of the chapter.

1. What is at the heart of the basic meter movement?
2. What is the purpose of hairsprings in the basic meter movement?
3. The pointer in the basic meter movement moves in relation to the _____ value of current through it.
4. What is used by the basic meter movement so that it can measure AC current?
5. Who first used the moving-coil meter in a measuring instrument?

10.2 OTHER METER MOVEMENTS

10.2.1 Electrodynamic Meter Movement

An *electrodynamic movement* uses the same basic operating principle as the D'Arsonval meter movement, except that the permanent magnet is replaced by fixed coils. The moving coil to which the indicator is attached is suspended between two field coils and connected in series with these coils. The same current flows through each.

Current flow in either direction through the three coils causes a magnetic field to exist between the field coils. A turning force is created on the moving coil and does not change direction, even if the current is reversed. Since reversing the current direction does not reverse the turning force, this type of meter can be used to measure both AC and DC if the scale is changed accordingly. The most important application of the electrodynamic principle of operation is in the *wattmeter*, which is used to measure power.

10.2.2 Moving-Vane Meter Movement

The *moving-vane meter movement* (sometimes called the moving-iron movement) is a commonly used movement for AC meters. The moving-vane meter operates on the principle of magnetic repulsion between like poles. The current to be measured flows through a coil, producing a magnetic field that is proportional to the strength of the current. Suspended in this field are two iron vanes. One is in a fixed position; the other, attached to the meter pointer, is movable. The magnetic field magnetizes these iron vanes with the same polarity regardless of the direction of current flow in the coil. Because like poles repel, the movable vane pulls away from the fixed vane, moving the meter pointer to indicate a value of current.

10.2.3   Hot-Wire and Thermocouple Meter Movements

Hot-wire (taut-band) and *thermocouple meter movements* both use the heating effect of current flowing through a resistance to cause meter deflection. Each uses this effect in a different manner. Because their operation depends only on the heating effect of current flow, they may be used to measure both DC and AC.

The hot-wire (taut-band) meter-movement deflection depends on the expansion of a high-resistance wire caused by the heating effect of the wire itself as current flows through it. A resistance wire is stretched taut between the two meter terminals, with a thread attached at a right angle to the center of the wire. Current flow heats the wire, causing it to expand. This motion is transferred to the meter pointer through the thread and a pivot.

A thermocouple meter consists of a resistance wire across the meter terminals, which heats in proportion to the amount of current. Attached to this wire is a small thermocouple junction of two unlike metal wires that connect across a very sensitive DC meter movement (usually a D'Arsonval meter movement). As the current being measured heats the heating resistor, a small current is generated by the thermocouple junction. The current being measured flows through the resistance wire only, not through the meter movement itself. The pointer turns in proportion to the amount of heat generated by the resistance wire.

SECTION 10.2 SELF-CHECK	Answers are given at the end of the chapter.

1. What is the difference between an electrodynamic movement and the basic D'Arsonval meter movement?
2. Hot-wire and thermocouple meters use the _____ effect of current in order to cause meter deflection.
3. The electrodynamic meter is used in the _____.

10.3   AMMETERS

An *ammeter* is a device that measures current. The analog DC ammeter is almost always a Weston-type instrument. The moving coil of a Weston instrument typically carries currents not larger than 30 μA. To be able to handle larger currents, a *shunt resistor* (a parallel-connected resistance) is placed across the moving-coil leads. Current going into the meter then divides, with part of the current going through the coil and the rest through the shunt. The shunt is a low-value resistance, usually made of manganin strips that are bonded to a heavy copper block. The use of manganin ensures a low-temperature coefficient of resistance (improved stability) and more accurate measurements.

The currents in the shunt and in the moving coil vary inversely with their resistances. This is expressed in equation form as

$$\frac{I_{SH}}{I_M} = \frac{R_M}{R_{SH}}$$

(10.1)

where R_{SH} = shunt resistance

R_M = meter resistance

I_{SH} = shunt current

I_M = meter current

EXAMPLE 10.1

What shunt resistance is required to extend the range of a 0–1-mA movement having an internal resistance of 50 Ω so that a total of 10 mA may be measured? This is illustrated in Figure 10.6.

Solution
$$\frac{I_{SH}}{I_M} = \frac{R_M}{R_{SH}}$$

$$R_{SH} = \frac{I_M R_M}{I_{SH}}$$

$$= \frac{(1 \times 10^{-3}\ A)(50\ \Omega)}{(9 \times 10^{-3})\ A}$$

$$= 5.56\ \Omega$$

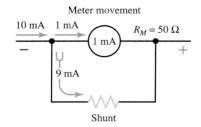

FIGURE 10.6
Circuit for Example 10.1.

Practice Problems 10.1   Answers are given at the end of the chapter.
1. What shunt resistance is required to extend the range of a 0–50-μA, 2-kΩ movement to 10 mA?
2. What shunt is required to extend the range of a 0–50-μA, 2-kΩ movement to 50 mA?

The ammeter is always placed *in series* to the component through which the desired current reading is to be made. The internal resistance of an ammeter is very small. This is because for virtually any value of current to be measured, a shunt resistance is always in parallel to the moving coil. Therefore, the total meter resistance is $R_M \| R_{SH}$. For the meter in Example 10.1, the total meter resistance is 50 Ω ∥ 5.56 Ω, or 5 Ω. Ideally, the internal resistance should be zero. If the meter resistance were appreciable, its insertion could change the resistance of the circuit, thereby changing the amount of circuit current. This is referred to as the *loading effect* of an ammeter.

10.3.1   Ammeter Placement

In complex electrical circuits, you are not always concerned with total circuit current. You may be interested in the current through a particular component or group of components. In any case, the ammeter is always connected in series with the circuit path you wish to test. The proper placement of an ammeter to measure currents in various portions of a circuit is illustrated in Figure 10.7.

10.3.2   Ammeter Sensitivity

Ammeter sensitivity is the amount of current (I_M) necessary to cause full-scale deflection (FSD) of the ammeter pointer. Typical FSD values are from about 10 μA to 30 mA. The corresponding meter-movement resistance (R_M) is from about 1 Ω to 2000 Ω. (R_M is the internal resistance of the wire in the moving coil.)

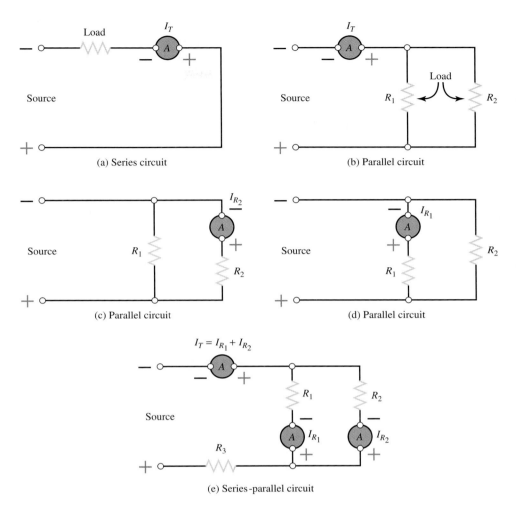

FIGURE 10.7
Proper ammeter connections: (a) measuring total current (I_T), series circuit; (b) measuring total current (I_T), parallel circuit; (c) measuring current through R_2 (I_{R_2}), parallel circuit; (d) measuring current through R_1 (I_{R_1}), parallel circuit; (e) measuring I_T, I_{R_1}, I_{R_2}, series-parallel circuit.

The smaller the amount of FSD current, the more sensitive the ammeter. For example, an ammeter that needs only 10 μA to deflect the pointer to FSD is much more sensitive than one that requires 10 mA. Sensitivity can be given for a meter movement, but the term *ammeter sensitivity* usually refers to the entire ammeter and not just the meter movement. An ammeter consists of more than just the meter movement.

10.3.3 Multirange Ammeter

A multirange ammeter can be constructed by using several values of shunt resistors and a rotary switch to select the desired range. This is illustrated in Figure 10.8(a). The switching mechanism is known as a *make-before-break* switch. As a new range is selected, the wide end of the moving contact on a switch connects to the next terminal before it loses contact with its present terminal. This is to ensure that there is a shunt across the coil of the meter at all times. Otherwise, an open circuit may occur, in which case the full current to be measured would flow through the coil of the meter, causing possible burnout.

Another method that is used to protect the ammeter coil from excessive current is through the use of an *Aryton shunt* (Figure 10.8(b)). By examining the diagram, note that no matter which range is selected, there is always a shunt or a combination of shunts in parallel to the meter. For some ranges, a shunt is in series to the meter as well.

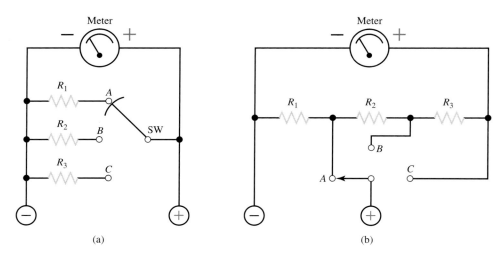

FIGURE 10.8
Multirange ammeters: (a) ammeter with individual shunts; (b) Aryton shunt.

10.3.4 Measuring Current

There are some important points to remember when using the ammeter to measure current in a circuit. Here are some rules to follow to ensure the proper use of an ammeter:

1. The ammeter must be connected in series to the path of the current. The circuit path must be opened and the ammeter inserted into the circuit at that point.
2. Always set the range selector to the highest scale and then reduce as needed.
3. Observe polarity. The positive lead of the ammeter (red) should be connected in the circuit so that it can be traced back to the positive side of the voltage supply, whereas the negative lead of the meter (black) should be connected in the circuit so that it can be traced back to the negative side of the voltage supply.
4. Most analog meters have an accuracy of ±3% of full scale. For this reason current readings should be as close to full scale as possible. For example, a 7-mA reading on a 10 mA scale has a maximum error of ±0.3 mA. The same reading on a 100-mA scale has a maximum error of ±3.0 mA. Thus the reading on the 10-mA scale could read from 6.7 to 7.3 mA, whereas the reading on the 100-mA scale could read from 4 to 10 mA.
5. Better-quality analog meters include a mirror along the scale. The meter is read so that the pointer and its mirror reflection appear as one. This eliminates the optical error called *parallax* caused by looking at the meter from the side.

**SECTION 10.3
SELF-CHECK**

Answers are given at the end of the chapter.

1. How are ammeters connected in a circuit?
2. What device enables the ammeter to measure multiple ranges of current flow?
3. Describe the term *ammeter sensitivity*.
4. What is *parallax* error?

10.4   VOLTMETERS

A current change causes a voltage change, and this relationship means that the moving-coil meter movement can be used to measure voltage as well. A 0–50-μA, 2000-Ω meter movement has a maximum voltage drop of 0.1 V. Thus, if 0.1 V is placed across the meter movement, a full-scale deflection of 50 μA occurs. With this setup, the meter can be used to measure any voltage from 0 to 0.1 V. Obviously, this setup has practical limitations. In order to measure higher voltages and extend the range, a resistance is placed *in series* with the movement to drop any excess voltage that the meter cannot safely handle. The series resistance is known as a *multiplier* and is usually connected internally. This is shown in Figure 10.9. The voltmeter is connected in parallel or across the component in order to measure voltage. With the use of additional resistances and a rotary switch, a practical multirange voltmeter can be made.

FIGURE 10.9
A voltmeter formed by connecting a multiplier resistor in series with meter movement.

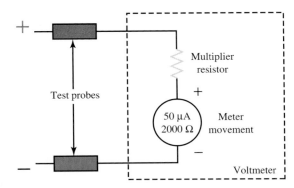

Test probes

Multiplier resistor

50 μA
2000 Ω

Meter movement

Voltmeter

To determine the value of an individual multiplier resistor, the rules of a series circuit are used. In equation form, this is expressed as:

$$R_{\text{mult}} = \frac{V_{\text{full-scale}}}{I_M} - R_M \tag{10.2}$$

where $V_{\text{full-scale}}$ = full-scale voltage at extended range

I_M = full-scale meter current

R_M = meter-movement resistance

EXAMPLE 10.2

What value of an individual multiplier resistance is needed to extend the range of the 50-μA, 2000-Ω meter movement to 10 V?

Solution $R_{\text{mult}} = \dfrac{10 \text{ V}}{50 \times 10^{-6} \text{ A}} - 2000 \text{ }\Omega$

$= 200{,}000 \text{ }\Omega - 2000 \text{ }\Omega$

$= 198{,}000 \text{ }\Omega$

$= 198 \text{ k}\Omega$

Practice Problems 10.2   Answers are given at the end of the chapter.
1. What multiplier is required to extend the range of a 0–50-μA, 2000-Ω meter movement to 50 V?
2. What multiplier is required to extend the range of a 0–50-μA, 2000-Ω meter movement to 100 V?

10.4.1 Multirange Voltmeters

Through the use of a range-selector switch, a multirange voltmeter can be made that switches the appropriate individual multiplier resistance in series with the meter movement. The schematic diagram for such a voltmeter is shown in Figure 10.10.

Instead of individual multipliers, each with its own nonstandard resistance value, the multipliers can be connected in a series string, as in Figure 10.11. Note, that for the most part these resistances can be obtained in standard resistance values. In order to calculate the specific resistances needed for a particular application, equation (10.2) is modified as follows:

$$R_{mult} = \frac{V_{full\text{-}scale}}{I_M} - (R_M + \text{all previous } R_{mult} \text{ values}) \qquad \textbf{(10.3)}$$

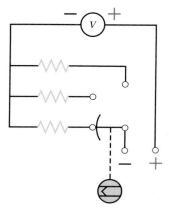

FIGURE 10.10
Multirange voltmeter utilizing individually connected multiplier resistors.

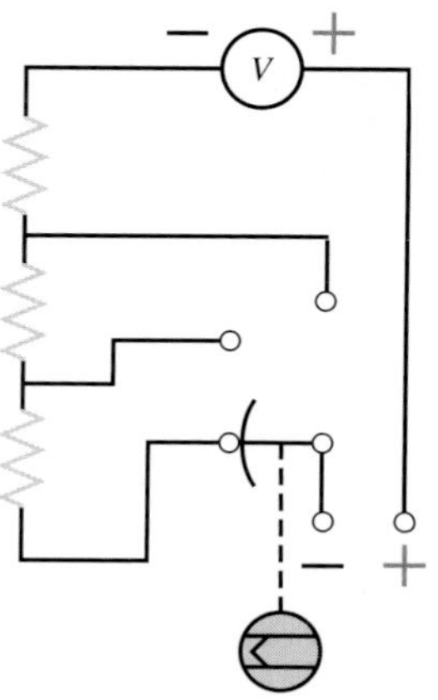

FIGURE 10.11
Multirange voltmeter utilizing a series string of multiplier resistors.

Here is an example problem to help clarify equation (10.3).

EXAMPLE 10.3

Determine the values of R_1, R_2, and R_3 in Figure 10.11 (the meter movement is a 50 μA, 3 kΩ) so that it has voltage ranges of 10 V, 50 V, and 100 V.

Solution
For the 10-V range:

$$R_1 = \frac{V_{full\text{-}scale}}{I_M} - R_M$$

$$= \frac{10 \text{ V}}{50 \times 10^{-6} \text{ A}} - 3 \times 10^3 \ \Omega$$

$$= 200 \times 10^3 \ \Omega - 3 \times 10^3 \ \Omega$$

$$= 197 \text{ k}\Omega$$

For the 50-V range:

$$R_2 = \frac{V_{full\text{-}scale}}{I_M} - (R_M + R_1)$$

$$= \frac{50 \text{ V}}{50 \times 10^{-6} \text{ A}} - (3 \text{ k}\Omega + 197 \text{ k}\Omega)$$

$$= 1 \text{ M}\Omega - 200 \text{ k}\Omega$$
$$= 800 \text{ k}\Omega$$

For the 100-V range:

$$R_3 = \frac{V_{\text{full-scale}}}{I_M} - (R_M + R_1 + R_2)$$

$$= \frac{100 \text{ V}}{50 \times 10^{-6} \text{ A}} - (3 \text{ k}\Omega + 197 \text{ k}\Omega + 800 \text{ k}\Omega)$$

$$= 2 \text{ M}\Omega - 1 \text{ M}\Omega$$

$$= 1 \text{ M}\Omega$$

Practice Problems 10.3   Answers are given at the end of the chapter.
1. Find R_1, R_2, and R_3 for a series-connected multirange voltmeter with a 10-μA, 2-kΩ meter movement connected like the one in Figure 10.11. The voltage ranges are 2.5 V, 30 V, and 100 V.

10.4.2   Voltmeter Sensitivity

Voltmeter sensitivity is expressed in ohms/volt (Ω/V). A voltmeter is considered more sensitive if it draws less current from the circuit. The sensitivity of a voltmeter varies inversely with the current required for full-scale deflection (FSD). Expressed mathematically,

$$\text{sensitivity} = \frac{1 \text{ V}}{I_{FSD}} \qquad\qquad\qquad \textbf{(10.4)}$$

where I_{FSD} = current required for FSD of the meter movement

The sensitivity of the 50-μA meter movement is simply the reciprocal of 50 μA, or 20,000 Ω/V (20 kΩ/V). Voltmeter sensitivity is usually indicated on the scale face and is higher for DC voltages than AC voltages. A general-purpose lab quality voltmeter should have a minimum sensitivity of 20 kΩ/V. Higher-value sensitivity voltmeters are available.

EXAMPLE 10.4

What is the sensitivity of a voltmeter that produces a FSD of 10 μA?

Solution   sensitivity = $1/10 \times 10^{-6}$
$\qquad\qquad\qquad\quad$ = 100,000 Ω/V
$\qquad\qquad\qquad\quad$ = 100 kΩ/V

Practice Problems 10.4   Answers are given at the end of the chapter.
1. What is the voltmeter sensitivity of a voltmeter with a 100-μA meter movement?
2. What is the voltmeter sensitivity of a voltmeter with a 1-mA meter movement?

10.4.3   Voltmeter Resistance

The multiplier resistance is usually a high value in order to limit the current flow through the meter movement. The voltmeter resistance changes as the multiplier is changed for each range. The voltmeter resistance is *the sum of the multiplier and meter-movement resistance* (R_M). It can also be found by multiplying the voltage range times the voltmeter sensitivity. Thus, on a 100-V range, a 20-kΩ/V voltmeter has a resistance of 2000 kΩ or 2 MΩ (100 $\times$ 20 kΩ). It should be noted that the voltmeter resistance has this value whether you read full scale or not. Further, it should be noted that only the voltmeter resistance varies with voltage range, the *sensitivity is constant* and *remains the same on all ranges*.

Since the voltmeter is placed in parallel, the higher the voltmeter resistance, the better. This ensures that little current is drawn by the voltmeter and circuit performance is not hindered by the application of the voltmeter.

10.4.4 Voltmeter Loading Effects

When the voltmeter resistance is not high enough, connecting it across a circuit component can change circuit resistance, which changes circuit current and voltage. The measured voltage decreases compared to the voltage without the voltmeter. This effect is called *voltmeter loading*, because additional current is drawn by the voltmeter. An ideal voltmeter has infinite resistance and no loading effects.

Voltmeter loading effects are illustrated in a series of diagrams in Figure 10.12. The effects are more pronounced in high-resistance circuits. Figure 10.12(a) shows the original circuit without the voltmeter attached. In Figure 10.12(b), a voltmeter with a 1-kΩ/V sensitivity is placed across R_2. The voltmeter range is selected as 100 V, the value of the voltage source. The voltmeter resistance equates to 100 × 1 kΩ/V, or 100 kΩ. This value loads R_2 and the circuit so that the equivalent R_2 resistance is lowered to 50 kΩ from its 100-kΩ value. The corresponding voltage across R_2 decreases from its normal value of 50 V to 33.3 V. This is shown in Figure 10.12(c).

FIGURE 10.12
Loading effects of a voltmeter: (a) original high-resistance circuit, normal voltage division;
(b) 1-kΩ/V voltmeter across R_2; (c) R_2 equivalent resistance is 50 kΩ, and R_2 voltage drop lowers to 33.3 V.

Lab Reference: The effects of voltmeter loading are demonstrated in Exercise 17.

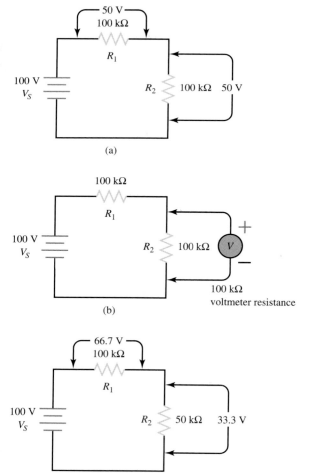

The loading effect is minimized by using a voltmeter with a resistance much greater than the resistance across which the voltage is measured (the 10-to-1 rule). This is illustrated in Figure 10.13. Here a 20-kΩ/V voltmeter is placed across a 100-kΩ resistor. The voltmeter range is again selected as 100 V, the value of the voltage source. Voltmeter resistance is 100 × 20 kΩ/V, or 2000 kΩ (2 MΩ). 2 MΩ in parallel with 100 kΩ results in an equivalent resistance practically equal to 100 kΩ. The loading effect is negligible, and the voltage division of the circuit remains unchanged.

Because the multiplier varies with the range selected, a higher range requires a higher multiplier resistance, thus increasing voltmeter resistance for less loading. A digital voltmeter has a resistance of 10 MΩ or more on all ranges and thus has practically no loading effect as a voltmeter.

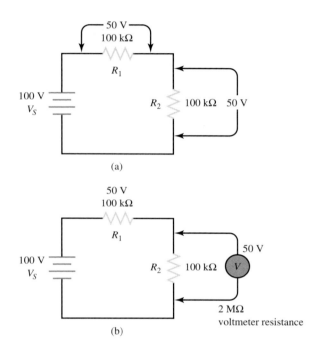

FIGURE 10.13
Negligible loading effects with a high-resistance voltmeter: (a) original high-resistance circuit, normal voltage division; (b) normal voltage division with high-resistance voltmeter connected to R_2.

Some rules to follow when measuring voltage are as follows:

1. Always set the range selector to the highest voltage range scale first and then reduce as needed.
2. Observe polarity as with the ammeter.
3. The voltmeter must be connected across or in parallel with the component voltage to be measured. No circuit breaks need be made.
4. As with the ammeter, voltage readings should be made as close to full scale as possible.
5. As with the ammeter, line up the pointer with its mirror reflection in order to prevent parallax error.

Practice Problems 10.5 Answers are given at the end of the chapter.
1. A voltmeter that has a 10-kΩ/V sensitivity and is on the 10-V range is placed across R_2 in Figure 10.14. What voltage drop is indicated by the voltmeter?

FIGURE 10.14
Circuit for Practice Problems 10.5, Problem 1.

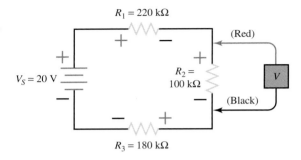

10.5 OHMMETERS

The basic meter movement can also be used to measure resistance. The resulting circuit is called an *ohmmeter*. In its basic form, the ohmmeter is nothing more than a meter movement, a battery, and a series resistance. The basic circuit of an ohmmeter is illustrated in Figure 10.15.

FIGURE 10.15
A basic ohmmeter.

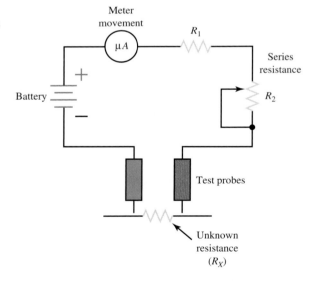

An ohmmeter forces a current to flow through an unknown resistance and then measures the resulting current. For a given voltage, the current is determined by the unknown resistance. That is, the amount of current measured by the meter is an indication of the unknown resistance. Thus, the scale of the meter movement is calibrated in ohms.

The purpose of the battery is to force current through the unknown resistance. Note that this is an *internal* voltage source. A circuit voltage source *is not used* when measuring resistance. The meter movement is used to measure the resulting current. The test probes provide a convenient method of connecting the ohmmeter to the unknown resistor (R_X). Fixed resistor R_1 limits the current through the meter to a safe level. Variable resistor R_2 is called the ZERO OHMS adjustment. Its purpose is to compensate for battery aging.

When using an ohmmeter, all voltage sources must be disconnected from the circuit under test. **Never use an ohmmeter on a circuit that has voltage applied to it.** When using the ohmmeter, it is first necessary to calibrate, or "zero," it. This is shown in Figure 10.16(a). When you short the leads of the ohmmeter together, the resistance value being measured is 0 Ω. The meter pointer deflects to FSD. This is the zero (0) indication of the meter. Notice that this is on the *right side* of the scale. At times, you may have to adjust the OHMS ADJUST potentiometer (R_2) until the meter reads full scale. This is known as calibrating the meter. This is necessary whenever the battery loses voltage through aging, or in the case of a multirange ohmmeter, when the range is changed. When the probes are apart, an open circuit exists; thus, there is no deflection of the meter, because no current can flow in the circuit. This point on the left end of the scale is calibrated as infinite ohms (∞). This is illustrated in Figure 10.16(b). When measuring other values of resistance, the meter deflects to points between zero and infinity. The meter scale is calibrated to indicate the value of resistance being measured. Additional ohmmeter ranges can be added with additional series resistances and a rotary switch. For higher resistance ranges, an additional (higher) battery voltage is used.

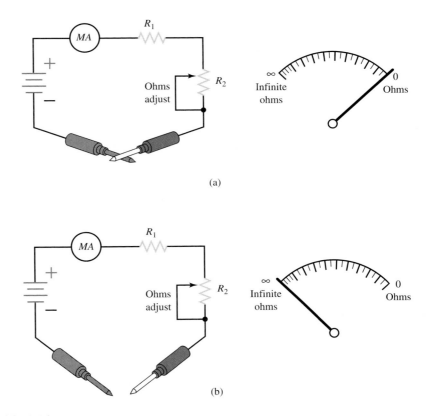

FIGURE 10.16
To zero the ohmmeter before using it: (a) short the leads together, adjusting for zero as necessary; (b) open the leads, no deflection; the left end of the scale is infinite ohms.

Notice that the ohmmeter scale is referred to as a *nonlinear* scale, as opposed to the voltmeter and ammeter scales, which are *linear.* Because zero is on the right end of the scale on an ohmmeter, this is also referred to as a *back-off scale.* Figure 10.17 shows the differences between nonlinear and linear scales. An ohmmeter scale is shown (utilizing a 0–1-mA movement and an internal 1.5-V battery) with relative resistance values at various portions of FSD (Figure 10.17(a)). Note that at every one-fourth FSD spacing, the resistance change is not uniform or linear. Thus, this scale is nonlinear, whereas the ammeter or voltage scale is linear (Figure 10.17(b)).

FIGURE 10.17
Nonlinear versus linear scale: (a) nonlinear ohmmeter scale; (b) linear scale for voltmeter or ammeter.

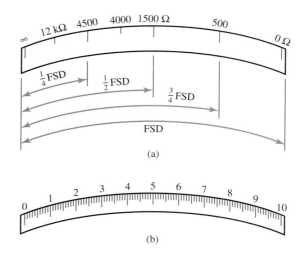

There are a few points to remember when measuring resistance with an ohmmeter:

1. Connect or short the test leads together and adjust the zero-adjust control so that the pointer is a 0 Ω on all scales. This must be repeated if you change ranges.
2. Ensure that power is removed or turned off to the circuit to be measured. Otherwise, damage could occur to the ohmmeter.
3. Connect the leads across the component and read off the resistance indicated on the scale, multiplying this value by 1, 10, 1000, etc., depending on the range selected.
4. If the component under test has a parallel connection of another component, an invalid reading may be obtained due to the parallel combination of resistances. To overcome this problem, isolate one end of the component under test from the circuit.

SECTION 10.5 SELF-CHECK

Answers are given at the back of the chapter.

1. The basic ohmmeter is made from a _____, a _____, and a _____.
2. The ohmmeter scale is _____.
3. A zero resistance is indicated on the _____ side of an ohmmeter.

10.6 THE ANALOG MULTIMETER (VOM)

The *analog multimeter* combines the ammeter, voltmeter, and ohmmeter into one unit. The name multimeter comes from the term multiple meter. It is also commonly called a *VOM* (volt-ohm-milliammeter). The VOM is a DC ammeter, an AC ammeter, a DC voltmeter, an AC voltmeter, and an ohmmeter all in one package. An example of a VOM is shown in Figure 10.18. Two well-known manufacturers of the VOM are the Simpson Co. and the Triplett Co.

Most multimeters employ a D'Arsonval meter movement and have a built-in rectifier for AC measurement. The lower portion of the meter contains the function switches and jacks. The function switch selects the appropriate circuitry inside the meter and also the desired range by changing the multiplier, shunt, etc. All the functions work just like the meters previously described.

The common (−) and positive (+) jacks are normally used for the meter probes; there are a number of other jacks present that can be used to extend the range of the instrument beyond those normally selected by the range switch. The RESET is a circuit breaker used to protect the meter movement. Not all VOMs have this form of protection and use internal fuses. When not in use, the multimeter's leads should be disconnected and it should be switched to the highest voltage scale.

FIGURE 10.18
Typical analog multimeters (VOM) (a Simpson model 260-8xi and a 260-8). Courtesy of Simpson Electric Co.

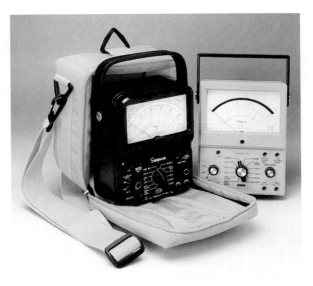

Table 10.1 summarizes the main points to remember when using the ammeter, voltmeter, or ohmmeter. These rules apply whether the meter is a single unit or one function of a VOM.

TABLE 10.1
Summary of analog meter use

Ammeter	Voltmeter	Ohmmeter
Power on in circuit	Power on in circuit	Power off in circuit
Connect in series	Connect in parallel	Connect in parallel, isolate as necessary
Low internal resistance	High internal resistance	Has internal battery
Has internal shunts, lower resistance on higher current ranges	Has internal multipliers, higher resistance on higher ranges	Higher battery voltage and more sensitive meter for higher ohms ranges
Be aware of parallax error	Be aware of parallax error	Be aware of parallax error

SECTION 10.6 SELF-CHECK

Answers are given at the end of the chapter.

1. What do the letters VOM stand for?
2. The reset function of a VOM is performed by a _____ _____.

10.7 THE DIGITAL MULTIMETER (DMM)

The *digital multimeter* (DMM) has become more popular than the VOM as a test instrument. This is because of its ease of use, decimal display, accuracy, and cost. A digital value of the measurement is displayed automatically with decimal point, polarity, and the unit for V, A, or Ω. Examples of DMMs are shown in Figure 10.19.

The heart of the DMM is an analog-to-digital (A/D) converter. It converts analog values at the input to an equivalent digital (binary) form. These values are processed by digital circuits to be shown on a liquid-crystal display (LCD) as decimal digits. The

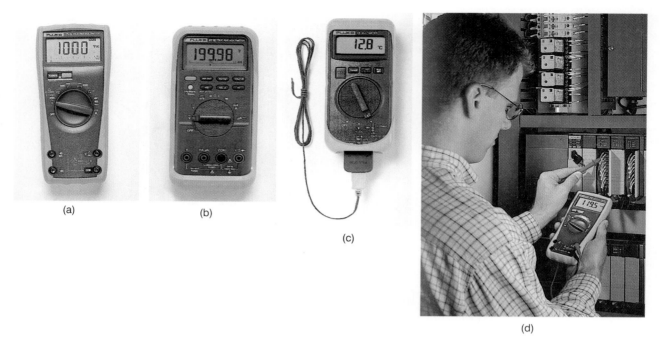

(a)   (b)   (c)   (d)

FIGURE 10.19
Examples of DMMs: (a)–(c) Examples of digital multimeters; (d) a technician using a DMM.
Courtesy of Fluke Corp. Reproduced with permission.

input resistance of the DMM is typically 10 MΩ on all ranges. This is high enough to eliminate voltage loading effects. For AC measurements a rectifier is used to convert the value to DC. The frequency range of the meter for AC measurements is limited to 45 to 1000 Hz. For higher-frequency measurements, a special meter is needed.

Other advantages of DMMs include automatic overrange indication, overload protection, and even automatic range selection. DMM accuracy refers to an indication of the maximum error that can be expected between the actual signal being measured and the reading indicated by the meter. *Resolution* is the ability to display the difference between values. In other words, it is a ratio of the minimum value that can be displayed to the maximum value that can be displayed on a given range. In most DMMs, the left, or most significant, digit is known as a half-digit because it can display only a 0 or 1. Consequently, a DMM with four digits is often called a $3\frac{1}{2}$-*digit DMM*. A $3\frac{1}{2}$-digit DMM can display 19.99 V, but 29.99 V is displayed as 30.0 V.

A drawback to the DMM is the fact that it must convert the signal being measured to a digital form utilizing a sample-and-hold technique. Thus, the display is not a real-time display of the actual value being measured. To compensate for this, many DMMs utilize a LCD bar graph across the bottom of the display. The bar graph display updates 25–40 times per second compared to the digital display updates of 2.5–4 times per second. The bar graph shows the relative magnitude of the input compared to the full-scale value of the range in use. This is useful when adjusting a circuit for a peak value or a minimum (null). The bar graph thus emulates the pointer on a VOM.

There are many other types of measurements that DMMs can do. These include frequency measurement, capacitance or inductance testing, and transistor h_{FE} testing. Another extremely useful feature that most DMMs have needs mentioning. This is a diode-test position that measures the junction voltages of semiconductor diodes and transistors. Current supplied by the DMM to a semiconductor junction under test is read as a voltage drop by the DMM. Normal values are 0.6–0.7 V for silicon semiconductors and 0.2–0.3 V for germanium semiconductors. A short-circuit junction reads 0 V, whereas an open junction is indicated by an over-range indication (too high to measure, "O.L" on many meters).

Table 10.2 compares analog and digital meters for various parameters.

TABLE 10.2
Selected comparisons of analog and digital meters

	Analog	**Digital**
Accuracy	Meter reading is more accurate on the right side of the scale.	Extremely accurate for fixed signals.
Reading errors	Errors in meter reading can result from parallax.	Unlikely due to digital display.
Range selection	Manual.	Automatic or manual.
Loading effects	Readings can be severely affected.	Minimal.
Cost	Over $100 for lab quality units.	Under $100 for hand-held and portables.

SECTION 10.7 SELF-CHECK

Answers are given at the end of the chapter.

1. The heart of the DMM is an _____.
2. A DMM input resistance is typically _____.
3. Besides the conventional testing of the VOM, what other parameters can DMMs measure?

■ SUMMARY

1. Circuit measurement is used to monitor the operation of an electrical or electronic device or to determine the reason a device is not operating properly.

2. A meter movement converts electrical energy into mechanical energy.

3. With an analog readout meter, the amount of pointer deflection across a scale is an analog, or similar, to the magnitude of the electrical property being measured.

4. The basic meter movement encompasses a permanent-magnet moving-coil movement.

5. The basic meter movement was first used by D'Arsonval and later by Weston.

6. A rectifier is used so that AC can be measured by a meter movement.

7. An ammeter measures current. It is placed in series to the component through which the desired current reading is to be made.

8. Shunts are used to extend the range of a meter movement for measuring current.

9. An ammeter internal resistance should be small—ideally zero.

10. Full-scale deflection (FSD) is the amount of current required to cause the pointer to deflect full scale.

11. The smaller the amount of FSD current, the more sensitive the ammeter.

12. Current readings should be made as close to full scale as possible.

13. A voltmeter can be made from a meter movement by placing a multiplier resistance in series to the meter movement.

14. Voltmeter sensitivity is expressed in ohms/volt. It is found by simply taking the reciprocal of the meter-movement current sensitivity.

15. Voltmeter resistance is the sum of the multiplier and the meter-movement resistance.

16. Only the voltmeter resistance varies with voltage range, the sensitivity remains the same on all ranges.

17. A voltmeter loads the circuit when its resistance is not high enough and is more pronounced in high-resistance circuits.

18. An ohmmeter has its own internal voltage source.

19. An analog VOM gets its name from volt-ohm-milliammeter.

20. A digital multimeter (DMM) has become popular because of its ease of use, decimal display, accuracy, and cost.

■ IMPORTANT EQUATIONS

$$\frac{I_{SH}}{I_M} = \frac{R_M}{R_{SH}} \qquad \text{(10.1)}$$

$$R_{mult} = \frac{V_{full\text{-}scale}}{I_M} - R_M \qquad \text{(10.2)}$$

$$R_{mult} = \frac{V_{full\text{-}scale}}{I_M} - (R_M + \text{all previous } R_{mult} \text{ values}) \qquad \text{(10.3)}$$

$$\text{sensitivity} = \frac{1 \text{ V}}{I_{FSD}} \qquad \text{(10.4)}$$

■ REVIEW QUESTIONS

1. Describe the basic meter movement.
2. What are the purposes of the hairspring and the pointer in a basic meter movement?
3. What was the first instrument that employed the moving-coil meter?
4. Describe what is needed in order for the meter movement to measure alternating current.
5. What is *damping* and how is it reduced?
6. Compare the internal makeup of the electrodynamic movement to the moving-vane movement.
7. How is pointer deflection accomplished in the hot-wire and thermocouple meter movements?
8. Describe the ammeter and its use.
9. What is the purpose of the shunt in an ammeter?
10. How is an ammeter connected in a circuit?
11. Define ammeter sensitivity.
12. What is the advantage of an Aryton shunt?
13. A voltmeter employs a basic meter movement and a _____ resistance that is connected in _____ with the movement.
14. What is the relationship between voltmeter sensitivity and meter FSD current?
15. What is *parallax error*?
16. What makes up voltmeter resistance?
17. What is the relationship of voltmeter sensitivity and the voltage range employed?
18. Describe voltmeter loading effects.
19. Why must the voltage source be removed when measuring resistance?
20. The ohmmeter scale is _____.
21. On most ohmmeter scales, zero is on the _____.
22. Describe the internal makeup of the analog VOM.
23. Describe how the digital DMM works.
24. What is DMM accuracy? Resolution?
25. What is a drawback to the DMM?
26. Compare features of the DMM to the VOM.

■ CRITICAL THINKING

1. Describe how the basic meter movement works and what makes the indicator deflect.
2. Why should readings be made as close to full scale as possible when using analog meters?
3. For negligible loading effects, should the ammeter be used on lower current ranges or higher current ranges? Why?
4. Describe the relationship between voltmeter sensitivity and voltmeter resistance.
5. Describe the relationship between DMM resolution and the number of digits displayed for various readings.

■ PROBLEMS

1. What shunt resistances are required to extend the range of a meter movement having $I_M = 1$ mA and $R_M = 50 \ \Omega$ to current ranges of 2 mA, 5 mA, 10 mA, 20 mA, 50 mA, and 100 mA, respectively?
2. Calculate the total meter resistance for each range in Problem 1.
3. What shunt resistances are required to extend the range of a meter movement having $I_M = 25 \ \mu A$ and $R_M = 5 \ k\Omega$ to current ranges of 250 μA, 1 mA, 5 mA, and 25 mA, respectively?
4. Calculate the total meter resistance for each range in Problem 3.
5. For a meter movement having $I_M = 20 \ \mu A$ and $R_M = 2.5 \ k\Omega$, calculate the multiplier resistors needed for the following voltage ranges: 1 V, 2.5 V, 10 V, 25 V, and 100 V.
6. For a meter movement having $I_M = 100 \ \mu A$ and $R_M = 1 \ k\Omega$, calculate the multiplier resistors needed for the following voltage ranges: 1 V, 2.5 V, 10 V, 50 V, and 100 V.
7. Calculate the voltmeter sensitivity for a voltmeter that utilizes a 20-μA meter movement.
8. Calculate the voltmeter sensitivity for a voltmeter that utilizes a 400-μA meter movement.
9. Calculate the total voltmeter resistance of a voltmeter on the 100-V range utilizing a 10-mA meter movement.
10. Calculate the total voltmeter resistance for the voltmeter in Problem 6 for each range listed.
11. What must be the current sensitivity of a meter movement if it is to be used to produce a 50-kΩ/V voltmeter?
12. An analog voltmeter has a sensitivity of 50 kΩ/V. What is the total voltmeter resistance if the voltage range of the voltmeter is 30 V? What is the current sensitivity of the same meter?
13. A series circuit consists of a 500-kΩ and 1-MΩ resistance. The voltage source is 12 V. Calculate the voltage across the 500-kΩ resistance with no meters connected.
14. Calculate the voltage across the 500-kΩ resistance of Problem 13 if a 20-kΩ/V voltmeter on the 5-V range is used.
15. Calculate the voltage across the 500-kΩ resistance of Problem 13 if a 20-kΩ/V voltmeter on the 50-V range is used.
16. Calculate the voltage across the 500-kΩ resistance of Problem 13 if a typical DMM is used to measure the voltage.

■ ANSWERS TO PRACTICE PROBLEMS

PRACTICE PROBLEMS 10.1

1. $10 \, \Omega$
2. $2 \, \Omega$

PRACTICE PROBLEMS 10.2

1. $998 \, k\Omega$
2. $1.998 \, M\Omega$

PRACTICE PROBLEMS 10.3

1. $R_1 = 248 \, k\Omega$, $R_2 = 2.75 \, M\Omega$, $R_3 = 7 \, M\Omega$

PRACTICE PROBLEMS 10.4

1. $10 \, k\Omega/V$
2. $1 \, k\Omega/V$

PRACTICE PROBLEMS 10.5

1. $2.22 \, V$

■ ANSWERS TO SELF-CHECKS

SECTION 10.1 SELF-CHECK

1. A permanent-magnet moving-coil movement
2. They return the coil to its initial position when there is no current flow, and they make electrical connections to the coil.
3. Average
4. A rectifier
5. D'Arsonval

SECTION 10.2 SELF-CHECK

1. The permanent magnet is replaced by fixed coils.
2. Heating
3. Wattmeter

SECTION 10.3 SELF-CHECK

1. In series
2. A shunt
3. Ammeter sensitivity is the amount of current necessary to cause full-scale deflection of the ammeter pointer.
4. An optical error caused by observing the pointer from the side instead of straight on.

SECTION 10.4 SELF-CHECK

1. In series
2. Ω/V
3. When the measured voltage decreases compared to the voltage without the voltmeter
4. $100 \, \mu A$

SECTION 10.5 SELF-CHECK

1. Meter movement, battery, series resistance
2. Nonlinear
3. Right

SECTION 10.6 SELF-CHECK

1. Volt-ohm-milliammeter
2. Circuit breaker

SECTION 10.7 SELF-CHECK

1. Analog-to-digital converter
2. $10 \, M\Omega$
3. Frequency, capacitance, inductance, transistor h_{FE} testing

CHAPTER 11

KIRCHHOFF'S LAWS, VOLTAGE, AND CURRENT DIVIDERS

OBJECTIVES

After completing this chapter, you will be able to:

1. Describe Kirchhoff's voltage law (KVL) and apply it to practical circuits.
2. Describe the effects of series-aiding and series-opposing voltage sources.
3. Calculate voltage drops utilizing the voltage-divider rule.
4. Describe Kirchhoff's current law (KCL) and apply it to practical circuits.
5. Calculate the amount of current in parallel branches utilizing the current-divider rule.
6. Design a loaded voltage divider and explain its application.

INTRODUCTION

Many types of circuits have components that are not in series, in parallel, or in series-parallel. Such circuits are called *complex* circuits. The pattern the resistances make is not easily recognizable. A complex circuit may have two or more voltage sources applied in different branches. Another example is the unbalanced Wheatstone bridge. If the general rules of series or parallel cannot be applied or Ohm's law is too cumbersome, a new method of circuit analysis becomes necessary. The application of Kirchhoff's laws, as described in this chapter, and the network theorems of Chapter 12 are examples of new methods applied to these complex circuits. As will be shown, any circuit can be analyzed using Kirchhoff's laws because they do not depend on circuits drawn with series or parallel connections.

11.1   KIRCHHOFF'S VOLTAGE LAW (KVL)

In 1847, G. R. Kirchhoff extended the use of Ohm's law by developing a simple concept concerning the voltages in a closed circuit path. Kirchhoff's voltage law (KVL) states:

> The algebraic sum of the voltage sources and the voltage drops in any closed path (loop) in a circuit is equal to zero.

Any closed path is a *loop* and may be defined as a continuous connection of branches where current flow leaving a point in one direction travels through a loop in one direction and returns to its point of origin.

To state the KVL another way, the voltage drops and voltage sources in a circuit are equal at any given moment in time. If the voltage sources are assumed to have one sign (positive or negative) at that instant and the voltage drops are assumed to have the opposite sign, then the result of adding the voltage sources and voltage drops is zero. Because the sum of the voltage drops equals the source voltage, KVL is another way of stating Equation 7.3, $V_S = V_1 + V_2 + V_3 + \cdots + V_n$.

Through the use of KVL, circuit problems can be solved that would be difficult and often impossible with the knowledge of just Ohm's law alone. (This is even more evident in the next chapter.) When KVL is properly applied, an equation can be set up for a closed loop, and the unknown circuit values can be calculated.

11.1.1   Application of Kirchhoff's Voltage Law

KVL can be written as

$$V_S + V_1 + V_2 + V_3 + \cdots + V_n = 0 \tag{11.1}$$

where V_1, V_2, etc., are the voltage drops around any closed-circuit loop.

To set up the equation for an actual circuit, the following procedure is used.

Lab Reference: The use of Kirchhoff's voltage law is demonstrated in Exercise 18.

1. Assume a direction of current through the circuit. (The correct direction is desirable but not necessary.)
2. Using the assumed direction of current, assign polarities to *all resistors and sources* through which the current flows. Remember that electrons flowing into a resistor make that end negative with respect to the other end. Where the current leaves a battery terminal is considered the negative side.
3. When tracing a loop, consider any voltage whose negative $(-)$ terminal is reached first as a negative term and any voltage whose positive terminal is reached first as a positive term. This method applies to *both voltage drops and voltage sources.*
4. Starting at any point in the circuit, trace around the circuit, writing down the amount and polarity of the voltage across each component in succession.
5. Place these voltages, with their polarities, into the equation, and solve for the desired quantity as necessary.

Utilizing a couple of examples, let's prove that KVL works and see how KVL can be used to solve for an unknown voltage.

EXAMPLE 11.1

Prove that KVL is true for the circuit in Figure 11.1.

Solution   Using the rules for KVL, assume a direction for current and indicate the polarity of all voltages (see Figure 11.2). Starting at point *A*, trace around the circuit in the direction of assumed current, writing down the amount and polarity of each voltage in succession:

$$(-12 \text{ V}) + (-8 \text{ V}) + (25 \text{ V}) + (-5 \text{ V}) = 0$$

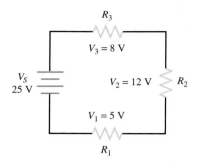

FIGURE 11.1
Circuit for Example 11.1.

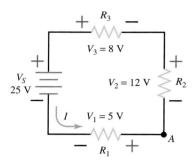

FIGURE 11.2
Solution for Example 11.1.

EXAMPLE 11.2

By utilizing KVL, find the unknown voltage drop across R_2 in Figure 11.3. Show the solution as Figure 11.4.

Solution Following the same procedures as in Example 11.1, we have

$$(-V_2) + (-7 \text{ V}) + (20 \text{ V}) + (-8 \text{ V}) = 0$$
$$-V_2 - 15 \text{ V} + 20 \text{ V} = 0$$
$$-V_2 + 5 \text{ V} = 0$$
$$-V_2 = -5 \text{ V}$$
$$V_2 = 5 \text{ V}$$

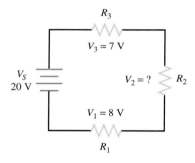

FIGURE 11.3
Circuit for Example 11.2.

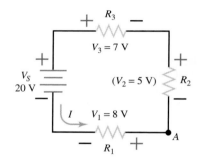

FIGURE 11.4
Solution for Example 11.2.

Practice Problems 11.1 Answers are given at the end of the chapter.
1. Using Figure 11.5, prove that KVL is true for the circuit.

FIGURE 11.5
Circuit for Practice Problems 11.1,
Problem 1.

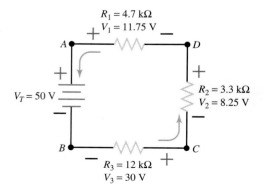

11.1.2 Series-Aiding and -Opposing Sources

In many practical applications (particularly those in Chapter 12) a circuit may contain more than one voltage source. Sources that cause current to flow in the same direction are considered to be series-aiding, and their voltages are additive. Sources that tend to force current in opposite directions are said to be series-opposing, and the effective source voltage is the *difference* between the opposing voltages. When two opposing sources are inserted into a circuit, current flow is in a direction determined by the larger source.

EXAMPLE 11.3

For the circuit in Figure 11.6 (having two sources), prove that KVL is true.

Solution   Following the same procedures as in Example 11.1, we have

$$(-5 \text{ V}) + (-5 \text{ V}) + (-5 \text{ V}) + (20 \text{ V}) + (-5 \text{ V}) = 0$$

See Figure 11.7.

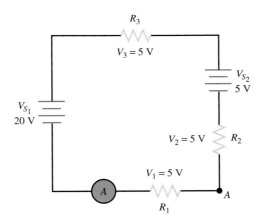

FIGURE 11.6
Circuit for Example 11.3.

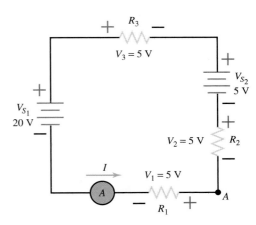

FIGURE 11.7
Solution for Example 11.3.

Practice Problems 11.2   Answers are given at the end of the chapter.
1. For the circuit in Figure 11.8, find V_2.

FIGURE 11.8
Circuit for Practice Problems 11.2,
Problem 1.

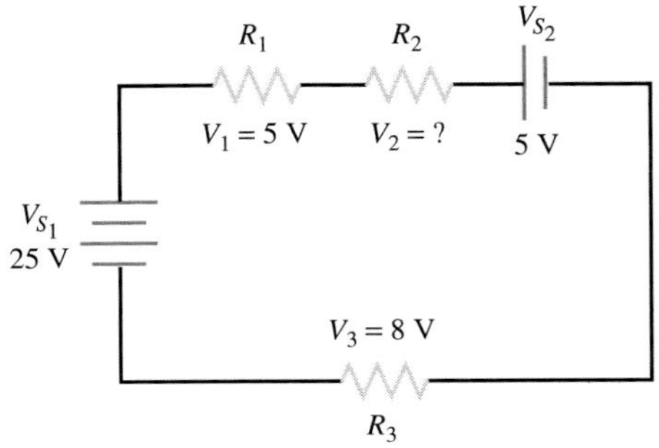

11.2 VOLTAGE DIVIDERS

Lab Reference: Voltage dividers are demonstrated in Exercise 18.

In a series circuit, the voltage across a given resistance has the same proportionality to the total voltage as the resistance has to the total resistance. The proportional method of determining a voltage drop is known as the *voltage-divider rule.* It is often easier to use the voltage-divider rule as calculating for current is not required. It may also allow the technician to approximate a voltage drop without the need for written calculations. This can save time when troubleshooting a circuit.

A useful application of the voltage-divider rule is to *voltage-divider circuits.* A voltage-divider circuit is used to get two or more different voltages from a common source voltage. Voltage dividers are popular in electronic circuits to provide the proper levels of voltage to transistors, integrated circuits, and the like.

Mathematically, the voltage-divider rule is stated in equation form as follows:

$$V_{RX} = \left(\frac{R_X}{R_T}\right)V_S \qquad (11.2)$$

where V_{RX} = voltage drop across a given resistance
R_X = given resistance
R_T = total resistance
V_S = source voltage

FIGURE 11.9
Potentiometer as a voltage divider.

The simplest type of voltage divider is the potentiometer. If a voltage source is placed across the potentiometer (Figure 11.9), an output voltage is obtained at the wiper (point *B*) and a reference. This output voltage represents a portion of the input voltage. The output voltage is taken either from point *A* to *B* or from point *B* to *C*. According to KVL, the sum of these two drops must equal the source voltage.

EXAMPLE 11.4

Use the voltage-divider rule to find the voltage drops across each resistor in the circuit shown in Figure 11.10.

FIGURE 11.10
Circuit for Example 11.4.

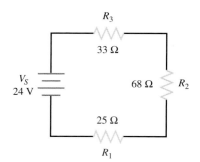

Solution $R_T = R_1 + R_2 + R_3$

$$V_{R_1} = \left(\frac{R_1}{R_T}\right)V_S$$

$$= \left(\frac{25 \ \Omega}{126 \ \Omega} \right) \times 24 \text{ V}$$

$$= 4.76 \text{ V}$$

$$V_{R_2} = \left(\frac{R_2}{R_T} \right) V_S$$

$$= \left(\frac{68 \ \Omega}{126 \ \Omega} \right) \times 24 \text{ V}$$

$$= 12.95 \text{ V}$$

$$V_{R_3} = \left(\frac{R_3}{R_T} \right) V_S$$

$$= \left(\frac{33 \ \Omega}{126 \ \Omega} \right) \times 24 \text{ V}$$

$$= 6.29 \text{ V}$$

According to KVL, the sum of these drops must equal V_S. You should check them to show that they do.

Practice Problems 11.3   Answers are given at the end of the chapter.
1. Find V_{R_1}, V_{R_2}, and V_{R_3} for the circuit in Figure 11.11.

FIGURE 11.11
Circuit for Practice Problems 11.3,
Problem 1. Find V_{R_1}, V_{R_2}, and V_{R_3}
using the voltage-divider rule.

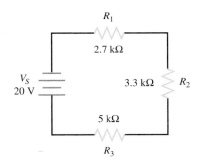

SECTION 11.2
SELF-CHECK

Answers are given at the end of the chapter.

1. Describe the task performed by a voltage divider.
2. What is the simplest type of voltage divider?

11.3   KIRCHHOFF'S CURRENT LAW (KCL)

The division of current in a parallel network follows a definitive pattern. This pattern is described by Kirchhoff's current law (KCL). The law states:

The algebraic sum of the currents entering and leaving any node is zero.

This law may be stated mathematically as

$$I_1 + I_2 + \cdots + I_n = 0 \tag{11.3}$$

where I_1, I_2, etc., are the currents entering or leaving the node.

Any equation based on Kirchhoff's current law is referred to as a *nodal equation*. Currents entering the node are considered to be positive ($+$) and the currents leaving the node are considered to be negative ($-$). When solving a problem using KCL, the currents must be evaluated with the proper polarity signs attached.

EXAMPLE 11.5

Solve for the value of I_3 in Figure 11.12.

Given $I_1 = 10$ A
$I_2 = 3$ A
$I_4 = 5$ A

Solution The currents are placed in the equation with the proper signs.

$$I_1 + I_2 + I_3 + I_4 = 0$$
$$10 \text{ A} + (-3 \text{ A}) + I_3 + (-5 \text{ A}) = 0$$
$$I_3 + 2 \text{ A} = 0$$
$$I_3 = -2.0 \text{ A}$$

I_3 has a value of 2 A, and the ($-$) sign shows it to be a current leaving the node.

FIGURE 11.12
Circuit for Example 11.5.

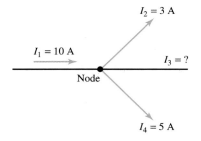

EXAMPLE 11.6

Using Figure 11.13, solve for the magnitude and direction of I_3.

Given $I_1 = 6$ A
$I_2 = 3$ A
$I_4 = 5$ A

Solution
$$I_1 + I_2 + I_3 + I_4 = 0$$
$$6 \text{ A} + (-3 \text{ A}) + I_3 + (-5 \text{ A}) = 0$$
$$I_3 - 2 \text{ A} = 0$$
$$I_3 = +2 \text{ A}$$

I_3 is 2 A, and its positive sign shows it to be a current entering the node.

FIGURE 11.13
Circuit for Example 11.6.

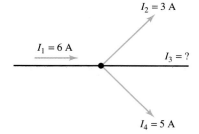

Practice Problems 11.4 Answers are given at the end of the chapter.

1. Using Figure 11.14 and the KCL, determine all unknown currents. Indicate the direction for all solved current utilizing an arrow indicator (for example, 5 A↑, 4 A →, or 3 A ←).

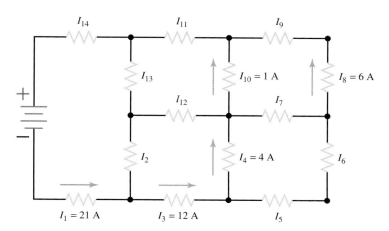

FIGURE 11.14
Circuit for Practice Problems 11.4, Problem 1.

| SECTION 11.3 SELF-CHECK | Answers are given at the end of the chapter.

1. State Kirchhoff's current law.
2. What is the polarity of current entering or leaving a node?

11.4 CURRENT DIVIDERS

It may be necessary to find individual branch currents in a parallel circuit when the total current and resistance are known but the source voltage is not. We know that currents divide inversely as the resistance of their paths; we can use this fact to determine the current division. Current always takes the path of least opposition, so current is always greatest through the path of least resistance. We may express this as the current-divider rule:

> The amount of current in one of two parallel branches is calculated by multiplying the total current by the opposite branch resistance divided by their resistive sum.

The equations for calculating currents through two branch circuits are

$$I_1 = \left(\frac{R_2}{R_1 + R_2}\right)I_T \tag{11.4}$$

$$I_2 = \left(\frac{R_1}{R_1 + R_2}\right)I_T \tag{11.5}$$

| EXAMPLE 11.7 | Using the current-divider rule, calculate the currents flowing through the branches of Figure 11.15.

Solution $I_1 = \left(\dfrac{R_2}{R_1 + R_2}\right)I_T \qquad I_2 = \left(\dfrac{R_1}{R_1 + R_2}\right)I_T$

$\qquad\qquad = \left(\dfrac{9}{6 + 9}\right) \times 50 \qquad = \left(\dfrac{6}{6 + 9}\right) \times 50$

$\qquad\qquad = 30 \text{ mA} \qquad\qquad\quad = 20 \text{ mA}$

FIGURE 11.15
Circuit for Example 11.7.

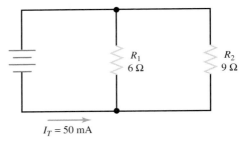

$I_T = 50$ mA

Note that equations (11.4) and (11.5) are valid only for a two-resistance parallel circuit. When a parallel circuit contains more than two resistances, the following equation is used to find an individual branch current:

$$I_X = \left(\frac{R_T}{R_X}\right)I_T \qquad\qquad\qquad \textbf{(11.6)}$$

where I_X = current through an individual branch
$\qquad R_X$ = resistance of an individual branch
$\qquad R_T$ = total resistance of the circuit

EXAMPLE 11.8

Using Figure 11.16 determine the current through R_2 using the current-divider equation.

Solution $R_T = \dfrac{1}{1/R_1 + 1/R_2 + 1/R_3}$

$\qquad\qquad = \dfrac{1}{1/510 \ \Omega + 1/270 \ \Omega + 1/430 \ \Omega}$

$\qquad\qquad = 125.2 \ \Omega$

$\qquad I_2 = \left(\dfrac{R_T}{R_2}\right)I_T$

$\qquad\qquad = \left(\dfrac{125.2 \ \Omega}{270 \ \Omega}\right) \times 300 \text{ mA}$

$\qquad\qquad = 139.1 \text{ mA}$

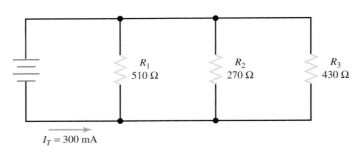

$I_T = 300$ mA

FIGURE 11.16
Circuit for Example 11.8.

Practice Problems 11.5 Answers are given at the end of the chapter.

1. Find the current in R_1 of Figure 11.17.

FIGURE 11.17
Circuit for Practice Problems 11.5, Problem 1.

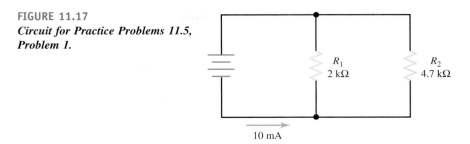

R_1
2 kΩ

R_2
4.7 kΩ

10 mA

2. Find the current in R_3 of Figure 11.18.

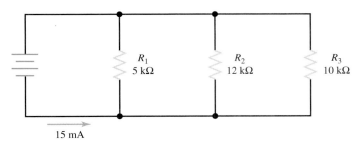

R_1
5 kΩ

R_2
12 kΩ

R_3
10 kΩ

15 mA

FIGURE 11.18
Circuit for Practice Problems 11.5, Problem 2.

**SECTION 11.4
SELF-CHECK**

Answers are given at the end of the chapter.

1. Currents divide _____ as the resistance of their paths.
2. Current always takes the path of _____ resistance.

11.5 LOADED VOLTAGE DIVIDERS

The voltage dividers shown in Section 11.2 are valid for a series-connected resistive circuit. This type of voltage divider provides two or more voltages obtained from a common voltage source. This type of circuit is called an *unloaded voltage divider*. However, if actual loads are placed across the voltage divider, the circuit becomes a series-parallel circuit, or a *loaded voltage divider*. Because of this configuration, the voltages and currents in the unloaded voltage divider change. Thus, to build a practical voltage divider we need to know the actual load requirements going in. In this way we can design the circuit to give us the proper values of voltage and current that the loads require. The number of taps, or sections, in the voltage divider depends on the number of loads required.

Figure 11.19 shows a practical loaded voltage divider. Three loads are connected to a common source, with each load requiring a different voltage and current for proper operation. The current through R_1 is called the *bleeder* current. It is the difference between the total current and the load current. Determining the bleeder current is the key to the start of the loaded voltage-divider design. This current does not flow through any of the loads; it is merely bled from the source. It is considered as a *loss* because it flows in the divider and not in a connected load. However, it does provide operational and safety requirements. Practical voltage sources (power supplies) have accumulated voltage charges on their filter capacitors that must be discharged before the technician performs routine maintenance or troubleshoots the circuit. Although high-voltage supplies are not as common as they used to be, there are high-current supplies that need to be discharged. The bleeder-current path provides for this.

Operationally, the amount of bleeder current that flows is a design compromise; if the current is too low, small variations in load current seriously affect the load voltage; if it is too high, large amounts of power are dissipated by the divider, resulting in an inefficient system. The bleeder current is generally about 10% to 25% of the total load current. (All examples in this chapter use a 10% value.)

FIGURE 11.19
A practical loaded voltage divider.

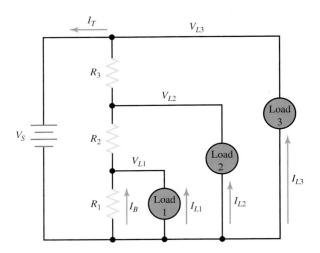

11.5.1 Designing a Loaded Voltage Divider

To design a loaded voltage divider, the load requirements must be known. This includes both their voltage and current requirements. By taking 10% of the total load current, we can ascertain the value of the bleeder current. The voltage drop across each divider resistor is found by the use of Kirchhoff's voltage law (KVL). The current through each divider resistor is found by the use of Kirchhoff's current law (KCL). The following example illustrates the design process using Kirchhoff's laws and Ohm's law.

EXAMPLE 11.9	Design a loaded voltage divider that will provide for the following loads. Load 1 requires 6 V and 20 mA, load 2 requires 9 V and 10 mA, and load 3 requires 12 V and 30 mA. The voltage source is 12 V. Assume a 10% bleeder current.

Solution For this circuit, three load voltages require three taps and three divider resistors.

$$I_{L_T} = I_{L_1} + I_{L_2} + I_{L_3} \quad \text{(Kirchhoff's current law)}$$
$$= 20\,\text{mA} + 10\,\text{mA} + 30\,\text{mA}$$
$$= 60\,\text{mA}$$
$$I_B = 10\% \text{ of } I_{L_T}$$
$$= 0.10 \times 60\,\text{mA}$$
$$= 6\,\text{mA}$$
$$I_T = I_B + I_{L_T} \quad \text{(Kirchhoff's current law)}$$
$$= 6\,\text{mA} + 60\,\text{mA}$$
$$= 66\,\text{mA}$$
$$V_1 = V_{L_1} = 6\,\text{V}$$

$$V_2 = V_{L_2} - V_{L_1} \quad \text{(Kirchhoff's voltage law)}$$
$$= 9\,V - 6\,V$$
$$= 3\,V$$

$$V_3 = V_{L_3} - V_{L_2} \quad \text{(Kirchhoff's voltage law)}$$
$$= 12\,V - 9\,V$$
$$= 3\,V$$

$$I_{R_1} = I_B = 6\,mA$$

$$I_{R_2} = I_{R_1} + I_{L_1} \quad \text{(Kirchhoff's current law)}$$
$$= 6\,mA + 20\,mA$$
$$= 26\,mA$$

$$I_{R_3} = I_{R_2} + I_{L_2} \quad \text{(Kirchhoff's current law)}$$
$$= 26\,mA + 10\,mA$$
$$= 36\,mA$$

$$R_1 = \frac{V_1}{I_{R_1}}$$
$$= \frac{6\,V}{6 \times 10^{-3}\,A}$$
$$= 1 \times 10^3\,\Omega$$
$$= 1\,k\Omega$$

$$R_2 = \frac{V_2}{I_{R_2}}$$
$$= \frac{3\,V}{26 \times 10^{-3}\,A}$$
$$= 115.4\,\Omega$$

$$R_3 = \frac{V_3}{I_{R_3}}$$
$$= \frac{3\,V}{36 \times 10^{-3}\,A}$$
$$= 83.3\,\Omega$$

See Figure 11.20.

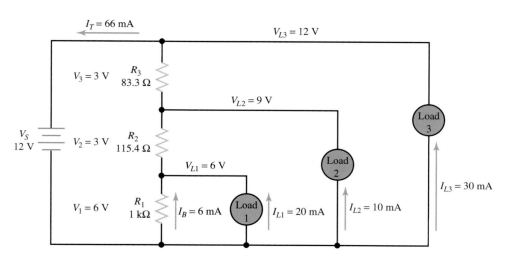

FIGURE 11.20
Solution to Example 11.9.

Using standard 5% resistors for a real circuit, we find that $R_1 = 1\ \text{k}\Omega$, $R_2 = 120\ \Omega$, and $R_3 = 82\ \Omega$. The next consideration is to find the power rating of the resistors in order to build a practical loaded voltage divider.

$$P_1 = I_{R_1}V_1$$
$$= (6 \times 10^{-3}\ \text{A}) \times 6\ \text{V}$$
$$= 36\ \text{mW}$$
$$P_2 = I_{R_2}V_2$$
$$= (26 \times 10^{-3}\ \text{A}) \times 3\ \text{V}$$
$$= 78\ \text{mW}$$
$$P_3 = I_{R_3}V_3$$
$$= (36 \times 10^{-3}\ \text{A}) \times 3\ \text{V}$$
$$= 108\ \text{mW}$$

For a safety factor, the power rating is doubled for each resistor. Using standard wattage values, R_1 requires a 72-mW (2×36-mW) resistor. The closest standard value is $\frac{1}{8}$ W (125 mW). R_2 requires 156 mW (2×78 mW). The closest standard value is $\frac{1}{4}$ W (250 mW). R_3 requires a 216-mW (2×108-mW) resistor. The closest standard value is $\frac{1}{4}$ W (250 mW).

Practice Problems 11.6   Answers are given at the end of the chapter.
1. Design a loaded voltage divider that provides for the following loads: load 1 is 50 V at 25 mA, load 2 is 70 V at 35 mA, and load 3 is 80 V at 30 mA ($V_S = 80$ V). Determine the values of R_1, R_2, and R_3. Assume a 10% bleeder current.

SECTION 11.5 SELF-CHECK

Answers are given at the end of the chapter.

1. What determines how many taps, or sections, are required by a loaded voltage divider?
2. Why is the bleeder current considered as a loss?
3. The bleeder current typically falls into what range of current?

■ SUMMARY

1. A complex circuit is one that contains resistances in arrangements that are not easily recognizable. In addition they may contain two or more sources in more than one branch.
2. Kirchhoff's voltage law (KVL) states that the algebraic sum of the voltage sources and the voltage drops in any closed path (loop) in a circuit is equal to zero.
3. In simple terms, KVL states that the sum of the voltage drops must equal the source voltage.
4. Sources that cause current to flow in the same direction are considered to be series-aiding, and their voltages are additive.
5. Sources that tend to force current in opposite directions are said to be series-opposing, and the effective source voltage is the difference between the opposing voltages.
6. A voltage-divider circuit is used to get two or more voltages from a common voltage source.
7. The simplest type of voltage divider is the potentiometer (pot).
8. Kirchhoff's current law (KCL) states that the algebraic sum of the currents entering and leaving any node is zero.
9. Currents entering a node are considered positive and those leaving a node are considered negative.
10. The current-divider rule can be used to solve for branch currents when only the total current and branch resistances are known.
11. The common series-connected voltage divider with loads attached is known as a loaded voltage divider.
12. The key to starting the design of a loaded voltage divider is to determine the bleeder current.
13. Both Kirchhoff's laws and Ohm's law are utilized in the design of a loaded voltage divider.
14. The normal bleeder current is generally about 10% to 25% of the total load current.
15. For a safety factor, the power rating is doubled for each resistor in the loaded voltage divider.

▪ IMPORTANT EQUATIONS

$$V_S + V_1 + V_2 + V_3 + \cdots + V_n = 0 \qquad (11.1)$$

$$V_{RX} = \left(\frac{R_X}{R_T}\right) V_S \qquad (11.2)$$

$$I_1 + I_2 + \cdots + I_n = 0 \qquad (11.3)$$

$$I_1 = \left(\frac{R_2}{R_1 + R_2}\right) I_T \qquad (11.4)$$

$$I_2 = \left(\frac{R_1}{R_1 + R_2}\right) I_T \qquad (11.5)$$

$$I_X = \left(\frac{R_T}{R_X}\right) I_T \qquad (11.6)$$

▪ REVIEW QUESTIONS

1. Describe a *complex* circuit. Give an example.
2. Describe Kirchhoff's voltage law (KVL).
3. How is KVL applied to resistive circuits?
4. Define the term *loop*.
5. How do you determine the polarity of each end of a resistor?
6. Describe series-aiding and series-opposing voltage sources.
7. Describe the voltage-divider rule.
8. Describe a voltage-divider circuit.
9. What is the simplest form of a voltage divider?
10. Where are voltage dividers used?
11. Describe Kirchhoff's current law (KCL).
12. Where is KCL used?
13. Describe the assignment of polarity to the currents entering or leaving a node.
14. Describe how current divides within a parallel circuit.
15. Describe the current-divider rule.
16. What is the difference between an unloaded voltage divider compared to a loaded voltage divider?
17. Describe the procedure to follow when designing a loaded voltage divider.
18. What does the term *bleeder current* refer to?
19. How is the level of bleeder current determined?
20. What function does the bleeder current provide?

▪ CRITICAL THINKING

1. Describe Kirchhoff's voltage law (KVL) and show how it can be used in complex circuits.
2. Describe Kirchhoff's current law (KCL) and show how it can be used in complex circuits.
3. Explain the role of bleeder current in a loaded voltage divider. Discuss the rationale for determining what level of bleeder current the circuit should use.

▪ PROBLEMS

1. Using Figure 11.21, determine V_3 by utilizing KVL.

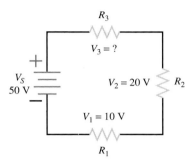

FIGURE 11.21
Circuit for Problem 1.

2. Using Figure 11.22, determine V_S by utilizing KVL.

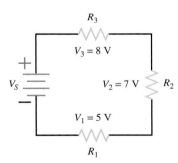

FIGURE 11.22
Circuit for Problem 2.

3. Using Figure 11.23, determine V_1 by utilizing KVL.

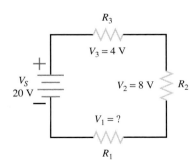

FIGURE 11.23
Circuit for Problem 3.

4. Using Figure 11.24, determine V_2 by utilizing KVL.

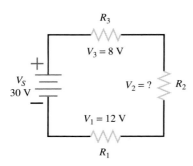

FIGURE 11.24
Circuit for Problem 4.

5. Using Figure 11.25, determine V_4 by utilizing KVL.

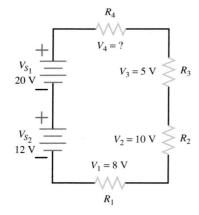

FIGURE 11.25
Circuit for Problem 5.

6. Using Figure 11.26, determine V_1 by utilizing KVL.

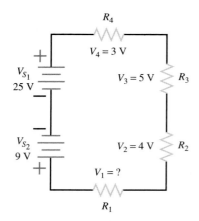

FIGURE 11.26
Circuit for Problem 6.

7. Using the voltage-divider rule find each resistor's voltage drop for the circuit in Figure 11.27.

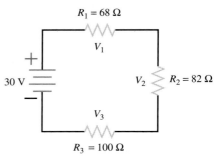

FIGURE 11.27
Circuit for Problem 7.

8. Using the voltage-divider rule, find each resistor's voltage drop for the circuit in Figure 11.28.

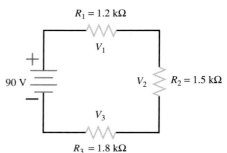

FIGURE 11.28
Circuit for Problem 8.

9. Using the voltage-divider rule, find each resistor's voltage drop for the circuit in Figure 11.29.

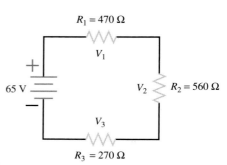

FIGURE 11.29
Circuit for Problem 9.

10. Using the voltage-divider rule, find each resistor's voltage drop for the circuit in Figure 11.30.

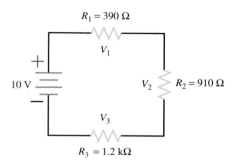

FIGURE 11.30
Circuit for Problem 10.

11. Using Figure 11.31, determine all unknown currents and indicate their direction with an arrow indicator (as in Practice Problems 11.4).

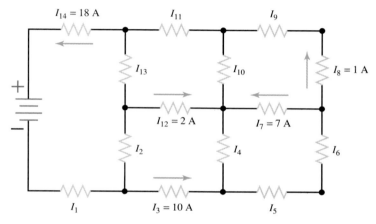

FIGURE 11.31
Circuit for Problem 11.

12. Using Figure 11.32, determine all unknown currents and indicate their direction with an arrow indicator (as in Practice Problems 11.4).

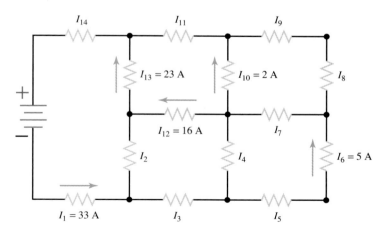

FIGURE 11.32
Circuit for Problem 12.

13. Using Figure 11.33, determine I_1 and I_2.

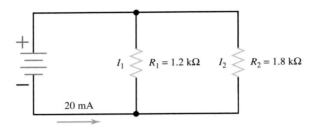

FIGURE 11.33
Circuit for Problem 13.

14. Using Figure 11.34, determine I_1 and I_2.

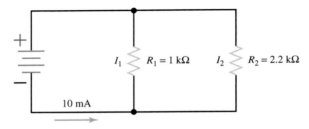

FIGURE 11.34
Circuit for Problem 14.

15. Using Figure 11.35, determine I_1, I_2, and I_3.

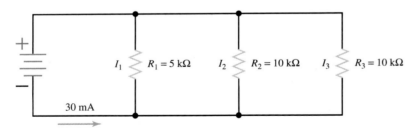

FIGURE 11.35
Circuit for Problem 15.

16. Using Figure 11.36, determine I_1, I_2, and I_3.

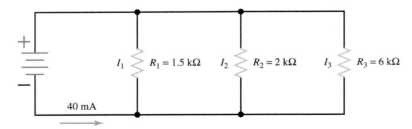

FIGURE 11.36
Circuit for Problem 16.

17. Design a loaded voltage divider with the following load requirements: load 1 is 10 V at 10 mA, load 2 is 12 V at 20 mA, and load 3 is 20 V at 15 mA. $V_S = 20$ V. Assume 10% bleeder current. Find R_1, R_2, and R_3.

18. Using 5% standard values, determine the closest values for the resistors in Problem 17.

19. Utilizing a safety factor by doubling the power rating, determine the proper wattage for each resistor in Problem 17.

20. Design a loaded voltage divider with the following load requirements: load 1 is 12 V at 9 mA, load 2 is 15 V at 6 mA, load 3 is 20 V at 20 mA, and load 4 is 30 V at 12 mA. $V_S = 30$ V.

■ ANSWERS TO PRACTICE PROBLEMS

PRACTICE PROBLEMS 11.1

1. From Point A (CCW), $(50) + (-30) + (-8.25) + (-11.75) = 0$.

PRACTICE PROBLEMS 11.2

1. $V_2 = 7$ V

PRACTICE PROBLEMS 11.3

1. $V_{R_1} = 4.9$ V, $V_{R_2} = 6$ V, $V_{R_3} = 9.1$ V

PRACTICE PROBLEMS 11.4

1. $I_2 = 9$ A↑, $I_5 = 8$ A→, $I_6 = 8$ A↑, $I_7 = 2$ A←, $I_9 = 6$ A←, $I_{12} = 5$ A←, $I_{13} = 14$ A↑, $I_{11} = 7$ A←, $I_{14} = 21$ A←

PRACTICE PROBLEMS 11.5

1. $I_1 = 7$ mA
2. $I_3 = 3.91$ mA

PRACTICE PROBLEMS 11.6

1. $R_1 = 5.6$ kΩ, $R_2 = 588$ Ω, $R_3 = 145$ Ω

■ ANSWERS TO SELF-CHECKS

SECTION 11.1 SELF-CHECK

1. A complex circuit is one where the pattern the resistors make is not easily recognizable as series, parallel, or series-parallel.
2. The sum of the voltage drops equals the source voltage.
3. Those that have currents flowing in same direction

SECTION 11.2 SELF-CHECK

1. To obtain two or more voltages from a common source
2. Potentiometer

SECTION 11.3 SELF-CHECK

1. KCL states that the algebraic sum of the currents entering and leaving any node is zero.
2. It is positive if entering a node and negative if leaving a node.

SECTION 11.4 SELF-CHECK

1. Inversely
2. Least

SECTION 11.5 SELF-CHECK

1. The number of loads required
2. Because it does not contribute to load current
3. 10–25% of load current (This text uses 10%.)

CHAPTER 12

NETWORK THEOREMS

OBJECTIVES

After completing this chapter, you will be able to:

1. Describe techniques that can be used to perform circuit analysis.
2. Describe the role that Kirchhoff's laws have in circuit analysis versus Ohm's law.
3. Differentiate between voltage sources and current sources and tell how to convert from one to the other.
4. State the merits of each of the following circuit-analysis tools: (a) superposition, (b) nodal analysis, (c) mesh current analysis, (d) Thevenin's theorem, (e) Norton's theorem, and (f) Millman's theorem.
5. Describe the importance of maximum power transfer.
6. Compare the load resistance required to accommodate maximum current, maximum voltage, or maximum power transfer.
7. Compare the level of efficiency achieved versus the degree of power transferred.

INTRODUCTION

Chapter 11 defined a complex circuit and demonstrated the use of Kirchhoff's laws to analyze those circuits. The primary purpose of circuit analysis is to determine the circuit response to a given voltage or current applied to the circuit. In many cases this involves determining the performance of an isolated portion of a complex circuit. This is done by replacing the remainder of the circuit with a simplified equivalent network. Thevenin's and Norton's theorems are examples of this technique.

Some circuit analysis can be done with just Ohm's law. However, most of the complex circuits in this chapter use the techniques of loop analysis and nodal analysis or the equivalent network, as previously described. These analyses are derived by the use of Kirchhoff's laws, which allow us to analyze circuits more quickly than with Ohm's law alone. Some circuits can't be analyzed by any method other than Kirchhoff's laws. The theorems found here represent the *real tools* used for circuit analysis.

The circuits in this chapter are defined as *linear* circuits. That is, they contain resistances driven by sources of *constant voltage* or *current* (defined in Section 12.1). In addition to being linear, the circuits must be *bilateral*. A bilateral component is one that has the same current flowing through it for opposite polarities of the voltage source.

Some analysis methods require the use of simultaneous linear equations. These circuits may require that we find two or three unknowns using two or three simultaneous equations. To solve simultaneous equations requires the use of intermediate algebra. The Sharp calculator has a built-in function to solve simultaneous equations.

12.1 VOLTAGE AND CURRENT SOURCES

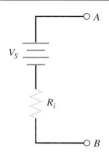

FIGURE 12.1
A voltage source with internal resistance. R_i is ideally zero.

12.1.1 Voltage Source

Before the various analysis methods can be demonstrated, it must be shown that a circuit can contain both a voltage source and a current source. You are familiar with voltage sources; the battery is an example. In an ideal condition, the battery produces a constant output voltage regardless of the value of a load resistance connected across its terminals. It makes no difference whether there is a heavy load (small load resistor) or a light load (large load resistor); the terminal voltage will remain constant.

In reality, every voltage source has a level of inefficiency caused by an internal resistance that generates heat. Under ideal conditions this internal resistance is assumed to be zero, but it is, in fact, several ohms. A voltage source with internal resistance is shown in Figure 12.1. The importance of this internal resistance is shown in the section on maximum power transfer later in this chapter.

With a load connected to the voltage source, the internal resistance of the source has a voltage drop. This, in turn, lowers the terminal voltage the source delivers to the load. This is illustrated in Figure 12.2. The smaller the load resistance, the greater the voltage drop across R_i.

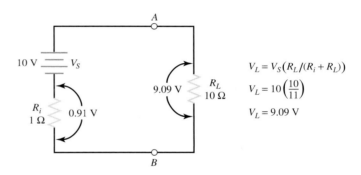

FIGURE 12.2
With a load connected to the voltage source, R_i has a voltage drop. This lowers the terminal voltage supplied to the load.

12.1.2 Current Source

Just as a voltage source has a certain voltage rating, the current source has a certain current rating. Likewise, just as a voltage source should deliver a constant voltage, an ideal current source should deliver a constant current, regardless of the value of load resistance connected across its terminals.

A current source can be thought of as a voltage source with an extremely high internal resistance (ideally infinite). In this way, the value of a load resistance connected across its terminals has little effect on the total resistance, and the load current remains constant. The symbol for a constant current source is shown in Figure 12.3. The arrow points in the direction of current (electron) flow. Solid-state devices are often analyzed with a current source rather than a voltage source. The only difference is that the arrow for solid-state devices indicates conventional current flow.

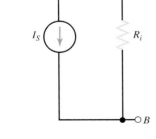

FIGURE 12.3
A current source. R_i is ideally infinite.

Practice Problems 12.1   Answers are given at the end of the chapter.
1. A 9-V battery develops a terminal voltage of 8.4 V when delivering a 1.2 A into a load resistance, R_L. Determine the values of R_i and R_L.

12.1.3 Source Conversions

A voltage source in series with a resistance can be converted to an equivalent current source in parallel with resistance, and vice versa. The sources are equivalent in the sense that when a load is connected between terminals *A* and *B* of either source, the same response is exhibited. However, without a load connected, the voltage-source circuit *does not* dissipate power, whereas the current-source circuit *does* dissipate power. Thus, the circuits *themselves* are not equivalent.

To convert the voltage source to an equivalent current source, simply divide the source voltage by its internal resistance. This calculation becomes the value of current for the current source. The required parallel resistance (for the current source) is the same as the initial series resistance of the voltage source. Figure 12.4 illustrates this process.

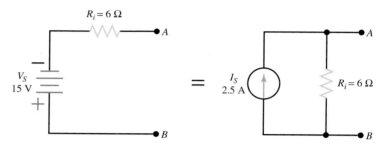

FIGURE 12.4
Voltage-source-to-current-source conversion.

**SECTION 12.1
SELF-CHECK**

Answers are given at the end of the chapter.

1. The ideal voltage source internal resistance is _____.
2. The ideal current source internal resistance is _____.

12.2 SUPERPOSITION THEOREM

The superposition theorem states that:

Lab Reference: The superposition theorem is demonstrated in Exercise 19.

> Each current and voltage in a linear circuit consisting of independent sources equals the algebraic sum of the currents and voltages produced by each source acting alone.

The term *independent sources* means that the terminal voltage of the voltage source does not depend upon the value of a current or a voltage anywhere in the circuit. Because power is not a linear quantity $(P = I^2R)$, you cannot use superposition directly to calculate power.

Circuit analysis by superposition can be done with Ohm's law. The following example demonstrates the superposition theorem.

EXAMPLE 12.1

Determine the current through R_1 of the circuit in Figure 12.5(a).

Solution   To begin, consider V_{S_1} (10-V source) acting alone. To do this, you must reduce all other sources to zero. This means any other voltage source is replaced by a short circuit (the voltage source's ideal internal resistance), as shown in Figure 12.5(b). This figure also shows that R_2 and R_3 are connected in parallel and have an equivalent resistance of 1.5 kΩ. Thus, the 10-V source circuit quickly reduces to the single-loop circuit of Figure 12.5(c). In this figure,

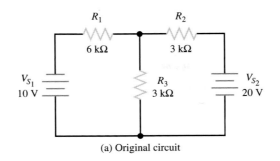

(a) Original circuit

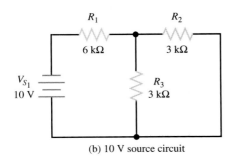

(b) 10 V source circuit

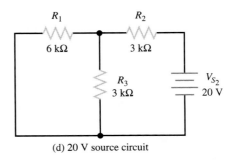

(d) 20 V source circuit

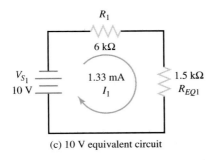

(c) 10 V equivalent circuit

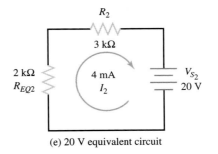

(e) 20 V equivalent circuit

FIGURE 12.5
Using the superposition theorem to solve a two-source circuit: (a) original circuit; (b) 10-V source circuit; (c) 10-V equivalent circuit; (d) 20-V source circuit; (e) 20-V equivalent circuit.

$$I_1 = \frac{V_{S_1}}{R_T}$$

$$= \frac{10 \text{ V}}{7.5 \text{ k}\Omega}$$

$$= \frac{10 \text{ V}}{7.5 \times 10^3 \text{ }\Omega}$$

$$= 1.33 \text{ mA} \rightarrow \qquad \text{(flowing clockwise)}$$

Next, consider the effects of V_{S_2} (20-V source) acting alone, as shown in Figure 12.5(d). As you can see, R_1 and R_3 connected in parallel and have an equivalent resistance of 2 kΩ. Figure 12.5(e) illustrates the resulting single-loop circuit. In this figure,

$$I_2 = \frac{V_{S_2}}{R_T}$$

$$= \frac{20 \text{ V}}{5 \text{ k}\Omega}$$

$$= \frac{20 \text{ V}}{5 \times 10^3 \text{ }\Omega}$$

$$= 4 \text{ mA}$$

Working backward from Figure 12.5(e) to 12.5(d), note that I_2 divides between R_1 and R_3. Using the current-divider rule,

$$I_{R_1} = \frac{I_2 R_3}{R_1 + R_3}$$

$$= \frac{(4 \text{ mA})(3 \text{ k}\Omega)}{9 \text{ k}\Omega}$$

$$= 1.33 \text{ mA} \rightarrow \qquad \text{(flowing clockwise)}$$

Thus, the actual current flowing through R_1 in the original circuit (Figure 12.5(a)) due to *both sources* is the algebraic sum of both clockwise currents.

$$I_{R_1} = 1.33 \text{ mA} + 1.33 \text{ mA}$$
$$= 2.66 \text{ mA} \rightarrow \qquad \text{(flowing clockwise)}$$

(Note that should either current flowing in R_1 be in the opposite direction to the other, I_{R_1} is simply the difference of the two.)

Practice Problems 12.2   Answers are given at the end of the chapter.
1. Using Figure 12.6, determine I_{R_2} using the superposition theorem.

FIGURE 12.6
Circuit for Practice Problems 12.1, Problem 1.

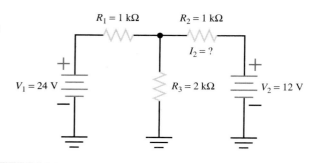

12.3   NODAL ANALYSIS

Nodal analysis is a technique where equations based on Kirchhoff's current law are written and unknown voltages are solved for. A node is a junction between two or more components. A *node voltage* is a voltage at a node with respect to a common reference point. The number of node voltages in a circuit is always less than the total number of nodes if one node is chosen as the reference or ground.

The reference, or ground, node is often chosen as the node connecting the largest number of components. The steps for solving circuit problems by nodal analysis are as follows:

1. Identify the number of nodes. The circuit in Figure 12.7(a) contains four nodes. They are identified as points *A*, *B*, *C*, and *D*.
2. Use one node as a reference node. In this case, node *B* is chosen as the reference. Points *C* and *D* are called dependent nodes because the voltage sources determine these voltages. Node *A* is referred to as an independent node because the size of the resistor

and the circuit configuration determine this node's voltage. The voltage between the independent node and the reference is the unknown voltage, V_X (Figure 12.7(b)).

3. Identify and assign a current direction at each independent node. The currents flowing into and out of node A are shown in Figure 12.7(c).

4. Use Kirchhoff's current law to write an equation for each independent node:

$$I_1 + I_2 + I_3 = 0$$

5. Use Ohm's law to express each current equation in terms of voltages:

$$I_1 = \frac{V_1}{R_1} = \frac{V_X - V_C}{R_1}$$

$$I_2 = \frac{V_2}{R_2} = \frac{V_X - V_D}{R_2}$$

$$I_3 = \frac{V_3}{R_3} = \frac{V_X}{R_3}$$

FIGURE 12.7
Using nodal analysis to solve a two-source circuit: (a) original circuit; (b) unknown voltage, V_X; (c) node currents.

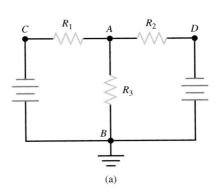

(a)

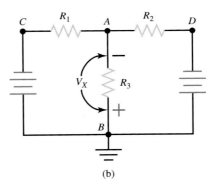

(b)

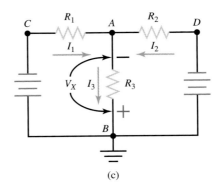

(c)

EXAMPLE 12.2

Use nodal analysis to find the voltage V_X for the circuit shown in Figure 12.8.

FIGURE 12.8
Circuit for Example 12.2.

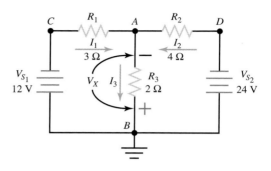

Solution   Apply KCL at node A:

$$I_1 + I_2 + I_3 = 0$$

Apply Ohm's law:

$$I_1 = \frac{V_1}{R_1} = \frac{V_X - V_C}{R_1} = \frac{V_X - 12}{3}$$

$$I_2 = \frac{V_2}{R_2} = \frac{V_X - V_D}{R_2} = \frac{V_X - 24}{4}$$

$$I_3 = \frac{V_3}{R_3} = \frac{V_X}{2}$$

Substitute the values obtained in the Ohm's law calculation into the KCL equation:

$$\frac{V_X - 12 \text{ V}}{3} + \frac{V_X - 24 \text{ V}}{4} + \frac{V_X}{2} = 0$$

Multiply both sides of the equation by 12:

$$4V_X - 48 \text{ V} + 3V_X - 72 \text{ V} + 6V_X = 0$$

Combine like terms and solve for V_X:

$$13V_X - 120 \text{ V} = 0$$

$$V_X = \frac{120 \text{ V}}{13}$$

$$= 9.23 \text{ V}$$

Because the circuit in Figure 12.8 has just one independent node, the solution involved only one unknown. Should a circuit have two independent nodes, two equations would be generated (with two unknowns). This solution requires the use of simultaneous linear equations. These equations are demonstrated in the next section. A circuit with two or more independent nodes is more easily solved by mesh current analysis (in light of the fact that the Sharp calculator can readily solve simultaneous linear equations).

Practice Problems 12.3   Answers are given at the end of the chapter.
1. Using nodal analysis, determine V_{R_3} in Figure 12.9.

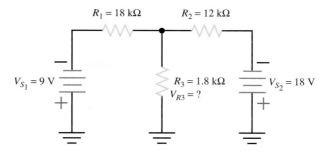

FIGURE 12.9
Circuit for Practice Problems 12.3, Problem 1.

SECTION 12.3 SELF-CHECK	Answers are given at the end of the chapter.
	1. Nodal analysis is based on _____.
	2. What is a node?

12.4 MESH-CURRENT ANALYSIS

A *mesh* is the simplest possible closed path. It differs from a loop in that a mesh has only one path. The current is assumed to flow around a mesh without dividing. Thus, the current does not divide at a branch point. A mesh current is an *assumed one,* whereas a branch current is the *actual* current path. However, when the mesh currents are known, all the individual currents and voltages can be easily determined. This is its advantage over other analysis methods.

This analysis differs from nodal analysis in that equations are written using Kirchhoff's voltage law to solve for unknown currents. The number of meshes equals the number of mesh currents, which is the number of equations required. Usually two or three equations are generated (having a commensurate number of unknowns). Solving these equations requires the use of simultaneous linear equations.

The steps for solving circuit problems using mesh-current analysis are as follows:

Lab Reference: Mesh currents are used to solve for unknowns in Exercise 20.

1. Assume the paths for the mesh currents. For consistency, each path is usually taken in the same direction. Generally, a clockwise direction is employed (and used throughout this chapter). Two mesh currents with labels I_A and I_B are shown for the circuit in Figure 12.10.

2. The first equation is generated by adding the voltage drops (*IR* drops) going around a mesh in the same direction as its mesh current. Any voltage drop produced by its own mesh current is considered positive and additive to the other voltage drops in the same mesh. The resistors within a mesh are considered to be in series.

3. Any resistance common to two meshes has two opposite mesh currents. It also has two opposing voltage drops. For one mesh it is positive, and it is negative for the opposing current of the adjacent mesh. R_3 in Figure 12.10 meets these criteria.

4. The first equation is completed by assigning the polarity of any voltage sources within the mesh current path. If the mesh current flows into the positive terminal first, as for V_1 in Figure 12.10, its quantity is considered positive. Should it flow into a negative terminal first, its quantity is considered negative. Should there be more than one voltage source within a mesh current path, a single voltage value is assigned using the series-aiding or series-opposing totals of the sources.

5. The second equation is generated following Steps 2 through 4 for the second mesh-current path.
6. After all needed equations are generated, the unknown mesh currents are solved using simultaneous linear equation rules. This may be by addition or subtraction of the equations, substitution, or the use of determinants.

When a mesh has no voltage source, the algebraic sum of the voltage drops within that mesh is zero. If the solution for a mesh current comes out negative, it means that the actual current for the mesh is in the opposite direction from the assumed current flow.

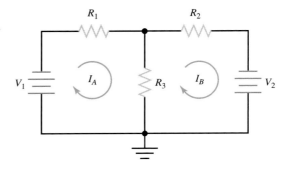

FIGURE 12.10
Mesh-current circuit showing mesh currents I_A and I_B.

EXAMPLE 12.3

Determine V_3 in the circuit of Figure 12.11. (*Note:* This is the same circuit used for nodal analysis.)

FIGURE 12.11
Circuit for Example 12.3.

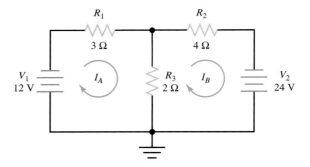

Solution Generate two equations based on I_A and I_B. Note that the equations generated are in the form of Kirchhoff's voltage law. That is, the sum of individual voltage drops equals the source voltage ($I_R + I_R + I_R = V_S$).

$$(3A + 2A) - 2B = 12$$

or
$$5A - 2B = 12 \quad \text{(first equation)}$$

$$-2A + (2B + 4B) = -24$$

or
$$-2A + 6B = -24 \quad \text{(second equation)}$$

Multiply the first equation by 3 and add the two equations.

$$\begin{array}{r} 15A - 6B = 36 \\ -2A + 6B = -24 \\ \hline 13A = 12 \end{array}$$

$$A = \tfrac{12}{13} = 0.923 \text{ A}$$

Substitute the value for *A* into either equation and solve for *B*.

$$15(0.923 \text{ A}) - 6B = 36 \text{ A}$$
$$13.85 \text{ A} - 6B = 36 \text{ A}$$
$$-6B = 22.15 \text{ A}$$
$$B = \frac{22.15 \text{ A}}{-6}$$
$$B = -3.69 \text{ A}$$

Because the current solution for *B* is negative, the assumed direction is wrong. Redraw the current path for I_B. V_3 is produced by two currents through R_3. Both I_A and I_B flow downward through R_3 and are additive. Find V_3 by Ohm's law.

$$\begin{aligned}
I_{R_3} &= I_A + I_B \\
&= 0.923 \text{ A} + 3.69 \text{ A} \\
&= 4.61 \text{ A} \\
V_3 &= I_{R_3} R_3 \\
&= 4.61 \text{ A} \times 2 \text{ }\Omega \\
&= 9.22 \text{ V}
\end{aligned}$$

Optional Calculator Sequence

The Sharp calculator has a built-in provision for solving simultaneous linear equations. This is labeled on the calculator as **3VLE** (and is accessed by selecting Mode 2), which stands for three-variable linear equations. Mesh currents meet this criterion. Note that in Example 12.3 two equations are needed. The Sharp calculator can accommodate up to three equations. If a third equation is not needed, simply use zero (0) for all terms representing third equation variables.

The calculator display prompts for each variable in the first equation as a1, b1, c1, and d1. a1, b1, and c1 represent the voltage drops; d1 represents the voltage source. In Example 12.3, a1 = 5, b1 = −2, c1 = 0, and d1 = 12. The prompts for the second equation are a2, b2, c2, and d2. Their respective values are a2 = −2, b2 = 6, c2 = 0, and d2 = −24. The calculator will prompt for the third equation. Since there is none, enter zero for all terms.

Should the resistances in a circuit be in kΩ, simply ignore the k when you generate the equations. Because of this shortcut, note that the solution for current flows are in mA (since V/kΩ = mA). By ignoring the k, it is easier to generate the equations and enter them on the calculator.

The calculator sequence for Example 12.3 using the Sharp calculator is as follows:

Step	Keypad Entry	Top Display Response	Step	Keypad Entry	Top Display Response
1	Mode	0-3	17	0	
2	2	a1?	18	= (enter)	d2?
3	5		19	24	
4	= (enter)	b1?	20	±	−24
5	2		21	= (enter)	a3?
6	±	−2	22	0	

(continued)

Step	Keypad Entry	Top Display Response	Step	Keypad Entry	Top Display Response
7	= (enter)	c1?	23	= (enter)	b3?
8	0		24	0	
9	= (enter)	d1?	25	= (enter)	c3?
10	12		26	0	
11	= (enter)	a2?	27	= (enter)	d3?
12	2		28	0	
13	±	−2	29	= (enter)	**x=** (bottom display) Answer **−0.923**
14	= (enter)	b2?	30	= (enter)	**y=** (bottom display) Answer **−3.69**
15	6		31	= (enter)	det= (bottom display) 26
16	= (enter)	c2?			

The value for I_A is $x = 0.923$ A, whereas the value for I_B is $y = -3.69$ A (the same answers arrived at manually). Note that in Step 31 the bottom display response is det = 26. This represents the constant used when solving simultaneous equations by determinants. The Sharp calculator solves simultaneous equations by the use of determinants. Because of this fact, the manual use of determinants is not done in this text.

Should the voltage drops of the other resistors in Figure 12.11 need to be known, the use of mesh currents furnishes the data needed to solve for all current values in the circuit. To find V_1 in Figure 12.11, it can be seen that only I_A flows through it. Thus,

$$V_1 = I_A R_1$$
$$= 0.923 \text{ A} \times 3 \text{ } \Omega$$
$$= 2.77 \text{ V}$$

To find V_2, it can be seen that only I_B flows through R_2. Thus,

$$V_2 = I_B R_2$$
$$= 3.69 \text{ A} \times 4 \text{ } \Omega$$
$$= 14.76 \text{ V}$$

It is left as an exercise for the student to prove that the voltages in either mesh are valid for KVL.

EXAMPLE 12.4

Solve for V_2 in the circuit of Figure 12.12. (Note this circuit requires three equations.)

Solution

$$7A - 3B + 0C = -18$$
$$-3A + 11B - 6C = 6$$
$$0A - 6B + 14C = 10$$

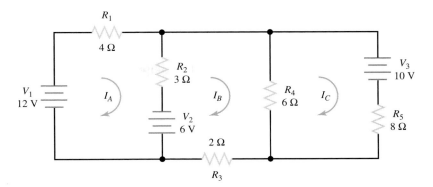

FIGURE 12.12
Circuit for Example 12.4.

Using the Sharp calculator, as in Example 12.3,

$$x = -2.42 \text{ A}$$
$$y = 0.36 \text{ A}$$
$$z = 0.87 \text{ A}$$

Remember that I_A is represented by the value of x, I_B by the value of y, and I_C by the value of z. The circuit is shown in Figure 12.13, with the actual currents labeled. Note that the direction for I_A has been corrected. As can be seen from this figure, R_2 has two series-aiding currents flowing through it. Thus,

$$I_{R_2} = I_A + I_B$$
$$= 2.42 \text{ A} + 0.36 \text{ A}$$
$$= 2.78 \text{ A}\uparrow \quad \text{(flowing upward)}$$
$$V_2 = I_{R_2}R_2$$
$$= 2.78 \text{ A} \times 3 \text{ }\Omega$$
$$= 8.34 \text{ V}$$

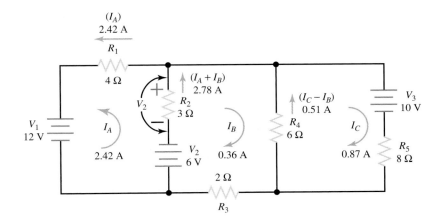

FIGURE 12.13
Solution for Example 12.4: solving for V_2. R_2 has two aiding currents flowing through it.

Practice Problems 12.4   Answers are given at the end of the chapter.
1. Using mesh currents, determine the correct loop equation for I_A and I_B for the circuit in Figure 12.14.
2. Using mesh currents, determine V_{R_1} for the circuit in Figure 12.15.
3. Using mesh currents, determine V_{R_3} for the circuit in Figure 12.16.

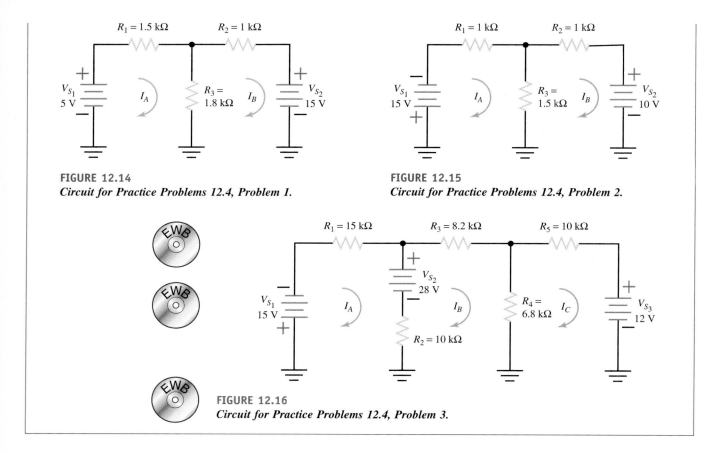

FIGURE 12.14
Circuit for Practice Problems 12.4, Problem 1.

FIGURE 12.15
Circuit for Practice Problems 12.4, Problem 2.

FIGURE 12.16
Circuit for Practice Problems 12.4, Problem 3.

SECTION 12.4 SELF-CHECK	Answers are given at the end of the chapter.

1. Define the term *mesh*.
2. The direction of current flow in a mesh is _____ .
3. Mesh equations are based on _____ .
4. Mesh equations require the use of _____ _____ _____ .

12.5 THEVENIN'S THEOREM

Thevenin's theorem (named after M. L. Thevenin, a French engineer) is one of the most important tools used to simplify the task of circuit analysis. This is especially true if you are mainly interested in obtaining the electrical response associated with a single component (such as a load resistance) and the load can vary in value.
   Thevenin's theorem states that:

Lab Reference: Various attributes of Thevenin's theorem are demonstrated in Exercise 21.

Everything in the original circuit, except the load, may be replaced by an equivalent circuit. The equivalent circuit consists of a series combination of a voltage source and a resistance.

   We are interested only in the responses produced in the load resistor connected to terminals *a* and *b* (see Figure 12.17). Everything to the left of terminals *a* and *b* is replaced by an equivalent voltage source (V_{TH}) and an equivalent resistance (R_{TH}). Thus, instead of having to reconfigure series and parallel combinations each time the load is varied, an equivalent circuit (sans the load) is created that consists of a simple series circuit.

The rules used in Thevenin's theorem are as follows:

To calculate V_{TH}:

1. Remove the load from the circuit.
2. Calculate the voltage between the terminals where the load was connected. This voltage is called the open-circuit voltage and equals V_{TH}.

To calculate R_{TH}:

1. Remove the load.
2. Reduce all sources to zero. This means voltage sources are replaced by short circuits (their ideal internal resistance) and any current sources by open circuits (their ideal internal resistance).
3. Calculate the resistance between the terminals where the load was connected (terminals a and b). This resistance equals R_{TH}.

The following examples illustrate how the rules are applied.

EXAMPLE 12.5

Calculate the voltage drop across R_L and the current through R_L in the circuit in Figure 12.17(a).

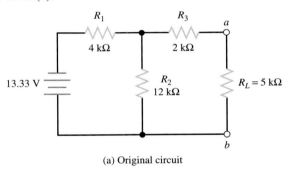

(a) Original circuit

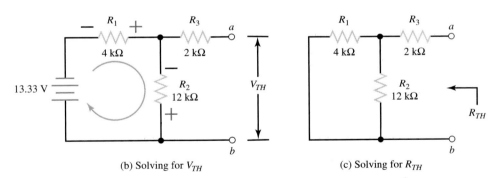

(b) Solving for V_{TH}

(c) Solving for R_{TH}

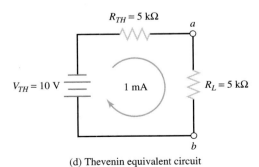

(d) Thevenin equivalent circuit

FIGURE 12.17
Circuits for Example 12.5: (a) original circuit; (b) solving for V_{TH}; (c) solving for R_{TH};
(d) Thevenin equivalent circuit.

Solution Remove the load as shown in Figure 12.17(b). The same current flows through R_1 and R_2, because they are in series. No current flows through R_3. The voltage between terminals a and b is V_{TH}. As can be seen in Figure 12.17(b), this is the same as V_2.

$$V_{TH} = V_2 = \frac{V_S(R_2)}{R_1 + R_2}$$

$$= \frac{13.33 \text{ V} (12 \text{ k}\Omega)}{(4 \text{ k}\Omega + 12 \text{ k}\Omega)}$$

$$= 10 \text{ V}$$

To calculate R_{TH}, remove the load and replace the voltage source with a short circuit, as shown in Figure 12.17(c). The resistance between terminals a and b in this figure is R_{TH}.

$$R_{TH} = R_{ab} = R_3 + R_1 \| R_2$$

$$= 2 \text{ k}\Omega + \frac{4 \text{ k}\Omega(12 \text{ k}\Omega)}{4 \text{ k}\Omega + 12 \text{ k}\Omega}$$

$$= 2 \text{ k}\Omega + 3 \text{ k}\Omega$$

$$= 5 \text{ k}\Omega$$

Figure 12.17(d) illustrates the Thevenin equivalent circuit with the 5-kΩ load resistor connected between terminals a and b. From this figure,

$$I_L = \frac{V_{TH}}{R_{TH} + R_L}$$

$$= \frac{10 \text{ V}}{10 \text{ k}\Omega}$$

$$= 1 \text{ mA}$$

$$V_L = I_L R_L$$

$$= 1 \text{ mA} \times 5 \text{ k}\Omega$$

$$= 5 \text{ V}$$

Should the load resistance change to 10 kΩ, simply reapply the series circuit rules and calculate V_L as

$$I_L = V_{TH}/R_{TH} + R_L \qquad \text{(with } R_L = 10 \text{ k}\Omega)$$

$$= 10 \text{ V}/15 \text{ k}\Omega$$

$$= 0.67 \text{ mA}$$

$$V_L = I_L R_L$$

$$= 0.67 \text{ mA} \times 10 \text{ k}\Omega$$

$$= 6.7 \text{ V}$$

(Note that the voltage-divider rule could be employed, as shown in Example 12.6, instead of solving via Ohm's law.)

EXAMPLE 12.6

Determine V_L of the load resistor in Figure 12.18(a). (You should recognize this circuit as an unbalanced Wheatstone bridge.)

Solution To determine V_{TH}, remove the load as shown in Figure 12.18(b). Note that once the load is removed, the two 10-kΩ resistors are connected in series, as are the 3-kΩ and 9-kΩ resistors. Applying the voltage-divider rule,

$$V_a = (20\ \text{V})\left(\frac{10\ \text{k}\Omega}{20\ \text{k}\Omega}\right) = 10\ \text{V}$$

$$V_b = (20\ \text{V})\left(\frac{3\ \text{k}\Omega}{12\ \text{k}\Omega}\right) = 5\ \text{V}$$

In Figure 12.17(b), $V_{TH} = V_{ab} = V_a - V_b$, so

$$V_{TH} = 10\ \text{V} - 5\ \text{V} = 5\ \text{V}$$

(Note that V_a is more positive than V_b.)

To determine R_{TH}, remove the load and reduce V_S to zero, as shown in Figure 12.18(c). This circuit is redrawn in Figure 12.18(d) to help visualize the equivalent resistance

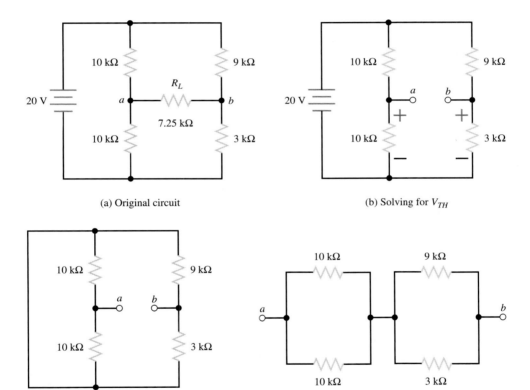

(a) Original circuit

(b) Solving for V_{TH}

(c) Solving for R_{TH}

(d) $R_{TH} = 10\ \text{k}\Omega \,\|\, 10\ \text{k}\Omega + 9\ \text{k}\Omega \,\|\, 3\ \text{k}\Omega$

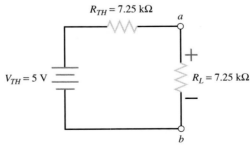

(e) Thevenin equivalent circuit

FIGURE 12.18
Circuits for Example 12.6: (a) original circuit; (b) solving for V_{TH}; (c) solving for R_{TH};
(d) $R_{TH} = 10\ k\Omega \| 10\ k\Omega + 9\ k\Omega \| 3\ k\Omega$; (e) Thevenin equivalent circuit.

between terminals a and b. Thus,

$$R_{ab} = R_{TH} = 10\ \text{k}\Omega\ \|\ 10\ \text{k}\Omega\ +\ 9\ \text{k}\Omega\ \|\ 3\ \text{k}\Omega$$
$$= 5\ \text{k}\Omega\ +\ 2.25\ \text{k}\Omega$$
$$= 7.25\ \text{k}\Omega$$

Figure 12.18(e) illustrates the Thevenin equivalent circuit. To find V_L using the voltage-divider rule,

$$V_L = \frac{V_{TH}(R_L)}{R_{TH} + R_L}$$
$$= \frac{(5\ \text{V})(7.25\ \text{k}\Omega)}{7.25\ \text{k}\Omega + 7.25\ \text{k}\Omega}$$
$$= 2.5\ \text{V}$$

Practice Problems 12.5 Answers are given at the end of the chapter.
1. Using Thevenin's theorem, find V_L in Figure 12.19.

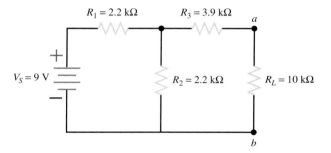

FIGURE 12.19
Circuit for Practice Problems 12.5, Problem 1.

2. Using Thevenin's theorem, find V_L in Figure 12.20.

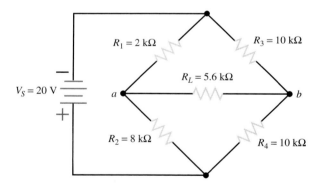

FIGURE 12.20
Circuit for Practice Problems 12.5, Problem 2.

SECTION 12.5 SELF-CHECK

Answers are given at the end of the chapter.

1. The equivalent Thevenin circuit is a _____ connected voltage source and a resistance.
2. True or false: Thevenin's theorem is especially useful if the load resistance varies in value.

12.6   NORTON'S THEOREM

Norton's theorem (named after E. L. Norton, a scientist at Bell Labs) is used for simplifying a network in terms of currents instead of voltages. It is the complement of Thevenin's theorem. Norton's theorem is stated as follows:

> Everything in the original circuit, except the load, may be replaced by an equivalent circuit. The equivalent circuit consists of a parallel combination of a current source and a resistance.

Everything to the left of the load (terminals a and b) is replaced by an equivalent current source (I_N) and an equivalent parallel resistance (R_N). The rules used in Norton's theorem are as follows.

To calculate I_N:

1. Remove the load from the circuit and replace it with a short circuit.
2. Calculate the current through the short circuit. This current is called the short-circuit current and equals I_N.

To calculate R_N:

1. Remove the load.
2. Reduce all sources to zero. This means voltage sources are replaced by short circuits and current sources are replaced by open circuits.
3. Calculate the resistance between the terminals where the load was connected. This resistance equals R_N.

The following example illustrates how the rules are applied.

EXAMPLE 12.7

Determine I_L, the current flowing through the load in the circuit of Figure 12.21. Note that this is the same figure used in Example 12.5 for Thevenin's theorem.

Solution   Remove R_L and replace it with a short (Figure 12.21(b)). To find the current in the short circuit (I_N), determine both R_T and I_T and then use the current-divider rule to get I_N.

$$R_T = R_1 + R_2 \| R_3$$
$$= 4\text{ k}\Omega + 12\text{ k}\Omega \| 2\text{ k}\Omega$$
$$= 4\text{ k}\Omega + 1.71\text{ k}\Omega$$
$$= 5.71\text{ k}\Omega$$

$$I_T = \frac{V_S}{R_T}$$
$$= \frac{13.33\text{ V}}{5.71\text{ k}\Omega}$$
$$= 2.33\text{ mA}$$

$$I_N = I_T\left(\frac{R_2}{R_2 + R_3}\right)$$
$$= (2.33\text{ mA})\left(\frac{12\text{ k}\Omega}{12\ \Omega + 2\text{ k}\Omega}\right)$$
$$= 2.33\text{ mA}\left(\frac{12\text{ k}\Omega}{14\text{ k}\Omega}\right)$$
$$= 2.00\text{ mA}$$

To get R_N, open the load, replace V_S with a short circuit, and find the resistance seen looking into terminals a and b (Figure 12.21(c)).

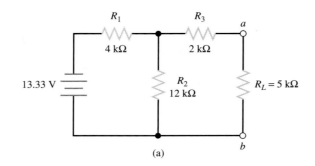

(a)

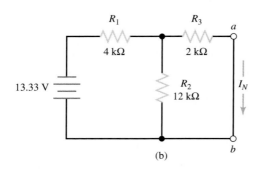

(b)

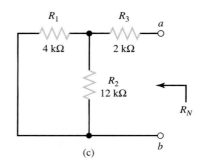

(c)

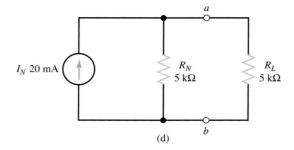

(d)

FIGURE 12.21
Circuits for Example 12.7: (a) original circuit; (b) finding I_N; (c) finding R_N; (d) Norton's equivalent circuit.

$$R_N = R_{ab} = R_3 + R_1 \| R_2$$

$$= 2 \text{ k}\Omega + \frac{(4 \text{ k}\Omega)(12 \text{ k}\Omega)}{4 \text{ k}\Omega + 12 \text{ k}\Omega}$$

$$= 2 \text{ k}\Omega + 3 \text{ k}\Omega$$

$$= 5 \text{ k}\Omega$$

To get I_L, apply the current-divider rule:

$$I_L = I_N \left(\frac{R_L}{R_N + R_L} \right)$$

$$= 2 \text{ mA}(5 \text{ k}\Omega/10 \text{ k}\Omega)$$

$$= 1.0 \text{ mA}$$

The Norton equivalent circuit for this example is shown in Figure 12.21(d). Note that R_N and R_{TH} are equal for the same circuit.

Practice Problems 12.6   Answers are given at the end of the chapter.
1. Using Norton's theorem, find I_L for the circuit in Figure 12.22.

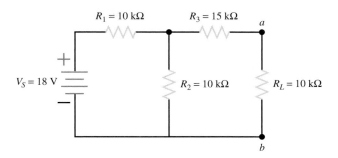

FIGURE 12.22
Circuit for Practice Problems 12.6, Problem 1.

12.6.1   Thevenin-Norton Conversions

A Thevenin's equivalent circuit can easily be converted to a Norton's equivalent circuit, and vice versa. The conversion process is illustrated in Figure 12.23.

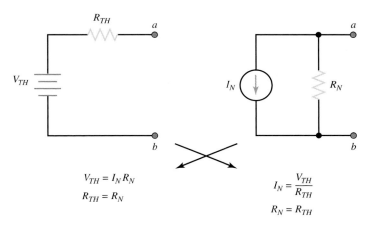

FIGURE 12.23
Thevenin-Norton conversions.

Answers are given at the end of the chapter.

1. Norton's theorem is the _____ of Thevenin's theorem.
2. A Norton's equivalent circuit consists of a _____ connected current source and a resistance.

12.7   MILLMAN'S THEOREM

Millman's theorem applies only to sources directly connected in parallel. This theorem provides a shortcut for finding the common voltage across any number of parallel branches with different voltage sources. A typical example is shown in Figure 12.24(a). (This circuit was previously solved utilizing other methods.) For all the branches, V_{AB} represents the common voltage across the branches. Figure 12.24(b) represents the same circuit redrawn for clarity to show each branch connected as a parallel circuit.

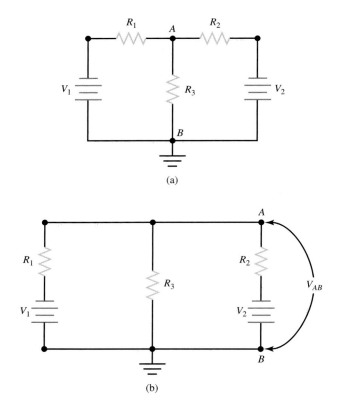

FIGURE 12.24
Using Millman's theorem to solve for V_{AB}: (a) original circuit; (b) circuit redrawn for clarity.

Finding V_{AB} gives the net effect of all the sources in determining the voltage at A with respect to B (ground). To calculate this voltage:

$$V_{AB} = \frac{\dfrac{V_1}{R_1} + \dfrac{V_2}{R_2} + \dfrac{V_3}{R_3} + \ldots + \dfrac{V_n}{R_n}}{\dfrac{1}{R_1} + \dfrac{1}{R_2} + \dfrac{1}{R_3} + \ldots + \dfrac{1}{R_n}} \tag{12.1}$$

The formula is derived from converting voltage sources to current sources and combining the results. The numerator, with V/R terms, is the sum of the parallel current sources. The denominator, with $1/R$ terms, represents the sum of the parallel conductances. Current divided by conductance is the same as current multiplied by resistance, or voltage, and thus represents V_{AB}.

EXAMPLE 12.8

Find V_{AB} (the same as V_3) for the circuit in Figure 12.25. This is the same circuit as in Figure 12.7 and was previously solved using both nodal and mesh analysis.

Solution   Substitute the circuit values into Equation 12.1.

$$V_{AB} = \frac{\dfrac{-12\text{ V}}{3\ \Omega} + \dfrac{0\text{ V}}{2\ \Omega} + \dfrac{-24\text{ V}}{4\ \Omega}}{\dfrac{1}{3\ \Omega} + \dfrac{1}{2\ \Omega} + \dfrac{1}{4\ \Omega}}$$

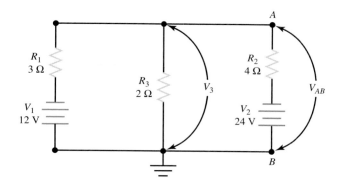

FIGURE 12.25
Circuit for Example 12.8.

$$= \frac{(-4 + 0 - 6)\ \text{V}/\Omega}{\dfrac{1}{3\ \Omega} + \dfrac{1}{2\ \Omega} + \dfrac{1}{4\ \Omega}}$$

$$= \frac{-10\ \text{V}/\Omega}{1.08\ \Omega}$$

$$= -9.23\ \text{V}$$

The negative sign indicates that point *A* is negative with respect to point *B*.

Practice Problems 12.7   Answers are given at the end of the chapter.
1. Using Millman's theorem, find V_{AB} for the circuit in Figure 12.26.

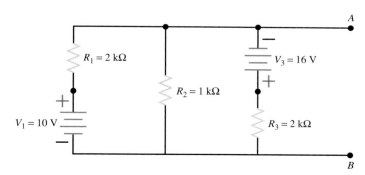

FIGURE 12.26
Circuit for Practice Problems 12.7, Problem 1.

SECTION 12.7 SELF-CHECK	Answers are given at the end of the chapter. **1.** Millman's theorem can be used only for _____ connected voltage sources.

12.8   MAXIMUM POWER TRANSFER THEOREM

Maximum power is transferred from the source to the load when the resistance of the load is equal to the internal resistance of the source. This theory is illustrated in Table 12.1 and

Lab Reference: The maximum
power transfer theorem is demonstrated, along with graphing from
measured quantities, in Exercise
22.

the circuit and graph of Figure 12.27. When the load resistance is 5 Ω, matching the source
internal resistance, a maximum power of 500 W is developed in the load.

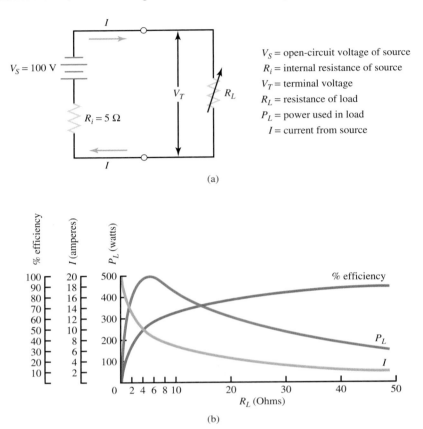

V_S = open-circuit voltage of source
R_i = internal resistance of source
V_T = terminal voltage
R_L = resistance of load
P_L = power used in load
I = current from source

(a)

(b)

FIGURE 12.27
*Maximum power transfer theorem: (a) circuit with variable load resistance; (b) graph of load
power (P_L), current (I), and % efficiency. (Note that at maximum load power, the efficiency is
50%.)*

The efficiency of power transfer (ratio of output power to input power) from the
source to the load increases as the load increases in resistance. The efficiency approaches
100% as the load resistance approaches a relatively large value compared with that of the
source, because less power is lost due to I^2R (the current has a small value). The efficiency of power transfer is 50% at the maximum power transfer point (when the load resistance equals the internal resistance of the source). The efficiency of power transfer approaches zero when the load resistance is relatively small compared with the internal
resistance of the source. This is also demonstrated on the chart.

When the need arises for both high efficiency and maximum power transfer, it is
resolved by a compromise between maximum power transfer and high efficiency. Where
the amounts of power involved are large and the efficiency is important, the load resistance is made large relative to the source resistance so that the losses are kept small. In
this case, the efficiency is high.

Where the problem of matching a source to a load is important, as in communications circuits (amplifier to speaker, transmission line to antenna, and the like), the transfer of a powerful signal is more important than a high efficiency. In such cases, the power
transfer is the maximum the source is capable of supplying when the source resistance
equals the load resistance. At this time the efficiency is about 50%.

The importance of having good efficiency is relative. The load cannot perform any
work if it doesn't receive current. Thus, a highly efficient circuit in a practical sense ac-
complishes little if no current is supplied to the load.

TABLE 12.1
Maximum power transfer chart

R_L (Ω)	V_T (V)	I (A)	P_L (W)	Eff. (%)
0	0	20	0	0
1	16.7	16.7	278.9	16.7
2	28.6	14.3	409	28.6
3	37.5	12.5	468.8	37.5
4	44.4	11.1	492.8	44.4
5	50	10	500	50
6	54.5	9.1	496	54.5
7	58.3	8.3	483.9	58.3
8	61.6	7.7	474.3	61.6
9	64.3	7.1	456.5	64.3
10	66.7	6.7	446.9	66.7
20	80	4	320	80
30	85.7	2.9	248.5	85.7
40	88.9	2.2	195.6	88.9
50	90.9	1.9	172.7	90.9

In the final analysis, the need of the load dictates the type of transfer that is made (regardless of efficiency). If a high voltage is required by the load, its resistance must be high compared to the source. If maximum power in the load is required, its resistance must match the source. If a high current is required by the load, its resistance must be small compared to the source.

**SECTION 12.8
SELF-CHECK**

Answers are given at the end of the chapter.

1. When does maximum transfer of power occur in a circuit?
2. The efficiency at maximum power transfer is about _____.
3. For high voltage transfer, the load resistance should be _____.
4. For high current transfer, the load resistance should be _____.

12.9   CHOOSING A THEOREM TO USE: A SUMMARY

Before you think that brain overload is imminent and that every theorem expressed in the chapter has to be mastered, let's review the merits of each theorem.

1. The superposition theorem is based solely on Ohm's law and the rules of series and parallel circuits. Advanced math skills are not needed. However, it is cumbersome and time consuming to analyze the circuit twice or more, once for each source acting alone.
2. Nodal analysis uses Kirchhoff's current law. The equations can be quickly generated, but the solution solves for just one unknown voltage and involves fractions. Advanced

math skills are not needed unless the circuit has two or more independent nodes; then the use of simultaneous equations is required.

3. Mesh analysis uses Kirchhoff's voltage law. The equations can be quickly generated and the solution solves for *all currents;* thus all circuit values can be found as well (using Ohm's law). Advanced math skills are needed, but with the availability of calculators such as the Sharp, the solutions to simultaneous equations are easy.

4. Thevenin's theorem involves the use of an equivalent voltage source and a series resistance circuit. Advanced math skills are not needed. The circuit values that are solved are useful for just the load. However, in designs where multiple load values occur, this theorem is very useful.

5. Norton's theorem involves the use of an equivalent current source and a parallel resistance circuit. It is simply the complement of Thevenin's theorem and more suited to engineering problems (particularly solid-state devices).

6. Millman's theorem does not require advanced math skills. However, it is limited to the form of the circuit. The circuit must have parallel connected sources.

7. The maximum power transfer theorem isn't technically a theorem but rather the source and load resistance requirements for transferring current, voltage, or power. It is an extremely important tool for anyone working in electronics.

Every text and every instructor has favorite tools for doing circuit analysis. That's why this chapter on circuit analysis is so important. In the author's opinion, you need proficiency in just three theorems in the chapter (particularly for students in technician-level programs). These include the mesh current theorem, Thevenin's theorem, and the maximum power transfer theorem. Because of the capabilities of the Sharp calculator (and its cost of under $20), mesh currents are easy to master, even for three equations. When utilizing this theorem, all the currents are known; therefore, all voltages and powers can be easily determined. The use of simplification of circuits with multiple load values makes Thevenin's theorem very useful. Finally, the maximum power transfer theorem is so important to circuit designs, it must be known by the technician. With a good understanding of these three theorems, virtually all circuits the technician is required to analyze can easily be done.

SECTION 12.9 SELF-CHECK

Answers are given at the end of the chapter.

1. Which theorems require the use of simultaneous linear equations?
2. Which theorems are complements of one another?

■ SUMMARY

1. Circuits can be analyzed having both voltage sources and current sources.

2. Voltage sources have an internal resistance that is small (ideally zero). Current sources have an internal resistance that is high (ideally infinite).

3. The superposition theorem states that each current and voltage in a linear circuit consisting of independent sources equals the algebraic sum of the currents and voltages produced by each source acting alone.

4. Nodal analysis is a technique by which equations based on KCL are written and unknown voltages are solved for.

5. A mesh is the simplest possible closed path and has only one path. A mesh current is an assumed one, whereas a branch current is the actual current path.

6. Mesh current analysis utilizes KVL and requires the use of simultaneous linear equations.

7. Simultaneous linear equations are solved by using addition, subtraction, substitution, or determinants.

8. Thevenin's theorem states that everything in the original circuits, except the load, is replaced by an equivalent circuit. The equivalent circuit consists of a series combination of a voltage source and a resistance.

9. Thevenin's theorem is useful for circuits that contain load values that vary.

10. Norton's theorem is the complement of Thevenin's theorem. The equivalent circuit consists of a parallel combination of a current source and a resistance.

11. Thevenin-to-Norton and Norton-to-Thevenin conversions are easily accomplished via Ohm's law.

12. Millman's theorem provides a shortcut for finding the common voltage across any number of parallel branches with different voltage sources.

13. Maximum power is transferred to the load when the source internal resistance and the load resistance are equal.

14. For voltage transfer, the load resistance should be large compared to the source resistance.

15. For current transfer, the load resistance should be small compared to the source resistance.

16. In the final analysis, the need of the load dictates the type of transfer and the level of efficiency that is made.

17. Nodal analysis containing two or more independent nodes and mesh-current analysis require the use of simultaneous linear equations.

■ IMPORTANT EQUATIONS

$$V_{AB} = \frac{\dfrac{V_1}{R_1} + \dfrac{V_2}{R_2} + \dfrac{V_3}{R_3} + \ldots + \dfrac{V_n}{R_n}}{\dfrac{1}{R_1} + \dfrac{1}{R_2} + \dfrac{1}{R_3} + \ldots + \dfrac{1}{R_n}} \qquad (12.1)$$

■ REVIEW QUESTIONS

1. The circuit analysis tools utilized in this chapter require a linear circuit. What is a linear circuit?

2. Define a bilateral circuit.

3. Describe a voltage source.

4. Describe a current source.

5. Which source dissipates power even with no load attached?

6. State the superposition theorem.

7. What is meant by an independent source?

8. Describe nodal analysis methods.

9. What are dependent versus independent nodes in nodal analysis?

10. When does nodal analysis require the use of simultaneous linear equations?

11. Define a mesh.

12. What is the difference between a mesh current and an actual current?

13. What does it mean when the solution to a mesh current is a negative quantity?

14. Describe Thevenin's theorem.

15. For what type of circuits is Thevenin's theorem most useful?

16. Describe Norton's theorem.

17. How do Thevenin's and Norton's theorems compare?

18. Describe Millman's theorem.

19. Millman's theorem can be used only for _____ connected voltage sources.

20. What are the requirements for maximum power transfer between source and load?

21. What determines the type of transfer that is made between source and load?

22. How does the level of efficiency compare to the type of transfer that is made between source and load?

■ CRITICAL THINKING

1. Show how mesh currents can be used to solve for unknowns in an unbalanced Wheatstone bridge circuit by finding I_L for the circuit in Figure 12.28.

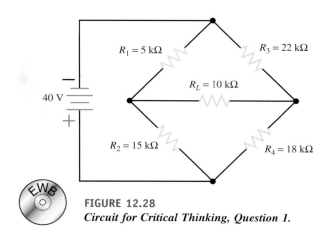

FIGURE 12.28
Circuit for Critical Thinking, Question 1.

2. Compare and contrast the features of the six circuit-analysis methods (exclude the maximum power theorem).

3. Compare and contrast the values of source resistance versus load resistance as to the type of transfer that is made and the level of efficiency achieved for that transfer.

■ PROBLEMS

1. A 1.5-V cell develops a terminal voltage of 1.25 V while delivering 250 mA into a load. Determine R_i and R_L.

2. Using Figure 12.29, convert the voltage source to its equivalent current source.

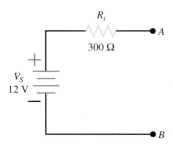

FIGURE 12.29
Circuit for Problem 2.

3. Using Figure 12.30, convert the current source to its equivalent voltage source.

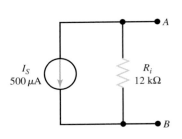

FIGURE 12.30
Circuit for Problem 3.

4. Using Figure 12.31, solve for V_{R_1} using the superposition theorem.

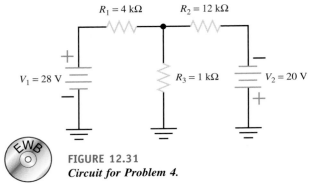

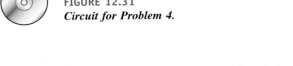

FIGURE 12.31
Circuit for Problem 4.

5. Using Figure 12.32, solve for V_{R_2} using nodal analysis.

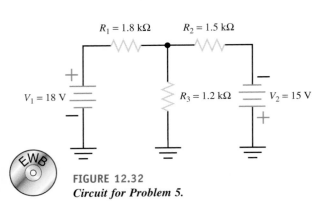

FIGURE 12.32
Circuit for Problem 5.

6. Using Figure 12.33, determine the correct loop equations for I_A and I_B.

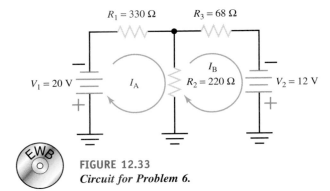

FIGURE 12.33
Circuit for Problem 6.

7. Using Figure 12.33, determine V_{R_1}, V_{R_2}, and V_{R_3}.
8. Using Figure 12.34, determine V_{R_1}, V_{R_2}, and V_{R_3} using mesh currents.

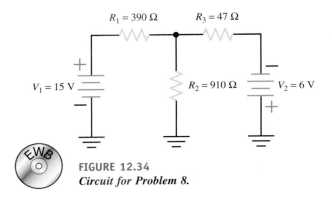

FIGURE 12.34
Circuit for Problem 8.

9. Using Figure 12.35, determine V_L using Thevenin's theorem.

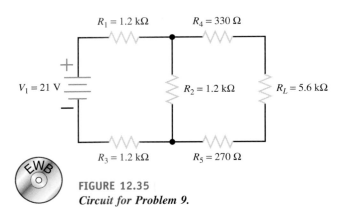

FIGURE 12.35
Circuit for Problem 9.

10. Using Figure 12.36, determine V_L using Thevenin's theorem.

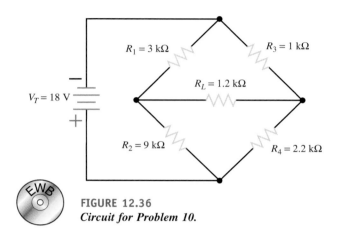

FIGURE 12.36
Circuit for Problem 10.

11. Using Figure 12.37, determine I_L using Norton's theorem.

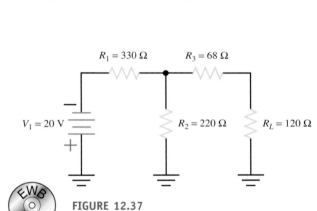

FIGURE 12.37
Circuit for Problem 11.

12. Using Figure 12.38, determine V_{AB} using Millman's theorem.

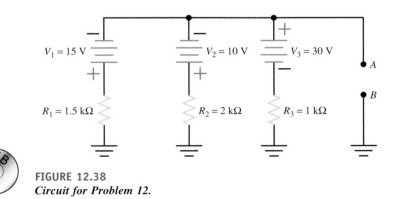

FIGURE 12.38
Circuit for Problem 12.

13. A 10-V source has an internal resistance of 100 Ω. Calculate I_L, V_L, P_L, P_T, and the efficiency power transfer for the following R_L values: 10 Ω, 25 Ω, 50 Ω, 100 Ω, 200 Ω, 500 Ω, and 1 kΩ.

For the remaining problems, determine all required unknowns using any theorem you choose.

14. Using Figure 12.39, find V_{R_2}.

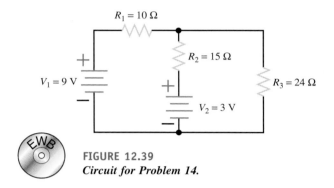

FIGURE 12.39
Circuit for Problem 14.

15. Using Figure 12.40, find V_{R_3}.

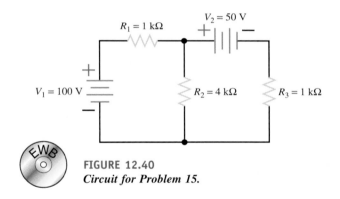

FIGURE 12.40
Circuit for Problem 15.

16. Using Figure 12.41, find V_{R_1}.

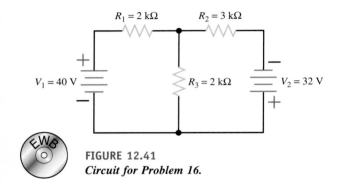

FIGURE 12.41
Circuit for Problem 16.

17. Using Figure 12.42, find V_{R_5}.

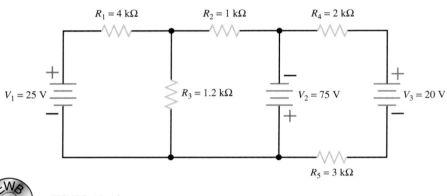

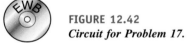

FIGURE 12.42
Circuit for Problem 17.

18. Using Figure 12.43, find V_{R_3}.

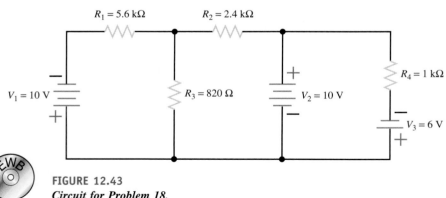

FIGURE 12.43
Circuit for Problem 18.

19. Using Figure 12.44, find V_{R_4}.

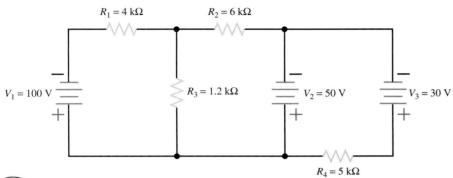

FIGURE 12.44
Circuit for Problem 19.

20. Using Figure 12.45, find V_L.

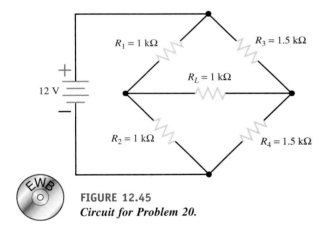

FIGURE 12.45
Circuit for Problem 20.

21. Using Figure 12.46, find V_L.

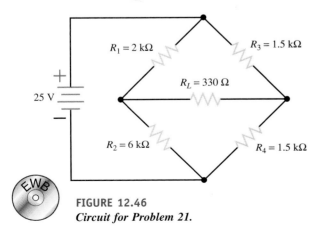

FIGURE 12.46
Circuit for Problem 21.

22. Using Figure 12.47, find V_L.

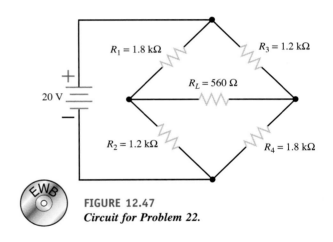

FIGURE 12.47
Circuit for Problem 22.

23. Using Figure 12.48, find V_L.

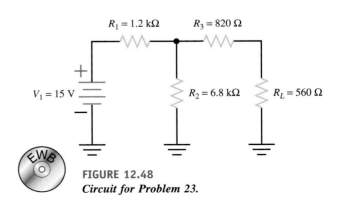

FIGURE 12.48
Circuit for Problem 23.

24. Using Figure 12.49, find V_L.

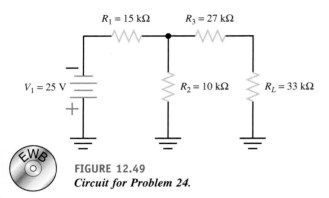

FIGURE 12.49
Circuit for Problem 24.

25. Using Figure 12.50, find V_{AB}.

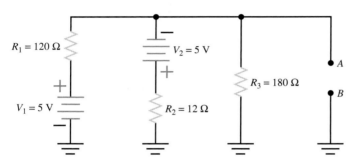

FIGURE 12.50
Circuit for Problem 25.

■ ANSWERS TO PRACTICE PROBLEMS

PRACTICE PROBLEM 12.1

1. $R_i = 0.5 \, \Omega$, $R_L = 7 \, \Omega$

PRACTICE PROBLEM 12.2

1. $I_{R_2} = 2.4$ mA

PRACTICE PROBLEM 12.3

1. $V_{R_3} = 2.88$ V

PRACTICE PROBLEMS 12.4

1. $A = 3.3A - 1.8B = -5$, $B = -1.8A + 2.8B = 15$
2. $V_{R_1} = 13.13$ V
3. $V_{R_3} = 2.67$ V

PRACTICE PROBLEMS 12.5

1. $V_L = 3$ V
2. $V_L = 2.8$ V

PRACTICE PROBLEM 12.6

1. $I_L = 0.3$ mA

PRACTICE PROBLEM 12.7

1. $V_{AB} = -1.5$ V

■ ANSWERS TO SELF-CHECKS

SECTION 12.1 SELF-CHECK

1. Zero
2. Infinite

SECTION 12.2 SELF-CHECK

1. Ohm's law
2. No, because power is nonlinear (I^2R)
3. The terminal voltage of the voltage source does not depend upon the value of a current or a voltage anywhere in the circuit.

SECTION 12.3 SELF-CHECK

1. KCL

2. A node is a junction of two or more components.

SECTION 12.4 SELF-CHECK

1. A mesh is a simplest possible closed path.
2. Assumed
3. KVL
4. Simultaneous linear equations

SECTION 12.5 SELF-CHECK

1. Series
2. T

SECTION 12.6 SELF-CHECK

1. Complement
2. Parallel

SECTION 12.7 SELF-CHECK

1. Parallel

SECTION 12.8 SELF-CHECK

1. When the source resistance equals the load resistance
2. 50%
3. High
4. Small

SECTION 12.9 SELF-CHECK

1. Nodal (with two independent nodes) and mesh currents
2. Thevenin's and Norton's theorems

CHAPTER 13

CONDUCTORS, CELLS, AND SUPPLEMENTAL COMPONENTS

OBJECTIVES

After completing this chapter, you will be able to:

1. Describe the various types of conductors.
2. Compare and contrast conductors via size, material, and resistivity.
3. Evaluate the differences between positive, negative, and zero temperature coefficients.
4. Calculate conductor resistance and the change in resistance at higher temperatures.
5. Utilize an American wire gauge table.
6. Define the terms *wire, cable,* and *conductor insulation.*
7. Describe superconductivity and its applications in electrical circuits.
8. Compare and contrast circuit-control–utilizing switches.
9. Compare and contrast circuit-protection–utilizing fuses and circuit breakers.
10. Compare and contrast various attributes of both primary and secondary cells.

INTRODUCTION

In the previous chapters you learned primarily how resistors comprised various types of electrical circuits. The resistor value and type of interconnection provided the majority of the operating characteristics for an electrical circuit. There are other components essential to an electrical circuit as well. Many of these components are described in this chapter.

13.1   CONDUCTORS

Conductors are the means used to tie or connect most circuit components together. There are many factors that determine the type of electrical conductor used to connect components. Some of these factors are the physical size of the conductor, the type of material used for the conductor, and the electrical characteristics of the conductor.

13.1.1   Conductor Sizes

To compare the resistance and size of one conductor with that of another, a standard or unit size must be established. A convenient unit of measurement, as far as the *diameter* of a conductor is concerned, is the *mil* (0.001 (one-thousandth) in.). A convenient unit of conductor *length* is the *foot.* The standard unit of size in most cases is the *mil-foot.* This standard references a wire with a unit size if it has a diameter of 1 mil and a length of 1 ft.

13.1.2   Circular Mil

The *circular mil* is the standard unit of measurement of a round wire's cross-sectional area (Figure 13.1). This unit of measurement is found in American and English wire tables. The diameter of a round conductor (wire) used to conduct electrical current may be only a fraction of an inch. Therefore, it is convenient to express this diameter in mils to avoid the use of decimals. For example, the diameter of a wire is expressed as 25 mils instead of 0.025 in. The area, in circular mils, of a round conductor is obtained by squaring the diameter (also in mils). Thus, a wire having a diameter of 25 mils has an area of 25^2, or 625, circular mils.

FIGURE 13.1
Cross-sectional areas of conductors.

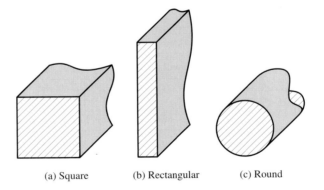

(a) Square   (b) Rectangular   (c) Round

13.1.3   Circular Mil-Foot

A *circular mil-foot,* as shown in Figure 13.2, is actually a unit of volume. It is a unit conductor 1 ft in length and having a cross-sectional area of 1 circular mil. Because it is considered a unit conductor, the circular mil-foot is useful in making comparisons between wires that are made of different metals. For example, a basis of comparison of the resistivity of various substances can be made by determining the resistance of a circular mil-foot of each of these substances.

FIGURE 13.2
Circular mil-foot.

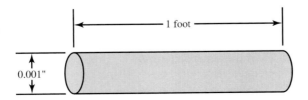

1 foot

0.001"

Answers are given at the end of the chapter.

1. What are the standard units of diameter and length for a conductor?
2. What is the standard unit of measurement for a round wire's cross-sectional area?
3. Describe the term *circular mil-foot.*

13.2 SPECIFIC RESISTANCE

Specific resistance, or resistivity, is the resistance in ohms offered by a unit volume (the circular mil-foot) of a substance to the flow of electric current. Resistivity is the reciprocal of conductivity. A substance that has a high resistivity will have a low conductivity, and vice versa.

Many tables of specific resistance are based on the resistance in ohms of a volume of a substance 1 ft in length and 1 circular mil in cross-sectional area. The temperature at which the resistance measurement is made is also specified. If the kind of metal of which a conductor is made is known, the specific resistance of the metal may be obtained from a table. The table may also show the temperature coefficient (α) of the substance. The temperature coefficient is used to determine how the resistance of a material varies with a change in temperature. (Temperature coefficient is more fully explained in Section 13.3.) The temperature coefficient and specific resistances of some common metals are given in Table 13.1.

TABLE 13.1
Specific resistances and temperature coefficient of common metals

Metal	Circular mil-foot (Ω at 20°C)	Temperature Coefficient (per °C, α)
Silver	9.8	0.004
Copper (drawn)	10.37	0.004
Gold	14.7	0.004
Aluminum	17.02	0.004
Tungsten	33.2	0.005
Nickel	52	0.005
Steel (soft)	95.8	0.003
Nichrome	660	0.0002

The resistance of a conductor of a uniform cross section varies directly as the product of the length and the specific resistance of the conductor and inversely as the cross-sectional area of the conductor. Expressed as an equation, the resistance in ohms of a conductor may be calculated as

$$R = \frac{\rho L}{A}$$

(13.1)

where R = resistance (Ω)
   ρ = specific resistance (Ω/circular mil-ft)
     (ρ is the Greek letter *rho*)
   L = length (ft)
   A = cross-sectional area (circular mils)

EXAMPLE 13.1

What is the resistance of 1000 ft of copper wire having a cross-sectional area of 10,400 circular mils (no. 10 wire) at a temperature of 20°C?

Solution

Given $\rho = 10.37$ (from Table 13.1)

$L = 1000$ ft

$A = 10,400$ circular mils

$R = \dfrac{\rho L}{A}$

$= \dfrac{(10.37 \times 1000 \text{ ft})}{10,400 \text{ circ. mils}}$

$= 1 \, \Omega$ (approximately)

Practice Problems 13.1 Answers are given at the end of the chapter.

1. What is the resistance of 100 ft of tungsten having a cross-sectional area of 1020 circular mils at a temperature of 20°C?

**SECTION 13.2
SELF-CHECK**

Answers are given at the end of the chapter.

1. Define the term *specific resistance*.
2. Define the term *temperature coefficient*.

13.3   TEMPERATURE COEFFICIENT

The resistance of pure metals—such as silver, copper, and aluminum—increases as the temperature increases. This is called a *positive temperature coefficient (PTC)* and is shown in Table 13.1. The increase in temperature causes additional vibration by the electrons of a conductor. As electrons in an electric current attempt to pass through the conductor, there is a greater chance for the energy field of one electron to interact with another. This, in turn, creates a frictional resistance that increases the resistance of the conductor. However, the resistance of some alloys—such as constantan and manganin—changes very little as the temperature changes.

Room temperature (20–25°C) is usually used as the reference when specifying specific resistance. As has been stated, the resistance of pure metals increases with an increase in temperature. The increase in resistance can be approximated using Equation 13.2, which utilizes the temperature coefficients given in Table 13.1.

$$R_T = R_o + R_o \, (\alpha \Delta t) \tag{13.2}$$

where R_T = the higher resistance at the higher temperature
R_o = the resistance at 20°C
α = temperature coefficient (α is the Greek letter *alpha*)
Δt = temperature rise above 20°C

EXAMPLE 13.2

Determine the resistance of a length of copper wire having a resistance of 10 Ω at 20°C when the temperature rises to 100°C.

Solution $R_T = R_o + R_o \, (\alpha \Delta t)$
$= 10 \, \Omega + 10 \, \Omega \, (0.004 \times 80)$
$= 10 \, \Omega + 3.2 \, \Omega$
$= 13.2 \, \Omega$

Practice Problems 13.2   Answers are given at the end of the chapter.
1. Determine the resistance of a length of copper with a resistance of 15 Ω at 20°C when the temperature rises to 80°C.

Semiconductor substances (silicon, germanium) exhibit a *negative temperature coefficient* (*NTC*). A negative value means less resistance at higher temperatures. A zero temperature coefficient (0TC) was shown to exist with constantan and manganin. These substances can be used for precision wirewound resistors, which do not change value as the temperature varies.

SECTION 13.3 SELF-CHECK

Answers are given at the end of the chapter.
1. Define *positive temperature coefficient*.
2. What temperature is used as the reference when specifying specific resistance?

13.4   WIRE MEASUREMENT

Wires are manufactured in numbered sizes. The sizes are designated utilizing American wire gauge (AWG) numbers. The AWG number is the most commonly used type of wire measurement. Tables for sizing conductors, either solid or stranded, and the material they are made from, such as copper or aluminum, are published by the National Bureau of Standards. One such table is shown in Table 13.2. The wire diameters become smaller as the gauge numbers become larger. The largest wire size shown in the table is AWG 0000, or no. 0000 (read as "4 aught"), and the smallest is AWG 40, or no. 40. An increase in three gauge numbers multiplies the resistance by a factor of 2 and decreases the area by the same factor. Many of the numbers in the table are rounded off for convenience. In addition to showing AWG numbers, the table shows values for diameter in mils, cross-sectional area in circular mils, and Ω/1000 ft at 25°C.

TABLE 13.2
Standard solid copper (AWG)

Gauge No.	Diameter, mils	Area, Circular Mils	Ohms per 1000 ft at 25°C	Gauge No.	Diameter, mils	Area, Circular Mils	Ohms per 1000 ft at 25°C
0000	460	212,000	0.050	19	36.0	1,290	8.21
000	410	168,000	0.630	20	32.0	1,020	10.4
00	365	133,000	0.795	21	28.5	810	13.1
0	325	106,000	0.100	22	25.3	642	16.5
1	289	83,700	0.126	23	22.6	509	20.8
2	258	66,400	0.159	24	20.1	404	26.2
3	229	52,600	0.201	25	17.9	320	33.0
4	204	41,700	0.253	26	15.9	254	41.6
5	182	33,100	0.319	27	14.2	202	52.5
6	162	26,300	0.403	28	12.6	160	66.2

Gauge No.	Diameter, mils	Area, Circular Mils	Ohms per 1000 ft at 25°C	Gauge No.	Diameter, mils	Area, Circular Mils	Ohms per 1000 ft at 25°C
7	144	20,800	0.508	29	11.3	127	83.4
8	128	16,500	0.641	30	10.0	101	105.0
9	114	13,100	0.808	31	8.9	79.7	133.0
10	102	10,400	1.02	32	8.0	63.2	167.0
11	91	8,230	1.28	33	7.1	50.1	211.0
12	81	6,530	1.62	34	6.3	39.8	266.0
13	72	5,180	2.04	35	5.6	31.5	335.0
14	64	4,110	2.58	36	5.0	25.0	423.0
15	57	3,260	3.25	37	4.5	19.8	533.0
16	51	2,580	4.09	38	4.0	15.7	673.0
17	45	2,050	5.16	39	3.5	12.5	848.0
18	40	1,620	6.51	40	3.1	9.9	1,070.0

In typical applications, hookup wires for electronic circuits and breadboards with current levels less than 1 A are generally about no. 22 or 24 gauge. The 110-V AC house wiring utilizes either no. 12 or 14, depending on current levels, which can go up to 20 A.

Figure 13.3 shows a wire gauge used to measure wires ranging from number 0 to number 36. A gauge number adjacent to a slot indicates the wire size. The slot has parallel sides. It is within the slot that the wire measurement is taken.

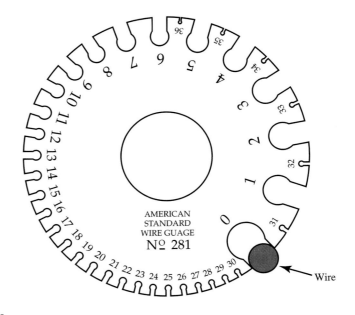

FIGURE 13.3
Wire gauge.

SECTION 13.4
SELF-CHECK

Answers are given at the end of the chapter.

1. What are AWG numbers used for?
2. What gauge wire should be used for hookup wire for breadboards?
3. What is the area in circular mils for no. 12 copper wire?
4. Where should wire measurement be taken utilizing the circular wire gauge shown in Figure 13.3?

13.5 CONDUCTOR TYPES AND CONNECTORS

A wire is a slender rod or filament of drawn metal. The definition restricts the term to what would ordinarily be understood as *solid wire*. If a wire is covered with insulation, it is called an insulated wire. Although the term *wire* properly refers to the metal, it is generally understood to include the insulation as well.

A conductor is a wire suitable for carrying an electric current. A stranded conductor is a conductor composed of a group of wires or of any combination of groups of wires. The wires in a stranded conductor are usually twisted together and not insulated from each other. A stranded conductor has greater flexibility than a solid wire conductor.

A cable is either a stranded conductor (single-conductor cable) or a combination of conductors insulated from one another (multiple-conductor cable). The term cable is a general one, and in practice it usually applies only to the larger sizes of conductors. A small cable is more often called a stranded wire, or cord (i.e., lamp cord). Cables are either bare or insulated. Electronic circuitry also utilizes conductors in the form of copper traces on printed circuit board assemblies. Some examples of conductors are shown in Figure 13.4.

FIGURE 13.4
Examples of conductors.

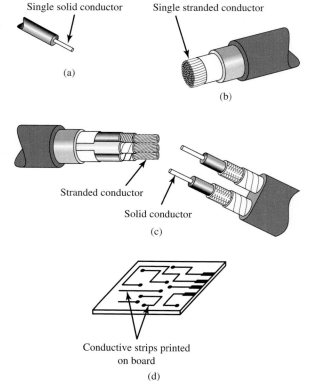

Single solid conductor

Single stranded conductor

(a)

(b)

Stranded conductor

Solid conductor

(c)

Conductive strips printed
on board

(d)

Electronic circuitry requires wires and cables to connect from one point to another. Various types of connectors are available to facilitate this connection. A sample of connectors is shown in Figure 13.5. Those in Figure 13.5(a) make temporary connections via clipping action. Common plugs are shown in Figure 13.5(b). The phone plug ($\frac{1}{4}$-in. diameter) also comes in miniature sizes (mm). The banana plug's mate is shown in Figure 13.5(d) as the banana jack. Likewise, the RCA plug's mate is the RCA jack in Figure 13.5(d). The coaxial plug and socket (Figure 13.5(b) and (d)) become the coaxial cable, an extremely common conductor used in electronic circuitry.

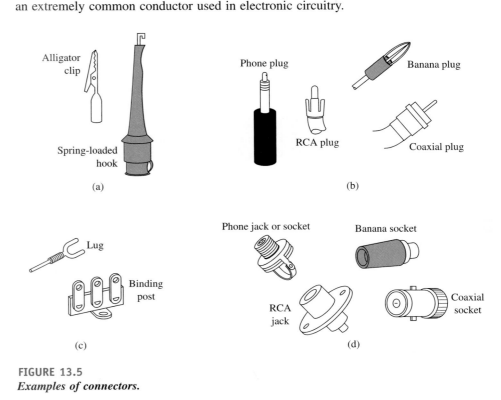

FIGURE 13.5
Examples of connectors.

SECTION 13.5 SELF-CHECK

Answers are given at the end of the chapter.

1. Define the term *wire.*
2. Define the term *conductor.*
3. What is the difference between a cable and a stranded conductor?
4. List some common connectors.

13.6 CONDUCTOR INSULATION

In general, current-carrying conductors must not be allowed to come in contact with one another, their supporting hardware, or personnel working near them. To accomplish this, conductors are coated or wrapped with various materials. These materials have such a high resistance that they are, for all practical purposes, nonconductors. They are generally referred to as *insulators,* or *insulating materials.*

Two fundamental properties of insulating materials (for example, rubber, glass, asbestos, or plastic) are insulation resistance and dielectric strength. These two terms have entirely different and distinctive properties.

Insulation resistance is the resistance to current leakage through the insulation materials. Insulation resistance can be measured by means of a "megger" (an ohmmeter capable of measuring megohms of resistance) without damaging the insulation. Clean, dry insulation having cracks or other faults may show a high value of insulation resistance but would not be suitable for use.

Dielectric strength is the ability of the insulator to withstand potential difference. (A table of dielectric strengths is shown in Table 19.2.) It is usually expressed in terms of the voltage at which the insulation fails because of the electrostatic stress caused by a potential difference. Maximum dielectric strength values can be measured only by raising the voltage of a test sample until the insulation breaks down.

Common substances that are used for insulation are rubber, plastics, varnished cambric, extruded polytetrafluoroethylene, fluorinated ethylene propylene (FEP), and enamel. The job of any insulator is to confine the current to its conductor.

SECTION 13.6 SELF-CHECK

Answers are given at the end of the chapter.

1. What is conductor insulation?
2. Define the term *insulation resistance.*
3. Define the term *dielectric strength.*

13.7 SUPERCONDUCTIVITY

Superconductors are materials that undergo a transformation at reduced temperature that gives them unique electrical properties. The most important of these properties is the ability to carry electrical currents with significantly reduced resistance. With significantly reduced resistance, electrical signals are not dissipated in the form of heat, so that all kinds of electrical and electronic devices and components become far more efficient.

13.7.1 Low-Temperature Superconductivity (LTS)

The first superconductor, mercury, was discovered in 1911 by a Dutch physicist named Onnes. This phenomenon had been restricted to metals and alloys with a transition temperature (the temperature when superconductivity occurs) of less than 23 K (kelvins). The Kelvin temperature scale is measured in degrees Celsius from absolute zero ($-273.15°C$). 0 K is equal to absolute zero. (Absolute zero is the hypothetical point at which a substance has no molecular motion and no heat.) Table 13.3 shows the relative temperature scale associated with superconductors.

TABLE 13.3
Relative temperature scale for superconductors

Kelvin (K)	Celsius (°C)	Fahrenheit (°F)
160	−113	−171
93	−180	−292
77	−196	−302
50	−223	−370
36	−237	−393
20	−253	−423
4	−269	−452

13.7.2 High-Temperature Superconductivity (HTS)

In 1986, superconductivity was discovered in ceramic materials. This precipitated an onrush of ceramic-based superconductors with transition temperatures as high as 130 K. The ceramic-based materials are commonly known as high-temperature superconductors (HTS), whereas the metallic and alloy materials are called low-temperature superconductors (LTS). HTS uses nitrogen (a liquid at 77 K) as a coolant that is inexpensive and commonplace compared to liquid helium, the original LTS coolant. Commercial applications of superconductivity use available mechanical refrigerators (known as cryogenic refrigerators) to provide the requisite cooling properties.

Superconductor applications include uses in the electric power industry. More cost-effective motors and generators are possible with HTS materials. Power cables made of HTS can carry 2 to 10 times more power than conventional cables in smaller or equal-sized cables. A new product, *superconducting magnetic energy storage* (SMES), will be used by electric utilities to control power fluctuations.

The transportation industry utilizes strong magnetic fields created by HTS coils to produce levitation by magnetism. This is the basic principle underlying magnetically levitated (maglev) trains.

The medical industry utilizes magnetic coils based on LTS materials for magnetic resonance imaging (MRI) equipment. The eventual use of HTS materials will enhance the cost-benefit aspect of this application.

**SECTION 13.7
SELF-CHECK**

Answers are given at the end of the chapter.

1. Define the term *superconductivity.*
2. What material was the first superconductor?
3. Define the term *HTS.*
4. Name some common uses for superconductivity.

13.8 CIRCUIT CONTROL: SWITCHES

Circuit control, in its simplest form, is the application and removal of power. This is also expressed as turning a circuit on and off or opening and closing a circuit. Circuit control is required so that if a circuit develops problems that could damage the equipment or endanger personnel, it is possible to remove the power from the circuit.

Circuit-control devices have many different shapes and sizes, but most circuit-control devices are either *switches, solenoids,* or *relays.* (Solenoids and relays are covered in Chapter 15.)

13.8.1 Switch Types

There are thousands and thousands of switch applications found in the home and in industry. Some switches operate by the touch of a finger and many others are operated automatically.

A *manual switch* is a switch that is controlled by a person. It is one that you turn on and off. Examples of common manual switches are a light switch, the ignition switch of an automobile, and the on/off switch of a consumer product.

An *automatic switch* is one that is controlled by a mechanical or electrical device rather than human action. Two examples of automatic switches are the thermostat and the remote-controlled switch of a consumer product.

Toggle, push-button, rocker, slide, rotary, snap-acting, microswitch, and DIP are examples of different types of switches.

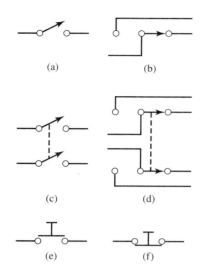

FIGURE 13.6
Examples of common switches:
(a) single-pole, single-throw;
(b) single-pole, double-throw;
(c) double pole, single-throw;
(d) double-pole, double-throw;
(e) normally open push-button;
(f) normally closed push-button.

13.8.2 Multicontact Switches

Switches are sometimes used to control more than one circuit or to select one of several possible circuits. An example of a switch controlling more than one circuit is the AM/FM selector on a radio. The selection of one of several circuits can be done by the range switch of a VOM. These switches are called *multicontact switches* because they have multiple contacts.

13.8.3 Poles and Throws

Multicontact switches are usually classified by the number of poles and throws they have. *Poles* are shown on a schematic as those contacts through which current enters (or leaves) the switch; they are connected to the moveable contacts. Each pole may be connected to another part of the circuit through the switch by *throwing* the switch (moveable contacts) to another position. This action provides an individual conduction path through the switch for each pole connection. The number of throws indicates the number of different circuits that can be controlled by each pole. By counting the number of points where current enters (or leaves) the switch (from the schematic symbol or the switch itself) you can determine the number of poles. By counting the number of different points each pole can connect with, you can determine the number of throws. Figure 13.6 illustrates several common switches.

13.8.4 Rotary Switches

A *rotary switch* is a multicontact switch with the contacts arranged in a full or partial circle. Instead of a push button or a toggle, the mechanism used to select the contact moves in a circular motion and must be turned or rotated. Rotary switches can be manual or automatic switches.

Some rotary switches are made with several layers or levels. The arrangement makes possible the control of several circuits with a single switch. These switches are also known as *wafer switches,* where each layer is known as a *wafer.* A wafer switch is shown in Figure 13.7.

FIGURE 13.7
Wafer switch.

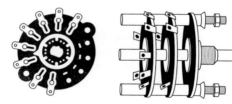

SECTION 13.8 **SELF-CHECK**	Answers are given at the end of the chapter.
	1. What is circuit control?
	2. What are common circuit control devices?
	3. What is the difference between a *pole* and a *throw*?
	4. What is the difference between a rotary switch and a wafer switch?

13.9 CIRCUIT PROTECTION: FUSES AND CIRCUIT BREAKERS

Circuit-protection devices are used to stop current flow or open a circuit. They are used to protect a circuit from the dangers of a direct short, excessive current, or excessive heat.

To provide the proper protection, a circuit-protection device *must always be connected in series* with the circuit it is protecting.

13.9.1 Fuses

A *fuse* is the simplest circuit-protection device. It derives its name from the Latin word *fusus,* meaning "to melt." The earliest type was simply a bare copper wire between two connections. The wire was smaller than the conductor it was protecting and, therefore, would melt before the conductor it was protecting was harmed. After changing from copper to other metals, tubes or enclosures were developed to hold the melting element. The enclosed fuse made possible the addition of a filler material that helps to contain the arc that occurs when the element melts.

For low-power uses, the filler material is not required. A simple glass tube is used. The use of a glass tube gives the added advantage of being able to see when a fuse is open.

13.9.2 Fuse Types

There are basically two types of fuses, *plug-type fuses* and *cartridge fuses.* The plug-type is rapidly being replaced by the circuit breaker. Both types employ a single wire or a ribbon as the fuse element (the part of the fuse that melts). The condition (good or bad) of some fuses can be determined by visual inspection. The condition of other fuses can be determined only with an ohmmeter. Fuse examples are shown in Figure 13.8.

FIGURE 13.8
Examples of fuses: (a) cartridge-type; (b) plug-type.

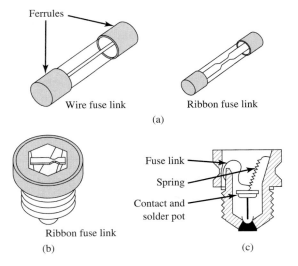

13.9.3 Fuse Ratings

Fuses are rated by current, voltage, and time-delay characteristics. To select the proper fuse, you must understand the meaning of each of the fuse ratings.

Current Rating

The current rating of a fuse is a value expressed in amperes that represents the maximum current the fuse will allow without opening. The current rating of a fuse is always indicated on the fuse.

Voltage Rating

The voltage rating of a fuse is *not* an indication of the voltage the fuse is designed to withstand while carrying current. The voltage rating indicates the ability of the fuse to quickly extinguish the arc after the fuse element melts and the maximum voltage the open fuse will block. In other words, once the fuse has opened, any voltage less than the voltage rating of the fuse is not able to "jump" the gap of the fuse.

Time-Delay Rating

Various electrical and electronic circuits require different levels of protection. In some of these circuits it is important to protect against temporary or transient current increases. Sometimes the device being protected is very sensitive to current and cannot withstand any increase in current. In these cases, the fuse must open very quickly if the current increases.

Fuses are time-delay rated to indicate the relationship between the current through the fuse and the time it takes for the fuse to open. The three time-delay ratings are delay, standard, and fast.

Delay   A *delay*, or *slow-blowing, fuse* has a built-in delay that is activated when the current through the fuse is greater than the current rating of the fuse. This fuse will allow temporary increases in current (surges) without opening.

Standard   *Standard fuses* have no built-in time delay. They are also designed to not be very fast acting. Standard fuses are sometimes used to protect against direct shorts only. A standard fuse can be used in any circuit where surge currents are not expected and a very fast opening of the fuse is not needed.

Fast   *Fast fuses* are designed to open very quickly when the current through the fuse exceeds the current rating of the fuse. Fast fuses are used to protect devices that are very sensitive to increased current.

Figure 13.9 shows the differences between delay, standard, and fast fuses. The figure shows that if a 1-A-rated fuse has 2 A of current through it (200% of the rated value), a fast fuse would open in about 0.7 s, a standard rated fuse would open in about 1.5 s, and a delay-rated fuse would open in about 10 s.

FIGURE 13.9
Time required for fuse to open.

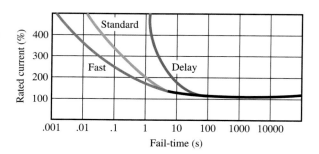

13.9.4   Identification of Fuses

Figure 13.10 shows the commercial designations (old and new) for fuses. The old designation uses a combination of letters and numbers (three in all) that indicates the style and time-delay characteristics. The time-delay characteristics must be looked up in a manufacturer's catalog. The 3AG fuse is a glass-bodied fuse, $\frac{1}{4}$ in. $\times$ $1\frac{1}{4}$ in., and has a standard time-delay rating.

The new commercial designation is the same except for the style portion of the coding. The new system uses letters only. In the example shown, 3AG in the old system becomes AGC in the new system. Because *C* is the third letter of the alphabet, it is used instead of the 3 used in the old system.

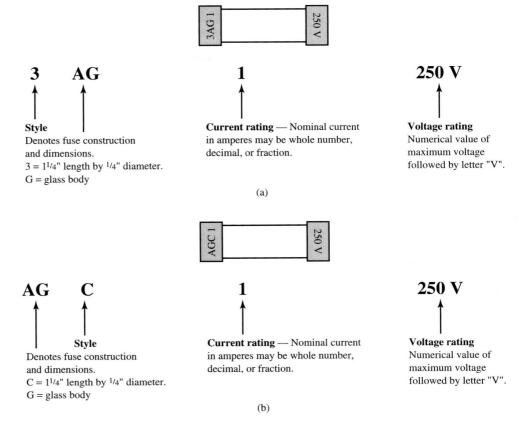

3   AG

Style
Denotes fuse construction
and dimensions.
3 = 1¹/4" length by ¹/4" diameter.
G = glass body

1

Current rating — Nominal current
in amperes may be whole number,
decimal, or fraction.

250 V

Voltage rating
Numerical value of
maximum voltage
followed by letter "V".

(a)

AG   C

Style
Denotes fuse construction
and dimensions.
C = 1¹/4" length by ¹/4" diameter.
G = glass body

1

Current rating — Nominal current
in amperes may be whole number,
decimal, or fraction.

250 V

Voltage rating
Numerical value of
maximum voltage
followed by letter "V".

(b)

FIGURE 13.10
*Commercial designations for fuses: (a) old commercial designations; (b) new commercial
designations.*

13.9.5   Circuit Breakers

A *circuit breaker* is a circuit-protection device that, like a fuse, stops current in the circuit if there is a direct short, excessive current, or excessive heat. Unlike a fuse, a circuit
breaker is reusable. Instead of replacing the circuit breaker, you reset it.

Circuit breakers are multielement devices. The most important element is a trip element. A *trip element* is the part of the circuit breaker that senses an overload condition
and causes a circuit breaker to trip, or break the circuit. The trip element uses thermal,
magnetic, or thermal-magnetic trip mechanisms. The thermal trip element employs a
bimetallic element that bends (and opens the circuit) when heated by excessive current.
This device acts as a delay fuse and protects against a small overload and excessive temperature. The magnetic trip element uses an electromagnet in series with the circuit load.
Excessive current causes an increase in the magnetic field strong enough to open the contacts (and the circuit) within the circuit breaker. This device will trip instantly. When both
types of protection are desired, a thermal-magnetic trip circuit breaker is used.

**SECTION 13.9
SELF-CHECK**

Answers are given at the end of the chapter.

1. A circuit-protection device must always be connected in _____ with the circuit it is protecting.
2. What is the most common type of fuse?

3. How are fuses rated?
4. What are the three time-delay ratings?
5. How does a circuit breaker differ from a fuse?

13.10 BATTERIES

As we saw in Chapter 4, a *cell* is a device in which chemical energy is converted to electrical energy. This process is called *electrochemical action.* A battery can be made up of just one cell or a combination of cells, which increases the potential difference.

The voltage across the electrodes depends on the elements from which the electrodes are made and the composition of the electrolyte. The current that a cell delivers depends upon the resistance of the entire circuit, including that of the cell itself. The internal resistance of the cell depends upon the size of the electrodes, the distance between them in the electrolyte, and the resistance of the electrolyte. The larger the electrodes and the closer together they are in the electrolyte (without touching), the lower the internal resistance of the cell and the more current the cell is capable of supplying to the load. In simpler terms, the cell's physical size and type of electrolyte determines its current capacity.

Cells are distinguished by either being a *primary* or a *secondary* type. Chemical action in the primary cell eats away one of the electrodes. When this happens, the cell must be discarded. A secondary cell is one in which the electrodes and the electrolyte are altered by the chemical action. These cells may be restored to their original condition by forcing an electric current through them in the direction opposite to that of discharge. Secondary cells are rechargeable; primary cells are not.

13.10.1 Primary Cells

Zinc-Carbon Cell

A common primary cell has carbon mixed with manganese dioxide and zinc electrodes. This cell is also known as the Leclanché cell, named after its inventor. The electrolyte is a solution of ammonium chloride and zinc chloride. The electrolyte is usually in the form of a paste or gel (hence the name *dry cell,* also associated with this type). This cell is shown in Figure 13.11.

The current flow through the load is the movement of electrons from the negative electrode of the cell (zinc) to the positive electrode (carbon). This causes fewer electrons in the zinc and an excess of electrons in the carbon. Hydrogen ions (bubbles released from the ammonia) being positively charged are attracted to the negative charge on the carbon electrode. This process is called *polarization* and decreases the voltage output while raising the internal resistance of the cell. However, manganese dioxide (a depolarizer) added to the carbon releases oxygen, which combines with the hydrogen bubbles to form water. In time, the *depolarizer* loses its effectiveness causing an increase in internal resistance and a need for battery replacement. The zinc electrode is eaten away during the discharge process, so that in time battery replacement is also necessary.

Note in Figure 13.11 that the zinc electrode (bottom of can) is labeled negative and the carbon electrode (top of can) is labeled positive. This represents the direction of current flow *outside* the cell from negative to positive.

Alkaline Cell

The alkaline cell (alkaline-manganese dioxide) has a higher energy output than the zinc-carbon cells. In comparison to the zinc-carbon cell, the alkaline cell delivers up to 10

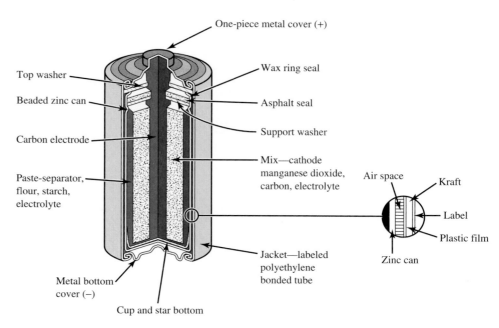

FIGURE 13.11
Cutaway view of carbon-zinc cell.

times the ampere-hour (Ah) capacity at high and continuous drain conditions. It also has a secure seal that provides excellent resistance to leakage and corrosion. It is the most popular primary cell in consumer use today.

The components of the alkaline-manganese battery are a zinc anode, a manganese dioxide cathode, and a highly conductive potassium hydroxide electrolyte. Depending on configuration (single or multicell), capacities are from 34 mAh to 15,000 mAh. Open-circuit voltage ranges from 1.5 to 1.6 V, depending on cathode formulation.

Lithium Cell

Lithium can be used as both a primary and secondary cell. As a primary cell it uses manganese dioxide to form the cathode. Lithium metal forms the anode, with a conductive organic electrolyte. These cells range in capacity from 76 mAh to 500 mAh. The open-circuit voltage is typically 3.2 V, compared to 1.5 V for most zinc cells.

Zinc Air Cell

The zinc air cell uses a powdered zinc anode, a highly conductive potassium hydroxide electrolyte, and atmospheric oxygen as the cathode reactant. Capacities are up to 1.5 Ah. The practical open circuit voltage is 1.4 V, with a flat discharge curve.

Zinc–Silver Oxide Cell

The zinc–silver oxide cell is noted for supplying energy at relatively high current drains. They are ideal for miniature devices where space is limited. The silver cell uses an amalgamated zinc anode, silver oxide as the cathode material, and a potassium hydroxide electrolyte. It is commonly manufactured in a button-cell configuration (Figure 13.12). The capacity ranges from 14 mAh to 180 mAh. The open-circuit voltage of the silver cell is 1.6 V with a flat discharge curve.

FIGURE 13.12
Zinc–silver oxide button cell.

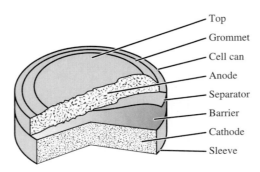

Top
Grommet
Cell can
Anode
Separator
Barrier
Cathode
Sleeve

13.10.2   Secondary Cells

A *secondary cell* is a storage cell that can be recharged by reversing the internal chemical reaction. A primary cell must be discarded after it has been completely discharged. The lead-acid cell is the most common type of storage cell.

Sealed Lead-Acid (SLA) Cell

The *flooded* version of the SLA cell is found in automobiles and is one with which you are probably somewhat familiar. Most portable equipment uses the *sealed* version. The SLA is commonly used when high power is required, weight is not critical, and cost must be kept low. The typical current range is 2 Ah to 30 Ah. The SLA is designed to be kept at a full charge level and has shallow discharge abilities. Uses include wheelchairs, UPS (uninterrupted power supplies) units, and emergency lighting. It does not suffer from memory problems (explained with the NiCd cell). Because of high lead content, the SLA is not environmentally friendly.

Nickel Cadmium (NiCd) Cell

For electronic circuits, a NiCd (Figure 13.13) represents a first-generation technology. It remains a popular choice for applications such as portable radios, cell phones, laptop computers, power tools, and video cameras. It prefers a fast charge over slow charge. A NiCd does not like to be pampered by sitting in chargers for days and being used only occasionally for brief periods. It performs best when it is periodically discharged to 1 V per cell (full discharge). If this periodic full discharge is omitted, the NiCd gradually loses performance due to *crystalline formation,* also referred to as *memory.*

Originally, the word memory was derived from *cyclic memory,* meaning that a NiCd battery remembers how much discharge was required on previous discharges. Improvements in battery technology have virtually eliminated this phenomenon. The problem with the modern NiCd is not cyclic memory, but the aforementioned crystalline formation. The active materials of the cell are present in crystalline form, and when the memory phenomenon occurs, these crystals grow, forming a spike or treelike spike crystals. These crystals gradually cause the NiCd to lose performance, and in severe cases they may puncture the separator, causing high self-discharge.

A full discharge is not required for each charge. A full discharge to 1 V per cell once a month is sufficient to keep the crystal formation under control. Such a discharge-charge cycle is referred to as an *exercise.*

Nickel Metal-Hydride (NiMH) Cell

The NiMH is a second-generation technology heralded as the shining star that will solve today's battery problems. This assumption may be overoptimistic, but the NiMH cell does have several advantages over the NiCd. It has more capacity than a standard NiCd, is less affected by memory, and is environmentally safe.

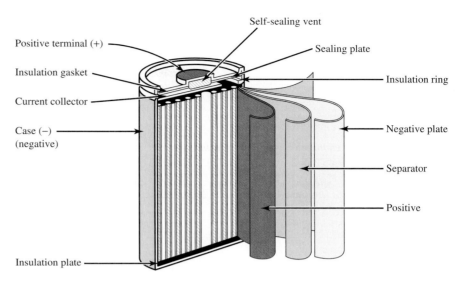

FIGURE 13.13
Drawing of a NiCd cell.

On the down side, it is rated for only 500 charge-discharge cycles. (In comparison, the NiCd can accept several thousand full discharge-charge cycles.) It cannot accept as quick a fast charge as the NiCd. The recommended discharge current is 0.2 C (one-fifth of the rated capacity).

Lithium Cell (Rechargeable)

The lithium cell is a third-generation technology and is the most talked about battery chemistry. Its energy density is the highest among commercial batteries, and the self-discharge is very low. There are several types of lithium cells of interest. The most promising chemistries are the lithium-ion (Li-Ion) and the lithium-polymer (Li-Polymer).

The Li-Ion provides an energy density that is twice that of the NiCd at comparable load currents. It deep discharges well, and behaves much like a NiCd as far as charge-discharge characteristics are concerned. It is safe because no metallic lithium is used. At the present time, it remains among the most expensive commercial batteries.

The Li-Polymer provides three times the energy density as NiCd. It features a very low self-discharge. With respect to charge-discharge characteristics it closely resembles the SLA. Shallow rather than deep discharging is preferred. When commercially available, it may remain fairly expensive, but the manufacturing costs will be lower than that of the Li-Ion.

Secondary Alkaline Cell

The secondary alkaline cell is relatively new. It has been heavily promoted, but its success will be dictated by the results it produces when used in everyday applications.

The idea of recharging an alkaline is not new. They have been used in flashlights for commercial aircraft for more than 15 years. The difference between a rechargeable and a standard alkaline cell lies in the energy density and the number of times the battery can be recharged. Whereas the standard alkaline offers maximum energy density, the rechargeable alkaline provides a higher cycle count but compromises on the energy density. It can be recycled a limited number of times. The longevity is a direct function of the depth of discharge; the deeper the discharge, the fewer cycles the battery endures. This type of secondary cell is used in light-load applications because of its relatively high internal resistance.

All rechargeable batteries vary in terms of performance and longevity. Nothing affects these aspects more than battery maintenance. The better commercial chargers employ many techniques to monitor the charge process. This includes "intelligent chargers" that employ a microcontroller to monitor battery voltage. Intelligent chargers ensure that proper battery charging is being exercised.

Table 13.4 summarizes the pros and cons of the six rechargeable cells just described.

TABLE 13.4
Pros and cons of the most commonly used rechargeable batteries

Battery Type	NiCd	NiMH	SLA	Li-Ion	Li-Polymer	Secondary Alkaline
Energy density (Wh/kg)	50	75	30	100	175	80 (initial)
Cycle life (typical)	1500	500	200–300	300–500	150	10 (to 65%)
Fast-charge time	$1\frac{1}{2}$ h	2–3 h	8–15 h	3–6 h	8–15 h	3–4 h
Self-discharge	Medium	High	Low	High	Very low	Very low
Cell voltage (nominal)	1.25 V	1.25 V	2 V	3.6 V	2.7 V	1.5 V
Load current	Very high	Medium	Low	High	Low	Very low
Exercise req (days)	30	90	180	NA	NA	NA
Cost	Low	Medium	Very low	Very high	High	Very low
Cost per cycle ($)	0.04	0.16	0.10	0.25	0.60	0.50
In commercial use since	1950	1970	1970	1990	≈1997	1990

1. *Energy density* is measured in watt-hours per kilogram (Wh/kg).
2. *Cycle life* indicates the typical number of charge-discharge cycles before the capacity decreases from the nominal 100% to 80% (65% for reusable alkaline).
3. *Fast-charge time* is the time required to fully charge an empty battery.
4. *Self-discharge* indicates the self-discharge rate when the battery is not in use. Moderate refers to 1–2% capacity loss per day.
5. *Cell voltage* multiplied by the number of cells provides the battery terminal voltage.
6. *Load current* is the maximum recommended current the battery can provide. High refers to a discharge rate of 1 C. The C-rate is a unit by which charge and discharge times are scaled. If discharged at 1 C, a 1000-mAh battery provides a current of 1000 mA.
7. *Exercise requirement* indicates the frequency the battery needs exercising to achieve maximum service life.
8. *Battery cost* is the estimated commercial price of a commonly available battery.
9. *Cost per cycle* indicates the operating cost derived by taking the average price of a commercial battery and dividing it by the cycle count.
10. *In commercial use since* is the approximate year the battery became commercially available.

SECTION 13.10 SELF-CHECK

Answers are given at the end of the chapter.

1. What determines cell voltage?
2. What is the difference between a primary and a secondary cell?
3. What is the most popular primary cell used for consumer products?
4. What causes modern NiCd memory problems?
5. Which cell has the highest cycle life?

■ SUMMARY

1. Conductors are the means used to tie or connect most circuit components together.

2. A convenient unit for a conductor's diameter is the mil (0.001 in.), for length is the foot, and for cross-sectional area is the circular mil.

3. The circular mil-foot is useful in making comparisons between wires that are made of different metals.

4. Specific resistance is the resistance in ohms offered by a circular mil-foot of a substance to the flow of electric current.

5. Pure metals have a positive temperature coefficient (PTC). That is, their resistance increases with an increase in temperature.

6. Wire is manufactured in numbered sizes known as the American wire gauge (AWG) number system. The larger the number, the smaller the wire diameter.

7. A wire is a slender rod or filament of drawn metal. A conductor is a wire suitable for carrying an electric current.

8. A cable is either a stranded conductor or a combination of conductors insulated from one another. It usually applies only to the larger sizes of conductors.

9. Connectors are used to facilitate the connection of wire or cables from one point to another.

10. Conductors are usually coated with an insulating material.

11. Two properties of insulating materials are insulation resistance and dielectric strength.

12. Superconductors exhibit superconductivity at very low temperatures. Superconductivity allows a conductor to carry an electric current at significantly reduced resistance.

13. High-temperature superconductors (HTS) were discovered in ceramic based materials in 1986.

14. Superconductor applications include use in the electric power industry for motors, generators, and power cables; in transportation for maglev trains; and in medical fields for magnetic resonance imaging (MRI).

15. A switch in its simplest form is used for the application and removal of power. A multicontact switch has a number of poles and throws used to control more than circuit.

16. Fuses and circuit breakers are used for circuit protection.

17. Fuses are rated by current, voltage, and time-delay characteristics. Fuse time-delay ratings include delay (slow-blow), standard, and fast.

18. A circuit breaker provides the same protection as a fuse, but instead of replacing the circuit breaker, you reset it.

19. Cell voltage is determined by electrode material and the electrolyte composition.

20. Primary cells cannot be recharged; they must be replaced when discharged. The most common primary cell for consumer applications is the alkaline cell.

21. Secondary cells are storage cells that can be recharged by reversing the internal chemical reaction.

22. NiCd memory problems occur because the cell has not been discharged to 1 V per cell once a month. Crystal formation thus occurs that inhibits the cell performance.

23. The most talked about battery chemistry is the lithium cell. It has extremely high energy density. However, it will be fairly expensive as they become commercially available.

24. The key to secondary cell performance is maintenance. This can be facilitated by utilizing intelligent chargers.

■ IMPORTANT EQUATIONS

$$R = \frac{\rho L}{A} \tag{13.1}$$

$$R_T = R_o + R_o\left(\alpha\Delta t\right) \tag{13.2}$$

■ QUESTIONS

1. How are conductors used?

2. What parameters are used to compare the resistance and size of one conductor with another?

3. Describe the common unit size for the measurement of conductors.

4. Define the term *specific resistance.*

5. Define the term *temperature coefficient.*

6. Why do pure metals have a positive temperature coefficient?

7. Describe the AWG number system for conductor sizes.

8. Differentiate the differences between a wire and a conductor.

9. What is a stranded conductor? Does it have any advantages?

10. Compare and contrast a cable and a cord.

11. What are connectors used for?

12. Define the term *insulation resistance.*

13. Define the term *dielectric strength.*

14. List some common substances used for conductor insulation.

15. What is a superconductor?

16. Define superconductivity.

17. What are the differences between LTS and HTS superconductors?

18. Why is the discovery of HTS potentially so important?

19. Describe superconductor applications.

20. What are the common circuit-control devices?

21. What is the difference between a manual and an automatic switch?

22. What are multicontact switches?

23. Compare and contrast *poles* and *throws.*

24. What is the difference between a rotary switch and a wafer switch?

25. How must circuit-protection devices be connected in a circuit?

26. Describe the earliest form of a fuse.

27. What is the filler in a fuse used for?

28. How are fuses rated?

29. What does the voltage rating of a fuse refer to?

30. Explain three forms of fuse time delays.

31. Give an example of the old versus new commercial designations for fuses.

32. Compare and contrast a circuit breaker and a fuse.

33. What is the most important advantage of a circuit breaker? Describe the differences in these types.

34. What determines cell voltage and capacity?

35. Compare and contrast the differences between primary and secondary cells.
36. What is the most commonly used element for primary cell electrodes?
37. Describe how a simple zinc-carbon cell works.
38. What is cell polarization?
39. What do the polarity indicators on a cell indicate?
40. Discuss why the alkaline is more popular than zinc-carbon cells.
41. Discuss the attributes of the various secondary cells.
42. What is secondary cell memory usually associated with NiCd cells? How does one prevent it?
43. Discuss the attributes of proper battery maintenance.
44. Define the term *energy density*.
45. Define the term *C-rate*.

■ CRITICAL THINKING

1. Why do pure metals exhibit a positive temperature coefficient (PTC)?
2. Explain how superconductivity can be used to allow power cables to carry 2 to 10 times more power than conventional cables in smaller or equal-sized cables.
3. What was the original memory problem associated with NiCd batteries and how does it differ from crystalline formation?

■ PROBLEMS

1. What is the resistance of 5000 ft of copper wire having a cross-sectional area of 10,400 circular mils at a temperature of 20°C?
2. What is the resistance of 500 ft of aluminum wire having a cross-sectional area of 10,400 circular mils at a temperature of 20°C?
3. What is the resistance of 10,000 ft of copper wire having a cross-sectional area of 10,400 circular mils at a temperature of 20°C?
4. What is the resistance of 1000 ft of no. 12 copper wire at a temperature of 25°C?
5. What is the resistance of 1000 ft of no. 20 tungsten wire at a temperature of 25°C?
6. Determine the resistance of a length of copper wire having a resistance of 5 Ω at 20°C when the temperature rises to 90°C.
7. Determine the resistance of a length of aluminum wire having a resistance of 20 Ω at 20°C when the temperature rises to 60°C.
8. How much resistance increase occurs to a length of copper wire having a resistance of 24 Ω when the temperature rises from 20°C to 120°C?
9. What is the resistance of 2000 ft of no. 20 tungsten wire when the temperature rises from 25°C to 100°C?
10. What is the resistance of 10,000 ft of no. 14 copper wire when the temperature rises from 25°C to 120°C?

■ ANSWERS TO PRACTICE PROBLEMS

PRACTICE PROBLEM 13.1

1. 3.25 Ω

PRACTICE PROBLEM 13.2

1. 18.6 Ω

■ ANSWERS TO SECTION SELF-CHECKS

SECTION 13.1

1. For the diameter, it is the mil; for length, it is the foot.
2. Circular mil
3. It is a unit of volume for a conductor 1 ft in length and having a cross-sectional area of 1 circular mil.

SECTION 13.2

1. Specific resistance is the resistance in ohms offered by a unit volume of a substance to the flow of electric current.
2. The temperature coefficient is used to determine how the resistance of a material varies with a change in temperature.

SECTION 13.3

1. PTC means that the resistance of a substance increases with a rise in temperature.
2. Room temperature (20°C or 25°C)

SECTION 13.4

1. To designate wire sizes
2. No. 22 or 24
3. 6530 circular mils
4. Within the slot

SECTION 13.5

1. A wire is a slender rod or filament of drawn metal.
2. A conductor is a wire suitable for carrying an electric current.
3. A cable generally refers to larger sizes of conductors.
4. Phone plug and jack, banana plug and jack, and coaxial plug and socket.

SECTION 13.6

1. The coating or wrapping on a conductor
2. Insulation resistance is the resistance to current leakage through the insulating materials.
3. Dielectric strength is the ability of the insulator to withstand potential difference.

SECTION 13.7

1. Superconductivity is the property of a superconductor to carry electrical current with significantly reduced resistance.

2. Mercury

3. High-temperature superconductors discovered in 1986 in ceramic-based materials

4. In the electric power industry for motors, generators, and power cables; in the transportation industry in maglev trains; in the medical industry in magnetic resonance imaging (MRI)

SECTION 13.8

1. The application and removal of power

2. Switches, solenoids, or relays

3. Poles are those contacts through which current enters (or leaves) the switch. A throw is a contact each pole can connect with.

4. A rotary switch is a multicontact switch. A rotary switch made with several layers or levels is known as a wafer switch.

SECTION 13.9

1. Series

2. Cartridge-type

3. By current, voltage, and time-delay characteristics

4. Delay, standard, and fast

5. It is similar in application with the exception that instead of replacing, you reset it.

SECTION 13.10

1. Cell voltage is determined by electrode elements and the composition of the electrolyte.

2. A primary cell cannot be recharged; a secondary cell can.

3. Alkaline

4. Crystalline formation due to not discharging the cell once a month to 1 V per cell.

5. NiCd

PART III

MAGNETISM AND
MAGNETIC DEVICES

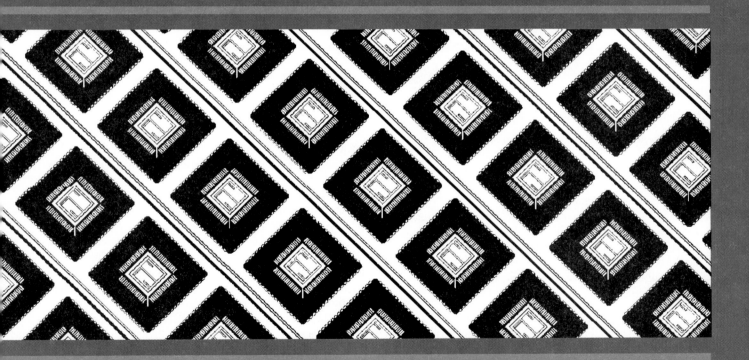

CHAPTER 14

MAGNETISM AND MAGNETIC QUANTITIES

OBJECTIVES

After completing this chapter, you will be able to:

1. Trace the background of magnetism and describe its nature.
2. Compare Weber's theory of magnetism to the domain theory.
3. Define common magnetic terms.
4. Identify the units of measurement for magnetic flux.
5. Describe the classifications of magnetic materials.
6. Compare permanent and temporary magnets and their applications.
7. Calculate the unknowns in problems related to magnetic circuits.

INTRODUCTION

In order to properly understand the principles of electricity, it is necessary to study magnetism and its effects on electrical equipment. Magnetism and electricity are so closely related that the study of either subject would be incomplete without at least a basic knowledge of the other.

Magnetism is generally defined as that property of a material that enables it to attract pieces of iron. A material possessing this property is known as a *magnet.* The word originated around the sixth century with the ancient Greeks, who found stones possessing this characteristic. The stones actually contained oxides of iron and were called *magnetite* by the Greeks. It was also noted that when suspended from a string, these stones would always point in one direction only. The Chinese, among others, used these stones as compasses around A.D. 200. They were called "leading stones," or *lodestones,* as they now are known.

In 1269 Petrus Peregrinus studied magnets and is credited with creating the term *poles* to define where the invisible magnetic lines terminated in a magnet. In 1600 William Gilbert was the first to study magnets seriously. He characterized the two poles of a magnet, that Earth is a magnet, that iron can be magnetized, and that magnetism can be destroyed by heating. In 1820, Hans Christian Oersted discovered that any current-carrying conductor generates a magnetic field. The interrelationship between electricity and magnetism became known as *electromagnetism.* This principle allows us to make an artificial magnet known as an *electromagnet.*

Table 14.1 lists several magnetic quantities and their respective SI units, along with symbols for each. These units are described further later in this chapter.

TABLE 14.1
SI units and symbols for magnetic quantities

Quantity	SI Symbol	SI Unit	SI Symbol
Flux	Φ	Weber	Wb
Flux density	B	Tesla	Wb/m^2 or T
Magnetomotive force	$\mathcal{F}$	Ampere-turn	A·t
Reluctance	$\mathcal{R}$	Ampere-turn/weber	A·t/Wb
Field intensity	H	Ampere-turns/meter	A·t/m

A magnetic circuit has many similarities to an electric circuit. In a magnetic circuit, magnetomotive force ($\mathcal{F}$) is analogous to the electric circuit quantity of voltage (V). Similarly, flux (Φ) is analogous to current (I), and reluctance ($\mathcal{R}$) is analogous to resistance (R). In an electric circuit it is possible to calculate the units of voltage, current, or resistance, providing sufficient data are given. Likewise, in a magnetic circuit, it is possible to calculate the units of magnetomotive force, flux, and reluctance, providing sufficient data are given.

14.1 THE NATURE OF MAGNETISM

14.1.1 Magnetic Definitions

Before describing the nature of magnetism, some terms need to be defined. Some of these terms are used in describing magnetic classifications. Many of the following terms are described in more detail later in the chapter. *Permeability* refers to the ease with which a material can pass magnetic lines of force. *Reluctance* is the opposition that a material offers to the magnetic lines of force. *Temporary magnets* easily lose most of their magnetic strength after the magnetizing force has been removed. The amount of magnetism that remains in a temporary magnet is referred to as its *residual magnetism*. Magnets made from substances such as hardened steel and certain alloys that retain a great deal of their magnetism are called *permanent magnets*. These materials are relatively difficult to magnetize because of their high reluctance. They also have low permeability. A temporary magnet produced from a material with low reluctance has a high permeability.

14.1.2 Weber's Theory

A popular theory of magnetism considers the molecular alignment of the material. This is known as *Weber's theory*. This theory assumes that all magnetic substances are composed of tiny molecular magnets, each with a north and south magnetic pole and with a surrounding magnetic field. The two poles form a *dipole*. Any unmagnetized material has the magnetic forces of its molecular magnets neutralized by adjacent molecular magnets, thereby eliminating any magnetic effect. A magnetized material has most of its molecular magnets lined up so that the north pole of each molecule points in one direction, and the south pole faces in the opposite direction. A material with its molecules thus aligned has one effective north pole and one effective south pole.

If the magnet is split in half, each half has both a north and a south pole. Further subdividing just yields more magnets having the same pole direction as the original magnet.

Another part of this theory states that any means used to disarrange the orderly array of the dipoles, such as jarring or heating, nullifies the magnetic properties of the material.

14.1.3 Domain Theory

A more modern theory of magnetism is based on the electron spin principle. From the study of atomic structure it is known that all matter is composed of vast quantities of atoms, with each atom containing one or more orbital electrons. The electrons are considered to orbit in various shells and subshells, depending upon their distance from the nucleus. It is now known that the electron also revolves on its axis as it orbits the nucleus of an atom. The phenomenon of magnetism seems to be associated with both the spinning and orbiting activities of electrons.

An electron carries a negative electrical charge. The spinning effect of the electron creates a magnetic field, or *magnetic moment.* The polarity of the magnetic field is determined by the direction of the electron spin. The strength of the magnetic field of an atom depends on the number of electrons spinning in each direction. If an atom has equal numbers of electrons spinning in opposite directions, the magnetic fields surrounding the electrons neutralize, or cancel, each other. The atom is unmagnetized. However, if more electrons spin in one direction than the other, the magnetic fields do not completely cancel, and the atom is said to be magnetized.

Using iron as an example, with an atomic number of 26, we know that it has 26 electrons that surround the nucleus. It is in the third shell (M shell) that the magnetic effects are generated. Fourteen electrons are in this shell. Of the 14 electrons, 9 spin in one direction and 5 spin in the other. This results in a net magnetic field caused by the spin of the 4 nonneutralized electrons, as illustrated in Figure 14.1.

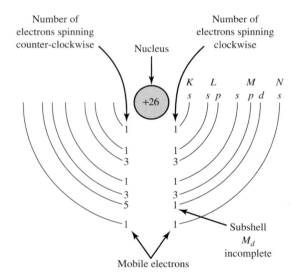

FIGURE 14.1
Iron atom.

Another fact is that in a given magnetic material atoms do not act independently but are bound together in groups called *domains.* This forms an interaction that is sustained even when the material is thermally agitated. However, above a certain temperature, the atomic alignment breaks down and the material becomes nonmagnetic. The temperature at which this occurs is known as the *Curie point.* For iron, this is 770°C. Above this temperature, iron loses its magnetic characteristics. When cooled below this temperature, iron can become magnetized again.

14.1.4 How Magnets Are Produced

An unmagnetized material can become magnetized by insertion into the coils of a temporary magnet known as an electromagnet (Section 14.5). The magnetic field of the electromagnet aligns the dipoles in the unmagnetized material so that they are in the same direction. Another way to produce magnetism is to stroke an unmagnetized material with another material that is already magnetized. Both processes utilize what is called *magnetic induction*. By stroking the unmagnetized material in *one direction only,* the dipoles of the unmagnetized material become aligned.

**SECTION 14.1
SELF-CHECK**

Answers are given at the end of the chapter.

1. Name a natural magnet.
2. What is a *dipole*?
3. The spinning effect of an electron creates a magnetic _____.
4. Describe magnetic induction.

14.2   MAGNETIC FIELDS

The space surrounding a magnet where magnetic forces act is known as a *magnetic field.* The pattern of a magnetic field can be obtained by performing an experiment with iron filings. A piece of glass is placed over a bar magnet and the iron filings are then sprinkled on the surface of the glass. If the glass is gently tapped, the iron particles align themselves to the magnetic field surrounding the magnet. The filings form a definitive pattern, as illustrated in Figure 14.2.

**FIGURE 14.2
*Pattern formed by iron filings.***

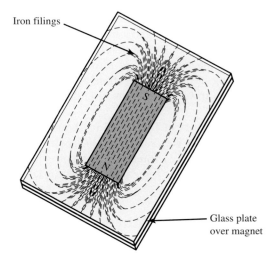

Iron filings

Glass plate
over magnet

14.2.1   Lines of Force

To further describe the magnetic phenomena, lines are used to represent the force existing in the area surrounding a magnet. These lines, called *magnetic lines of force,* do not actually exist but are imaginary lines used to illustrate and describe the pattern of the magnetic field. These lines are also known as *flux lines.* Flux lines (Figure 14.3) are assumed to emanate from the north pole of the magnet, pass through the surrounding space, and enter the south pole. Inside the magnet, the lines travel from south to north, thus completing a closed loop.

FIGURE 14.3
Bar magnet showing lines of force (flux lines).

Lab Reference: Magnetic lines of force and their patterns are demonstrated in Exercise 23.

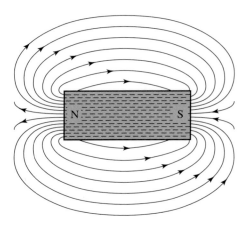

14.2.2 The Maxwell

One *maxwell* (Mx) is equal to one magnetic flux line. A flux line is also noted by the Greek symbol phi, Φ. The maxwell is named for James Clerk Maxwell (1831–1879), an important Scottish physicist who greatly contributed to our understanding of electrical and magnetic field theories by their unification.

14.2.3 The Weber

A *weber* is a larger unit of magnetic flux. Most magnets produce thousands of flux lines. The weber (Wb) represents 1×10^8 Mx, or individual flux lines. The microweber is a more practical unit and equals 10^{-6} Wb: $1 \ \mu$Wb is equal to 100 Mx, or flux lines. This unit is named after Wilhelm Weber (1804–1890), a German physicist.

The characteristics of magnetic lines of force can be summarized as follows:

1. Magnetic lines of force are continuous and always form a closed loop.
2. Magnetic lines of force never cross one another.
3. Parallel magnetic lines of force traveling in the same direction repel one another. If traveling in opposite directions they tend to unite with each other and form into single lines traveling in a direction determined by the magnetic poles creating the lines of force.
4. Magnetic lines of force tend to shorten themselves. Therefore, the magnetic lines of force existing between two unlike poles cause the poles to be pulled together.
5. Magnetic lines of force pass through all materials, both magnetic and nonmagnetic.
6. Magnetic lines of force always enter or leave magnetic material at right angles to the surface.

14.2.4 Magnetic Attraction and Repulsion

When the poles of two magnets are brought close together, there is a force of attraction or repulsion generated between them. Further, the force of attraction or repulsion varies with the strength of the magnets and their distance from each other. This principle is similar to Coulomb's law for electric charges. The following summarizes these facts:

1. Like magnetic poles repel each other.
2. Unlike magnetic poles attract each other.
3. The attraction or repulsion between magnets varies directly with the product of their strength and inversely with the square of the distance between them.

The mutual attraction or repulsion of the poles produces a more complicated pattern than that of a single magnet. These magnetic lines of force can be roughly illustrated by the use of iron filings, as before. This is shown in Figure 14.4.

FIGURE 14.4
Magnetic poles in close proximity.

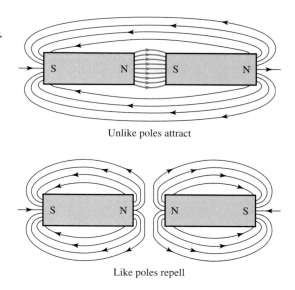

Unlike poles attract

Like poles repell

**SECTION 14.2
SELF-CHECK**

Answers are given at the end of the chapter.

1. What are *flux lines*?
2. How many flux lines does a maxwell represent? A weber?
3. State the magnetic law of attraction and repulsion.

14.3   CLASSIFICATION OF MAGNETIC MATERIALS

14.3.1   Ferromagnetism

Iron, cobalt, nickel (and various alloys of these materials) are called *ferromagnetic materials*. Ferromagnetic substances show a permanent, spontaneous magnetizing effect in the absence of a magnetizing field. The magnetization occurs due to the combined effects of spinning electrons and their associated magnetic moments. This, in turn, causes domains to form that have the same direction of magnetization. The domain magnetization is the greatest at absolute zero. It is reduced by temperature, being zero at its Curie temperature. Above the Curie temperature, it behaves as a paramagnetic (defined in Section 14.3.2) material, and the ability to possess permanent magnetism disappears.

An additional weak magnetizing field is all that is needed to further align the domains in a ferromagnetic material, so that the material exhibits a strong permanent magnetic field after alignment. They are used as permanent magnets and electromagnets. They exhibit high permeability factors up to 5000, with permalloy having a permeability upwards of 100,000. Contrast this to air, which is assigned a permeability of 1.

Ferromagnetic materials are generally put in two categories: those with high retentivities, called *hard,* and those with large initial permeabilities, called *soft.* Hard magnets make up what are popularly called *permanent magnets.* Magnetically soft materials find application in such devices as lifting electromagnets and the power transformers that convert standard 60-Hz commercial electric power from one voltage to another.

14.3.2   Paramagnetism

Paramagnetic materials include aluminum, platinum, manganese, and chromium. They have a mild attraction to a magnetic field. Their permeability is slightly more than 1.

14.3.3 Diamagnetism

Diamagnetic materials include bismuth, antimony, copper, zinc, mercury, gold, and silver. Their permeability is less than 1. They can become weakly magnetized, but their magnetic field is in the opposite direction from the magnetizing field.

SECTION 14.3 SELF-CHECK

Answers are given at the end of the chapter.

1. Name some common ferromagnetic materials.
2. True or false: Above the Curie temperature, a ferromagnetic material loses its magnetic characteristics.
3. A material with a permeability of less than 1 is _____ .

14.4   TYPES OF MAGNETS

All magnets are artificial, with the exception of magnetite and the planet Earth. Various types of artificial magnets are described next.

14.4.1   Permanent Magnets

A *permanent magnet* (PM) is made from a piece of hardened steel or other special alloy; when magnetized by the presence of an external magnetic field, it retains its magnetism (high retentivity), even when the magnetizing force is removed. The retained magnetism does not become exhausted with use, as the magnetic properties are determined by the structure of the internal atoms. Any loss of magnetism is accomplished only through high temperatures, a physical shock, or a strong demagnetizing force.

Application for Permanent Magnets

A common PM material is *alnico,* an alloy of aluminum, nickel, and iron, with cobalt, copper, and titanium added to produce varying grades. Alnico grade V is often used for PM loudspeakers. The loudspeaker uses a moving coil and a PM. The PM establishes a strong, stationary magnetic field. The moveable coil is placed around part of the magnet and in its magnetic field. This is illustrated in Figure 14.5. The coil is forced to move back and forth by the varying magnetic field caused by the current it receives from the amplifier. This current varies at an audio rate. The coil is alternately attracted and repelled by the stationary field of the PM.

Attached to the moving coil is a large cone, or diaphragm. As the coil vibrates, the cone vibrates, setting the air around the cone in motion at the same rate. This produces a sound-pressure variation that the ear interprets as the original sound.

FIGURE 14.5
A loudspeaker.

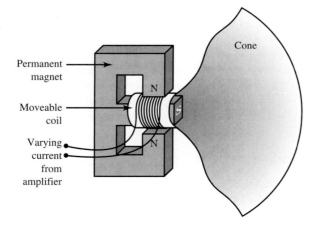

14.4.2 Temporary Magnets

A temporary magnet is a piece of soft iron that is magnetized while in the presence of a magnetizing field but demagnetizes the moment the iron is taken away from the magnetizing force or the force is no longer present. Soft iron possesses low retentivity.

Electromagnets

An electromagnet is a temporary magnet that consists of insulated wire wrapped in the form of a coil (Figure 14.6). A magnetic field is produced only when a direct current flows through the coil. The strength of the magnetic field is affected by the number of turns of the wire, the core material, and the magnitude of the current.

Soft iron is used as a core material in order to concentrate the flux lines surrounding the coil. This results in a stronger magnetic field.

FIGURE 14.6
A coiled current-carrying conductor becomes an electromagnet.

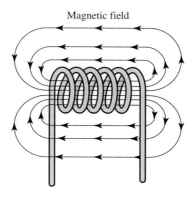

Magnetic field

Applications for Electromagnets

If the current to the coil is controlled by a switch, then the magnetism produced can be turned on or off. Applications include lifting magnets, buzzers, bells, and relays. A *relay* is an electromagnetic switch whose contacts are opened or closed by electromagnetism. Figure 14.7 shows how the relay can be used to switch the antenna between the receiver and transmitter sections. When the transmit switch is closed, the relay coil is energized. The magnetic field from the coil pulls the contact of the armature down. This completes the path for the transmitter to be connected to the antenna, while opening the contact on the receiver.

FIGURE 14.7
A relay used as a transmit-receive switch.

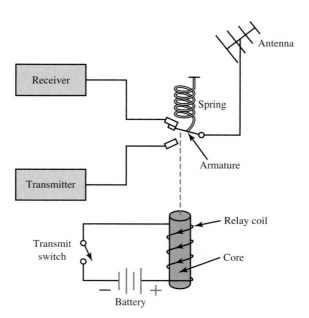

Another common application is in magnetic tape recording. Particles of iron oxide coat a mylar tape. The recording head is an electromagnet (coil) that produces a magnetic field proportional to the current through it. The flux preferentially travels through the core. Thus, the core is deliberately broken at an air gap. Flux in the air gap creates a *fringing field* that extends some distance from the core. As the tape passes the air gap of the head, small areas of the coating become magnetized by induction. The ferromagnetic surface becomes permanently magnetic. This is illustrated in Figure 14.8. On playback, the moving magnetic tape produces variations in electric current within the recording head.

As the current in the recording head varies, the amount of magnetism induced into the oxide coating also varies. This is known as *analog signal processing*. (Analog signal processing was described in Chapter 1.)

FIGURE 14.8
Magnetic tape recording. The fringing field in the air gap magnetizes the ferromagnetic surface on the tape.

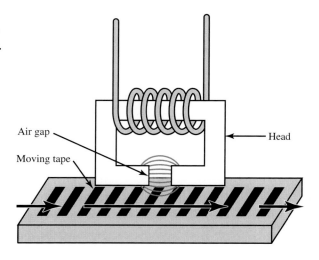

An example of *digital signal processing* utilizing the effects of magnetism can be demonstrated via a computer hard disk. Most modern hard disks are 3.5 in. in diameter or smaller. They consist of thin aluminum platters with a magnetic coating. Data are written and read by read/write heads that float over the platter on a microscopic cushion of air. The illustration in Figure 14.9 shows three platters with six read/write heads that move synchronously.

FIGURE 14.9
A computer hard disk showing three platters and six read/write heads.

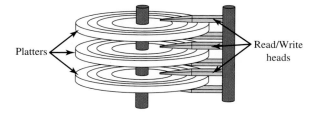

A read/write head consists of a tiny electromagnet. When a disk rotates under the read/write head, the head can either read existing data or write new data:

■ If current is applied to the coil, the head becomes magnetic. This magnetism induces and thus orients the micromagnets within the magnetic coating in the track. This is known as the *write mode*.
■ If the head moves along the track without current applied to the coil, it senses the micromagnets within the magnetic coating in the track. This magnetism induces current in the coil. These flashes of current represent the data on the disk. This is known as the *read mode*.

In order for the head to digitally interpret the data, the magnetic domains of the platter are initially arranged to all point in the same direction. To record data, the head reverses the magnetic polarity of a domain. This is interpreted as the digit 1, whereas an unchanged polarity is interpreted as the digit 0 (Figure 14.10). The digits 1 and 0 represent the numbers used by a digital system that is referred to as a *binary number system.*

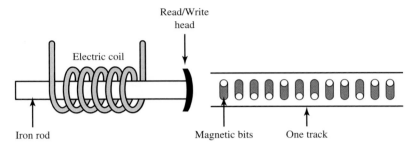

FIGURE 14.10
The read/write head of a computer hard disk. The magnetic domains (bits) are interpreted as a digital 1 or 0.

14.4.3   Ferrites

Ferrites are nonmetallic materials that exhibit the ferromagnetic properties of iron. They have very high permeabilities but values less than that of iron. Their permeability ranges from 50 to 3000. They are made of powdered iron oxide and ceramic materials using a sintering process. (Sintering is a process whereby bonded metal particles are shaped and partially fused by pressure and heating below the melting point of the metal.) Because each iron oxide particle is encapsulated in ceramic, the ferrite is an electrical insulator. Their specific resistance is 10^5 Ω/cm.

Applications for Ferrites

Common applications for ferrites include using them as magnetic cores. Whether in a fixed or variable form, ferrite cores are much more efficient than iron, especially when the current alternates at a very high frequency. Eddy currents (Chapter 15) cannot form in the insulated ferrite material; thus, I^2R losses are minimal. Ferrites are used at microwave frequencies (generally above 1 GHz), where they exhibit a phenomenon known as *precession.* Precession is analogous to a one-way valve for very high frequency currents. Applications include filters and oscillators.

Ferrites also come in the form of small toroid (doughnut-shaped) beads. When strung on conductors, the beads effectively block any high-frequency currents from passing through. This is known as *filtering.*

**SECTION 14.4
SELF-CHECK**

Answers are given at the end of the chapter.

1. What is the difference between a permanent magnet and a temporary magnet?
2. The air gap in a magnetic tape head creates a _____ _____.
3. Magnetic tape recording is an example of _____ _____ _____.
4. The computer hard disk is an example of _____ _____ _____.
5. Why are ferrites electrical insulators?

14.5   **MAGNETIC SHIELDING**

There is no known insulator for magnetic flux. If a nonmagnetic material is placed in a magnetic field, there is no appreciable change in flux; that is, the flux penetrates the nonmagnetic material. For example, a glass plate placed between the poles of a horseshoe magnet has no appreciable effect on the field, although glass itself is a good insulator to electric current.

If a magnetic material (for example, soft iron) is placed in a magnetic field, the flux is redirected (concentrated) to take advantage of the greater permeability of the soft iron. Notice also that when the same flux lines pass through a piece of plastic (nonmagnetic material), no concentration or redirection occurs. This is demonstrated in Figure 14.11.

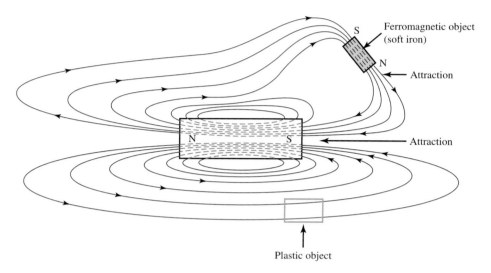

FIGURE 14.11
Permeability effects: Soft iron redirects and concentrates the flux lines. The flux lines are un-affected by the plastic object.

The sensitive mechanisms of electric instruments and meters can be unduly influenced by stray magnetic fields. The stray magnetic fields can cause errors in their readings. Because instrument mechanisms cannot be insulated from magnetic fields, it is necessary to employ some means of redirecting the flux around the instrument. This is accomplished by placing a soft-iron case, called a *magnetic screen,* or *shield,* about the instrument. The instrument is effectively shielded, because magnetic flux is more readily established in the soft-iron case.

SECTION 14.5 SELF-CHECK	Answers are given at the end of the chapter. 1. True or false: There is no known insulator for magnetic flux. 2. _____ _____ is used as a shield for magnetic flux.

14.6 MAGNETIC QUANTITIES

14.6.1 Magnetomotive Force (mmf)

Current passing through a coil causes a magnetic field or flux lines to be produced. *Magnetomotive force* (mmf), symbolized by $\mathscr{F}$, is the magnetic pressure that produces the flux. (Remember, this is analogous to the electrical pressure supplied by voltage in an electric circuit.) The amount of mmf produced is the product of the magnitude of current in amperes (A) and the number of turns (N) of the coil. It is measured in A·t, an SI unit of measurement. The equation for mmf is

$$\mathscr{F} = NI \qquad \qquad \text{(14.1)}$$

where $\mathscr{F}$ = magnetomotive force (ampere − turns (A·t))
N = number of turns of the coil
I = current (A)

EXAMPLE 14.1

Calculate the mmf for a coil with 3000 turns and a 10-mA current.

Solution $\mathscr{F} = NI$
$= 3000 \text{ turns} \times 10 \times 10^{-3} \text{ A}$
$= 30 \text{ A·t}$

Practice Problems 14.1   Answers are given at the end of the chapter.
1. A coil with 40,000 turns must produce 1000 A·t of mmf. How much current is necessary?
2. Calculate the A·t for a coil with 120 turns and 50 mA of current.

14.6.2   Reluctance $(\mathscr{R})$

Reluctance $(\mathscr{R})$ was previously defined as the opposition that a material offers to the magnetic lines of force. This is equivalent to resistance in an electrical circuit. Thus, the equation for calculating reluctance is the magnetic equivalent of Ohm's law. This is shown as

$$\mathscr{R} = \frac{\mathscr{F}}{\Phi} \qquad \qquad \text{(14.2)}$$

where $\mathscr{R}$ = reluctance (A·t/Wb)
$\mathscr{F}$ = magnetomotive force (A·t)
Φ = flux (Wb)

EXAMPLE 14.2

What is the reluctance when the mmf for a coil is 200 At and the flux is 50×10^{-6} Wb?

Solution $\mathscr{R} = \dfrac{\mathscr{F}}{\Phi}$

$= \dfrac{200 \text{ A·t}}{25 \times 10^{-6} \text{ Wb}}$

$= 8 \times 10^{6} \text{ A·t/Wb}$

Practice Problems 14.2   Answers are given at the end of the chapter.
1. A magnetic material has a total flux of 32 μWb with an mmf of 40 A·t. Calculate the reluctance.
2. A coil has an mmf of 250 A·t and a reluctance of 1.25×10^{6} A·t/Wb. Calculate the total flux.

14.6.3   Flux Density (B)

Flux density (B) is equal to the number of magnetic lines of flux (Φ) per square meter. The SI unit of flux density, Wb/m², is known as the tesla (T). The mathematical relationship between flux density and lines of force is expressed as

$$B = \frac{\Phi}{A}$$

(14.3)

where B = flux density (T)
   Φ = total field flux (Wb)
   A = cross-sectional area (m²)

EXAMPLE 14.3

An iron core with a cross-sectional area of 0.5 m² has a total flux of 800 μWb. Calculate the flux density in the core material.

Solution $B = \dfrac{\Phi}{A}$

$$= \frac{800 \times 10^{-6}\,\text{Wb}}{0.5\,\text{m}^2}$$

$$= 1.6 \times 10^{-3}\,\text{T}$$

$$= 0.0016\,\text{T}$$

Practice Problems 14.3   Answers are given at the end of the chapter.
1. How much is B in teslas for a flux of 400 μWb through an area of 0.25 m²?
2. The flux density in an iron core is 2 T. If the area of the core is 50 cm², calculate the total number of flux lines in the core.

14.6.4   Field Intensity (*H*)

The ampere-turns of mmf specifies the magnetizing force, but the intensity of the magnetic field depends on the length of the coil (electromagnet). A longer coil has more air space between the turns and between the ends of the coil and, thus, a weaker magnetic field.

   Field intensity (*H*) is a measurement of the mmf needed to establish a specified flux density in a unit length of the coil. This term applies equally to a permanent magnet's length, because both produce the same kind of a magnetic field.

   For a coil the field intensity may be expressed as

$$H = \frac{\mathscr{F}}{l}$$

(14.4)

where H = field intensity (A·t/m)
   $\mathscr{F}$ = applied mmf (A·t)
   l = length of coil (m)

EXAMPLE 14.4

Calculate the field intensity generated by a coil that has 200 T, 3 A of current, and is 0.05 m long.

Solution   First we must find $\mathscr{F}$.

$$\mathscr{F} = NI \quad \text{(from equation (14.1))}$$

$$= 200\,\text{turns} \times 3\,\text{A}$$

$$= 600\,\text{A·t}$$

$$H = \frac{\mathscr{F}}{l}$$

$$= \frac{600\,\text{A·t}}{0.05\,\text{m}}$$

$$= 12{,}000\,\text{A·t/m}$$

Practice Problems 14.4   Answers are given at the end of the chapter.
1. Calculate the field intensity generated by a coil that has 100 T, 2.5 A of current, and length 1.5 cm.
2. How long is a coil having a field intensity of 10,000 A·t/m, 75 T, and a current of 1 A?

14.6.5   Permeability (μ)

Permeability (μ) was previously defined as the ease with which a material passes flux lines. It is the electrical equivalent of conductance. Permeability (μ) is measured in henrys per meter (H/m). (The unit henry (H) is named in honor of American physicist Joseph Henry (1797–1878), who did extensive research on electromagnetism.) A high permeability factor indicates that a magnetic field can easily be established within a material. At the same time we can deduce that the material has a low reluctance ($\mathcal{R}$).

For *practical* purposes air and other nonmagnetic materials are assigned a permeability of unity (1). Magnetic materials such as iron, nickel, steel, and their alloys have a value much greater than unity. Numerical values of μ for different materials are assigned by comparing their permeability with that of air or a vacuum. This is known as the relative permeability (μ_r) and has no units.

Absolute permeability (μ_o) is the actual permeability of free space (air or a vacuum) and is the *standard SI reference value*. Free space has a permeability of $4\pi \times 10^{-7}$ H/m, or 1.26×10^{-6} H/m. The SI permeability (μ) of a material is the product of the relative permeability times the absolute permeability. It is defined by the following equation:

$$\mu = \mu_r \times \mu_o \tag{14.5}$$

where μ = permeability (H/m)
μ_r = relative permeability (no units)
μ_o = absolute permeability (1.26×10^{-6} H/m)

In SI units, the permeability (μ) of a material is also given by

$$\mu = \frac{B}{H} \tag{14.6}$$

where μ = permeability (H/m or T/(A·t/m))
B = flux density (T)
H = field intensity (A·t/m)

EXAMPLE 14.5

Find μ when $\mu_r = 3000$.

Solution $\mu = \mu_r \times \mu_o$
$= 3000 \times 1.26 \times 10^{-6}$ T/(A·t/m)
$= 378 \times 10^{-3}$ T/(A·t/m)

EXAMPLE 14.6

A coil has a field intensity, H, of 400 A·t/m. The μ_r of the core is 400. Calculate B in teslas.

Solution $\mu = \mu_r \times \mu_o$
$= 400 \times 1.26 \times 10^{-6}$ T/(A·t/m)
$= 504 \times 10^{-6}$ T/(A·t/m)

$$B = \mu \times H$$
$$= 504 \times 10^{-6} \text{ T/(A·t·m)} \times 400 \text{ A·t/m}$$
$$= 0.201 \text{ T}$$

Practice Problems 14.5   Answers are given at the end of the chapter.
1. Find μ when the relative permeability of a material is 500.
2. Find B in teslas when $\mu_r = 100$ and $H = 150$ A·t/m.

Table 14.2 lists the permeability figure in henrys per meter and the relative permeability of several materials.

TABLE 14.2
Permeabilities of various materials

Material	Permeability, μ (H/m)	Relative Permeability, μ_r (No Units)
Air or vacuum	1.26×10^{-6}	1
Nickel	6.28×10^{-5}	50
Cobalt	7.56×10^{-5}	60
Cast iron	1.1×10^{-4}	90
Machine steel	5.65×10^{-4}	450
Transformer iron core	6.9×10^{-3}	5,500
Silicon iron	8.8×10^{-3}	7,000
Permalloy	0.126	100,000
Supermalloy	1.26	1,000,000

**SECTION 14.6
SELF-CHECK**

Answers are given at the end of the chapter.

1. Define mmf.
2. Define reluctance.
3. The SI unit of flux density is the _____.
4. What factor affects the intensity of the magnetic field?
5. What is the permeability of free space?

14.7   THE *B-H* HYSTERESIS CURVE

The two most important magnetic quantities are flux density (B) and the magnetizing force (H). Their relationship can be graphed as the *B-H* hysteresis curve (Figure 14.12) with iron as the magnetic material. B and H are directly proportional terms, with H the independent variable and B the dependent variable. The curve between points O and A represents the magnetization curve. The induced magnetization eventually saturates at point A. Magnetic saturation occurs when a further magnetizing force (H) does not produce an increase in the flux density (B). Saturation develops when all the magnetic dipoles have become aligned.

As the magnetizing force reduces to zero, the flux density (B) falls to point R and does not follow the original curve. To reduce the flux to zero, a coercive force (an opposite-polarity magnetizing force), $-H$, must be applied, tending to establish magnetic flux in the opposite direction. The flux density (B) now increases in direction $-B$. At point Q is another saturation point, albeit in the opposite direction. If the magnetizing force is again reversed, the resulting magnetizing curve moves to point U and then up to point A once again. From this point on, the curve repeats itself. Note that the increasing and decreasing magnetization curves are separated from each other. It can be said that the flux density (B) *lags* the field intensity (H). The term *lagging* is called *hysteresis*. Hysteresis occurs because all the magnetic dipoles cannot reverse their alignment when the magnetizing force reverses. There is a net residual magnetism in each direction on the curve. An internal molecular friction prevents all the dipoles from reversing their direction when the magnetizing field reverses. The energy wasted in heat as the dipoles lag the magnetizing force is called the *hysteresis loss.*

FIGURE 14.12
The B-H hysteresis curve.

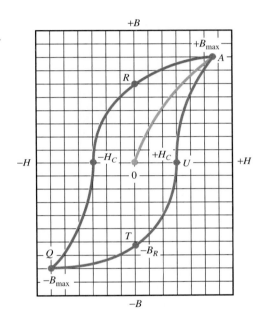

In order to determine the hysteresis characteristics of a magnetic material, the magnetizing force is provided by an AC current, which by nature reverses periodically. The losses are greater as the reversals increase with respect to time (a higher frequency). A large part of the magnetizing force is used just to overcome the internal friction of the dipoles. Heat is produced in this process. The faster the magnetizing force changes, the greater the hysteresis effect. For steel and hard magnetic materials, the hysteresis losses are even higher than in a soft magnetic material like iron. Thus, the hysteresis curve or loop has a specific width in relation to the type of magnetic material. The wider the hysteresis loop, the greater the hysteresis loss.

SECTION 14.7
SELF-CHECK

Answers are given at the end of the chapter.

1. What is magnetic saturation?
2. The term *lagging* is also called _____.
3. Describe hysteresis loss.
4. Hysteresis losses are _____ proportional to the _____ of an AC source.

14.8 THE HALL EFFECT

The *Hall effect* was discovered by E. H. Hall in 1879. The effect is noticed when a magnetic field brought close to a gold strip carrying current causes a voltage to be produced. The voltage produced from the magnetic field is called a *Hall voltage*. The Hall effect sensor, named in his honor, is used to measure the *flux density* (*B*) of a magnetic field. Semiconductors are used today instead of gold.

Figure 14.13 shows the Hall effect on a commonly used indium-arsenide (InAs) semiconductor probe and the relationships between the applied magnetic field, the current flowing through the semiconductor, and the resulting Hall voltage that is produced. A magnetic field applied at right angles to the semiconductor causes a separation in charge carriers. The redistribution of these carriers produces a Hall voltage that is induced in a direction perpendicular to the current and the magnetic field. The magnitude of the Hall voltage is proportional to the product of the magnetic field and current. By holding the current constant, the Hall voltage varies directly with the strength of the magnetic field or flux density.

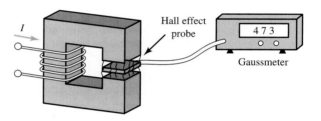

FIGURE 14.13
An application for the Hall effect. The gaussmeter measures flux density as a function of the voltage produced at the Hall effect probe.

In order to have enough sensitivity to measure minute amounts of flux density, the Hall effect sensor is calibrated in gauss rather than in tesla. The *gauss* (G), named in honor of Karl F. Gauss (1777–1855), a German mathematician, is a cgs unit of flux density. It is a very small unit of measurement. (The *tesla* (T), an SI unit of flux density measurement, is 10,000 (10^4) times the gauss.) A gaussmeter is an instrument that indicates the magnetic flux density, which is directly proportional to the Hall voltage produced by the semiconductor.

SECTION 14.8 SELF-CHECK

Answers are given at the end of the chapter.

1. Describe the Hall effect.
2. In order to measure a small amount of flux density, the _____ unit of measurement is used.

■ SUMMARY

1. Magnetism is generally defined as that property of a material that enables it to attract pieces of iron.
2. Lodestones are made from magnetite.
3. Permeability refers to the ease with which a material can pass magnetic lines of force.
4. Reluctance is the opposition that a material offers to the magnetic lines of force.
5. The amount of magnetism that remains in a temporary magnet is referred to as its residual magnetism.
6. Magnets made from substances such as hardened steel and certain alloys that retain a great deal of their magnetism are called permanent magnets.
7. Weber's theory stipulates that a magnetized substance has tiny molecules, each with a north and south pole (dipoles) lined up so that the north pole of each molecule points in the same direction.

8. The domain theory stipulates that magnetism is developed because of electron spin within the atomic structure of the material.

9. Magnetic induction is a process whereby an unmagnetized material can become magnetized by the effect of a magnetic field from a magnetized material.

10. Magnets possess magnetic lines of force known as flux lines that emanate from the north to south pole of the magnet.

11. One maxwell is equal to one magnetic flux line.

12. A weber represents 1×10^8 individual flux lines.

13. Like magnetic poles repel each other; unlike magnetic poles attract each other.

14. Iron is an example of a ferromagnetic material.

15. A common PM is made from alnico, an alloy of aluminum, nickel, and iron along with cobalt, copper, and titanium added to produce varying grades.

16. An electromagnet is a temporary magnet that consists of insulated wire wrapped in the form of a coil.

17. Electromagnets are used in lifting magnets, buzzers, bells, and relays.

18. Magnetic tape recording is an example of an analog signal process.

19. The data on a computer hard disk is an example of a digital signal process.

20. Ferrites are nonmetallic materials that exhibit the ferromagnetic properties of iron.

21. Ferrites have high permeabilities but are an electrical insulator.

22. Ferrites are used as magnetic cores in high frequency applications.

23. There is no known insulator for magnetic flux.

24. Soft iron is used as a magnetic shield for sensitive instruments that cannot be exposed to magnetic flux.

25. Magnetomotive force (mmf) is the magnetic pressure that produces the flux.

26. Flux density (B) is equal to the number of magnetic lines of flux (Φ) per square meter.

27. Field intensity (H) is a measurement of the mmf needed to establish a certain flux density in a unit length of the coil.

28. The relationship between B and H is graphed as the B-H hysteresis curve.

29. Hysteresis refers to the lagging effect between flux density and field intensity.

30. The energy wasted in heat as the dipoles lag the magnetizing force is called the hysteresis loss.

31. A Hall effect sensor is used to measure the flux density of a magnetic field.

■ IMPORTANT EQUATIONS

$$\mathscr{F} = NI \tag{14.1}$$

$$\mathscr{R} = \frac{\mathscr{F}}{\Phi} \tag{14.2}$$

$$B = \frac{\Phi}{A} \tag{14.3}$$

$$H = \frac{\mathscr{F}}{l} \tag{14.4}$$

$$\mu = \mu_r \times \mu_o \tag{14.5}$$

$$\mu = \frac{B}{H} \tag{14.6}$$

■ REVIEW QUESTIONS

1. Define magnetism.

2. Name a natural magnetic material.

3. For what applications were the first magnets used?

4. What is the difference between a temporary magnet and a permanent magnet?

5. Describe Weber's theory.

6. Describe the domain theory.

7. What is the Curie point?

8. Describe the process of magnetic induction.

9. What are flux lines?

10. Describe the conditions for attraction or repulsion between poles of two magnets.

11. Why is iron considered a ferromagnetic material?

12. What are the differences between hard and soft magnets?

13. Describe the nature of a permanent magnet.

14. Describe an application for a permanent magnet.

15. Describe the nature of a temporary magnet.

16. What does the fringing field emanating from a magnetic recording head do?

17. What are ferrites?

18. Where are ferrites used?

19. How does soft iron form a magnetic shield?

20. Define the term *magnetomotive force* (mmf).

21. Define reluctance.

22. Describe flux density.

23. Describe field intensity.

24. What is the permeability of free space?

25. Describe the effects of magnetic hysteresis.

26. Describe the use of a Hall effect sensor.

27. Why is a gaussmeter calibrated in units of gauss?

28. Describe an application for a temporary magnet.

■ CRITICAL THINKING

1. Explain how a spinning electron produces a magnetic field.

2. Using the characteristics of magnetic lines of force, explain how an attraction exists between unlike poles of two magnets.

3. Compare and contrast the characteristics of magnetic recording in an analog signal process and a digital signal process.

4. Describe how a B-H curve is formed and what affects the width of the curve.

5. Describe how the Hall effect is used by a Hall sensor in a gauss-meter.

■ PROBLEMS

1. Calculate the mmf for a coil with 20,000 turns and a 2.0-mA current.

2. Calculate the A·t for a coil with 120 turns and 75 mA of current.

3. How much current is necessary to produce 8000 A·t in a coil with 20,000 turns?

4. How much current is necessary to produce 400 A·t in a coil with 10,000 turns?

5. How many turns must a coil have to produce 800 A·t if it has 50 mA of current through it?

6. Calculate the reluctance of a magnetic material that has a mmf of 50 A·t and a total flux of 40 μWb.

7. A coil has an mmf of 100 A·t and a reluctance of 40×10^6 A·t/Wb. Calculate the total flux in μWb.

8. How much mmf would be required to produce a flux of 1000 μWb if the reluctance of the material is 30×10^6 A·t/Wb?

9. Calculate the reluctance when 1000 μWb of flux is produced by an mmf of 5000 A·t.

10. What is the reluctance of a material that produces 500 μWb of flux when the applied mmf is 400 A·t?

11. Find B when the total flux is 1000 μWb and the area is 0.6 m^2.

12. If $B = 3$ T, find the total flux when a core area is 0.33 m^2.

13. Find B when the total flux is 5000 μWb and the area of a coil is 0.5 cm^2.

14. Calculate H for a coil that is 25.4 mm long with 1.5 A of current and 400 T.

15. How long is a coil having an H of 5000 A·t/m, 80 T, and a current of 0.75 A?

16. Find μ when $\mu_r = 500$.

17. Find μ_r when $\mu_r = 45 \times 10^{-6}$ H/m.

18. Find μ_r of a core when $H = 50$ A·t/m and produces a B of 0.063 T.

19. Find μ when $B = 0.2$ T and $H = 100$ A·t/m.

20. Find μ when $B = 0.84$ T and $H = 200$ A·t/m.

■ ANSWERS TO PRACTICE PROBLEMS

PRACTICE PROBLEMS 14.1

1. 25 mA

2. 6 A·t

PRACTICE PROBLEMS 14.2

1. 1.25×10^6 A·t/Wb

2. 200 μWb

PRACTICE PROBLEMS 14.3

1. 0.016 T

2. 10×10^{-3} Wb

PRACTICE PROBLEMS 14.4

1. 16.67×10^3 A·t/m

2. 7.5×10^3 m

PRACTICE PROBLEMS 14.5

1. 630×10^{-6} T/(A·t/m)

2. 0.019 T

■ ANSWERS TO SELF-CHECKS

SECTION 14.1 SELF-CHECK

1. Magnetite

2. The two poles of a tiny molecular magnet

3. Field

4. When the field from a magnetic source aligns the dipoles of an unmagnetized material

SECTION 14.2 SELF-CHECK

1. Imaginary lines used to represent the force existing around a magnet

2. 1 Mx = 1 flux line, 1 Wb = 10^8 flux lines

3. Like poles repel, unlike poles attract.

SECTION 14.3 SELF-CHECK

1. Iron, cobalt, nickel

2. True

3. Diamagnetic

SECTION 14.4 SELF-CHECK

1. A permanent magnet retains its magnetism even when the magnetizing force is removed. A temporary magnet does not.

2. Fringing field

3. Analog signal processing

4. Digital signal processing

5. Each iron oxide particle is encapsulated in ceramic.

SECTION 14.5 SELF-CHECK

1. True

2. Soft iron

SECTION 14.6 SELF-CHECK

1. The magnetic pressure that produces flux

2. The opposition that a material offers to the magnetic lines of force

3. The tesla, T

4. The length of the coil

5. 1.26×10^{-6} H/m or T/(A·t/m)

SECTION 14.7 SELF-CHECK

1. Magnetic saturation occurs when a further magnetizing force does not produce an increase in the flux density.
2. Hysteresis
3. The energy wasted in heat as the dipoles lag the magnetizing force is called the hysteresis loss.
4. Directly, frequency

SECTION 14.8 SELF-CHECK

1. A magnetic field brought close to a semiconductor carrying current causes a voltage to be produced.
2. Gauss

CHAPTER 15

ELECTROMAGNETIC INDUCTION

OBJECTIVES

After completing this chapter, you will be able to:

1. Utilize the left-hand rule to determine associated parameters with conductors and coils.
2. Describe Faraday's law for electromagnetic induction.
3. State the factors that affect the amount of induced voltage.
4. Calculate the amount of induced voltage from quantities given.
5. Describe Lenz's law.
6. Describe the makeup of both DC and AC generators.
7. Compare and contrast the characteristics of DC and AC generators.
8. Describe any losses associated with generator components.
9. Describe the makeup of both DC and AC motors.

INTRODUCTION

The electron plays a very important role in magnetism. We know that a spinning electron and its orbital rotation produce a magnetic moment. From this, one can conclude that the prime link between electricity and magnetism is *motion*. Anytime a charged particle moves, a magnetic field is generated. Consequently, current flow, which is the movement of electrons, produces a magnetic field. This fact was discovered by Oerstad. The interrelationship between electricity and magnetism was defined as electromagnetism in the previous chapter.

In 1831, the English physicist Michael Faraday further explored Oerstad's work in electromagnetism. He found that the process could be reversed. Faraday observed that if a conductor were passed through a magnetic field, a voltage would be induced across the conductor. If the conductor were part of a complete circuit, a current could flow as well. This phenomenon is referred to as *electromagnetic induction*. Properties of this term and additional applications for electromagnetic induction are covered in this chapter.

15.1   ELECTROMAGNETISM'S LEFT-HAND RULE

15.1.1   Left-Hand Rule for Conductors

We know that a magnetic field is generated by a current-carrying conductor. The relationship between the direction of the magnetic lines of force around a conductor and the direction of current in the conductor is determined by means of the left-hand rule. (The left-hand rule makes the assumption that the current flow in the conductor is electron flow. If conventional current flow is used, the right-hand rule is invoked.) If you grasp the conductor in your left hand with the thumb extended in the direction of electron flow ($-$ to $+$), your fingers point in the direction of the magnetic lines of force. This is illustrated in Figure 15.1.

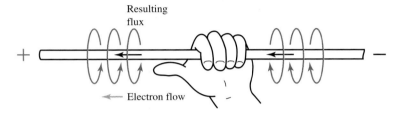

FIGURE 15.1
The left-hand rule for a conductor. The thumb points in the direction of current flow, and the fingers indicate the direction of the magnetic lines of force.

Notice that the magnetic lines are circular, because the field is symmetrical with respect to the wire in the center. Also, the magnetic field with circular lines of force is in a plane perpendicular (at right angles) to the current in the wire. The field strength decreases inversely as the square of the distance from the conductor. In addition, the current does not need to be within a conductor. An electron beam, produced within various electronic devices, has an associated magnetic field. In all cases, the magnetic field has circular lines of force in a plane perpendicular to the direction of motion of the current. If the direction of the current changes, the resulting magnetic field reverses as well.

When parallel conductors carry current in the same direction, the magnetic lines of force combine and increase the strength of the field, as shown in Figure 15.2(a). Parallel

FIGURE 15.2
Magnetic field around two parallel conductors.

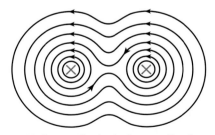

(a) Currents flowing in the same direction

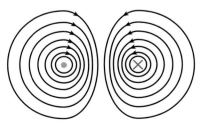

(b) Currents flowing in opposite directions

conductors carrying currents in opposite directions are shown in Figure 15.2(b). Notice that the field around one conductor is opposite in direction to the field around the other. The resulting flux lines oppose each other in the space between the wires, thus deforming the field around each conductor. From this we can conclude that two conductors carrying current in the same direction have the fields of the two conductors aiding each other. Conversely, if the two conductors are carrying current in opposite directions, the fields about the conductors repel each other. For indicating purposes, an arrow is usually used to show the direction of current in electrical circuits. Currents shown flowing away from the observer use a cross (the tail end of an arrow), whereas currents shown flowing toward the observer use a dot (the head of an arrow).

15.1.2 Left-Hand Rule for Coils

Bending a straight conductor into the shape of a loop has two effects (Figure 15.3(a)). First, the magnetic field lines are more dense inside the loop. Their number is the same as in a straight conductor, but now they are concentrated in a smaller space. Also, the lines inside the loop are all in the same direction. This is illustrated in Figure 15.3(b).

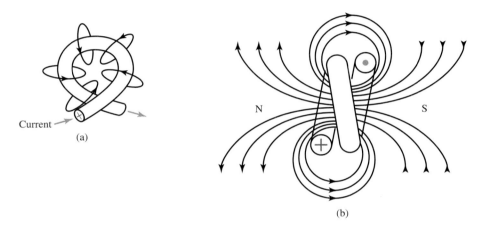

FIGURE 15.3
Flux lines around a loop of wire: (a) concentrated flux lines within the loop; (b) lines are in the same direction.

Notice that this action also creates north and south poles. In fact, the loop of wire has all the properties of a permanent magnet. The difference is that you can vary the intensity of the magnetic field by changing the amount of current through the loop. Another way to increase the intensity of the magnetic field is to add more turns of wire. This forms a *coil*. By adding an iron core, even further strengthening of the magnetic field occurs.

To determine the magnetic polarity, use the left-hand rule, as illustrated in Figure 15.4. By grasping the coil with the fingers of the left hand curled around the coil in the direction of the electron flow, the thumb points in the direction of the north pole.

The magnetic polarity depends on the direction of the current flow and the direction of the windings. The orientation of polarity of the voltage source determines the direction of the current flow. The direction of the windings is a manual adjustment. If either of the aforementioned changes directions, the polarity of the magnetic poles reverses as well.

SECTION 15.1 SELF-CHECK

Answers are given at the end of the chapter.

1. Describe how to use the left-hand rule for conductors.
2. When using the left-hand rule for coils, the thumb points in the direction of the _____ pole.

FIGURE 15.4
The left-hand rule for coils: (a) north pole to the right;
(b) current reverses, north pole to the left.

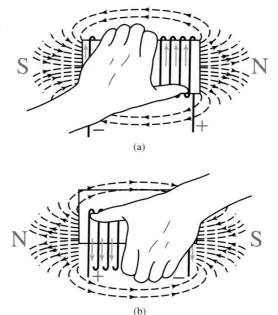

(a)

(b)

Lab Reference: The left-hand
rule for coils is demonstrated in
Exercise 23.

THE LAWS OF INDUCED VOLTAGE

15.2.1 Faraday's Law

Faraday discovered that a voltage is induced by magnetic flux cutting the turns of a coil. A magnetic linkage exists between the flux and the coil. This is known as *flux linkage*. The amount of induced voltage depends on the number of turns on the coil and how fast the flux moves across the conductor. Either the flux or the conductor can move; the important point is that there is *relative* motion between the two.

The voltage that is associated with a changing magnetic field is referred to as an *induced emf*, or an *induced voltage*. Induced voltages that are developed by mechanical motion between conductors and magnetic fields are called *generated voltages*. The induced voltage is generated because a separation in charge (electrons) occurs as the magnetic field cuts through the coil.

There are three basic factors that determine the magnitude of an induced emf:

1. The number of turns in the coil
2. The strength (number of flux lines) of the magnetic field
3. The relative speed (rate of change) between the coil and the magnetic field

It is very important to realize here that the flux must be moving (that is, expanding or collapsing) in order for there to be an induced voltage across the ends of the coil. A stationary magnetic field cannot induce a voltage in a stationary coil.

Mathematically, this may be expressed as

$$V_{\text{ind}} = N\left(\frac{\Delta\Phi}{\Delta t}\right) \tag{15.1}$$

where V_{ind} = voltage induced (V)
N = number of turns of the coil
$\Delta\Phi$ = change in flux (Wb)
Δt = change in time (s)

(Notice the symbol Δ in $\Delta\Phi$ and Δt represents *a change in*. Some texts use *d* in place of Δ.)

The ratio $\Delta\Phi/\Delta t$ represents the rate of change and is constant, assuming the flux is changing at a linear rate. For induced voltage, only this ratio is important, not the actual value of flux.

EXAMPLE 15.1

A magnetic flux of 1000 μWb cuts across a coil of 2000 turns in 0.1 s. Calculate the induced voltage.

Solution: $V_{\text{ind}} = (2000 \text{ turns}) \dfrac{1000 \times 10^{-6} \text{ Wb}}{0.1 \text{ s}}$

$= 20 \text{ V}$

Practice Problems 15.1 Answers are given at the end of the chapter.
1. A magnetic field cuts across a coil having 2000 turns at the rate of 2000 μWb/s. Calculate the induced voltage.
2. A coil of 1800 turns has an induced voltage of 100 V. Calculate the rate of flux change ($\Delta\Phi/\Delta t$) in Wb/s.
3. A coil of 20,000 turns has a change in flux of 200 Mx/s. Calculate the induced voltage. (Note the change in flux is in Mx, not Wb.)

15.2.2 Lenz's Law

About the same time that Faraday was experimenting with electromagnetic induction, a German physicist, Heinrich Lenz, was doing similar experiments. He concluded that the current induced in a coil due to a change in the magnetic flux is such that it opposes the cause producing it. This is known as *Lenz's law*. Another way of stating it is to say that the induced emf in a circuit causes a current to flow in such a direction that the magnetic field established by the current reacts in such a way as to stop or oppose the motion that generated the emf.

Common sense tells us that this must be the case. If the induced emf causes a current to flow and that current, in turn, generates a magnetic field that aids the original motion that produced the emf, we would have a perpetual motion machine.

**SECTION 15.2
SELF-CHECK**

Answers are given at the end of the chapter.

1. Describe Faraday's law.
2. What factors affect the magnitude of induced voltage?
3. True or false: A stationary magnetic field can induce a voltage.
4. Describe Lenz's law.

15.3 AC GENERATORS

15.3.1 AC Generator Characteristics

The *AC generator,* or *alternator,* is an example of a device that converts mechanical energy into electrical energy using the principles of electromagnetic induction. The amount of voltage generated depends on the following:

1. The strength of the magnetic field
2. The angle at which the conductor cuts the magnetic field
3. The speed at which the conductor is moved
4. The length of the conductor within the magnetic field

The polarity of the voltage depends on the direction of the magnetic lines of flux and the direction of movement of the conductor. To determine the direction of current flow in a given situation, the left-hand rule for generators is used.

The rule is explained in the following manner. Extend the thumb, forefinger, and middle finger of your left hand at right angles to one another, as shown in Figure 15.5. Point your thumb in the direction the conductor is being moved. Point your forefinger in the direction of the magnetic flux (from north to south). The middle finger then points in the direction of current flow in an external circuit to which the voltage is applied.

FIGURE 15.5
The left-hand rule for generators.

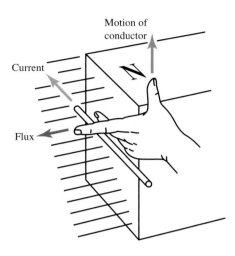

15.3.2 The Elementary AC Generator

An elementary AC generator (Figure 15.6) consists of a wire loop placed so that it can be rotated in a stationary magnetic field. This produces an induced emf in the loop. Sliding contacts (brushes) connect the loop to an external circuit load in order to pick up or use the induced emf.

The pole pieces (marked N and S) provide the magnetic field. The pole pieces are shaped and positioned as shown to concentrate the magnetic field as close as possible to the wire loop. The loop of wire that rotates through the field is called the *armature*. The ends of the armature loop are connected to rings called *slip rings*. They rotate with the armature. The *brushes,* usually made of carbon with leads attached to them, ride against the rings. The generated voltage appears across these brushes.

FIGURE 15.6
An elementary generator.

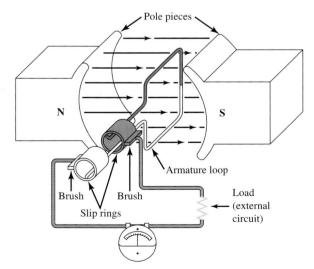

A voltage (potential difference) is produced in the following manner. The armature loop is rotated in a clockwise manner (Figure 15.7). Its initial, or starting, position is shown in *A.* (This will be considered the 0° position.) At 0° the armature loop is perpendicular to the magnetic field. The black-and-white conductors are moving parallel to the magnetic field; thus, they do not cut any lines of force. The induced emf is zero, as the meter shows at *A*. As the armature loop rotates from position *A* to *B*, the conductors cut through more and more lines of force at a continually increasing angle. At 90° (position *B*), they are cutting through a maximum number of lines of force and at a maximum angle. The result is that between 0° and 90°, the induced emf in the conductors builds up from zero to a maximum value. Observe that from 0° to 90°, the black conductor *cuts down* through the field. At the same time the white conductor *cuts up* through the field. The induced emfs in the conductors are series-aiding, the result of the same relative motion between the armature and the field causing more flux lines to be cut. The meter at position *B* reads a maximum value.

As the armature loop continues rotating from position *B* (90°) to position *C* (180°), the conductors that were cutting through a maximum number of lines of force at position *B* now cut through fewer lines. At position *C*, they are again moving parallel to the magnetic field. Thus, from 90° to 180°, the induced voltage decreases to zero. However, the polarity of the induced voltage has remained the same. This is shown by *A* through *C* on the graph. As the loop starts rotating beyond 180°, from position *C* through *D*, and back to position *A*, the voltage is in the direction opposite to that shown from position *A*, *B*, and *C*. The magnitude of the voltage is the same as it was from *A* to *C* except for its reversed polarity (as shown by meter deflection in *D*). The voltage output waveform for one complete revolution of the loop is shown on the graph in Figure 15.7 and is a single-phase voltage.

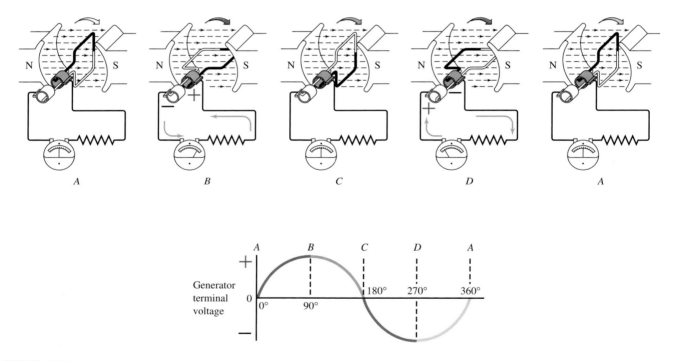

FIGURE 15.7
Output voltage of an elementary generator during one revolution.

15.3.3 The Practical AC Generator

All electrical generators depend on the principle of magnetic induction. The part of a generator that produces the magnetic field is called the *field*. The part in which the voltage is induced is called the *armature*. For relative motion to take place between the armature and the magnetic field, all generators must have two mechanical parts—a rotor and a stator. The *rotor* is the part that rotates; the *stator* is the part that remains stationary. In AC generators, the armature may be either the rotor or stator. Both rotating-armature and rotating-field alternators are available (Figure 15.8(a) and (b)). The rotating-field alternator has a stationary armature winding that is directly connected to the load. This type is usually used in high-voltage applications because it doesn't have high AC voltage on its slip rings. (High-voltage on slip rings can cause arcover.)

FIGURE 15.8
Types of ac generators: (a) rotating armature alternator; (b) rotating field alternator.

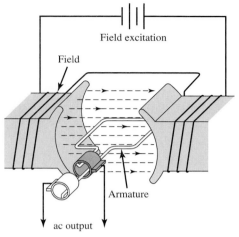

(a) Rotating armature alternator

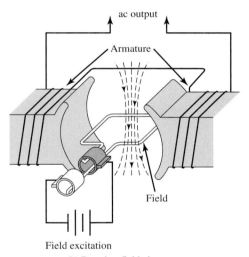

(b) Rotating field alternator

A practical alternator uses an electromagnet to provide a stationary magnetic field (stator). Figure 15.9 shows the stator windings used in the rotating-field alternator.

The stator windings may be wired to provide single- or three-phase voltage outputs. The three-phase is very common because of its efficiency. It contains three single-phase windings spaced so that the voltage induced in any one phase is displaced by 120° from the other two (Figure 15.10(a)).

Rather than have six leads coming out of the three-phase alternator, the stator coils are joined together in either wye or delta connections, as shown in Figure 15.10(b). This allows for a convenient connection to three-phase motors or power-distribution transformers.

In the wye-connected alternator, the total voltage across any two of the three line leads is the vector sum of the individual phase voltages. Each line voltage is 1.73 times one of the phase voltages. Because the windings form only one path for current flow between phases, the line and phase currents are equal.

FIGURE 15.9
Stationary armature windings.

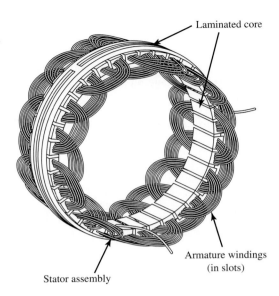

In the delta-connected alternator, the line voltages are equal to the phase voltages, but each line current is equal to 1.73 times the phase current. Because of this flexibility and increased current or voltage output, the three-phase alternator is the most-used type of alternator.

FIGURE 15.10
Three-phase alternator or transformer connections: (a) windings A, B, and C produce voltages that are displaced 120°; (b) windings connected in either delta or wye configuration.

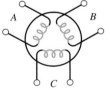

Three-phase alternator

(a)

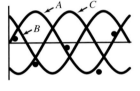

Delta connected   Wye connected

(b)

15.4 DC GENERATORS

15.4.1 The Basic DC Generator

A *basic DC generator* results when you replace the slip rings of an elementary generator with a two-piece *commutator,* changing the output voltage from AC to pulsating DC (Figure 15.11). The commutator mechanically reverses the armature-loop connections to the external circuit. This occurs at the same instant that the polarity of the voltage in the armature loop reverses.

FIGURE 15.11
A basic dc generator.

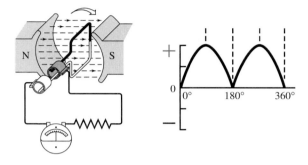

15.4.2 The Practical DC Generator

A *practical DC generator* adds additional coils (windings) on the armature. The commutator is also changed accordingly. The number of segments must equal the number of armature coils. They also use more than one pair of magnetic poles. The increased number of magnetic poles provides a stronger magnetic field (greater number of flux lines). Utilizing both, the additional coils cut through greater flux, which increases the magnitude of the output voltage. In addition, the output voltage has a higher average value (with less ripple voltage), as illustrated in Figure 15.12.

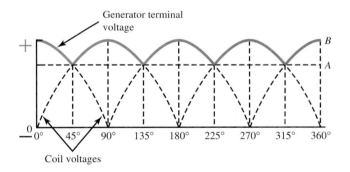

FIGURE 15.12
A multiple-cell armature decreases the ripple voltage in the output and increases the output voltage.

The field windings of a generator are supplied with energy by the generator itself or from an outside source. A *self-excited generator* is one that supplies its own field excitation. This is possible only if the field-pole pieces have retained some residual magnetism.

Self-excited generators are classified according to the type of field connection they use. These fields may be connected to the armature in a series, shunt, or compound-wound arrangement. These arrangements give correspondingly different output voltage–load current proportions. The compound generator is used in many applications because it provides a relatively constant voltage under varying load conditions. The various types of windings and graphs of their output voltage versus load current are shown in Figure 15.13.

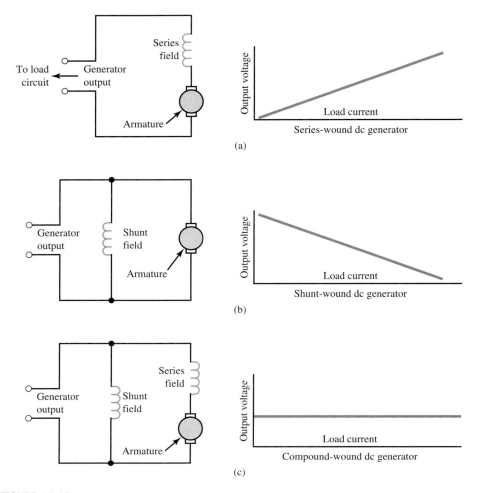

FIGURE 15.13
Self-excited generators and output voltage-load current proportions: (a) series field; (b) shunt field; (c) compound field.

15.5 GENERATOR LOSSES

In the armature, certain forces act to decrease the efficiency. These forces are considered losses and may be defined as follows:

1. I^2R, or copper losses in the winding
2. Eddy current loss in the core material
3. Hysteresis loss (a sort of magnetic friction)

15.5.1 Copper Losses

The power lost in the form of heat in the armature winding is known as *copper loss*. Heat is generated anytime current flows through a resistance. Copper loss is minimized in armature windings by using large-diameter wire to reduce its resistance.

15.5.2 Eddy Current Losses

The core of an armature is made from soft iron. Any conductor has currents induced in it when it is rotated in a magnetic field. The currents induced in the armature core are called *eddy currents*. The power dissipated in the form of heat as a result of the eddy currents is considered a loss.

Eddy currents, just like any other electrical currents, are affected by the resistance of the material in which the currents flow. The resistance of the armature is inversely proportional to its cross-sectional area. By laminating the core and insulating the laminations from one another, the eddy currents are kept appreciably small, because a lamination has an increased resistance due to its smaller cross-sectional area. Most generators use armatures with laminated cores to reduce eddy current losses.

15.5.3 Hysteresis Losses

Hysteresis loss is a heat loss caused by the magnetic properties of the armature. Because of armature rotation, the magnetic dipoles of the armature keep changing direction. This molecular friction produces heat. This heat is transmitted to the armature winding and causes the armature resistance to increase.

To compensate for hysteresis losses, heat-treated silicon-steel laminations are used in most DC generator armatures. The heat treatment causes an annealing process that reduces the hysteresis loss to a low value.

**SECTION 15.5
SELF-CHECK**

Answers are given at the end of the chapter.

1. Describe a copper loss.
2. How are eddy currents reduced in an armature?
3. How is hysteresis loss reduced in an armature?

15.6 DC MOTORS

The motor does the opposite of a generator; it converts electrical energy into mechanical energy. It utilizes many of the same elements contained in a generator. There is a definite relationship between the direction of the magnetic field, the direction of current in the conductor, and the direction in which the conductor tends to move. This relationship is called the *right-hand rule* for motors (Figure 15.14).

To find the direction of motion of a conductor, extend your thumb, index finger, and middle finger on the right hand so they are at right angles to each other. With your forefinger pointing in the direction of the magnetic flux (north to south) and your middle finger pointing in the direction of current flow in the conductor, then your thumb points in the direction of conductor motion. Thus, in a simple DC motor, current flow through the armature coil causes it to act as a magnet. The armature poles are attracted to field poles of opposite polarity, causing a torque (thrust or force) on the armature that causes it to rotate.

FIGURE 15.14
Right-hand rule for motors.

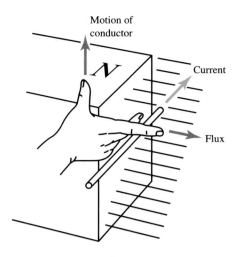

Figure 15.15 illustrates how torque is produced when a current-carrying conductor is placed in a magnetic field. In part (A) of the figure, the two fields below the conductor oppose, whereas above the conductor the two fields aid. This interaction of magnetic fields generates a force that pushes the conductor down and out of the magnetic field. The resulting motion forms the basis of operation for the DC motor. In part (b) of the figure, the current in the conductor is reversed, and the resulting torque is likewise reversed.

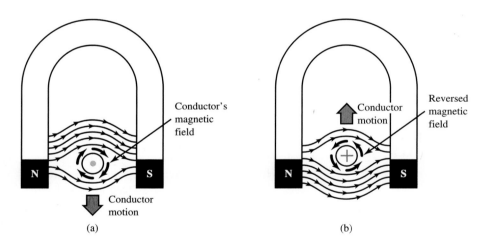

FIGURE 15.15
Torque results when a current-carrying conductor is placed in a magnetic field: (a) current flow is toward the observer, torque is downward; (b) current flow is away from observer, torque is upward.

15.6.1 Types of DC Motors

Series Motors

In *series motors,* the field windings are connected in series with the armature coil. The field strength varies with changes in armature current. When its speed is reduced by a load, the series motor develops greater torque. Its starting torque is greater than other types of DC motors. Its speed varies widely between full-load and no-load.

Shunt Motors

In *shunt motors,* the field windings are connected in parallel (shunt) across the armature coil. The field strength is independent of the armature current. Shunt-motor speed varies only slightly with changes in load, and the starting torque is less than that of other types of DC motors.

Compound Motors

In *compound motors,* one set of field windings is connected in series with the armature, and one set is connected in parallel. The speed and torque characteristics are a combination of both series and shunt motors.

SECTION 15.6 SELF-CHECK	Answers are given at the end of the chapter. 1. A motor converts _____ energy into _____ energy. 2. The _____-_____ rule is used for motors. 3. True or false: In shunt motors, speed varies only slightly with changes in load.

15.7 AC MOTORS

Because of the availability of AC power, AC motors are quite commonplace. In general, AC motors cost less than DC motors. Most AC motors do not use brushes and commutators, which tends to reduce maintenance costs.

An AC motor is particularly well suited to constant-speed applications. This is because its speed is determined by the frequency of the AC voltage applied to the motor terminals and by the number of poles it has. The DC motor is better suited for variable speed applications.

15.7.1 Types of AC Motors

There are generally three types of AC motors: series, synchronous, and induction AC motors.

Series AC Motors

Series AC motors are nearly identical to series DC motors. Special construction techniques allow series motors to be used as universal motors for either AC or DC. Rotating fields are developed by applying multiphase voltages to stator windings, which consist of multiple field coils. This rotating magnetic field causes the rotor to be pushed and pulled because of interaction between it and the rotor's own field.

Synchronous Motors

Synchronous motors are specifically designed to maintain constant speed, with the rotor synchronous to the rotating field. Synchronous motors require modification (such as squirrel-cage windings) to be self-starting.

Induction Motors

Induction motors are the most commonly used of all electric motors due to their simplicity and low cost. Induction motors may be single-phase or multiphase. They do not require an electrical rotor connection. Split-phase motors with special starting windings and shaded-pole motors are types of induction motors. Split-phase motors are designed to use inductance, capacitance, or resistance to cause a phase-shift to develop a starting torque. For greater starting torque, the capacitor-start split-phase induction motor is used.

Figure 15.16 illustrates two types of split-phase induction motors. Notice that in both types a *starting* winding is employed. As soon as the motor reaches its operating speed, a centrifugal switch opens the starting winding circuit. The other coil, the *running* winding, remains in the circuit.

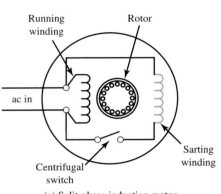

(a) Split-phase induction motor

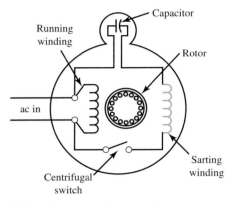

(b) Capacitor start, split-phase induction motor

FIGURE 15.16
Schematic drawing of motor circuits: (a) split-phase induction motor; (b) capacitor-start, split-phase induction motor.

**SECTION 15.7
SELF-CHECK**

Answers are given at the end of the chapter.

1. An AC motor is particularly well suited for _____ _____ applications.
2. The DC motor is better suited for _____ _____ applications.
3. Split phase induction motors employ a _____ winding in order to create a starting torque.

■ SUMMARY

1. The prime link between electricity and magnetism is motion (charges in motion).
2. A magnetic field is generated by a current-carrying conductor.
3. In a conductor, the left-hand rule is used to determine the direction of the magnetic lines of force.
4. In a coil, the left-hand rule is used to determine the direction of the north pole.
5. The magnetic polarity of the coil depends on the direction of current flow and the direction of the windings.
6. Faraday discovered that a voltage is induced by magnetic flux cutting the turns of a coil.
7. There must be relative motion between the flux and the turns of a coil for an induced voltage to be produced.
8. The induced voltage is generated because a separation in charge (electrons) occurs as the magnetic field cuts through the coil.
9. The magnitude of induced emf is determined by the number of turns in the coil, the strength of the magnetic field, and the relative speed between the coil and the magnetic field.
10. Lenz concluded that the current induced in a coil due to a change in the magnetic flux is such that it opposes the cause producing it.
11. An alternator is an example of a device that converts mechanical energy to electrical energy using the principles of electromagnetic induction.
12. An alternator uses slip rings to provide a means for the generated voltage to be taken from the device.
13. In a generator, the part that produces the magnetic field is called the field, whereas the part in which the voltage is induced is called the armature.
14. Stator coils in a three-phase alternator are either connected in a wye or delta configuration.
15. A wye configuration produces a higher output voltage; a delta configuration produces a higher output current.
16. To form a basic DC generator, the slip rings are replaced with a commutator.
17. The output of the commutator is a pulsating DC waveform.
18. A self-excited generator is one that supplies its own field excitation.

19. The fields of self-excited generators are configured in either a series, shunt, or compound-wound arrangement.

20. The compound generator is popular because it provides a relatively constant voltage under varying load conditions.

21. Generator losses are caused by copper losses, eddy current losses, and hysteresis losses.

22. A motor converts electrical energy into mechanical energy.

23. The right-hand rule for motors is used to determine the relationship between the direction of the magnetic field, the direction of the current in the conductor, and direction in which the conductor tends to move.

24. The field windings in motors are connected in series, shunt, or compound connections.

25. An AC motor is particularly suited for constant-speed applications, whereas a DC motor is better suited for variable-speed applications.

26. The three types of AC motors are series, synchronous, and induction.

27. Induction motors are the most commonly used of all electric motors.

28. A multiphase induction motor requires a phase shift to develop a starting torque.

■ IMPORTANT EQUATIONS

$$V_{ind} = N\left(\frac{\Delta\Phi}{\Delta t}\right) \qquad (15.1)$$

■ REVIEW QUESTIONS

1. What is the prime link between magnetism and electricity?
2. Describe the process of electromagnetic induction.
3. What is the left-hand rule for conductors?
4. Describe the magnetic lines of force emanating from a current-carrying conductor.
5. What is the left-hand rule for coils?
6. What determines the magnetic polarity of a current-carrying coil?
7. Describe Faraday's law.
8. What is flux linkage?
9. What is the important point between flux and a conductor in order for induction to take place?
10. What factors determine the amount of induced emf?
11. Describe Lenz's law.
12. The alternator converts _____ energy into _____ energy.
13. What factors control the amount of voltage generated by an alternator?
14. Describe the components of an elementary AC generator.
15. Through what components is voltage extracted from the output of an alternator?
16. What is the difference between a rotor and a stator?
17. Which type of alternator is used in high-AC-voltage applications?

18. The practical alternator used a(n) _____ to provide a stationary magnetic field.
19. How does a three-phase voltage output differ from a single-phase output?
20. What component is used to convert an alternator to a DC generator?
21. The output of a DC generator is _____ _____.
22. Why is the compound generator more commonly used?
23. Describe copper loss and how to circumvent it.
24. Describe eddy current loss and how to circumvent it.
25. Describe hysteresis loss and how to circumvent it.
26. The motor converts _____ energy into _____ energy.
27. How are field windings connected in a DC motor?
28. What are the three types of AC motors?
29. What are AC motors particularly well suited for? DC motors?
30. Which AC motor is the most widely used of all electric motors?

■ CRITICAL THINKING

1. Explain how Lenz's law must be true.
2. How do the factors that affect the voltage generated by an alternator determine the amount of voltage?
3. Describe in detail how AC is produced by the rotation of an armature within a magnetic field.
4. Compare and contrast the wye- and delta-connected types of stator coils in alternators.
5. Describe the operation of an induction motor with respect to the types of phase-shifting components, why a phase shift is needed, and what happens after the motor reaches running speed.

■ PROBLEMS

1. The magnetic flux changes from 10,000 μWb to 4000 μWb in 2.0 ms. How much is $\Delta\Phi/\Delta t$?
2. The magnetic flux changes from 1000 μWb to 10,000 μWb in 3 μs. If a coil has 400 turns, determine the amount of induced voltage.
3. The magnetic flux around a coil is 100 μWb. If the coil has 30,000 turns, what is the induced voltage if the flux remains stationary?
4. A magnetic flux of 200 Mx cuts across a coil having 10,000 turns in 1 ms. Calculate the induced voltage.
5. A coil has 100 V induced when the rate of flux change is 10,000 μWb/s. How many turns are in the coil?

■ ANSWERS TO PRACTICE PROBLEMS

PRACTICE PROBLEMS 15.1

1. 4 V
2. 55,555 μWb/s
3. 40 mV

■ ANSWERS TO SECTION SELF-CHECKS

SECTION 15.1 SELF-CHECK

1. If you grasp the conductor in your left hand with your thumb extended in the direction of electron flow, your fingers point in the direction of the magnetic lines of force.
2. North

SECTION 15.2 SELF-CHECK

1. A voltage is induced by magnetic flux cutting the turns of a coil.
2. The number of turns, the strength of the magnetic field, and the relative speed between the coil and the magnetic field
3. False
4. The current induced in a coil due to a change in the magnetic flux is such that it opposes the cause producing it.

SECTION 15.3 SELF-CHECK

1. Mechanical, electrical
2. Slip rings
3. A rotor is the part of the generator that rotates, whereas the stator is the part that remains stationary.
4. Voltage
5. Current

SECTION 15.4 SELF-CHECK

1. Pulsating DC
2. A commutator
3. Residual magnetism
4. Compound

SECTION 15.5 SELF-CHECK

1. The power lost in the form of heat in the armature winding
2. By laminating the core
3. By using heat-treated silicon steel laminations

SECTION 15.6 SELF-CHECK

1. Mechanical, electrical
2. Right-hand
3. True

SECTION 15.7 SELF-CHECK

1. Constant-speed
2. Variable-speed
3. Starting

PART IV

AC AND TRANSFORMERS

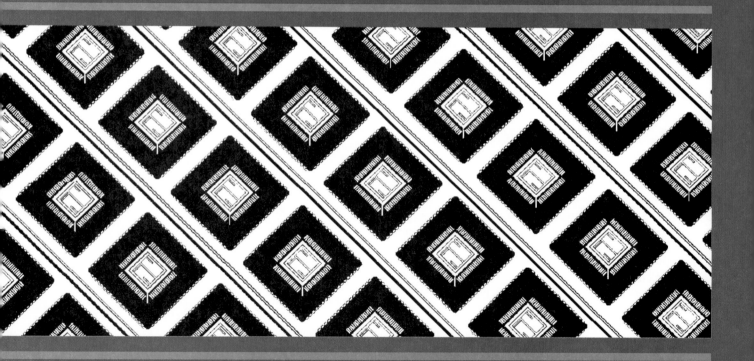

CHAPTER 16

ALTERNATING CURRENT (AC)

OBJECTIVES

After completing this chapter, you will be able to:

1. Describe why AC is the preferred form for power transmission.
2. Differentiate how AC is used for power transfer versus information transfer.
3. Compare and contrast the various amplitude levels associated with a sine wave's maximum (peak) level.
4. Calculate various amplitude values associated to a sine wave's maximum value.
5. Describe the terms *frequency, period,* and *wavelength* and calculate various unknowns.
6. Describe the phase relationships associated with AC waveforms and the use of phasor diagrams.
7. Compare and contrast nonsinusoidal waveforms.
8. Describe harmonic frequencies and show how they relate to nonsinusoidal waveform analysis.

INTRODUCTION

Up to this point, all the components and circuits we have studied have had a direct current (DC) source. A DC source is usually a battery or its electronic equivalent, the power supply. *Pure DC* is used to describe current that flows in one direction and always has the same value. In Chapter 15 we saw that a DC generator produced a DC voltage that varied in amplitude, known as *pulsating DC*. However, because this form of current does not change direction, it is still considered DC.

The commercial use of electricity began with DC generators and Thomas Edison, the famous American inventor, around the turn of the twentieth century. When a commercial DC system is used, the voltage is generated at the level (amplitude or value) required by the load. If a resistor is used to drop excess voltage to a load, power is dissipated as heat in the resistor and energy is wasted.

Transmission from a generating station over a long distance is done through wires. The amount of power lost due to the resistance of the wire is I^2R. Over long distances, this power loss can be appreciable. The DC power loss can be reduced by transmitting it at high voltage levels with correspondingly low current levels. Mathematically this is demonstrated by the following:

$$P = IV$$
$$= 10 \text{ A} \times 100 \text{ V DC} = 1 \text{ kW}$$
$$= 1 \text{ A} \times 1 \text{ kV DC} = 1 \text{ kW}$$
$$= 0.1 \text{ A} \times 10 \text{ kV DC} = 1 \text{ kW}$$

The transmission of DC at higher voltage levels is not a practical solution, because the loads would be required to work at the higher voltage levels.

Practically all commercial electric power companies generate and distribute alternating current (AC) because of the disadvantages related to transmitting DC. Unlike DC voltages, alternating voltages can be stepped up or down in amplitude by a device called a *transformer*. (The transformer is the subject of a later chapter.) The use of transformers permits an efficient transmission of electrical power over long-distance lines. At the power station, the transformer output power is a ratio of high voltage and low current levels. The low current value keeps I^2R losses to an acceptable level. At the consumer end of the transmission line, the voltage is stepped down by a transformer to the value required by the load. This is illustrated in Figure 16.1.

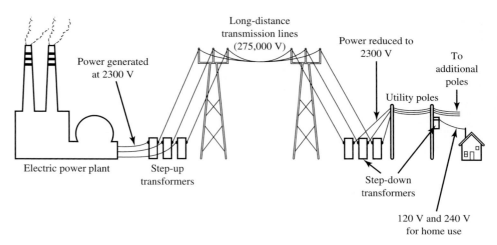

FIGURE 16.1
A typical electrical power–distribution system.

Alternating current is a very descriptive term, because AC has an amplitude that varies over time (*time-varying*). In addition, its amplitude reverses polarity each half-cycle (or at regular intervals). Figure 16.2 illustrates one cycle of AC as it appears on an oscilloscope screen.

FIGURE 16.2
One cycle of AC as it appears on an oscilloscope screen.

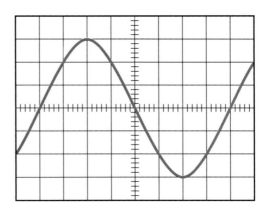

Practical *electronic circuits* require DC. To accommodate them, the AC supplied by the power station is converted to DC via rectifier circuits to form a DC power supply. Rectifiers and power supplies, as well as other forms of solid-state devices, are the subject of electronic devices texts.

AC is used for both power transfer and information transfer. Power transfer refers to the generation, transmission, and distribution of AC. Information transfer refers to how an AC waveform can convey something meaningful to the end user. Examples are signals or messages used in communication circuits (telephone, radio, television, satellites, and the like). *Electronic equipment manages the flow of information, whereas electrical equipment manages the flow of power.* These two uses of AC help define the differences between electronics and electricity.

As technicians, engineers, or scientists, we must know how AC is measured and used. To aid in this understanding, we identify three common attributes (amplitude, frequency, and phase quantities). These and additional characteristics of AC are the subject of this chapter.

16.1 THE SINE WAVE

The most basic and most used form of alternating current is called the *sinusoidal waveform,* or simply the *sine wave.* It is produced by an AC generator or by an electronic oscillator. It is a naturally occurring signal as well. The vocal cords, parts of musical instruments, and the pendulum are examples of devices that produce waveforms that resemble the sine wave or multiples of the sine wave. The term *sine* also refers to the fact that a sine wave's amplitude changes at the same rate as the sine, a trigonometric function (this is demonstrated later in the chapter).

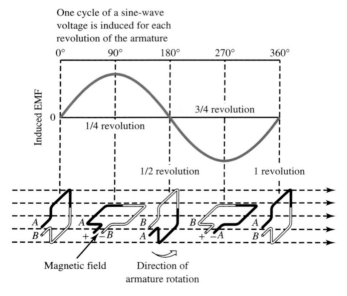

FIGURE 16.3
Generating a sine wave.

Section 15.3 showed how AC is produced via an AC generator. Figure 16.3 reviews how this is done. Note that one complete revolution of the armature produces a *cycle* of a sine-wave voltage. Because the amplitude of the sine wave is not constant, there are a variety of ways used to describe it. These are defined as follows.

16.1.1 Peak and Peak-to-Peak Values

Figure 16.4 shows the positive alternation of a sine wave (one half-cycle) and a DC waveform occurring simultaneously. Also note that the DC starts and stops at the same moment as the positive alternation and that both waveforms rise to the same maximum value. However, the DC values are greater than the corresponding AC values at all points except the point at which the positive alternation passes through its maximum value. At this point the DC and AC values are equal. This point on the sine wave is referred to as the maximum, or *peak, value* (*p,* or *pk, value*).

FIGURE 16.4
Maximum, or peak, value.

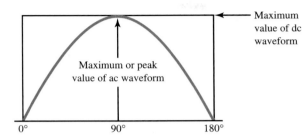

During each complete cycle of AC, there are always two maximum, or peak, values, one for the positive half-cycle and the other for the negative half-cycle. The difference between the peak positive value and the peak negative value is called the *peak-to-peak* (*pp,* or *pk-pk*) *value* of the sine wave (Figure 16.5). This value is twice the maximum, or peak, value of the sine wave and is often utilized for waveform measurements of AC voltages using an oscilloscope.

FIGURE 16.5
Peak and peak-to-peak values.

Lab Reference: The measurement of peak and peak-to-peak values is demonstrated in Exercise 25.

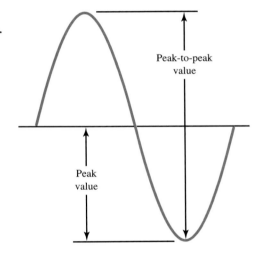

Mathematically, these values may be expressed as

$$V_p = 0.5V_{pp} \tag{16.1}$$

$$I_p = 0.5I_{pp} \tag{16.2}$$

where V_p and I_p denote peak values.

$$V_{pp} = 2V_p \tag{16.3}$$

$$I_{pp} = 2I_p \tag{16.4}$$

where V_{pp} and I_{pp} denote peak-to-peak values.

16.1.2 Average Value

Graphically, the *average value* of a waveform equals the area under the waveform divided by the base of the waveform. The true average value of a sine wave is zero, because the waveform has the same amount of positive (above-zero) area and negative (below-zero) area. Thus, the net area is zero, which makes the true average zero over the complete cycle. However, the only real difference in each alternation is the fact that the current reverses direction. To an AC load, the direction in which the current flows through the device is of little consequence. It is the work done, or heat generated, that is important.

For AC, the average value is taken as the average of all the instantaneous values during *one alternation (half-cycle) only*. This computation (derived by calculus) shows that one alternation of a sine wave has an average value equal to

$$V_{avg} = (63.7\%)V_p, \quad \text{or} \quad 0.637V_p \qquad (16.5)$$

$$I_{avg} = (63.7\%)I_p, \quad \text{or} \quad 0.637I_p \qquad (16.6)$$

Figure 16.6 shows the relationship between average and effective (rms) values.

FIGURE 16.6
Average and effective values.

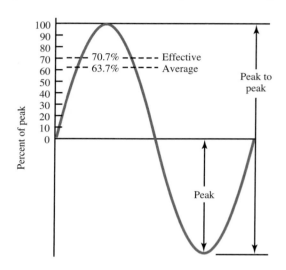

The term DC value is often used to mean average value, because a DC instrument provides a reading equal to the true average if it is used to measure an AC waveform. For this reason, a DC voltmeter reads 0 V if it is used to measure a sine-wave voltage.

EXAMPLE 16.1

Determine the average value of a positive alternation whose maximum amplitude is 200 V.

Solution $V_{avg} = 0.637V_p$
$= 0.637 \times 200 \text{ V}$
$= 127.4 \text{ V}$

Practice Problems 16.1   Answers are given at the end of the chapter.
1. Determine the average value of a sine wave whose peak current is 20 mA.
2. Determine the peak value of a sine wave whose average voltage is 20 V.

16.1.3 Effective (RMS) Value

The *effective value* is the most commonly used value for an AC waveform. This is the value of AC that has the same heating effect on a resistance as an equal value of DC. Expressed another way, the effective value is the *equivalent DC heating value* of the AC. For example, an AC sine-wave voltage whose peak is 170 V has an effective value of 120 V. When this AC source is connected to a certain resistance, it dissipates the same amount of electrical energy (heat) as a DC voltage of 120 V connected to the same resistance.

Because the heating effect varies as the square of the current (I^2R) or the square of the voltage (V^2/R), the effective value must be based on the squares of the instantaneous values and then the average taken. Thus, *rms* refers to a root-mean-square value. It is found by squaring the instantaneous values, adding these values, dividing by the number of values, and, finally, taking the square root of that value. Mathematically this equates to

$$V_{rms} = (70.7\%)V_p, \quad \text{or} \quad 0.707V_p \tag{16.7}$$

$$I_{rms} = (70.7\%)I_p, \quad \text{or} \quad 0.707I_p \tag{16.8}$$

It is often necessary to convert from rms to the peak value. This can be done as follows:

$$V_p = \left(\frac{1}{0.707}\right)V_{rms}, \quad \text{or} \quad 1.414V_{rms} \tag{16.9}$$

$$I_p = \left(\frac{1}{0.707}\right)I_{rms}, \quad \text{or} \quad 1.414I_{rms} \tag{16.10}$$

Using the peak-to-peak value as the unknown, we get

$$V_{pp} = 2.828V_{rms} \tag{16.11}$$

$$I_{pp} = 2.828I_{rms} \tag{16.12}$$

EXAMPLE 16.2

Determine the V_{rms} if the peak value of a waveform is 350 V.

Solution $V_{rms} = 0.707V_p$
$= 0.707 \times 350 \text{ V}$
$= 247.45 \text{ V}$

EXAMPLE 16.3

Determine V_{pp} when the rms of a waveform is 120 V.

Solution $V_{pp} = 2.828V_{rms}$
$= 2.828 \times 120 \text{ V}$
$= 339.36 \text{ V}$

Practice Problems 16.2 Answers are given at the end of the chapter.
1. Find the rms value of a sine wave whose peak voltage is 200 V.
2. Find the peak value of a sine wave whose rms is 120 V.
3. Find the rms value of a sine wave whose pk-pk value is 300 V.

Figure 16.7 shows the mathematical relationships between the various values of a sine wave of voltage or current in a handy table format. It shows equations 16.1 through 16.12 as well as some equations not demonstrated. Use this table as a reference whenever you are asked to do conversions involving the various amplitudes of an AC waveform.

Practice Problems 16.3 Answers are given at the end of the chapter.
1. Convert the following into peak values: (a) 15 V rms, (b) 200 V avg, (c) 200 V pp, (d) 120 V rms.
2. Convert the following into rms values: (a) 50 V pp, (b) 100 mA avg, (c) 220 V pp, (d) 75 V pk.

One final note: the rms value is so meaningful because of its DC equivalency that *all AC values are given in rms values unless otherwise specified.* This is true for component labeling, schematic diagrams, text, and the like.

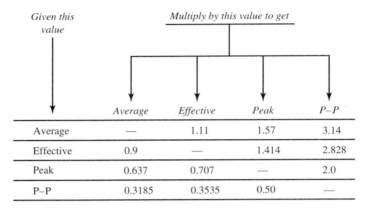

	Average	Effective	Peak	P–P
Average	—	1.11	1.57	3.14
Effective	0.9	—	1.414	2.828
Peak	0.637	0.707	—	2.0
P–P	0.3185	0.3535	0.50	—

FIGURE 16.7
Relationships between the various values of a sine wave of voltage or current.

16.1.4 Instantaneous Values

The *instantaneous value* of an alternating voltage is the value of voltage at a particular instant. The instantaneous value can be any value between and including the two extremes of zero and the plus or minus peak value. There are an infinite number of instantaneous values between zero and either peak value. Any instantaneous value is related to the sine of the angle of rotation of the armature of an AC generator at a particular instant and can be found by

$$v_i = V_p \sin \theta \qquad\qquad (16.13)$$

where $\quad v_i =$ instantaneous value (V)
$\qquad V_p =$ peak value (V)
$\quad \sin \theta =$ sine of angle θ

(Remember that the peak value occurs at an angle of 90° and 270°, because sin 90° is equal to 1.)

EXAMPLE 16.4

What is the instantaneous value of a waveform having a V_p of 150 V at angle of 30°?

Solution $\quad v_i = V_p \sin \theta$
$\qquad\qquad = (150 \text{ V}) \sin 30°$
$\qquad\qquad = 75 \text{ V}$

Note that the solution requires using the sine function. All scientific calculators have trig functions.

Optional Calculator Sequence

Using the Sharp calculator, you must enter the equation just as it is written because this calculator uses direct algebraic logic. The Sharp calculator solution to this example is as follows:

Step	Keypad Entry	Top Display Response
1	150	
2	sin	150sin
3	30	150sin
4	=	150sin30=
	Answer	75
		(bottom display)

Note that the calculator must be in the *degree mode* (not in rad or grad). On the Sharp calculator there is a degree key next to the percent key. To change modes, simply press this key until **deg** appears on the top of the LCD display.

EXAMPLE 16.5

Determine the instantaneous value when V_p is 200 V and the angle is 345°.

Solution $v_i = V_p \sin \theta$
$= (200 \text{ V})\sin 345°$
$= -51.76 \text{ V}$

Note that the instantaneous value is negative. This is because the angle is greater than 180°; thus it occurs within the negative alternation. All scientific calculators insert the negative sign automatically when the angle is greater than 180° and less than 360°.

Practice Problems 16.4 Answers are given at the end of the chapter.
1. Calculate the value of the instantaneous voltage for a sine wave with a 150-V pk at the following angles: (a) 30°, (b) 160°, (c) 245°, (d) 315°.
2. Calculate the peak value for a sine wave that has an instantaneous value of 45 V at 150°.

SECTION 16.1 SELF-CHECK

Answers are given at the end of the chapter.

1. The most basic and most used form of alternating current is the _____ _____.
2. AC has an amplitude that varies over _____.
3. AC is used for both _____ transfer and _____ transfer.
4. Define rms voltage.
5. All AC values are given in _____ unless otherwise specified.
6. For an angle greater than 180°, the instantaneous value of a sine wave has a _____ quantity.

16.2 FREQUENCY AND PERIOD

16.2.1 Frequency

A *cycle* is one positive and one negative alternation of alternating current. If an AC generator produces one complete cycle of AC during one revolution per second (360° of a sine wave), the frequency at which the AC is produced is 1 Hz. The number of complete cycles of AC completed each second is referred to as the *frequency*. Frequency is also the number of repetitions of any periodic wave in a unit of time. A *periodic* wave is one that repeats. It is not limited to sine waves; other waveforms are possible (as shown in Figure 16.9). Frequency is always measured and expressed in hertz (cycles per second), in honor of the German physicist Heinrich Hertz. In Figure 16.8(a) the frequency shown is 2 Hz, whereas in Figure 16.8(b) the frequency is four times higher (8 Hz), because there are more cycles occurring in the same amount of time.

In the United States, the power utilities have chosen 60 Hz as the generating frequency with 120 V (rms) as the household voltage. Lower frequencies cause flickering in

Lab Reference: The measurement of period and the calculation of frequency is demonstrated in Exercise 25.

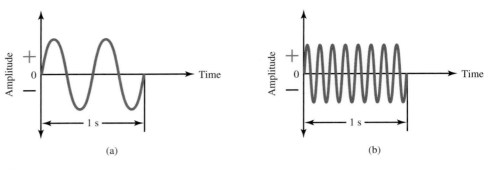

FIGURE 16.8
Frequency of a sine wave: (a) 2 cycles/s = 2 Hz; (b) 8 cycles/s = 8 Hz.

electric lights, and higher frequencies cause unacceptable power losses. England and most of Europe use an AC power line frequency of 50 Hz with 240 V to the home.

Frequency is an important term to understand, because most AC electrical equipment requires a specific frequency for proper operation (power transfer). In electronic circuits (information transfer), frequency is also important. Several bands of frequencies have been assigned for different aspects of the communications field. A representative sample of some common frequencies is shown in Table 16.1.

TABLE 16.1
Representative sample of common frequencies

Frequency	Application
60 Hz	Power-generating stations (U.S.)
20–20,000 Hz	Audio equipment
535–1605 kHz	AM radio broadcast band
26.965–27.405 MHz	CB radio band
54–60 MHz	Channel 2 TV band
88–108 MHz	FM radio broadcast band
4–6 GHz	C-band (satellite communications)
12–14 GHz	Ku-band (satellite communications)

16.2.2 Period

The *period* is the amount of time it takes a waveform to complete one cycle. The symbol T is used to represent the period of a waveform and is measured in seconds. The period of a waveform is inversely proportional to its frequency (they are reciprocals). For example, the U.S. household frequency of 60 Hz has a period of $\frac{1}{60}$ second, or 16.67 ms. The calculations are

$$T = \frac{1}{f} \tag{16.14}$$

$$f = \frac{1}{T} \tag{16.15}$$

EXAMPLE 16.6

Determine the period of a waveform with a frequency of 3500 kHz.

Solution $T = \dfrac{1}{f}$

$$= \frac{1}{3500 \times 10^3 \text{ Hz}}$$
$$= 0.286 \times 10^{-6} \text{ s}$$
$$= 0.286 \text{ } \mu\text{s}$$

EXAMPLE 16.7

What is the frequency of a waveform whose period is 120 μs?

Solution $f = \dfrac{1}{T}$

$$= \frac{1}{120 \times 10^{-6} \text{ s}}$$
$$= 8.33 \times 10^3 \text{ Hz}$$
$$= 8.33 \text{ kHz}$$

Practice Problems 16.5 Answers are given at the end of the chapter.
1. Determine the period for the following frequencies: (a) 1 kHz, (b) 50 kHz, (c) 2 MHz, (d) 4 GHz.
2. Determine the frequency for the following periods: (a) 2 ms, (b) 16.67 ms, (c) 12.5 μs, (d) 0.002 s.

Figure 16.9 shows the period of some typical periodic waveforms. Note that the period is from a point to where the same point repeats. Saying it another way, it is

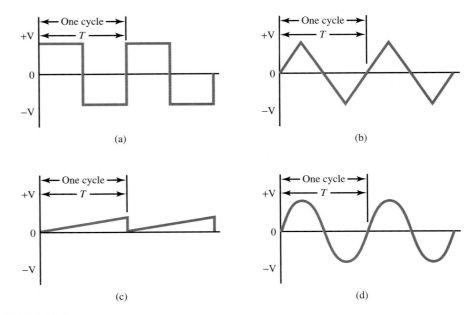

FIGURE 16.9
Typical periodic waveforms: (a) square wave; (b) triangular wave; (c) sawtooth wave; (d) sine wave.

between two successive points that have the same value and direction. For AC waveforms, this includes one positive and one negative alternation. Notice in Figure 16.9(c) that this is a DC waveform (because it does not change direction). DC waveforms *can* have a period and, thus, a frequency. (You will note this again in the study of DC power supplies, which are typically covered in a devices course.)

**SECTION 16.2
SELF-CHECK**

Answers are given at the end of the chapter.

1. A periodic wave is one that _____.
2. The frequency range for Channel 2 in the TV band is _____.
3. Define the term *period*.
4. Frequency and period are _____ proportional.
5. True or false: A DC waveform can have a frequency.

16.3   WAVELENGTH

The period was previously defined as the time it takes a waveform to complete one cycle. The distance traveled by the waveform during this time is referred to as the *wavelength* (Figure 16.10). Distance is simply velocity times time ($d = vt$). Because frequency is the reciprocal of time, the distance formula may be rewritten as $d = v/f$. For electronic circuits, wavelength (distance) is given the Greek symbol lambda, λ. Thus, in general, wavelength is found by

$$\lambda = \frac{v}{f} \qquad (16.16)$$

where λ = wavelength (m)
       v = velocity (m/s)
       f = frequency (H)

**FIGURE 16.10
*Wavelength.***

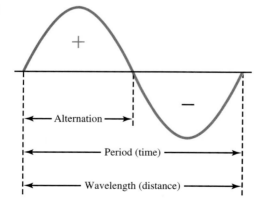

In order to calculate the wavelength, we need to know the *type* of wave that is propagating and the *velocity* of that wave. The velocity of a wave is different for an electromagnetic wave in free space versus a sound wave in free space.

16.3.1   Electromagnetic (Radio) Waves

Radio waves travel at the speed of light, or about 3.0×10^8 m/s. For convenience, this may also be expressed as 3.0×10^{10} cm/s or 3.0×10^{11} mm/s. This velocity is valid when the radio wave is propagating through free space (air or a vacuum). The higher the frequency, the shorter the wavelength, which is why a shortwave radio receiver is designed to receive high-frequency signals.

EXAMPLE 16.8

Determine the corresponding wavelengths for the following radio wave frequencies: 790 kHz, 96.7 MHz, and 6.0 GHz.

Solution $\lambda = \dfrac{v}{f}$ (here $v = 3 \times 10^8$ m/s)

$$= \frac{3.0 \times 10^8 \text{ m/s}}{790 \times 10^3 \text{ Hz}}$$

$$= 379.7 \text{ m} \qquad \text{(for 790 kHz)}$$

$$\lambda = \frac{3.0 \times 10^8 \text{ m/s}}{96.7 \times 10^6 \text{ Hz}}$$

$$= 3.1 \text{ m} \qquad \text{(for 96.7 MHz)}$$

$$\lambda = \frac{3.0 \times 10^8 \text{ m/s}}{6.0 \times 10^9 \text{ Hz}}$$

$$= 0.05 \text{ m} \qquad \text{(for 6.0 GHz)}$$

(Note that the wavelength is correspondingly shorter as the frequency increases.)

Practice Problems 16.6   Answers are given at the end of the chapter.
1. Find the wavelength for the following electromagnetic waves: (a) 60 Hz, (b) 1 kHz, (c) 102.7 Mhz, (d) 1 GHz.

16.3.2   Sound Waves

The velocity of sound waves is very low compared to radio waves. Sound waves travel through air at about 340 m/s ($\approx$1100 ft/s), depending on air temperature.

EXAMPLE 16.9

What is the wavelength in meters of the sound waves produced by a bass speaker at a frequency of 200 Hz?

Solution $\lambda = \dfrac{v}{f}$ (here $v = 340$ m/s)

$$= \frac{340 \text{ m/s}}{200 \text{ Hz}}$$

$$= 1.7 \text{ m}$$

Practice Problems 16.7   Answers are given at the end of the chapter.
1. Determine the wavelength in meters for a sound wave of 1 kHz.
2. Determine the wavelength in feet for a sound wave of 440 Hz.

SECTION 16.3 SELF-CHECK

Answers are given at the end of the chapter.

1. Define the term *wavelength*.
2. What is the velocity of a radio wave in mm/s?
3. The symbol for wavelength is _____.

16.4   PHASE RELATIONSHIPS

We have seen how AC sine waves can differ in frequency and amplitude. The third difference is in phase. *Phase* is the number of electrical degrees that one wave leads or lags another wave. The phase of a sine wave is *always relative to another sine wave of the same frequency.* The amplitude of the two waves has no bearing on the phase relationship.

Figure 16.11 shows four sine waves. Their amplitude and frequency are the same. The only difference is the phase variation between them. It is possible for one sine wave to lead or lag another sine wave by any number of degrees, except 0° or 360°. When either of these angles occurs, the two waves are said to be *in phase.* Using the waveform in *A* of Figure 16.11 as the reference, *B* is exactly in phase with *A*. The waveform in *C* crosses the zero line 90° later than the reference waveform in *A*, so the waveform in *C* is said to lag the waveform in *A* by 90°. The waveform in *D* crosses the zero line 180° after the reference, so it lags the waveform in *A* by 180°. *The terms lead or lag are relative.* It could also be said that *A* leads *D* by 180°. Note that a 180° phase angle is also termed as 180° *out of phase* because the two waveforms in question are exact opposites.

FIGURE 16.11
Phase relationships between AC sine waves.

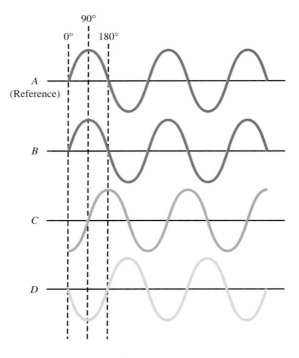

The values for the cosine function are the same as the values of the sine with the exception that for the cosine, the angle is shifted by 90° from that of the sine value. When one wave is shifted by 90° from another, it is known as a *cosine wave.* The phase angle of 90° for current and voltage waveforms has many applications in sine-wave AC circuits with components having inductance or capacitance. The cosine wave is illustrated in Figure 16.12. This figure illustrates that a cosine wave *leads* a sine wave by 90°, *or* a sine wave *lags* a cosine wave by 90°.

16.4.1   Phasor Diagrams

The AC waveforms shown so far in this section are called *time domain expressions.* These are the same waveforms that would appear on the oscilloscope if it had multiple channel inputs. Time domain diagrams utilize *magnitude values graphed in relation to time.* As is shown in Chapter 17, the oscilloscope displays the amplitude of a wave with respect to time.

To compare phases of alternating voltages, it is much more convenient to use phasor diagrams corresponding to the voltage waveforms. A *phasor* is a quantity that has both

FIGURE 16.12
Displacement between cosine and sine waves.

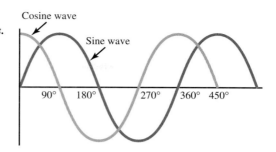

magnitude and direction. (DC quantities are known as *scalar* quantities, because scalar values have magnitude only.) Phasors are also called *frequency domain expressions,* because they contain amplitude and phase information. A *phasor diagram* shows the magnitude of a waveform relative to phase.

Figure 16.13(a) illustrates three time domain voltages sketched on the same scale. If you wish to compare the phase relationships of each voltage to the others, it is much easier and more apparent with the phasor diagram of Figure 16.13(b). The angle of the arrow with respect to the horizontal axis indicates the phase angle. A *reference waveform* (required) is always shown on the horizontal axis, representing 0°. The length of the phasor arrow represents the magnitude. A comparison of the two diagrams in Figure 16.13 shows that

$$V_1 \text{ lags } V_2 \text{ by } 45°$$

$$V_1 \text{ leads } V_3 \text{ by } 45°$$

$$V_2 \text{ leads } V_1 \text{ by } 45°$$

$$V_2 \text{ leads } V_3 \text{ by } 90°$$

$$V_3 \text{ lags } V_1 \text{ by } 45°$$

$$V_3 \text{ lags } V_2 \text{ by } 90°$$

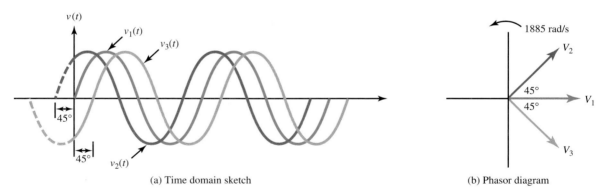

(a) Time domain sketch

(b) Phasor diagram

FIGURE 16.13
Time domain and phasor diagrams for three AC sine waves. It is much easier to compare phase relationships with phasor diagrams than with time domain sketches: (a) time domain sketch; (b) phasor diagram.

SECTION 16.4
SELF-CHECK

Answers are given at the end of the chapter.

1. Define the term *phase*.
2. Phase comparisons are done on waves having the same _____.
3. What is the difference between a sine wave and a cosine wave?
4. A phasor has both _____ and _____.
5. The length of the phasor arrow represents _____.

16.5   NONSINUSOIDAL WAVEFORMS

For power transfer, the sine wave is the most common waveform, but for information transfer, not all waveforms need be sinusoidal. In fact, the most common information waveform used today is a nonsinusoidal waveform called the *pulse waveform.* One reason for this is that it is a routine practice to take the sine-wave voice signal and convert it to a pulse wave utilizing an analog-to-digital converter. Once converted to digital, the pulse wave is processed with a digital signal-processing circuit.

Any waveform that is not a sine or cosine wave is a *nonsinusoidal waveform.* Other examples include the square (a special case of the pulse waveform), sawtooth, and triangular waveforms. These waveforms are produced by television signals, music, digital data, and so forth. Besides information transfer, they are used in electronic test equipment, in the analysis of the frequency response of amplifiers, as clock and signal sources in digital logic circuits, and in amplifier distortion analysis.

16.5.1   Pulse Wave

The pulse waveform is shown in Figure 16.14. A *pulse* is a signal of relatively short duration. The popular compact disc (CD) player (both audio and video) uses pulses via laser light. This conversion employs digital modulation techniques. An ideal pulse is a rectangle whose amplitude changes quickly from one value to another. The difference between amplitude levels $(+V$ and $-V)$ is called the *pulse amplitude,* and the length of time the signal stays at the new (high) value is called the *pulse width* (also known as *ON* time). The length of time the pulse stays low (t_L) is also known as *OFF* time. These are also shown in Figure 16.14.

FIGURE 16.14
Pulse waveform.

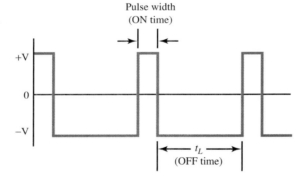

A real pulse takes a certain amount of time to reach its maximum amplitude (peak value) and return to its starting value. The amount of time a pulse takes to go from a low value to a higher value is called the *rise time.* Correspondingly, the amount of time a pulse takes to go from a high value to a lower value is called the *fall time.* For standardization, the levels of 10% and 90% are used as the reference points in order to measure either the rise or fall time. The rise time is the difference in time between the 90% point and the 10% point, whereas the fall time is the difference in time between the 10% point and the 90% point. For pulse-width measurements, the 50% level is used. This is shown in Figure 16.15.

The frequency at which the pulses occur is known as the *pulse repetition rate (PRR)* or *pulse repetition frequency (PRF).* The period is known as the *pulse repetition time (PRT).* A group of consecutive pulses is called a *pulse train.* If all the pulses in a pulse train are identical in shape and evenly spaced, the signal is considered to be a periodic (repeating) waveform. Microprocessors often employ periodic waveforms. These waveforms are considered to have a specific *signature;* thus, *signature analyzers* are available

FIGURE 16.15
Various references for pulse measurements.

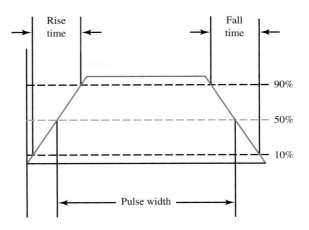

for testing microprocessor circuits. Signature analyzers have a memory capability for storing the pulse train (signature) of various test points. This allows for comparisons between signals.

When the pulse width of a pulse is multiplied by the repetition rate, a quantity known as the duty cycle is obtained. The *duty cycle* is simply a ratio of the pulse width to the period of one cycle (which is the same as multiplying the pulse width by the PRR). It may be expressed as a numerical ratio or a percent. It is found by

$$DC = \frac{PW}{T} \qquad \qquad (16.17)$$

where   DC = duty cycle as a ratio
                    (multiply by 100 to convert to a percentage)
        PW = pulse width (s)
          T = period (s)

EXAMPLE 16.10

Determine the duty cycle when the pulse width is 200 μs and the period is 1000 μs.

Solution $DC = \dfrac{PW}{T}$

$\qquad\quad = \dfrac{200\ \mu s}{1000\ \mu s}$

$\qquad\quad = 0.20, \quad \text{or} \quad 20\%$

Practice Problems 16.8   Answers are given at the end of the chapter.
1. Calculate the duty cycle for a pulse that has a PW of 5 μs and a PRR of 50 kHz.
2. If the DC is 0.33, what is the PW of a pulse having a period of 100 μs?

A higher duty cycle simply means that the load receives more average voltage (energy) and thus is on longer. As an example, a variable duty cycle DC pulse feeding a DC motor causes the motor to vary its speed (rpm). This is also known as *pulse-width modulation (PWM)*.

The average value of the pulse wave is found by multiplying the duty cycle by the peak-to-peak voltage of the waveform and adding the baseline. The baseline represents the lowest (minimum) value of the pulse waveform voltage. Because the pulse can either be an AC or DC waveform, the baseline can be negative, positive, or zero. The average value is calculated from

$$V_{avg} = \text{baseline} + (DC)(V_{pp})$$ **(16.18)**

where
V_{avg} = the average DC voltage
baseline = V_{min} of the pulse waveform
DC = duty cycle as a ratio
$V_{pp} = V_{max} - V_{min}$

EXAMPLE 16.11

Determine the average value of the pulse waveform shown in Figure 16.16.

FIGURE 16.16
Waveform for Example 16.11.

Lab Reference: Various attributes of nonsinusoidal waveforms are demonstrated in Exercise 26.

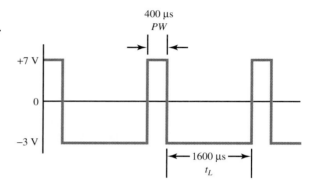

Solution
$$T = PW + \text{time pulse is low}$$
$$= 400 \ \mu s + 1600 \ \mu s = 2000 \ \mu s$$

$$DC = \frac{PW}{T}$$

$$= \frac{400 \ \mu s}{2000 \ \mu s}$$

$$= 0.20$$

$$V_{pp} = 7 \ V - (-3 \ V) = 10 \ V$$

$$V_{avg} = \text{baseline} + (DC)(V_{pp})$$

$$= -3 \ V + (0.20)(10 \ V)$$

$$= -1.0 \ V$$

Practice Problems 16.9   Answers are given at the end of the chapter.
1. For the circuit in Figure 16.17, determine V_{avg}.

FIGURE 16.17
Circuit for Practice Problems 16.9, Problem 1.

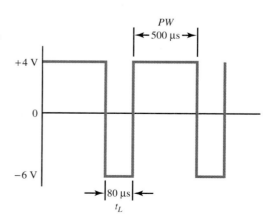

16.5.2 Square Wave

If the duty cycle is exactly 50%, the pulse is considered to be a *square wave* (a *symmetrical pulse*). This is shown in Figure 16.18(a). A square wave is a pulse having a pulse width that is one-half its period. Another way of stating this is to say that the on and off times are equal. Note that the square wave need not be an AC waveform with a zero line in the middle of the wave. In digital circuits it is common to have the zero line at either the top or bottom of the waveform; thus, the wave is considered a DC waveform. A very common form of digital logic (known as TTL logic) has zero as the bottom of the wave, with 5 V as the top, or higher, level.

FIGURE 16.18
Other nonsinusoidal waveforms.

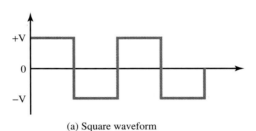

(a) Square waveform

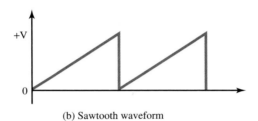

(b) Sawtooth waveform

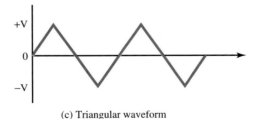

(c) Triangular waveform

16.5.3 Sawtooth Wave

A DC sawtooth wave is shown in Figure 16.18(b). It consists of two ramp voltages that occur in different amounts of time. As with other waveforms, the sawtooth can be considered an AC or DC wave, depending on where the zero line is located. Sawtooth waves are used in oscilloscopes, television receivers, and electronic triggering circuits.

16.5.4 Triangular Wave

Figure 16.18(c) shows a triangular wave. The triangular wave is defined as a periodic ramp waveform where the voltage level changes from one level to another at a constant rate. Both the positive-going and negative-going ramps are linear (direct proportionality). The waveform is considered to be symmetrical.

**SECTION 16.5
SELF-CHECK**

Answers are given at the end of the chapter.

1. The most common nonsinusoidal waveform is the _____ wave.
2. Define the term *nonsinusoidal waveform.*

3. What are the references for measuring rise and fall time?
4. Define the term *duty cycle.*
5. A square wave is a _____ pulse.
6. Where are sawtooth waves used?

16.6 HARMONIC FREQUENCIES

Consider a 100-Hz square wave, a nonsinusoidal waveform. Its fundamental repetition rate is 100 Hz. Exact whole-number multiples of the fundamental frequency are called *harmonics.* The second harmonic is 200 Hz, the third is 300 Hz, and so on. It should be noted that the series of harmonics have decreasing amplitudes proportional to their frequencies. Thus, the second harmonic has one-half the amplitude of the fundamental, the third harmonic has one-third the amplitude of the fundamental, and so on.

In the early 1800s, the French mathematician Fourier discovered that nonsinusoidal periodic waves can be represented by the sum of a constant value and a harmonic series of sinusoids. Stated another way, nonsinusoidal waveforms comprise a fundamental sine wave and a series of odd, even, or all harmonics of the fundamental sine wave. In addition, some nonsinusoidal waves contain only sine terms, whereas some contain cosine terms. The point to be emphasized is that each nonsinusoidal wave is unique and can be represented by a unique Fourier representation (a required study for engineering students). Figure 16.19 shows how a square wave corresponds to a fundamental sine wave and the odd harmonics of the fundamental. The series from Figure 16.19(a) through (e) shows how the square wave develops as more odd harmonics are added to the fundamental.

In practical terms, harmonics are useful in analyzing distorted sine waves or nonsinusoidal waveforms. An amplifier must amplify a complex input signal without affecting the signal's harmonic content. It is the harmonic content that makes one source of sound different from another with the same fundamental frequency. Harmonics are either deliberately created, such as in waveshaping circuits, or are a cause of unwanted interference, such as distortion. Because the magnitude of the harmonics of a sine wave decreases as the frequency increases, only the first few harmonics need be considered when examining the effects of harmonics in electronic components and circuits.

16.6.1 Units for Frequency Multiples

A common unit for frequency multiples is the *octave.* An octave has a frequency range of 2:1. The octave (not to be confused with octal) comes from music studies, where there are eight consecutive notes between frequencies two times apart in value.

Another common unit for frequency multiples is the *decade.* The decade has a frequency range of 10:1. When graphing frequency as a variable on semilog graph paper, it is common to plot the frequency as decade multiples across the horizontal axis of the graph paper.

SECTION 16.6 SELF-CHECK

Answers are given at the end of the chapter.

1. What are harmonic frequencies?
2. The third harmonic has _____ the frequency and _____ the amplitude of the fundamental frequency.
3. An octave has a frequency range of _____.
4. A decade has a frequency range of _____.

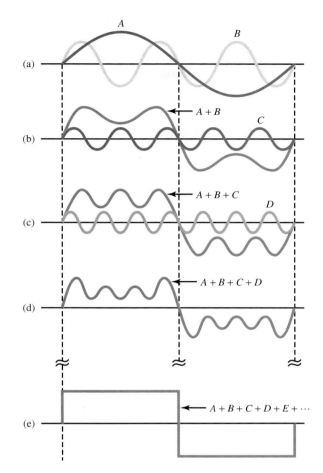

FIGURE 16.19
Square-wave composition is harmonically related: (a) fundamental and third harmonic; (b) fifth harmonic is added to fundamental and third; (c) seventh harmonic is added to fundamental, third, and fifth; (d) results of addition of fundamental, third, fifth, and seventh; (e) pure square wave produced when infinite odd harmonics are added to fundamental.

■ SUMMARY

1. Practically all commercial electric power utilities generate and distribute AC because of the disadvantages related to transmitting DC.

2. Alternating current has an amplitude that varies over time. In addition, its amplitude reverses polarity each half-cycle.

3. AC is used for both power transfer and information transfer.

4. Electronic equipment manages the flow of information, whereas electrical equipment manages the flow of power.

5. The most basic and used form of AC is the sine wave.

6. The average value of a waveform equals the area under the waveform divided by the base of the waveform.

7. The average value of a sine wave is zero.

8. The effective value (rms) is the most commonly used value for AC.

9. The effective value is the equivalent DC heating value of the AC.

10. All AC values are given in rms value unless otherwise specified.

11. The instantaneous value of an alternating voltage is the value of voltage at one particular instant.

12. The instantaneous value is related to the sine of the angle or rotation of the armature of an AC generator at a particular instant.

13. The number of cycles of AC completed each second is referred to as the frequency.

14. A periodic waveform is one that repeats.

15. The period is the amount of time it takes a waveform to complete one cycle.

16. DC waveforms can have a period and, thus, a frequency.

17. The distance traveled by a waveform during one cycle of time is referred to as the wavelength.

18. The velocity of electromagnetic (radio) waves is approximately 3.0×10^8 m/s.

19. The velocity of sound waves is approximately 340 m/s.

20. Phase is the number of electrical degrees that one wave leads or lags another wave.

21. The terms lead and lag are relative; there must be a reference waveform.

22. A cosine wave is shifted by 90° from the sine wave.

23. Time domain diagrams utilize magnitude values graphed in relation to time.

24. A phasor is a quantity that has both magnitude and direction.

25. Any waveform that is not a sine or cosine wave is a nonsinusoidal waveform.

26. A pulse is a signal of relatively short duration.

27. The length of time a pulse stays at the high value is called the pulse width.

28. A square wave is a pulse with a duty cycle of 50%.

29. Exact whole-number multiples of a fundamental frequency are called harmonics.

30. An octave has a frequency range of 2:1, and a decade has a frequency range of 10:1.

■ IMPORTANT EQUATIONS

$$V_p = 0.5V_{pp} \tag{16.1}$$

$$I_p = 0.5I_{pp} \tag{16.2}$$

$$V_{pp} = 2V_p \tag{16.3}$$

$$I_{pp} = 2I_p \tag{16.4}$$

$$V_{avg} = (63.7\%)V_p, \quad \text{or} \quad 0.637V_p \tag{16.5}$$

$$I_{avg} = (63.7\%)I_p, \quad \text{or} \quad 0.637I_p \tag{16.6}$$

$$V_{rms} = (70.7\%)V_p, \quad \text{or} \quad 0.707V_p \tag{16.7}$$

$$I_{rms} = (70.7\%)I_p, \quad \text{or} \quad 0.707I_p \tag{16.8}$$

$$V_p = \left(\frac{1}{0.707}\right)V_{rms} = 1.414V_{rms} \tag{16.9}$$

$$I_p = \left(\frac{1}{0.707}\right)I_{rms} = 1.414I_{rms} \tag{16.10}$$

$$V_{pp} = 2.828V_{rms} \tag{16.11}$$

$$I_{pp} = 2.828I_{rms} \tag{16.12}$$

$$v_i = V_p \sin \theta \tag{16.13}$$

$$T = \frac{1}{f} \tag{16.14}$$

$$f = \frac{1}{T} \tag{16.15}$$

$$\lambda = \frac{v}{f} \tag{16.16}$$

$$DC = \frac{PW}{T} \tag{16.17}$$

$$V_{avg} = \text{baseline} + (DC)(V_{pp}) \tag{16.18}$$

■ REVIEW QUESTIONS

1. Describe the term *pure DC*.
2. What causes power loss via transmission lines?
3. Why is AC preferred over DC for power transmission?
4. Describe an AC waveform.
5. What is the difference between power transfer and information transfer?
6. What is the difference between electronic equipment and electrical equipment?
7. Define the term *sine wave*.
8. Describe the ways that a sine wave can be described in terms of a voltage level.
9. Define the term *effective value*.
10. All AC values are labeled in _____ unless otherwise specified.
11. The instantaneous value of a sine wave at an angle greater than 180° is a _____ quantity.
12. Define the term *frequency*.
13. What is a periodic wave?
14. What is the frequency for C-band satellite communications?
15. Define the term *period*.
16. What is the wavelength of a waveform?
17. What is the velocity of radio waves?
18. Define the term *phase*.
19. What is a cosine wave?
20. Define the term *phasor*.
21. What does a phasor diagram show?
22. Define the term *nonsinusoidal waveform*.
23. What is a *pulse* waveform?
24. What is the difference between PRR and PRT?
25. What is the duty cycle of a waveform?
26. Compare and contrast the square, sawtooth, and triangle waveforms.
27. What are harmonic frequencies?
28. What is the amplitude of the third harmonic?
29. How is harmonic analysis used?
30. What is the difference between an octave and a decade?

■ CRITICAL THINKING

1. Describe why AC is preferred over DC for power transfer.
2. Describe the differences in how a sine wave's voltage can be characterized.
3. Describe harmonics, how they are used, and how they can describe nonsinusoidal waveforms.

■ PROBLEMS

1. Convert the following into rms values: (a) 40 V pp, (b) 1 A pk, (c) 50 V avg, (d) 30 V pp.
2. Convert the following into average values: (a) 25 mA rms, (b) 115 V rms, (c) 100 V AC, (d) 300 mA pp.
3. Convert the following into peak values: (a) 20 V rms, (b) 100 mA avg, (c) 1 kV pp, (d) 400 V rms.
4. Convert the following into peak-to-peak values: (a) 100 V avg, (b) 40 μV pk, (c) 150 mV rms, (d) 75 mV avg.
5. Calculate the instantaneous value for the following angles if

the peak value is 170 V: (a) 0°, (b) 60°, (c) 90°, (d) 150°, (e) 180°, (f) 210°, (g) 225°, (h) 315°, (i) 345°.

6. Calculate the instantaneous value for the following angles if the peak value is 120 mA: (a) 0°, (b) 45°, (c) 90°, (d) 120°, (e) 210°, (f) 270°, (g) 300°, (h) 330°, (i) 360°.

7. A sine wave of voltage has an instantaneous of 120 V when θ is 45°. Calculate the peak voltage.

8. A sine wave of voltage has an instantaneous of 15.5 V when θ is 75°. Determine the peak voltage and v_i at 150°.

9. A sine wave of voltage has $v_i = -26.8$ V at 250°. What is v_i at 80°?

10. Determine the period for the following frequencies: (a) 60 Hz, (b) 1 kHz, (c) 250 kHz, (d) 102.7 MHz, (e) 1.8 GHz.

11. Determine the frequency for the following periods: (a) 20 ms, (b) 8.33 ms, (c) 5 μs, (d) 40 μs, (e) 10 ns.

12. Calculate the wavelength in meters for electromagnetic waves having the following frequencies: (a) 60 Hz, (b) 1 kHz, (c) 125 kHz, (d) 900 MHz, (e) 2.4 GHz.

13. Calculate the wavelength in meters for sound waves having the following frequencies: (a) 40 Hz, (b) 220 Hz, (c) 400 Hz, (d) 840 Hz.

14. Calculate the frequency of a radio wave that has a wavelength of 2 m.

15. Calculate the period of a radio wave that has a wavelength of 10 cm.

16. Calculate the duty cycle of a pulse that has a PW of 25 μs and a period of 500 μs.

17. Calculate the duty cycle of a pulse that has a PW of 100 μs and a period of 133 μs.

18. A square wave has a PW of 4 ms. Calculate the PRT and PRR.

19. Calculate the average value of an AC square wave whose peak-to-peak value is 20 V.

20. Calculate the average value of the pulses shown in Figure 16.20(a)–(d).

21. What is the fifth harmonic of 5 kHz?
22. What is the second harmonic of 35 kHz?
23. What is the fourth even harmonic of 100 Hz?
24. What is the fourth octave above 12 kHz?
25. What is the frequency 3 decades below 1 MHz?

■ ANSWERS TO PRACTICE PROBLEMS

PRACTICE PROBLEMS 16.1
1. 12.7 mA
2. 31.4 V

PRACTICE PROBLEMS 16.2
1. 141.4 V
2. 169.7 V
3. 106.1 V

PRACTICE PROBLEMS 16.3
1. (a) 21.2 V (b) 314 V (c) 100 V (d) 169.7 V
2. (a) 17.7 V (b) 111 mA (c) 77.77 V (d) 53 V

PRACTICE PROBLEMS 16.4
1. (a) 75 V (b) 51.3 V (c) −135.9 V (d) −106.1 V
2. $V_{pk} = 90$ V

PRACTICE PROBLEMS 16.5
1. (a) 1 ms (b) 20 μs (c) 500 ns (d) 250 ps
2. (a) 500 Hz (b) 60 Hz (c) 80 kHz (d) 500 Hz

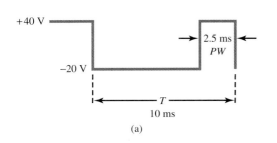
(a)

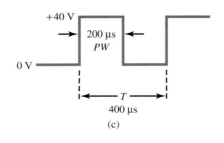
(c)

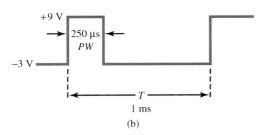
(b)

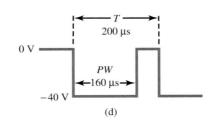
(d)

FIGURE 16.20
Circuit for Problem 20.

PRACTICE PROBLEMS 16.6

1. (a) 5×10^6 m (b) 300×10^3 m (c) 2.92 m (d) 0.3 m

PRACTICE PROBLEMS 16.7

1. 0.34 m
2. 2.5 ft

PRACTICE PROBLEMS 16.8

1. 0.25
2. 33 μs

PRACTICE PROBLEMS 16.9

1. 2.62 V

▪ ANSWERS TO SECTION SELF-CHECKS

SECTION 16.1

1. Sine wave
2. Time
3. Power, information
4. The rms voltage is the equivalent DC heating value of the AC.
5. rms
6. Negative

SECTION 16.2

1. Repeats
2. 54–60 MHz
3. The period is the amount of time it takes a waveform to complete one cycle.
4. Inversely
5. True

SECTION 16.3

1. The distance traveled by a waveform during one cycle is the wavelength.
2. 3×10^{11} mm/s
3. λ

SECTION 16.4

1. Phase is the number of electrical degrees that one wave leads or lags another wave.
2. Frequency
3. When one wave is shifted from another by 90° it is called a cosine wave.
4. Magnitude, direction
5. Magnitude

SECTION 16.5

1. Pulse waveform
2. Any waveform that is not a sine wave or cosine wave is a nonsinusoidal waveform.
3. 10% and 90%
4. The duty cycle is simply the ratio of pulse width to the period of one cycle.
5. Symmetrical
6. Sawtooth waves are used in oscilloscopes, television receivers, and electronic triggering circuits.

SECTION 16.6

1. Exact whole-number multiples of the fundamental frequency are called harmonics.
2. $3\times, \frac{1}{3}$
3. 2:1
4. 10:1

CHAPTER 17

AC TEST EQUIPMENT

OBJECTIVES

After completing this chapter, you will be able to:

1. Describe the operating characteristics of various types of test equipment.
2. Explain how an analog oscilloscope works.
3. Describe the constraints that affect oscilloscope performance.
4. Explain how a digital oscilloscope works.
5. Compare and contrast features of an analog versus a digital scope.
6. Describe the characteristics of oscilloscope passive probes.

INTRODUCTION

All AC test equipment comes in one of two forms; these are *signal-measuring instruments* or *signal-generating instruments.* In Chapter 10 we learned about basic meter movement and how it could be used to measure various quantities of DC. Both analog and digital forms of multimeters were discussed. To measure AC quantities, the basic meter movement employs the use of a rectifier to convert AC into pulsating DC. (This was shown in Chapter 10 as well.) The average value of the pulsating DC is translated into a rms value and is displayed on the multimeter analog scale. Other attributes of multimeters used for AC measurements are discussed in this chapter.

Another signal-measuring instrument and one of the most used pieces of test equipment is the oscilloscope. It can be used to measure both DC and AC values. It utilizes a *time domain representation,* with voltage plotted versus time. There are many characteristics of the oscilloscope to discuss. As with multimeters, they come in analog and digital formats. In addition to oscilloscope functions, the oscilloscope probe is discussed. The probe is a most important part of the oscilloscope and is often overlooked. It is the interface between the oscilloscope input and the circuit under test.

The spectrum analyzer is another signal-measuring instrument. It is found in most engineering labs, in communications labs, and as a portable unit for field use. It is similar to an oscilloscope, but it differs in that its display shows signal amplitude plotted as a function of frequency.

Signal-generator instruments and some specialty instruments are also covered in this chapter.

17.1 AC MULTIMETERS

17.1.1 Voltage and Current Measurements

There can be different scales to observe when using an analog multimeter for measuring AC voltage or current. (On some analog multimeters, the DC scale is in black, whereas the AC scale is in red.) Many technicians forget this fact and read the pointer off the DC scale. Examine Figure 17.1 closely (this is the same analog scale discussed in Chapter 10 and is found on the popular Simpson Model 260 VOM). There are two scales. The same range values are below the DC scale but above the AC scale. The scales are different because the AC scale is calibrated to indicate the rms value, not the average, or DC, value.

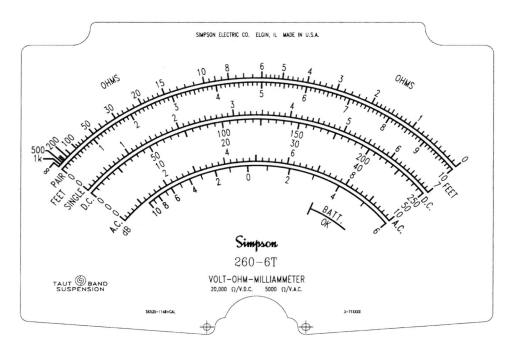

FIGURE 17.1
Note difference in AC and DC scales. Courtesy of Simpson Electric Co.

Another consideration when making analog AC measurements is that the voltmeter sensitivity of the AC meter is 5 kΩ/V (versus 20 kΩ/V for DC values). Depending on circuit values, loading can be more of a problem with AC measurements. Figure 17.2 illustrates the loading effects of the VOM in the AC volts position. Note that the correct reading across one of the 100-kΩ resistors should be 5 V rms rather than 2.5 V rms.

FIGURE 17.2
The loading effects of the VOM in the AC volts position:
(1) the VOM's sensitivity is 50 kΩ;
(2) R_2's resistance lowers to 33.3 kΩ (50 kΩ//100 kΩ);
(3) R_2's voltage drop is 2.5 V AC
($V_{R_2} = 33.3 \text{ k}\Omega/133.3 \text{ k}\Omega \times 10 \text{ V AC} = 2.5 \text{ V AC}$);
(4) R_2's voltage drop should be 5 V AC
($V_{R_2} = 100 \text{ k}\Omega/200 \text{ k}\Omega \times 10 \text{ V AC} = 5 \text{ V AC}$).

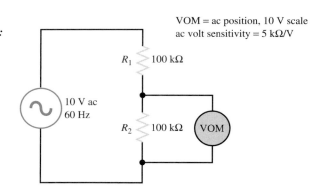

VOM = ac position, 10 V scale
ac volt sensitivity = 5 kΩ/V

R_1 100 kΩ

10 V ac
60 Hz

R_2 100 kΩ

VOM

17.1.2 Current Probes and Clamps

The ammeter requires that the current path be broken and the ammeter inserted across the break. To eliminate this problem, a current probe is available. The probe's jaw is opened and placed over a current-carrying conductor. A Hall effect device within the probe produces a voltage commensurate with the magnetic field surrounding the conductor. This is interpreted as a current value on the voltmeter scale.

If the probe and voltmeter are in one unit, it is called a *clamp meter*. Both DC and AC measuring probes are available (Figure 17.3). Figure 17.3(a) shows a typical clamp meter used by technicians and engineers, whereas Figure 17.3(b) shows a technician taking a line-power measurement with an industrial clamp meter. Industrial clamp meters are normally used to measure larger ranges of current.

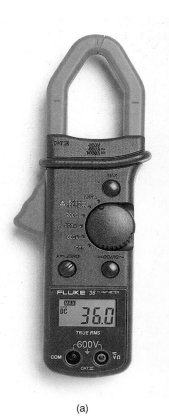

(a)

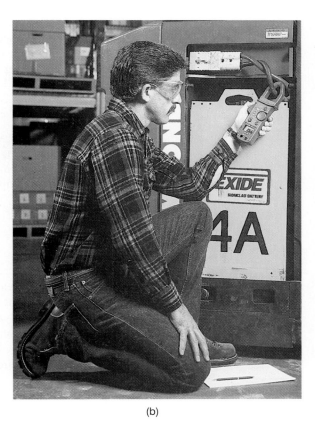

(b)

FIGURE 17.3
A clamp meter: (a) clamp meter used by technicians and engineers; (b) using an industrial clamp meter to measure line power. Courtesy of Fluke Corporation. Reproduced with permission.

17.1.3 True RMS Voltmeter

As we have already learned, an analog AC voltmeter rectifies the AC and then senses the average value. The average value is translated onto the scale as a rms value. (Remember, the rms of an AC is the most used value.) This method is valid for sine waves containing no distortion. But for distorted sine waves or nonsinusoidal waveforms, errors are introduced into the measurements. Distortion levels are quite common on AC signals.

A true rms value can be detected utilizing a thermal detector to measure a heating value. However, more modern digital meters use a digital calculation of the rms value. They obtain this by squaring the signal on a sample by sample basis, averaging it over a period of time, and then taking the square root of the result. Thus, a true rms is indicated.

17.1.4 Radio-Frequency (RF) Probes

Multimeters have a frequency limitation when measuring AC signals, which is generally about 2 kHz. Above this frequency, AC losses occur that are caused by the rectifier and the moving coil. RF probes are available for measuring higher frequencies.

These probes contain a capacitor to block any DC component and a rectifier to convert the AC to DC, which is then displayed as a rms value.

17.1.5 High-Voltage Probes

A typical voltmeter can measure voltages up to approximately 1000 V. For voltages higher than this, probes are available that contain an additional multiplier (sampling) resistor to drop the extra voltage. A common multiplier resistor uses a 100:1 dropping ratio. Thus the voltage to the meter is dropped by a factor of 100. The meter circuit of the high voltage probe is calibrated to display a value that is 100 times what is present across the probe's sampling resistor. These probes are well insulated to protect the user from a possible shock at higher voltages.

SECTION 17.1 SELF-CHECK	Answers are given at the end of the chapter. **1.** How do DC and AC analog scales differ? **2.** How is a true rms value determined? **3.** The frequency limit for an AC measurement is about _____. **4.** A common multiplier resistor for high-voltage probes uses a ×_____ dropping ratio.

17.2   OSCILLOSCOPE BASICS

The AC meters previously described provide a reasonably accurate measurement of current and voltage, but they do not allow you to "see" what an AC quantity looks like. To troubleshoot or analyze almost any electronic circuit, it is often necessary to know exactly what an AC waveform looks like. In many cases, it is necessary to know its peak and peak-to-peak values, its instantaneous values, phase relationships, and its frequency and period. These measurements, as well as others, can be performed by using a device known as an *oscilloscope.*

An oscilloscope, or *scope,* as it is commonly called, is one of the most important test instruments for measuring DC or AC quantities. The scope is effectively a voltmeter that optically displays peak-to-peak waveforms.

17.2.1   Basic Operation

A scope is capable of measuring an AC or DC voltage and displaying it in a graphical manner. The measured voltage appears as a picture on a screen that is similar to the type of screen used in a television set. The scope contains a number of controls that are used to adjust the amplitude (vertically) and the number of complete AC waveform cycles (horizontally) that are displayed. Scopes are calibrated so that the waveform presented on the screen can be analyzed and its most important characteristics determined.

A simplified block diagram of an oscilloscope is shown in Figure 17.4. The *vertical input* requires two connections to be made to measure AC or DC voltages. These terminals must be connected in parallel with the voltage source under test. The scope probe provides the interface between the vertical input terminals and the device under test. The probe contains a measuring tip and a ground-lead connector that allows for an easy interface to feed the voltage under test to the vertical input terminals.

Lab Reference: The basic operation of an oscilloscope is demonstrated in Exercise 24.

The heart of the scope is a *cathode-ray tube* (CRT), where the input signal is converted to an image on the screen of the CRT. The CRT generates an electron beam that is acted on by the vertical and horizontal amplifiers. The beam strikes a phosphor coating that illuminates, leaving an image that represents the signal under test.

A *vertical amplifier* is a multistage device that is the main factor in determining the sensitivity and bandwidth (discussed in the next section) of the scope. The *vertical sensitivity* is a measure of how much the electron beam deflects (across the *Y*-, or vertical, axis) for a specified input signal. The vertical amplifier stage contains an attenuator to match the high impedance of the scope probes (1 MΩ or 10 MΩ) to the low impedance of the vertical preamplifiers.

A *horizontal amplifier* provides the deflection voltages required to deflect the beam across the *X*-, or horizontal, axis of the CRT. A sawtooth (Chapter 16) waveform (sweep signal) is generated by the horizontal stage to provide a linearly increasing voltage that causes the beam to be deflected equal distances horizontally per unit of time. The rate at which the sawtooth wave rises establishes what is known as the *time base*.

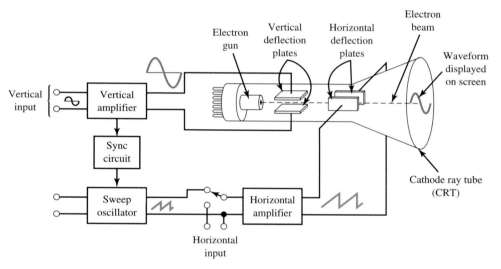

FIGURE 17.4
A basic oscilloscope.

Figure 17.5 shows how an AC input signal (waveform *A*) at the vertical input and a sawtooth waveform (waveform *B*) at the horizontal input must have an identical frequency for one cycle to be displayed by the CRT (waveform *C*). If the period of the sawtooth waveform differs from that of the input signal, then more or fewer cycles of the input signal are displayed on the CRT.

FIGURE 17.5

The scope display: Waveform B moves the beam left to right, whereas waveform A moves the beam up and down. Waveform C is the result.

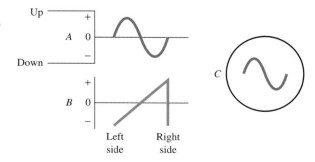

A *trigger control* is used to provide a stable waveform display. A trigger source receives a replica of the input signal, called the *sync pickoff,* to control triggering. For the waveform to be stable, the display must commence at exactly the instant that the input waveform is at its zero position. The trigger ensures that the sweep begins at the same point of a repeating signal, resulting in a clear picture. Figure 17.6 shows the difference between an untriggered display and a triggered display.

FIGURE 17.6
Triggering stabilizes a repeating waveform.

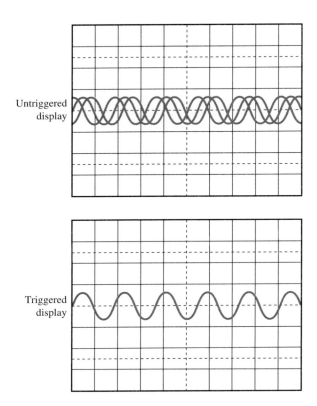

Untriggered display

Triggered display

Multichannel inputs (dual-trace, etc.) can be provided by a CRT supplying two electron beams, but for general-purpose instruments, cost precludes this. Most analog scopes provide extra channels by multiplexing through a single vertical channel. This is accomplished by using either a chop or an alternate mode.

The chop mode causes a scope to draw small parts of each signal by switching back and forth between them. The switching rate is too fast for the eye to notice, so the waveform looks whole. This mode is typically used with low-frequency signals requiring sweep speeds of 1 ms per division or less.

An alternate mode draws each channel alternately—the scope completes one sweep on channel 1, then one sweep on channel 2, a second sweep on channel 1, and so on. This mode is used with medium- to high-speed signals (higher frequency), when the sec/div scale is set to 0.5 ms or faster.

**SECTION 17.2
SELF-CHECK**

Answers are given at the end of the chapter.

1. The scope is equivalent to a _____.
2. The heart of a scope is the _____.
3. What is the trigger-control stage used for?
4. How do most analog scopes provide extra channel displays?

17.3 ANALOG OSCILLOSCOPES

The oscilloscope previously discussed is an *analog oscilloscope*. Analog scopes such as the 100-MHz, three-channel scope in Figure 17.7 work by directly applying a voltage being measured to an electron beam moving across the oscilloscope screen. AC voltage variations deflect the beam up and down proportionally. At the same time a horizontal sweep signal moves the beam horizontally. The two interactions cause a replica of the input signal to be displayed on the screen. This gives an immediate picture of the waveform. People often prefer analog oscilloscopes when it is important to display rapidly varying signals in *real time* (or as they occur). These types of scopes have been in use for many years. There are often the scope of choice for educational institutions because of cost.

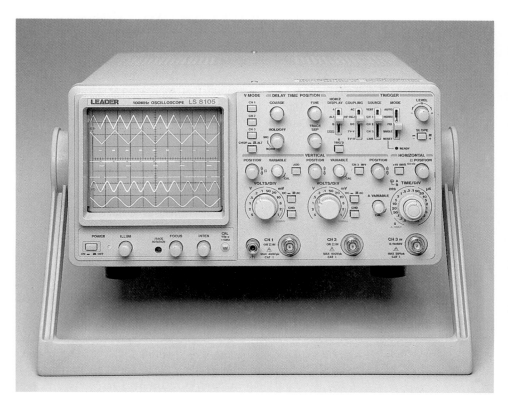

FIGURE 17.7
A 100-MHz, three-channel analog oscilloscope. Courtesy of Leader Instruments Corp.

17.3.1 Oscilloscope Constraints: Bandwidth (BW)

The vertical channels of an oscilloscope are designed for a broad bandpass, generally from some low frequency (DC) up to several megahertz. The difference between the high-frequency value (in MHz) and DC (0 Hz) is the oscilloscope's *bandwidth*. In today's technology, 100-MHz-bandwidth scopes are common. The bandwidth specification gives you an idea of the instrument's ability to handle high-frequency signals within a specified attenuation. Bandwidth is specified by listing the frequency at which a *sinusoidal* input signal has been attenuated to 0.707 of the middle frequencies. This is called the −3-dB point.

Oscilloscopes are generally used to describe waveforms that are rather complex in shape; you need to see not only the fundamental frequency but also other frequencies, which may be many times higher. Square waves, for instance, contain many odd harmonic frequencies that are more than 10 times the fundamental frequency of the signal. The *first guideline* to observe is that the scope bandwidth should be at least three times as high as the fundamental frequency of the fastest signal to be measured in order to display the signal accurately.

Bandwidth also affects the accuracy of both amplitude and timing measurements. The scope's frequency response should be fairly flat from DC to the -3-dB point. Beyond this frequency, the response drops by 6 dB/octave. The *second guideline* is to make amplitude measurements that are not dominated by the scope's frequency response. The -3-dB bandwidth of the scope should be at least 10 times the highest significant frequency in the signal to be measured. This guideline applies more to signals whose amplitude is dominated by a high-frequency component, such as a modulated RF, than to waveforms whose amplitude is dominated by a DC, or low-frequency, component.

The effect of bandwidth on timing measurements is a little less obvious. The frequency response of the scope acts as a low-pass filter (see Chapter 26 figures) in series with the measurement. You can see this effect when using the scope to measure rise time (see Figure 16.15). The measured rise time is a result of the signal's actual rise time and the scope's own rise time. The rise time is related to bandwidth. It is typically found by taking 0.35 divided by the bandwidth ($t_r = 0.35/\text{BW}$). The *third guideline* is that for timing measurements that approach the accuracy of the scope's time base, you need a scope that has a rise time at least 20 times faster than the signal you're measuring. For more normal use, a scope that is 5 times faster than the signal is acceptable.

Bandwidth isn't free. There are real costs associated with higher bandwidth. To keep the cost of bandwidth as low as possible, you need to consider what kind of signals you intend to measure before selecting which scope to use.

SECTION 17.3 SELF-CHECK

Answers are given at the end of the chapter.

1. Define the term *analog scope*.
2. An analog scope displays a signal in _____ _____.
3. How is a scope's bandwidth measured?

17.4 DIGITAL OSCILLOSCOPES

A *digital oscilloscope* samples the waveform and uses an analog-to-digital converter (ADC) to convert the voltage being measured into digital information. It then uses this digital information to reconstruct the waveform on the screen. The block diagram of a digital oscilloscope is shown in Figure 17.8.

When you connect a digital scope to a circuit, the vertical system adjusts the amplitude of the signal, just as with the analog scope. Next, the ADC in the acquisition system samples the signal at discrete points in time and converts the signal's voltage at these points to digital values called *sample points*. The horizontal system's sample clock determines how often the ADC takes a sample. The rate at which the clock "ticks" is called the sample rate and is measured in samples per second. Sampling methods usually fall into two types, real-time and equivalent-time methods.

The sampled points from the ADC are stored in memory as waveform points. More than one sample point may make up one waveform point. Together, the waveform points make up one waveform record. The number of waveform points used to make a waveform record is called the *record length*. The trigger system determines the start and stop points of the record. The display receives these record points after being stored in memory. Because of this memory capability, digital scopes are called *digital storage oscilloscopes* (DSOs).

Depending on the capabilities of the scope, additional processing of the sample points may take place, enhancing the display. Pretrigger may be available, allowing you to see events before the trigger point.

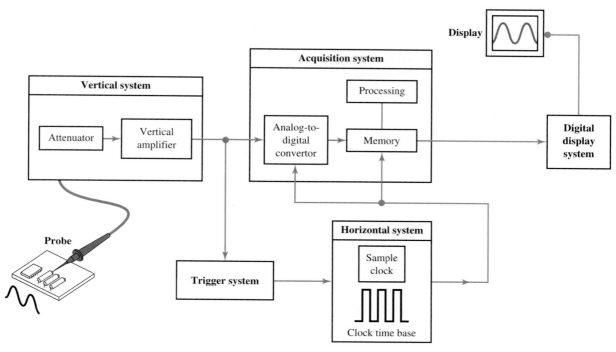

FIGURE 17.8
Block diagram of a digital oscilloscope.

Greatly simplified, the bandwidth of a digitizing oscilloscope defines the accuracy of the measurement, just as it does with analog oscilloscopes. The bandwidth requirement is the same as it is for an analog scope. The sampling speed determines the resolution of the display. Digital real-time sampling acquires all the samples of the waveform in a single trigger event up to the full bandwidth. Equivalent time sampling builds up a picture of the input signal over multiple acquisitions, with one or more samples gathered for each scope trigger. Therefore, equivalent-time sampling is useful only for signals that repeat, such as stable microprocessor clocks or communications test signals. For signals that don't repeat, digital real-time sampling is required.

Oversampling is required in order to achieve the highest signal fidelity and to ensure finding tricky fault conditions. Oversampling is sampling at more than twice the highest frequency in the signal to be measured. If you are looking strictly at sine waves, 5 times the highest frequency in the signal to be measured ensures accuracy. For square waves, however, a good rule of thumb is for the sampling rate to be 25 times the fundamental frequency.

Figure 17.9(a) shows a portable DSO that includes multimeter functions (the Fluke ScopeMeter). An industrial version of the scopemeter is shown in Figure 17.9(b). Here an industrial technician is shown using the scopemeter to monitor the line power.

17.4.1 The Digital Phosphor Oscilloscope (DPO)

New to the test-equipment arsenal is the digital phosphor oscilloscope, shown in Figure 17.10. The Tektronix DPO displays, stores, and analyzes data in real time, using three dimensions of signal information (amplitude, time, and the distribution of amplitude over time). The DPO faithfully emulates the best display attributes of the analog scope and provides the benefits of digital acquisition and processing as well. Like the DSO, the DPO uses a raster screen. But instead of a chemically coated phosphor CRT, it employs special parallel processing circuitry that emulates the intensity-graded trace of an analog scope. Thus, these scopes have the look of an analog scope but all the attributes of a DSO.

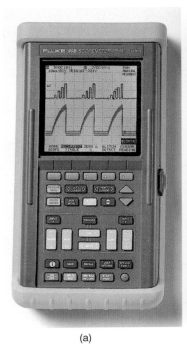
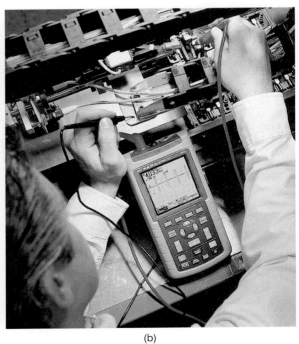

(a)   (b)

FIGURE 17.9
Portable DSO scopemeter: (a) this scopemeter includes multimeter functions; (b) scopemeter used by industrial technician to monitor line power. Courtesy of Fluke Corp. Reproduced with permission.

FIGURE 17.10
Tektronix TDS3054 DPO: the newest venture in DSO, the digital phosphor oscilloscope (DPO). Courtesy of Tektronix, Inc.

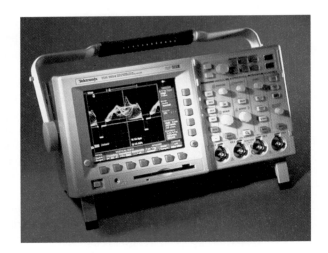

SECTION 17.4 SELF-CHECK	Answers are given at the end of the chapter.

1. The digital scope uses an _____ to convert the analog signal being measured into digital information.
2. The clock rate determines the _____ rate.
3. What are the two types of sampling methods employed by a digital scope?
4. Define the term *oversampling*.

17.5 ANALOG VERSUS DIGITAL: CHOOSING THE SCOPE TO USE

17.5.1 Analog Scopes

Here are some key points associated with analog scopes:

1. There are two major signal paths, horizontal and vertical.
2. Everything (including the CRT) must work at the same speed as the input signal.
3. All input channels are usually multiplexed through a single vertical path to the CRT.
4. The horizontal path is responsible for triggering.
5. The scope triggers on a voltage level and a rising or falling slope.
6. As input bandwidth goes up, the cost of the CRT also goes up, while reliability and accuracy of the CRT go down.

People like analog scopes because of the following:

1. Responsiveness: the analog scope responds instantly both to the input signal and to the control-panel adjustments.
2. People are familiar with knobs and switches right on the front panel of the scope. (Early digital scopes made you walk through menu after menu of softkeys to find even basic functions.)
3. A digital scope samples the input signal, which provides some delay in getting the signal on the display.
4. With analog scopes, you always have the option of picking up basic troubleshooting scopes at budget prices.

17.5.2 Digital Scopes

Here are some key points associated with digital scopes:

1. A digital scope captures data by sampling it, storing it in memory, and then reconstructing it on the screen.
2. The conceptual difference between this scheme and the way an analog scope works is that everything past the ADC doesn't need to work as fast as the signal.
3. In a digitizing scope, everything following the memory works at the same speed as the microprocessor, instead of having to work at the same speed as the incoming signal.

People like digital scopes because of the following:

1. The scope can process the waveform data and automatically measure parameters such as rise time, frequency, and time intervals.
2. Digital scopes have separate ADCs for multichannel displays. Information is captured and digitized simultaneously. By capturing data on all channels simultaneously, the time relationship among them is preserved.
3. Images can be stored on the display screen indefinitely.
4. A digitizing scope, because of microprocessor-based architecture, can provide an output port directly to a printer. (Analog scopes use a scope camera to capture images.)
5. Connection to a desktop computer is easy, allowing use in computer-automated test systems.
6. Waveforms can be stored in internal memory or on mass storage for future use and comparison.

The prices and speed of operation of DSOs have become competitive with analog scopes. The DSO can capture nonrepetitive signals, such as glitches. The DSO avoids the flicker problem analog scopes exhibit at slow sweep speeds and the fade problems analog scopes have at high sweep speeds.

In more and more electronic circuits, the processing circuits are going to a digital format whenever possible. This is also true in test equipment. The digital scope is a natural evolution of the analog storage scope. The major scope manufacturers are decreasing their analog scope products and increasing their digital scope products; in some cases they have stopped making analog scopes altogether.

17.6   OSCILLOSCOPE PROBES

Probes provide the interface between the scope's vertical input channel and the device under test (DUT). Probes are designed to not unduly influence the behavior of the circuit you are testing. The unintentional interaction of the probe and oscilloscope with the circuit being tested is called *circuit loading*. To minimize circuit loading, a 10X attenuator (passive) probe is commonly used.

Most scopes arrive with a passive probe as a standard accessory. Passive probes provide an excellent tool for general-purpose testing and troubleshooting. For more specific measurements or tests, many other types of probes exist. Two examples are active and current probes. The emphasis in this section is on passive probes, because this probe type allows for the most flexibility of use.

Most passive probes have some degree of attenuation factor, such as 10X or 100X. By convention, attenuation factors, such as for the 10X attenuator probe, have the X after the factor. In contrast, magnification factors, such as X10. have the X first.

The 10X (read as "times ten") attenuator probe minimizes circuit loading and is an excellent general-purpose probe. Circuit loading becomes more pronounced at higher frequencies, so this type of probe should be used for measuring signals above 5 kHz. Attenuation is caused primarily by the internal capacitance of the probe. The 10X attenuator probe improves measurement accuracy, but it also reduces the amplitude of the signal seen on the screen by a factor of 10.

Because it attenuates the signal, the 10X attenuator probe makes it difficult to look at signals less than 10 mV. The 1X probe is similar to the 10X attenuator probe, but it lacks the attenuation circuitry. With this circuitry, more interference is introduced to the circuit being tested. Use the 10X attenuator probe as your standard probe, but keep the 1X probe handy for measuring weak signals. Some probes have a convenient feature that

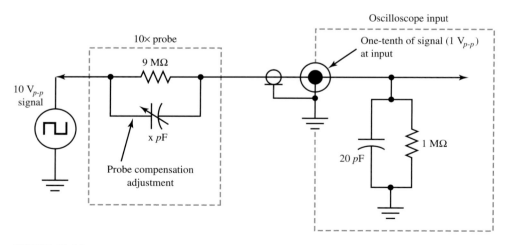

FIGURE 17.11
Typical probe-oscilloscope 10-to-1 divider network.

allows switching between 1X and 10X attenuation at the probe tip. If your probe has this feature, make sure you are using the correct setting before taking measurements. In addition, you must read from the proper 1X or 10X markings on the volts/div control on the scope.

The 10X probe works by balancing the probe's electrical properties against the scope's electrical properties. Before using a 10X probe you need to adjust this balance (called compensating the probe) for your particular scope. Figure 17.11 shows a simple diagram of the internal workings of a probe, its adjustment, and the input of an oscilloscope.

You should get into the habit of compensating the probe every time you set up your oscilloscope. A poorly adjusted probe can throw off your measurements, particularly amplitude measurements. Most scopes have a square-wave reference signal available at a terminal on the front panel for use to compensate the probe. If this low-frequency adjustment is not made properly, inaccuracies result in high-frequency measurements. A properly compensated probe produces a signal like the one shown in Figure 17.12(a); Figure 17.12(b) and (c) shows improperly compensated probes.

FIGURE 17.12
The effects of improper probe compensation: (a) properly compensated probe; (b) undercompensated probe; (c) overcompensated probe.

Lab Reference: Scope probe compensation is demonstrated in Exercise 27.

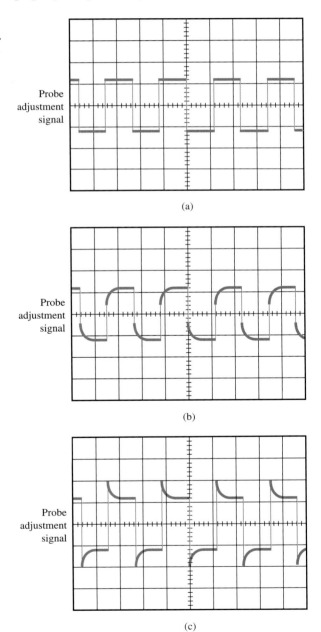

Probe adjustment signal

(a)

Probe adjustment signal

(b)

Probe adjustment signal

(c)

Measuring a signal requires two connections: the probe tip connection and a ground connection. Probes come with an alligator-clip attachment for grounding the probe to the circuit under test. In practice, you clip the grounding clip to a known ground in the circuit, such as the metal chassis of a stereo you are repairing or the ground strip on a protoboard, and then touch the probe tip to a test point in the circuit.

All probes are rated by bandwidth. In this respect, they are like scopes or other devices that are rated by bandwidth. In general, the probe's bandwidth should match the scope's bandwidth. This provides full scope bandwidth at the probe tip.

SECTION 17.6 SELF-CHECK

Answers are given at the end of the chapter.

1. What provides the interface between the scope's vertical input and the device under test?
2. Most passive probes have some degree of an _____ factor.
3. When should a 1× passive probe be used?
4. Each time you use a scope probe, you should _____ it.
5. True or false: The probe's bandwidth should match the scope's bandwidth.

17.7 COMMON TEST EQUIPMENT

17.7.1 Spectrum Analyzers

A *spectrum analyzer* is an instrument that graphically presents a plot of signal amplitude as a function of frequency for a selected portion of the spectrum. It is essentially a frequency-selective, peak-responding voltmeter calibrated to optically display the rms value of a sine wave. This is known as a *frequency domain* presentation. Signals are displayed as a spectrum on a CRT screen, with signal energy plotted on the vertical axis against frequency on the horizontal axis.

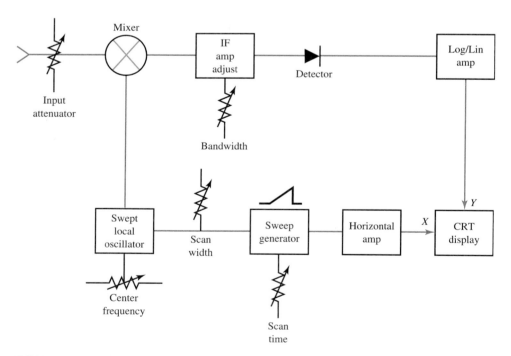

FIGURE 17.13
Simplified block diagram of spectrum analyzer.

A spectrum-analyzer display provides the following information: the presence or absence of signals, their frequencies (harmonic content), frequency drift, noise, relative amplitude of the signals, and the nature of modulation, if any, plus many other characteristics.

Figure 17.13 shows a simplified block diagram of a spectrum analyzer. Note that it has circuitry similar to both an oscilloscope and a superheterodyne (radio) receiver. Input frequencies are displayed as vertical deflections that are synchronized with the horizontal sweep, so that the lower frequencies appear at the start of the sweep and the higher frequencies appear at the upper end of the sweep.

The spectrum analyzer displays a square-wave signal as a series of odd harmonics decreasing in amplitude (Figure 17.14). This figure is indicative of how the spectrum analyzer displays amplitude as a function of frequency.

FIGURE 17.14
Spectrum analyzer display for a square wave.

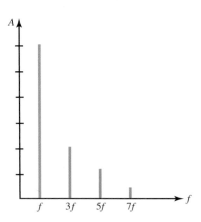

17.7.2 Frequency Counters

Oscilloscopes naturally measure the period of a signal. The frequency of the signal is then determined by calculation. Many high-end analog and most digital scopes have an additional feature that takes the period measurement, makes the frequency calculation, and provides a digital readout on the face of the CRT. Separate digital frequency counters are also available.

All digital frequency counters can measure the frequency of an unknown signal, and some have an extra feature in that they can measure and display the period of the signal as well.

Figure 17.15 shows a simplified block diagram of a *frequency counter.* The input signal is first conditioned into a series of pulses; then it is passed to the main gate. The frequency is measured by generating a gate time, consisting of a number of cycles of the reference clock, during which the input signal is counted. The frequency is calculated by dividing the number of cycles by the gate time.

Varying by cost, the range of frequencies that can be measured by a frequency counter is from DC to several gigahertz.

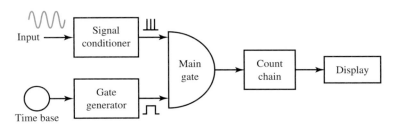

FIGURE 17.15
Simplified block diagram of a digital frequency counter.

17.7.3 Signal Generators

Signal generators (or signal sources) are designed to provide an alternating voltage at a specified frequency and amplitude. The heart of a signal generator is an electronic oscillator circuit that provides a tunable band of frequencies with amplitude control over the peak-to-peak voltage value of the signal.

Signal generators are generally available for two bands of frequency. These are the audio band and the RF (radio-frequency) band. The audio band is technically from 20 to 20 kHz, but most audio-signal generators provide a sine-wave output frequency as high as 1 MHz. In addition, these generators usually provide a square-wave output as well. The RF generator provides a sine-wave output with a frequency range from low kilohertz to several megahertz, depending on the model. A special RF generator feature involves the ability to provide a modulated RF output signal for testing AM and FM radio receivers.

The *function generator* is another common signal generator found in most labs. It provides a selectable sine, square, or triangular output with a range of frequencies from DC to several MHz. Many function generators have a DC offset control that is used to insert a DC level (either positive or negative) to the generator output if desired.

**SECTION 17.7
SELF-CHECK**

Answers are given at the end of the chapter.

1. The spectrum analyzer employs a _____ domain presentation.
2. The heart of the signal generator is an _____ _____.
3. For what two bands of frequencies are signal generators made available?

17.8 SPECIALTY TEST EQUIPMENT

There are a plethora of specialty test instruments available to help a technician in the diagnostics, test, and repair of special types of electronic equipment. Which of these instruments a particular lab may have is usually a function of budgets, technician expertise, and shop specialization. A small sampling of these instruments includes the following.

17.8.1 NTSC Color Bar Pattern Generator

The NTSC (National Television Standards Committee) signal generator is used to provide the test patterns for repairing and adjusting TV receivers. The NTSC standard is a 525-line, 30-frame-per-second signal standard. It is available as both a bench unit and a portable unit.

17.8.2 Wow-Flutter Meter

The wow-flutter meter is a direct-reading-type instrument designed for the measurement of wow, flutter, and drift characteristics. *Wow* and *flutter* are sound distortions consisting of a slow rise and fall of pitch, caused by speed variations in audio-video playback equipment. This meter is ideal for servicing VCRs, tape recorders, reel-to-reels, CDs, and other recording or playback equipment.

17.8.3 Automatic Distortion Meter

The automatic distortion meter can quickly measure distortion in the audio range. It is ideal for servicing radio, tape recorder, and audio products.

17.8.4   AM/FM Synthesized Signal Generator

The AM/FM synthesized signal generator displays RF signals with its built-in frequency counter, modulation levels, and memory. It is ideal for communications circuits, AM/FM receivers, and two-way radio transceiver servicing. One model can be used for FM stereo modulation.

**SECTION 17.8
SELF-CHECK**

Answers are given at the end of the chapter.

1. A common instrument used for the servicing of TVs is the _____.
2. Describe wow and flutter distortion.

■ SUMMARY

1. Any piece of AC test equipment is either a signal-measuring instrument or a signal-generating instrument.

2. The AC scale is a different scale from the DC scale on an analog meter.

3. The AC scale is calibrated to indicate the rms value.

4. For AC voltage measurements, the voltmeter sensitivity is lower than for DC values.

5. A current probe is clamped over a current-carrying conductor and utilizes a Hall effect sensor.

6. A common analog AC voltmeter displays an rms value as an adjusted average value that the meter actually measures.

7. A true rms voltmeter can be detected by measuring a heating value or by digital calculation of the rms value.

8. Multimeters have an AC frequency limitation of about 2 kHz.

9. Many high-voltage probes employ a multiplying resistor (sampling resistor) that drops the voltage by a factor of 100.

10. The oscilloscope is effectively a voltmeter that displays peak-to-peak waveforms.

11. The heart of a scope is a cathode-ray tube (CRT).

12. A vertical circuit causes an electron beam to move vertically and a horizontal circuit provides a sweep voltage to move the beam horizontally. The combination of the two results in a display that represents the input signal.

13. A trigger control is used to provide a stable waveform display.

14. Most analog scopes provide for multichannel displays through a multiplexing process called the chop or alternate mode.

15. The vertical channel circuitry of a scope affects the scope's bandwidth capability.

16. The scope's bandwidth should be at least three times as high as the fundamental frequency of the fastest signal under test for an accurately displayed signal.

17. A digital oscilloscope samples the waveform and uses an analog-to-digital converter to convert the voltage being measured into digital information.

18. Digital sampling methods employ real-time or equivalent-time methods.

19. In digital scopes, oversampling is required in order to achieve the highest signal fidelity.

20. People like analog scopes because of responsiveness, cost, and ease of use.

21. People like digital scopes because of signal storage, memory, and direct connections to computers.

22. Most analog scopes are shipped with a passive probe.

23. Most passive probes employ some degree of attenuation factor.

24. The 10× passive probe should be compensated each time it is used.

25. A spectrum analyzer displays the signal as a function of frequency (rather than time).

■ REVIEW QUESTIONS

1. All AC test equipment comes in what two forms?

2. What is the difference in scales for DC versus AC on analog multimeters?

3. Why do analog voltmeters suffer from greater circuit loading on the AC range?

4. How can one measure current without breaking the circuit?

5. What is actually measured when an AC voltmeter indicates a voltage value?

6. Describe a true rms voltmeter.

7. Describe how a high-voltage probe works.

8. What can be measured when utilizing an oscilloscope?

9. In simple terms, what is an oscilloscope?

10. The input signal under test is applied to the _____ input of the scope.

11. What provides the interface between the device under test and the vertical input terminals of the scope?

12. Over what does the vertical amplifier of the scope have control?

13. What is the function of the horizontal amplifier of the scope?

14. What does the trigger control do?

15. How do scopes provide for multichannel inputs?

16. How is the bandwidth of a scope defined?

17. What processes must be done in order for a digital scope to display a waveform on the screen?

18. What sampling methods are used in digital scopes?

19. Why are digital scopes called digital storage oscilloscopes?

20. What is oversampling?

21. What features of an analog scope do people prefer?

22. What features of a digital scope do people prefer?

23. What supplies the interface between the signal under test and the vertical input terminals of the scope?

24. Describe the makeup of a passive probe.

25. What does it mean to compensate a probe?

26. A spectrum analyzer utilizes a _____ domain presentation.

27. The spectrum analyzer provides for what type of information gathering?

28. The heart of a signal generator is an _____ _____.

29. What are the frequency ranges for most signal generators?

30. A function generator provides for what additional features?

31. What is the NTSC standard?

32. What are wow and flutter distortion?

■ CRITICAL THINKING

1. Describe in detail how an analog scope takes a signal under test and provides a display on the CRT.

2. Describe a scope's bandwidth capability and what guidelines to employ for greater accuracy in measurements.

3. Compare and contrast real-time versus equivalent-time sampling methods used in digital scopes.

4. Compare and contrast the features of an analog scope versus a digital scope.

5. Describe in detail the features of a passive probe (for example, attenuation, loading, when to use, and compensation).

■ ANSWERS TO SECTION SELF-CHECKS

SECTION 17.1 SELF-CHECK

1. The DC scale has its range values below it, whereas the AC scale has its range values above it.

2. By utilizing a thermal detector, or by digital calculation

3. 2 kHz

4. 100

SECTION 17.2 SELF-CHECK

1. A voltmeter that optically displays peak-to-peak waveforms

2. Cathode-ray tube (CRT)

3. To provide a stable waveform display

4. By multiplexing through a single vertical channel.

SECTION 17.3 SELF-CHECK

1. An analog scope directly applies a voltage under test to an electron beam moving across the oscilloscope screen.

2. Real time

3. The bandwidth extends to the frequency at which a sinusoidal input signal has been attenuated to 0.707 of the middle frequencies.

SECTION 17.4 SELF-CHECK

1. Analog-to-digital converter

2. Sampling

3. Real-time and equivalent-time methods

4. Oversampling is sampling at more than twice the highest frequency in the signal to be measured.

SECTION 17.5 SELF-CHECK

1. Same

2. Multiplexing

3. Printer

4. The DSO

SECTION 17.6 SELF-CHECK

1. The probe

2. Attenuation

3. For measuring weak signals

4. Compensate

5. True

SECTION 17.7 SELF-CHECK

1. Frequency

2. Electronic oscillator

3. Audio and RF bands

SECTION 17.8 SELF-CHECK

1. An NTSC color bar pattern generator

2. They are sound distortions consisting of a slow rise and fall of pitch, caused by speed variations in audio-video playback equipment.

CHAPTER 18

TRANSFORMERS

OBJECTIVES

After completing this chapter, you will be able to:

1. Describe the process of mutual induction.
2. Describe the components that make up a transformer.
3. Differentiate between no-load and load conditions to the transformer.
4. Describe a transformer's voltage and current ratio and compare it to a transformer's turns ratio.
5. Calculate voltages and currents based on transformer turns ratios and a given primary voltage.
6. Describe the types of transformer losses and the technique used to minimize these losses.
7. Describe a transformer's ratings.
8. Compare and contrast various types and applications of transformers.
9. Describe transformer faults.

INTRODUCTION

In Chapter 16 you learned that alternating current (AC) has certain advantages over direct current (DC). One important advantage is that when AC is used, the voltage and current levels can be increased or decreased by means of a transformer.

A *transformer* is a device that transfers electrical energy from one circuit to another with no physical connection between them. The electrical energy is always transferred without a change in frequency but may involve changes in magnitudes of voltage and current. A transformer's transfer of energy from one circuit to another is accomplished by the process of *mutual induction*. Mutual induction involves a varying magnetic field of one coil inducing a voltage into a nearby coil. A transformer utilizes two or more coils. (Mutual induction is more fully discussed in Section 20.4 in the chapter on inductance.)

We saw the effects of a varying magnetic field and induced voltage in Section 15.2. This section defined the basic factors that affected the magnitude of an induced emf and also specified that the flux must be moving for induction to take place. AC meets this criteria and is used as the source voltage for a transformer.

As you know, the amount of power used by the load of an electrical circuit is equal to the current in the load times the voltage across the load, or $P = IV$. If, for example, the load requires an input of 2 A at 10 V (20 W) and the source is capable of delivering only 1 A at 20 V, the circuit could not normally be used with this particular source. However, if a proper transformer is connected between the source and the load, the voltage can be decreased (stepped down) to 10 V and the current can be increased (stepped up) to 2 A. Notice in this example that the power remains the same. That is, 20 V times 1 A yields the same power as 10 V times 2 A.

A transformer that receives energy at one voltage and delivers it at the same voltage is called a *one-to-one transformer.* It is also known as an *isolation transformer,* because it does not transform the voltage-current ratios. It affords a physical separation between circuits that may be used for safety reasons. When the transformer receives energy at one voltage and delivers it at a higher voltage, it is called a *step-up transformer.* If the energy received by the transformer is delivered at a lower voltage, it is referred to as a *step-down transformer.* There are also transformers that provide for *impedance matching* between circuit elements. Various attributes of these types of transformers are described in this chapter.

18.1 BASIC TRANSFORMER OPERATION

18.1.1 Basic Transformer Elements

In its most basic form a transformer consists of

- A primary coil or winding,
- A secondary coil or winding,
- A core that supports the coils or windings.

While examining Figure 18.1 note that the primary winding is connected to a 60-Hz AC voltage source. As current flows in the primary winding, a magnetic field (flux) builds up (expands) and collapses (contracts) around it. The expanding and contracting magnetic field around the primary winding cuts the secondary winding and induces an AC voltage into that winding. This induced voltage causes AC to flow through the load. Current flows in the secondary if a complete path for current exists. The voltage may be stepped up or down depending on the design of the primary and secondary windings.

If DC is used as the primary source, no variation occurs in the current. Without current variation, there is no change in magnetic flux, no electromagnetic induction, and no induced secondary voltage.

FIGURE 18.1
Basic transformer action.

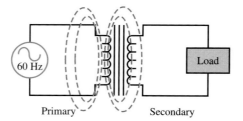

Primary   Secondary

Element Functions

The principal functions of the parts of a transformer described in Section 18.1.1 are as follows:

- The primary winding receives energy from the AC source and creates a changing magnetic flux.
- The secondary winding receives energy from the primary winding and delivers it to the load.
- The core material provides a path for the magnetic lines of flux.

18.1.2 Core Characteristics

The two windings are wound on some type of core material. In some cases the coils are wound on a cylindrical or rectangular cardboard form. In effect, the core material is air, and the transformer is called an *air-core transformer.* These transformers are used in high-frequency applications. Transformers used at low frequencies, such as 60 and 400 Hz, require a core of low-reluctance, high-permeability magnetic material, usually silicon iron or nickel alloys. This type of transformer is called an *iron-core transformer.* Silicon iron is used for power transformers, whereas nickel alloys are used for audio and telecommunication transformers. Ferrites may also be used as cores. Ferrite cores are used for low-power and high-frequency applications, including telecommunications. The shape of the ferrite core determines its application.

Core Composition

The composition of a transformer core depends on such factors as voltage, current, and frequency. Size limitations and construction costs are also factors to be considered.

The most popular and efficient transformer core is the *shell core,* as illustrated in Figure 18.2. As shown, each layer of the core consists of E- and I-shaped sections of metal. (Other shapes for the sections are also available, depending on the application.) These sections are butted together to form the laminations. The laminations are electrically insulated from each other (with varnish) and then pressed together to form the core.

FIGURE 18.2
Shell-type core construction.

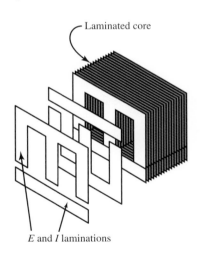

Laminated core

E and I laminations

18.1.3 Transformer Windings

In the transformer shown in the cutaway view of Figure 18.3(b), the primary consists of many turns of relatively small wire. It is wound in layers directly on a rectangular cardboard form. The wire is coated with varnish (enamel), so that each turn of the winding is insulated from every other turn. In transformers designed for high-voltage applications, sheets of an insulating material, such as mylar, are placed between the layers of windings to provide additional insulation.

When the primary winding is completely wound, it is wrapped in insulating paper or cloth. The secondary winding is then wound on top of the primary winding. After the secondary winding is complete, it too is covered with insulating paper. Next, the E and I sections of the shell core are inserted into and around the windings, as shown in Figure 18.3(a).

The leads from the windings are normally brought out through a hole in the enclosure of the transformer. Terminals are sometimes provided on the enclosure for connections to the windings.

FIGURE 18.3
(a) Exploded view of shell-type transformer construction;
(b) cutaway view of shell-type core with windings.

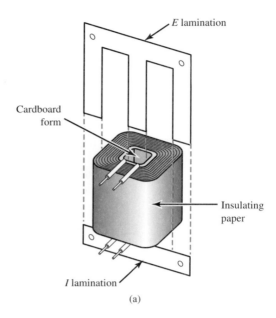

(a)

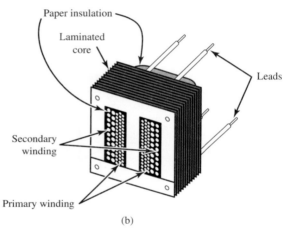

(b)

**SECTION 18.1
SELF-CHECK**

Answers are given at the end of the chapter.

1. True or false: If DC is used as the primary source, the load also has DC in it.
2. Transformers used at low frequencies require a high _____ core material.
3. A common core material for power transformers is _____ _____.
4. The most popular and efficient transformer core is the _____ core.
5. Insulation for each wire in the transformer is provided by coating with _____.

18.2 TRANSFORMER LOADING

18.2.1 No-Load Condition

A *no-load condition* is said to exist when a voltage is applied to the primary, but no load is connected to the secondary, as illustrated in Figure 18.4. Because of the open switch, there is no current flowing in the secondary winding. Even with the switch open, the AC voltage applied to the primary forms a small amount of primary current, called the *exciting current*. What this current does is essentially "excite" the coil of the primary to create a magnetic field. The level of primary current is kept small by what is known as *counter-emf.* As current flows in the primary, a magnetic field builds and expands outward. This changing flux cuts through the primary wire and induces a voltage in the opposite direction to the source (Lenz's law).

FIGURE 18.4
Transformer under a no-load condition.

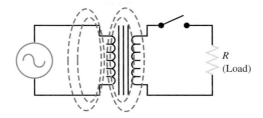

18.2.2   Load Condition

If the switch is closed in Figure 18.4, the load is connected across the secondary. A change in conditions occurs and the transformer acts differently. An increase in primary current occurs due to mutual inductance. *Mutual inductance* is the process whereby the primary induces a voltage in the secondary and a secondary current flows, which in turn generates a magnetic field that feeds back to the primary and weakens or cancels the primary counter-emf set up by the primary current. The flux links in both directions—for example, the primary flux links to the secondary, and the secondary flux links to the primary.

Now you can see why if a short occurs in the secondary, an increase in primary current also occurs. Because the primary is usually fused, the fuse blows. Even though the primary and secondary are physically separated, the flux linkage exists *both ways*. Thus, the primary current is affected by what happens in the secondary.

SECTION 18.2 **SELF-CHECK**	Answers are given at the end of the chapter.

1. Under no-load conditions the current in the primary is called an _____ _____.

2. Describe how the primary current is kept small under a no-load condition.

3. Under a load condition there is a(n) _____ in primary current.

18.3   TRANSFORMER RATIOS

18.3.1   Voltage Ratio and Turns

The total voltage induced into the secondary winding of a transformer is determined mainly by the ratio of the number of turns in the primary to the number of turns in the secondary (N_P/N_S) and by the amount of voltage applied to the primary.

As voltage is applied to the primary, flux lines cut both the primary and secondary. The voltage induced into the secondary is the same as the counter-emf induced into each turn in the primary. Because the counter-emf in the primary is equal (or almost equal) to the applied voltage, a proportion may be set up to express the value of the voltage induced in terms of the voltage applied to the primary and the number of turns in each winding. This proportion also shows the relationship between the number of turns in each winding and the voltage across each winding. This proportion is expressed by the equation

$$\frac{N_P}{N_S} = \frac{V_P}{V_S} \tag{18.1}$$

where N_P = number of turns in the primary
N_S = number of turns in the secondary
V_P = voltage applied to the primary
V_S = voltage induced in the secondary

Lab Reference: These voltages are measured as part of Exercise 28.

If any three of the quantities in equation (18.1) are known, the fourth quantity can be calculated. The equation can be transposed as needed to determine any unknown.

EXAMPLE 18.1

A transformer has 200 turns in the primary, 50 turns in the secondary, and 120 V applied to the primary. What is the voltage across the secondary (V_S)?

Solution $V_S = (V_P)\left(\dfrac{N_S}{N_P}\right)$

$= (120 \text{ V})\left(\dfrac{50}{200}\right)$

$= 30 \text{ V}$

EXAMPLE 18.2

There are 400 primary turns in a transformer. How many turns must be wound on the secondary to have a secondary voltage of 1 V if the primary voltage is 5 V?

Solution $N_S = \dfrac{V_S N_P}{V_P}$

$= \dfrac{1 \text{ V} \times 400 \text{ turns}}{5 \text{ V}}$

$= 80 \text{ turns}$

Practice Problems 18.1   Answers are given at the end of the chapter.
1. Determine the amount of secondary voltage when the primary voltage is 10 V AC, and there are 100 turns on the primary and 1000 turns on the secondary.
2. Determine the secondary turns when the secondary voltage is 12.6 V AC, the primary voltage is 120 V AC, and there are 200 primary turns.

If there is less voltage across the secondary than across the primary, then the secondary must have fewer turns than the primary. A transformer in which the voltage across the secondary is less than the voltage across the primary is called a *step-down transformer*. Step-down transformers are very common in electronic circuits, because most electronic circuits work at lower voltages than those supplied by the line source.

A transformer that has fewer turns in the primary than in the secondary produces a greater voltage across the secondary than the voltage applied to the primary. A transformer in which the voltage across the secondary is greater than the voltage applied to the primary is called a *step-up transformer*.

Turns Ratio

In many transformer problems, just the turns ratio and one voltage are given. The turns ratio is simply a ratio of primary and secondary turns. Either N_P/N_S or N_S/N_P can be selected as the ratio, depending on preference (the V_P/V_S ratio is adjusted accordingly to match the ratio selected). This text uses the N_P/N_S notation, as shown in equation (18.1). Regardless of which ratio is selected, the most important consideration is to examine secondary turns versus primary turns. If the secondary has more turns, the transformer is a step-up transformer, whereas a secondary having fewer turns is a step-down transformer.

A common way of indicating a step-down transformer turns ratio numerically is simply 10:1. This implies that the secondary has one-tenth ($\frac{1}{10}$) the number of turns of the primary. This notation follows directly from the N_P/N_S notation for the turns ratio, because the 10 in the ratio 10:1 represents the primary winding, and the 1 represents the secondary winding. Note that these numbers do not indicate the number of turns on the primary and secondary windings, just their ratio. Thus, a 10:1 transformer with a 120-V primary supplies 12 V to the secondary (assuming no losses).

EXAMPLE 18.3

A transformer has a turns ratio of 7:1. If 5 V is developed across the secondary, what is the voltage applied to the primary?

Solution $V_P = V_S\left(\dfrac{N_P}{N_S}\right)$

$$= (5\text{ V})\left(\frac{7}{1}\right)$$

$$= 35\text{ V}$$

EXAMPLE 18.4

A transformer's windings are labeled 1:5. If 120 V is applied to the primary, what is the secondary voltage?

Solution $V_S = V_P\left(\dfrac{N_S}{N_P}\right)$

$$= (120\text{ V})\left(\frac{5}{1}\right)$$

$$= 600\text{ V}$$

Practice Problems 18.2   Answers are given at the end of the chapter.
1. What is the turns ratio if $V_S = 24$ V and $V_P = 120$ V?
2. An 8:1 turns ratio for a transformer produces a $V_S = 15$ V. What is the primary voltage?

18.3.2   Current Ratio and Turns

The number of flux lines developed in a core is proportional to the magnetizing force (in ampere-turns) of the primary and secondary windings. Ampere-turns is a measure of the magnetomotive force developed by 1 A of current flowing in a coil of one turn. The flux that exists in the core of a transformer surrounds both the primary and secondary windings. Because the flux is the same for both windings, the ampere-turns in both windings must be the same.

Therefore,

$$I_P N_P = I_S N_S \tag{18.2}$$

where $I_P N_P$ = ampere-turns in the primary
$I_S N_S$ = ampere-turns in the secondary

By rearranging the equation you can see that the secondary current (I_S) is equal to the primary current (I_P) times the ratio N_P/N_S.

EXAMPLE 18.5

A transformer with 100 turns in the primary and 600 in the secondary has 3 A of current in the primary. What is the value of current in the secondary?

Solution $I_S = I_P\left(\dfrac{N_P}{N_S}\right)$

$$= (3\text{ A})\left(\frac{100}{600}\right)$$

$$= 0.5\text{ A}$$

EXAMPLE 18.6

A transformer with a turns ratio of 1:8 has 2 A of current in the primary. What is the value of secondary current?

Solution $I_S = I_P\left(\dfrac{N_P}{N_S}\right)$

$= (2\text{ A})\left(\dfrac{1}{8}\right)$

$= 0.25\text{ A}$

Practice Problems 18.3   Answers are given at the end of the chapter.
1. What is I_P when $I_S = 50$ mA, $V_S = 300$ V, and $V_P = 100$ V?
2. What is the turns ratio when $I_P = 1$ A and $I_S = 5$ A?

A transformer having fewer turns in the secondary than in the primary steps down the voltage and steps up the current. Thus, a 10:1 step-down transformer has 10 times the current in the secondary as in the primary. A 1:2 step-up transformer has one-half the current in the secondary as in the primary.

**SECTION 18.3
SELF-CHECK**

Answers are given at the end of the chapter.

1. What determines the voltage induced into the secondary of a transformer?
2. What is a step-down transformer?
3. If a secondary has more voltage than the primary, what type of transformer is it?
4. A transformer labeled 12.5:1 indicates that the primary has _____.

18.4   TRANSFORMER LOSSES

By combining equations (18.1) and (18.2) we can show that $V_P/V_S = I_S/I_P$. By cross-multiplying, we have $V_P I_P = V_S I_S$. The product of voltage and current is power; thus, with the exception of any power consumed by the transformer elements, all power delivered to the primary by the source is delivered to the load. The form (levels of voltage and current) of the power may change, but the power in the secondary nearly equals the power in the primary.

Practical power transformers, although highly efficient, are not perfect devices. Efficiencies range from 80% to nearly 100%, depending on transformer type. Most transformers have nearly a 100% efficiency; thus, most transformer problems do not take into account real losses and assume a 100% efficiency.

The power that is lost in a transformer is a combination of different types of losses. There are three types of losses that result in the undesirable conversion of electrical energy into heat energy. These losses were first described in Chapter 15 dealing with generators.

18.4.1   Copper Loss

Whenever current flows in a conductor, power is dissipated in the resistance of the conductor in the form of heat. The amount of power dissipated is directly proportional to the resistance of the wire and to the square of the current through it (I^2R). This loss is a result of the DC resistance in the primary and secondary windings. Copper loss can be reduced by using larger-diameter wire. Large-diameter wire is required for high-current windings, whereas small-diameter wire can be used for low-current windings.

18.4.2 Eddy-Current Loss

The core of a transformer is usually constructed of some type of ferromagnetic material. Whenever the primary of a ferromagnetic core transformer is energized by an AC source, a fluctuating magnetic field is produced. This magnetic field cuts the conducting core material and induces a voltage into it. (This process is the same as that which produces the counter-emf in the primary.) The induced voltage causes random currents to flow through the core. This in turn causes power to be dissipated in the form of heat. The random currents are called *eddy currents* and are undesirable.

To minimize the effects of eddy currents, transformer cores are *laminated.* Because the thin, electrically insulated laminations do not provide an easy path for current, eddy current losses are greatly reduced.

18.4.3 Hysteresis Loss

When a magnetic field is passed through a core, the core material becomes magnetized. To become magnetized, the domains within the core must align themselves with the external field. If the direction of the field reverses, the domains must rotate so that their poles are aligned with the new direction of the external field.

Power transformers normally operate from either 60 Hz or 400 Hz alternating current. Each tiny domain must realign itself twice during each cycle, or a total of 120 times a second when 60-Hz AC is used. The energy used to turn each domain is dissipated as heat within the iron core. This loss, called *hysteresis loss,* can be thought of as resulting from molecular friction. Hysteresis loss can be held to a small value by the proper choice of core materials during the manufacturing process.

SECTION 18.4 **SELF-CHECK**	Answers are given at the end of the chapter. **1.** How is copper loss minimized? **2.** Describe eddy currents. **3.** Power transformers normally operate from either _____-Hz or _____-Hz alternating current.

18.5   TRANSFORMER SYMBOLS, POLARITY INDICATORS, AND RATINGS

18.5.1 Symbols

Figure 18.5 shows some typical schematic symbols for transformers. The symbol for an air-core transformer is shown in Figure 18.5(a). Parts (b) and (c) show iron-core transformers. The bars between the coils are used to indicate an iron core. Additional connections are frequently made to the transformer windings at points other than the ends of the

FIGURE 18.5
Schematic symbols for various types of transformers:
(a) air-core transformer; (b) iron-core transformer;
(c) iron-core transformer, with multiple-secondary windings and center tap.

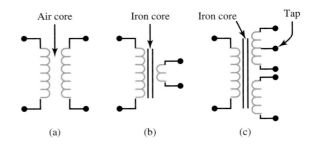

(a)   (b)   (c)

windings. These additional connections are called *taps*. When a tap is connected to the center of the winding, it is called a *center tap*. Figure 18.5(c) shows the schematic representation of a center-tapped iron-core transformer. Because the tap is in the center, the secondary voltage available is one-half from either end and the center tap as compared to the entire secondary voltage.

In addition to showing a center-tapped secondary, Figure 18.5(c) also has two distinct secondary windings, which is known as a *multiple-winding transformer*. This allows one transformer to provide two or more distinct secondary voltages. The primary side of a transformer can have more than one winding as well. This situation is common when the source can either be 120 V AC or 240 V AC. Both center-tapped and multiple-winding transformers were first used with vacuum-tube circuits and continue to be used in a variety of electronic circuits.

18.5.2 Polarity Indicators

The secondary voltage of a simple transformer may be either in phase or out of phase with the primary voltage, depending on the direction in which the windings are wound, the arrangement of the connections to the external circuit (load), and the reference points chosen for comparison. (V_P and V_S always rise and fall together with a resistive load; their phase relationship is determined by the reference points for comparison.)

Phase-indicating dots (dot convention) are used to indicate points on a transformer schematic symbol that have the same instantaneous polarity (points that are in phase). The use of phase-indicating dots is illustrated in Figure 18.6. In part (a) of the figure, both the primary and secondary windings are wound in the same direction. In this way, the top lead of the primary and the top lead of the secondary have the same polarity (and phase) with respect to a common reference. This is indicated by the dots on the transformer symbol being on the same side for both the primary and secondary.

Part (b) of the figure illustrates a transformer in which the primary and secondary are wound in opposite directions. Notice the top leads of the primary and secondary have opposite polarities. This is indicated by the dots being placed on opposite ends of the transformer symbol. Using a common reference, the secondary and primary voltages are now out of phase with each other.

FIGURE 18.6
Phase-indicating dots on transformers: (a) top leads have the same polarity; (b) top leads have opposite polarity.

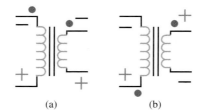

(a)   (b)

18.5.3 Ratings

When a transformer is to be used in a circuit, more than just the turns ratio must be considered. The voltage, current, and power-handling capabilities of the primary and secondary windings must also be considered.

The maximum voltage that can safely be applied to any winding is determined by the type and thickness of the insulation used. When a better (and thicker) insulation is used between the windings, a higher maximum voltage can be applied to the windings. This voltage is usually stamped on the body of the transformer and is the *rms value* of the AC voltage. Further, this is the rated value of voltage under full-load conditions.

The maximum current that can be carried by a transformer winding is determined by the diameter of the wire used for the winding. If current is excessive in a winding, a higher-than-ordinary amount of power is dissipated by the winding in the form of heat. This heat may be sufficiently high to cause the insulation around the wire to break down and to damage the transformer.

The current rating of a transformer is usually specified only for the secondary. This is because if the secondary current rating is not exceeded, the primary current cannot be exceeded either. Excessive secondary current can destroy both the secondary and primary windings of the transformer due to overheating.

The power-handling capacity of a transformer depends on its ability to dissipate heat. If the heat can safely be removed, the power-handling capacity of the transformer can be increased. This is sometimes accomplished by immersing the transformer in oil or by using cooling fins.

The power-handling capacity of a transformer is measured in *volt-amps* (V·A) rather than watts (W). This is because the power is not actually dissipated by transformer but by the load connected to the secondary. The watt is reserved for the actual dissipation of power in a resistive load, but the volt-ampere is called *apparent power,* because it is the power apparently used by the transformer. The product of $V_P I_P$ or $V_S I_S$ must not exceed the rated apparent power (V·A) of the transformer. If either side of the transformer has more power than the apparent power, the transformer will overheat and can be destroyed.

Transformers are also rated by frequency. Two common power-transformer frequencies are 60 Hz and 400 Hz. 400 Hz is often used for aircraft transformers because these transformers are much smaller and lighter than 60-Hz transformers having the same power rating. A transformer can be operated at a higher frequency than its rating but not at a lower frequency.

A typical power transformer could be labeled 24 V AC, 3 A, 60 Hz. This implies a 24-V rms secondary, a 3-A maximum secondary current, and 60-Hz frequency applications. (It is understood the primary voltage is 110–120 V AC.)

SECTION 18.5 SELF-CHECK

Answers are given at the end of the chapter.

1. What is a transformer *tap*?
2. What is a multiple-winding transformer?
3. What are phase-indicating dots used for?
4. What determines the maximum current that a transformer can handle?
5. The power-handling capacity of a transformer is measured in _____.

18.6 TYPES AND APPLICATIONS OF TRANSFORMERS

A transformer has many useful applications in an electrical circuit. A brief discussion of some of these applications should help in recognizing the importance of the transformer in electricity and electronics.

18.6.1 Power Transformers

Power transformers are used to supply voltages to various circuits in electrical equipment. The power-supply section of a circuit utilizes a power transformer. These transformers have two or more windings (multiple windings) wound on a laminated core. The number of windings and the turns per winding depend on the voltages that the transformer is to supply.

You can usually distinguish between the high-voltage and low-voltage windings in a power transformer by measuring the resistance of the windings. The low-voltage winding has fewer turns; it usually carries the higher current and therefore has the larger-diameter wire. This means that its resistance is less than the resistance of the high-voltage winding, which has more turns, normally carries less current, and therefore may be constructed of smaller-diameter wire.

The typical power transformer has several secondary windings, each providing a different voltage. A circuit employing a *multiple-winding transformer* is shown in Figure 18.7. The transformer is used in the power supply of an analog oscilloscope.

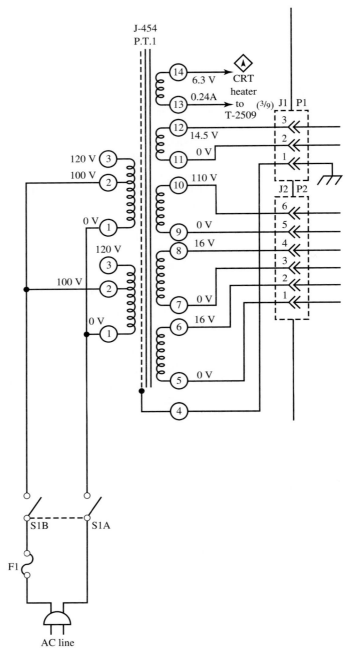

FIGURE 18.7
A multiple-winding transformer used in the power supply of an analog oscilloscope.

There are many types of power transformers. They range in size from the huge transformers weighing several tons (used in power substations of commercial utility companies) to very small ones weighing as little as a few ounces (used in electronic equipment).

18.6.2 Autotransformers

It is not necessary for the primary and secondary to be separate and distinct windings in a transformer. Figure 18.8 shows a schematic diagram of what is known as an *autotransformer*. This is also called a *variac*. Note that a single coil of wire is *tapped* to produce what is electrically a primary and secondary winding. The voltage across the secondary winding has the same relationship to the voltage across the primary that it would have if they were two distinct windings. The moveable tap in the secondary is used to select a value of output, either higher or lower than V_P, within the range of the transformer. That is, when the tap is at point A, V_S is less than V_P; when the tap is at point B, V_S is greater than V_P. Thus the secondary voltage is a variable AC voltage (variac).

FIGURE 18.8
Schematic diagram of an autotransformer (variac).

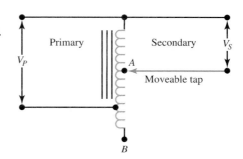

18.6.3 Intermediate-Frequency (IF) and Radio-Frequency (RF) Transformers

Intermediate-frequency (IF) and *radio-frequency* (RF) *transformers* are used in high-frequency circuits to couple (connect) circuits. Typical frequencies are in the AM and FM radio broadcast bands (535–1635 kHz and 88.1–107.9 MHz, respectively). The windings are wound on a tube of nonmagnetic material and have a special tunable powdered-iron core or contain only air as the core material. DC voltages are blocked by the transformers, because they can only couple the high-frequency AC signals between electronic circuits. The tunable part of the core is called the *slug*. These are color coded to indicate their place in a circuit.

18.6.4 Impedance-Matching Transformers

In Chapter 12 we saw that for maximum or optimum transfer of power between the source and load, it is necessary for the resistance of the source to match the load. A difference in AC circuits is that the limitation to current is not just from resistance. As will be shown in later chapters, there are other components that limit or oppose current as well. The term used to describe the general opposition to AC current is *impedance* (*Z*).

In electronic communication circuits, the concept of maximum power transfer is equally as important. This is particularly true between the transmitter and the antenna and the power amplifier and the speaker. Therefore, in these types of AC circuits, the impedance of one circuit must match the impedance of another if maximum power is to be transferred.

A common impedance-matching device is the transformer. A transformer is used to change, or transform, a secondary load impedance to a new value as seen by the primary. The secondary load impedance is said to be reflected back into the primary. Thus, it is known as a *reflected impedance*. The reflected impedance of the secondary may be stepped up or down by the product of the square of the turns ratio:

$$Z_P = \left(\frac{N_P}{N_S}\right)^2 \times Z_S$$

The more common form of this equation is

$$\frac{Z_P}{Z_S} = \left(\frac{N_P}{N_S}\right)^2 \qquad (18.3)$$

where Z_P = primary impedance, reflected from secondary

Z_S = secondary impedance

$\dfrac{N_P}{N_S}$ = turns ratio

EXAMPLE 18.7

Determine the reflected impedance into the primary when the load is 25 Ω and the turns ratio is 6:1.

Solution $Z_P = \left(\dfrac{N_P}{N_S}\right)^2 \times Z_S$

$$Z_P = \left(\frac{6}{1}\right)^2 \times 25\ \Omega$$

$$= 900\ \Omega$$

To obtain the proper turns ratio when the impedance is known, the following mathematical relationship for the transformer can be used:

$$\frac{N_P}{N_S} = \sqrt{\frac{Z_P}{Z_S}} \qquad (18.4)$$

EXAMPLE 18.8

An impedance-matching transformer is to be used to match the 8-Ω speaker to an amplifier impedance (R_i) of 800 Ω, as illustrated in Figure 18.9. Determine the required turns ratio, N_P/N_S, for the transformer.

Solution $\dfrac{N_P}{N_S} = \sqrt{\dfrac{Z_P}{Z_S}}$

$$= \sqrt{\frac{800\ \Omega}{8\ \Omega}}$$

$$= 10{:}1$$

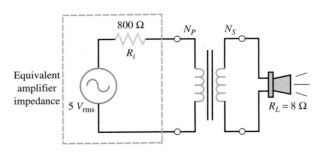

FIGURE 18.9
Circuit for Example 18.8.

The transformer in Example 18.8 must have 10 primary turns for every secondary turn for maximum power transfer to occur. This process is valid only if the load and source are assumed to be *pure resistances.* If either the source or load resistance is *complex,* meaning that it is made up of resistance and capacitance or inductance, then a different mathematical approach must be taken in order to match the impedances. (Complex impedances are discussed fully in Chapter 23.)

EXAMPLE 18.9

Determine the turns ratio needed in a transformer to match a 1-kΩ source resistance to a load resistance of 50 Ω, as illustrated in Figure 18.10(a). Calculate the power delivered to the load with a direct connection (no transformer) and with a matching transformer. The source voltage is 20 V. (Assume no losses in the transformer.)

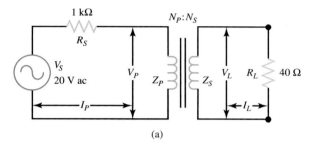

(a)

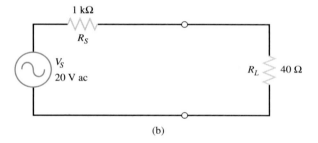

(b)

FIGURE 18.10
Transformer impedance-matching application: (a) circuit for Example 18.9; (b) direct connection, no transformer used.

Solution For a direct connection (Figure 18.10(b)):

$$V_L = \frac{R_L}{R_L + R_S} \times V_S$$

$$= \frac{40 \text{ }\Omega}{1040 \text{ }\Omega} \times 20 \text{ V}$$

$$= 0.77 \text{ V}$$

$$P_L = \frac{V_L^2}{R_L}$$

$$= \frac{(0.77 \text{ V})^2}{40 \text{ }\Omega}$$

$$= 14.8 \text{ mW}$$

For maximum power transfer, $Z_P = R_S$ and $Z_S = R_L$. The turns ratio is calculated as

$$\frac{N_P}{N_S} = \sqrt{\frac{Z_P}{Z_S}}$$

$$= \sqrt{\frac{1000 \text{ }\Omega}{40 \text{ }\Omega}}$$

$$= \sqrt{\frac{25}{1}}$$

$$= 5{:}1$$

Because $Z_P = R_S$ (for maximum power transfer to the primary) then V_P is found as

$$V_P = \frac{Z_P}{R_S + Z_P} \times V_S \quad \text{(the voltage-divider rule)}$$

$$= \frac{1000\ \Omega}{2000\ \Omega} \times 20\ \text{V}$$

$$= 10\ \text{V}$$

$Z_P = R_P = R_S$, so the primary current is calculated as

$$I_P = \frac{V_P}{R_P}$$

$$= \frac{10\ \text{V}}{1 \times 10^3\ \Omega}$$

$$= 10\ \text{mA}$$

$$P_P = I_P V_P$$

$$= (10 \times 10^{-3}\text{A}) \times 10\ \text{V}$$

$$= 100\ \text{mW}$$

In the transformer, $P_P = P_S = P_L$; therefore, P_L must also equal 100 mW. As proof, calculate the secondary current and solve for P_L.

$$I_S = \left(\frac{N_P}{N_S}\right)I_P \quad \text{(from equation (18.2))}$$

$$= 5 \times 10\ \text{mA}$$

$$= 50\ \text{mA}$$

$$P_L = I_S^2 R_L$$

$$= (50 \times 10^{-3}\text{A})^2 \times 40\ \text{V}$$

$$= 100\ \text{mW}$$

Notice how the transformer has been used as an impedance-matching device from the source to the load. There is approximately 6.8 times as much power delivered to the load when the matching transformer is used, compared to connecting the load directly across the source.

Practice Problems 18.4   Answers are given at the end of the chapter.
1. Determine Z_P when the turns ratio is 10:1 and $Z_S = 16\ \Omega$.
2. Determine the turns ratio when $Z_P = 3.2\ \text{k}\Omega$ and $Z_S = 8\ \Omega$.
3. Determine Z_S when $Z_P = 400\ \Omega$ and the turns ratio is 1:5.

18.6.5   Isolation Transformers

When a transformer has separate primary and secondary windings, the secondary load is not connected directly to the primary AC source. The secondary is said to be *floating* because it is no longer referenced to the ground on the primary side. This isolation is an advantage in reducing the chance for electric shock due to a "hot" ground. A hot ground is

Lab Reference: The application of DC to a transformer is demonstrated in Exercise 28.

also associated with an autotransformer, because it does not employ an isolated secondary. Care must be taken when using it.

Another advantage of an isolated secondary is the fact that any DC in the primary is not transferred to the secondary by transformer action. Sometimes a transformer with a 1:1 ratio is used just for isolation purposes.

**SECTION 18.6
SELF-CHECK**

Answers are given at the end of the chapter.

1. Low-voltage windings are usually _____ in diameter.
2. Describe an autotransformer.
3. The tunable part of an IF or RF core is called a _____.
4. What is reflected impedance?

18.7 TROUBLESHOOTING TRANSFORMER FAULTS

Because a transformer winding is essentially a coil, typical faults are either open circuits or short circuits. A common open occurs at the point where the wire of a winding attaches to a transformer terminal. This can be visually inspected and repaired by resoldering the terminal.

An ohmmeter is best used to test for an open. A reading of infinity indicates an open condition. A low resistance reading is normal, because the windings have some small DC resistance, depending on the number of turns. The windings usually have a rated ohmic value, and a value less than this (but not zero) indicates a partial short. A partial short changes the transformer turns ratio. In most cases, a partial short requires the replacement of the transformer. A short or a partial short is best tested by checking the Q of the primary or secondary coil. This method is fully described in Section 20.9.2 for inductors.

A short-circuited load condition causes an excessive current that can melt the varnish insulation, allowing bare wires to touch. This in turn causes further excessive current to flow. The windings then become shorted or partially shorted. Because the primary is usually fused, excessive secondary current should blow it. There is a distinctive smell when a transformer overheats, created by the melting varnish. There is no practical method for repairing a shorted winding. Consequently, these devices must be replaced if a short circuit has occurred.

A voltmeter can also be used to test transformers. An open in either winding is indicated by 0 V across the secondary.

**SECTION 18.7
SELF-CHECK**

Answers are given at the end of the chapter.

1. The typical faults in a transformer winding are either _____ or _____ circuits.
2. An ohmmeter is best used to test for a(n) _____.
3. An open in either winding is indicated by _____ V across the secondary.

■ SUMMARY

1. A transformer is a device that transfers electrical energy from one circuit to another with no physical connection between them.
2. The transfer of energy is accomplished by mutual induction.
3. The power remains the same on both sides of a transformer, but the *VI* ratio can be changed.
4. A basic transformer consists of a primary winding, a secondary winding, and a core that supports both windings.
5. Voltage is always induced in a secondary winding, but secondary current flows only if there is a path for current.

6. High-frequency applications use an air-core transformer.

7. The most popular and efficient transformer core is the shell core.

8. To provide electrical insulation, the windings of a transformer are coated with varnish.

9. Under no-load conditions, the primary creates a small current called the *exciting current.*

10. The primary current induces a voltage into the primary called *counter-emf.*

11. With a load connected, mutual inductance causes a secondary current that in turn causes an increase in primary current.

12. Flux links both ways in a transformer (from primary to secondary and from secondary to primary).

13. The turns ratio and primary voltage dictate the amount of induced secondary voltage.

14. A step-down transformer has less secondary voltage than primary voltage but more secondary current than primary current.

15. A 10:1 turns ratio means that the secondary has 10 times fewer windings than the primary.

16. Transformer losses are caused by copper losses, eddy currents, and hysteresis losses.

17. Additional connections to transformer windings are called *taps.*

18. Phase-indicating dots are used to indicate points on a transformer symbol that have the same polarity.

19. The power-handling capacity of a transformer is rated in volt-amps (V·A).

20. Power transformers are used to supply voltages to a power-supply circuit.

21. An autotransformer uses a single coil of wire whose secondary voltage is varied by a moveable tap.

22. IF and RF transformers are used to couple high-frequency circuits.

23. For maximum power transfer an impedance matching transformer is used to match impedance between the load and the source.

24. When using an isolation transformer, the secondary *floats,* because it is no longer referenced to the ground on the primary side.

25. An ohmmeter can be used to test for an open transformer winding.

■ IMPORTANT EQUATIONS

$$\frac{N_P}{N_S} = \frac{V_P}{V_S} \tag{18.1}$$

$$I_P N_P = I_S N_S \tag{18.2}$$

$$\frac{Z_P}{Z_S} = \left(\frac{N_P}{N_S}\right)^2 \tag{18.3}$$

$$\frac{N_P}{N_S} = \sqrt{\frac{Z_P}{Z_S}} \tag{18.4}$$

■ REVIEW QUESTIONS

1. What does a transformer do?

2. What elements make up the basic transformer?

3. What is the function of each of the elements that make up the basic transformer?

4. What are the types of cores associated with transformers?

5. The most popular and efficient transformer core is the _____ core.

6. Describe how primary and secondary windings are applied over core material.

7. Describe what happens on the primary side of a transformer under a no-load condition.

8. What causes an induced secondary voltage?

9. What determines the amount of voltage induced into a secondary winding?

10. What differentiates a step-up transformer from a step-down transformer?

11. How is a step-down transformer ratio indicated?

12. What are the types of losses that can occur in a transformer?

13. What is a tap? A center tap?

14. What is the purpose of phase-indicating dots?

15. What is done to ensure that a transformer can handle a higher voltage?

16. What is done to the secondary to allow it to handle higher current levels?

17. The power-handling capacity of a transformer is measured in _____.

18. Two common power-transformer frequencies are _____ Hz and _____ Hz.

19. Where are power transformers used?

20. Describe the autotransformer.

21. The secondary load impedance is _____ back to the primary.

22. Name an application for an isolation transformer.

23. What are the common types of transformer faults?

24. What piece of test equipment is best suited for testing for an open transformer winding?

25. How can a voltmeter be used to test transformers?

■ CRITICAL THINKING

1. Describe mutual inductance in detail and discuss the factors affecting secondary winding parameters.

2. Describe the types of cores available for transformer use and their applications.

3. Describe in detail what happens within the transformer in a no-load condition and in a load condition.

4. Describe how the turns ratio affects voltage and current values within a transformer.

5. Why isn't a transformer's power-handling capacity rated in watts?

■ PROBLEMS

1. There are 250 primary turns in a transformer. How many turns must be wound on the secondary to have a secondary voltage of 10 V if the primary voltage is 100 V?

2. There are 1000 secondary turns in a transformer. How many turns must be wound on the primary to have a secondary voltage of 20 V if the primary voltage is 200 V?

3. A transformer has a turns ratio of 8:1. If 10 V is developed across the secondary, what is the voltage applied to the primary?

4. A transformer has a turns ratio of 1:10. If 500 V is developed across the secondary, what is the voltage applied to the primary?

5. A transformer has a turns ratio of 1:5. If 200 V is applied to the primary, what is the secondary voltage?

6. A transformer with 200 turns in the primary and 500 in the secondary has 2 A of current in the primary. What is the value of current in the secondary?

7. A transformer has 250 mA in the primary. What is the secondary current if the turns ratio is 1:20?

8. A power transformer is connected to a 120-V AC line. Calculate the turns ratio needed for the following secondary voltages: (a) 6 V, (b) 10 V, (c) 15 V, (d) 24 V, (e) 36 V, (f) 48 V, (g) 120 V.

9. A power transformer with a voltage step-up ratio of 1:3 is connected to a 120-V AC line. What is the smallest load resistor that can be connected to the secondary without exceeding a power rating of 50 VA?

10. Solve for each unknown listed in each of the diagrams listed in Figure 18.11(a)–(f).

11. For the circuit in Figure 18.12, determine the turns ratio required for a maximum power transfer.

12. For Problem 11, determine the load power.

■ ANSWERS TO PRACTICE PROBLEMS

PRACTICE PROBLEMS 18.1

1. $V_S = 100$ V
2. $N_S = 21$

PRACTICE PROBLEMS 18.2

1. Turns ratio = 5:1
2. $V_P = 120$ V

PRACTICE PROBLEMS 18.3

1. $I_P = 150$ mA
2. Turns ratio = 5:1

PRACTICE PROBLEMS 18.4

1. $Z_P = 1600$ Ω
2. Turns ratio = 20:1
3. $Z_S = 10$ kΩ

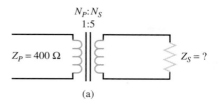

(a)

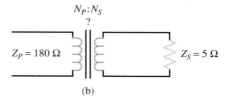

(b)

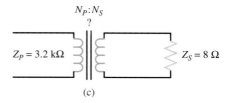

(c)

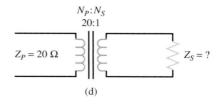

(d)

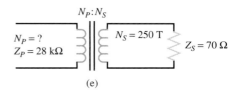

(e)

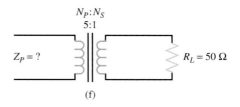

(f)

FIGURE 18.11
Circuits for Problem 10; solve for the unknowns listed.

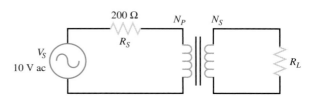

FIGURE 18.12
Circuit for Problem 11.

■ ANSWERS TO SECTION SELF-CHECKS

SECTION 18.1 SELF-CHECK

1. False
2. Permeability
3. Silicon iron
4. Shell
5. Varnish

SECTION 18.2 SELF-CHECK

1. Exciting current
2. From a counter-emf set up as the primary current builds a magnetic field within its windings
3. Increase

SECTION 18.3 SELF-CHECK

1. The turns ratio and V_P
2. One that has fewer turns on the secondary
3. Step-up transformer
4. 12.5 turns for every turn on the secondary

SECTION 18.4 SELF-CHECK

1. By using larger-diameter wire
2. Eddy currents are caused by induced core voltage setting up random currents within the core.
3. 60, 400

SECTION 18.5 SELF-CHECK

1. A tap is a connection to windings at points other than the ends of the windings.
2. A transformer with two or more primary or secondary windings
3. To indicate points on a transformer schematic that have the same instantaneous polarity
4. The diameter of the wire used for the winding
5. Volt-amps

SECTION 18.6 SELF-CHECK

1. Larger
2. It consists of a single coil of wire tapped to produce a primary and a secondary winding.
3. Slug
4. The secondary load impedance is reflected back into the primary.

SECTION 18.7 SELF-CHECK

1. Open, short
2. Open
3. 0

CAPACITORS, INDUCTORS, AND
TRANSIENT RESPONSE

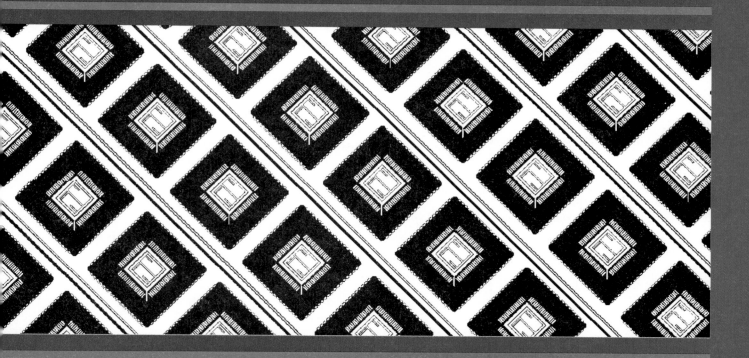

CHAPTER 19

CAPACITORS AND CAPACITANCE

OBJECTIVES

Upon completing this chapter, you will be able to:

1. Describe the characteristics of a capacitor.
2. Calculate capacitance and capacitance charging current.
3. Compare and contrast the various types of capacitors via the dielectric used.
4. Evaluate capacitor value codes.
5. Describe temperature compensation as associated with capacitors.
6. Compare and contrast capacitor faults.

INTRODUCTION

Up to this point we have concentrated on circuits containing only resistance. Resistance opposes the flow of current and converts or dissipates energy in the form of heat.

In simple terms, *capacitance* is the property of a dielectric to *store* electric charge. A device especially designed to have a certain value of capacitance is called a *capacitor*. A capacitor has the ability to store electrons and release them at a later time. A capacitor is analogous to a spring. A spring accumulates energy as its coils are condensed. A capacitor has also been called an *accumulator* and a *condensor*. The number of electrons that the capacitor can store for a given applied voltage is a measure of its capacitance and is the basis for equation (19.1). The capacitor stores energy in an *electrostatic field*. The energy is stored in such a way as to oppose any change in voltage. The unit of capacitance is the *farad*, F, named after Michael Faraday.

Capacitors have a number of important uses in electronics. These include filtering, coupling, and bypassing. In fact, it is virtually impossible to find an electronic circuit that doesn't use capacitors. In this chapter, you will learn about what a capacitor is, factors affecting capacitance, capacitor types, capacitor value coding, and capacitor faults. Our discussions are limited to DC circuits only. Capacitors and AC circuits are the subject of a later chapter.

CAPACITOR CHARACTERISTICS

19.1.1 Electrostatic Field

The electrostatic field and charges were discussed in Chapter 3. Charges were characterized utilizing Coulomb's law. We saw that any charged particle is surrounded by invisible lines of force, called *electrostatic lines of force*. These lines of force have the following characteristics:

- They are polarized from positive to negative.
- They radiate from a charged particle in straight lines and do not form closed loops.
- They have the ability to pass through any known material.
- They have the ability to distort the orbits of tightly bound electrons.

If two unlike charges are placed on opposite sides of an atom whose outermost (valence) electrons cannot escape from their orbits, the orbits of the electrons are distorted, as shown in Figure 19.1. Part (a) shows the normal orbit (circular), and part (b) shows the same orbit (elliptical) in the presence of charged particles. Since electrons have a negative charge, the positive charge attracts them, pulling them closer to it. The negative charge repels the electrons, pushing them further from it. It is this ability of an electrostatic field to attract and to repel charges that allows the capacitor to store energy.

FIGURE 19.1
Distortion of an electron orbit due to electrostatic force.

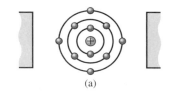

(a)

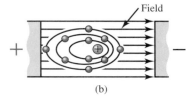

(b)

19.1.2 A Simple Capacitor

A simple capacitor is the parallel-plate capacitor, consisting of two conductors or electrodes separated by a dielectric material of uniform thickness. The conductors can be any material that conducts electricity easily. The dielectric material must be a poor conductor—that is, an insulator. This is shown in Figure 19.2.

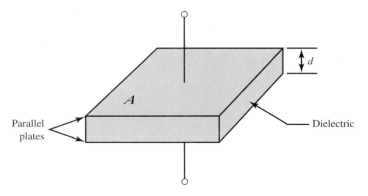

FIGURE 19.2
The parallel-plate capacitor.

19.1.3   Charging and Discharging a Capacitor

Figure 19.3(a) shows that each plate of an uncharged capacitor has the same number of free electrons before voltage is applied to the capacitor. Initially, there is no potential difference across the capacitor. In part (b) the switch is closed. The top plate becomes positively charged due to the repulsion of electrons within that plate by the distorted electron orbit and the attraction of these electrons to the positive terminal of the battery. The bottom plate becomes negatively charged due to the attraction of electrons into that plate by the distorted electron orbit. This produces a current flow, which is limited by resistor R_1.

Although a charge current does flow in the circuit, it does *not* flow through the capacitor. Electrons flow out of the capacitor's positive plate and into the negative plate via the battery and any circuit resistance. They cannot flow through the capacitor due to the insulating effects of the dielectric.

What the voltage source does is simply *redistribute* some electrons from one side of the capacitor to the other. This process is called *charging* the capacitor. This process continues until the potential difference across the capacitor terminals equals the source voltage. At this point the source voltage and the capacitor voltage are equal series-opposing voltages, and current ceases to flow in the circuit.

In part (c) of the figure, the switch has been opened, and the capacitor has retained its charge. This occurs because the dielectric material is an insulator, and the electrons in the bottom plate (negative charge) have no path to reach the top plate (positive plate). The distorted orbits of the atoms of the dielectric plus the electrostatic force of attraction between the two plates hold the positive and negative charges in their original position. Thus,

FIGURE 19.3
Charging the capacitor: (a) uncharged, no potential difference across capacitor; (b) charging the capacitor, the resistor limits the charging current; (c) capacitor retains charge.

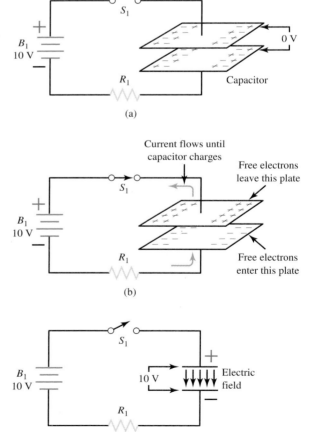

the energy that came from the battery is now stored in the electrostatic field of the capacitor.

The capacitor can be discharged by providing a path for current from one plate to the other (simply shorting the plates together is an easy method). A discharge current flows, which equalizes the number of free electrons on each plate. When the plates contain the same number of free electrons, there is no potential difference across the plates; hence the capacitor is completely discharged. At this point the valence electron orbits become circular again.

19.1.4 Capacitor Symbols

Figure 19.4 shows some common symbols for the capacitor. The symbol in Figure 19.4(a) looks very much like the parallel-plate model. The curved plate in (b) of the figure indicates that this plate is to be connected to a point closest to ground potential. Part (c) is used to indicate a variable capacitor. Part (d) is a more recently used symbol.

FIGURE 19.4
Circuit symbols for capacitors: (a) parallel-plate model; (b) curved plate is usually connected to lowest potential; (c) variable capacitor; (d) a more recent symbol.

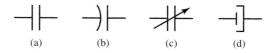

(a)　　(b)　　(c)　　(d)

19.1.5 Capacitance

For any given capacitor, the ratio of the charge on one plate to the potential difference across the plates is a constant. This constant is a property of the capacitor called its *capacitance*. Thus, capacitance is defined as

$$C = \frac{Q}{V} \qquad (19.1)$$

where　C = the capacitance (F)
　　　　Q = the charge per plate (C)
　　　　V = the potential difference across the capacitor (V)

Thus, 1 F = 1 C/V. A capacitor has a capacitance of 1 F if 1 C of charge must be deposited to raise the potential difference by 1 V across the capacitor. The farad is a very large amount of capacitance. The microfarad (μF) and picofarad (pF) are much more practical units of capacitance. Practical capacitors are available with capacitances ranging from 1 pF to more than 500,000 μF.

$$1 \text{ microfarad} = 1 \ \mu\text{F} = 1 \times 10^{-6} \text{ F}$$
$$1 \text{ picofarad} = 1 \text{ pF} = 1 \times 10^{-12} \text{ F}$$

EXAMPLE 19.1

Determine the capacitance when 50 μC of charge deposited on the plates raises the potential difference to 2 V.

Solution　$C = \dfrac{Q}{V}$

$$= \frac{50 \times 10^{-6} \text{ C}}{2 \text{ V}}$$

$$= 25 \times 10^{-6} \text{ F}$$

$$= 25 \ \mu\text{F}$$

EXAMPLE 19.2

A constant current of 4 mA charges a 10-μF capacitor in 1 s. How much voltage is across the capacitor?

Solution: $Q = It$

$$= 4 \times 10^{-3} \, \text{A} \times 1 \, \text{s}$$

$$= 4 \times 10^{-3} \, \text{C}$$

$$V = \frac{Q}{C}$$

$$= \frac{4 \times 10^{-3} \, \text{C}}{10 \times 10^{-6} \, \text{F}}$$

$$= 400 \, \text{V}$$

Practice Problems 19.1   Answers are given at the end of the chapter.
1. A constant current of 5 mA charges a 20-μF capacitor for 80 ms. What is the voltage across the capacitor?
2. How much charge is stored in a 10,000-μF capacitor with 40 V across it?
3. A capacitor stores 500-μC of charge with 200 V across its plates. Determine the value of C.

19.1.6   Capacitive Current Equation

We have seen that $Q = CV$. Assume that a small additional charge (ΔQ) raises the potential difference across the capacitor by a small amount (ΔV). Then, $\Delta Q = C\Delta V$. The symbol Δ (Greek delta) means a "change in." If these changes take place in a small amount of time (Δt), then $\Delta Q/\Delta t = C(\Delta V/\Delta t)$. Because $\Delta Q/\Delta t$ is the average current (i) during this short time interval (Δt),

$$i_c = C\left(\frac{\Delta V}{\Delta t}\right)$$

To emphasize that this is the current into or out of the capacitor (charge or discharge) and that we must consider the changing voltage across the capacitor, we have

$$i_c = C\left(\frac{\Delta V_c}{\Delta t}\right) \tag{19.2}$$

where i_c = current into or out of capacitor (A)
C = capacitance (F)
ΔV_c = change in voltage across capacitor (V)
Δt = time interval (s)

What is important about this equation is that the capacitor current varies *only* when the voltage across the capacitor is *changing*.

EXAMPLE 19.3

Determine the capacitor current when the voltage across a 4-μF capacitor changes from 16 to 24 V in 2 ms.

Solution $i_c = C\left(\dfrac{\Delta V_c}{\Delta t}\right)$

$$= (4 \times 10^{-6}\,\text{F})\left(\frac{24\,\text{V} - 16\,\text{V}}{2 \times 10^{-3}\,\text{s}}\right)$$

$$= 16\,\text{mA}$$

Practice Problems 19.2 Answers are given at the end of the chapter.

1. Determine the capacitor current when the voltage across a 10-μF capacitor changes from 10 to 20 V in 2.5 ms.

SECTION 19.1
SELF-CHECK

Answers are given at the end of the chapter.

1. Describe a simple capacitor.
2. When the capacitor is charging, the source is simply _____ some electrons from one side of the capacitor to the other.
3. True or false: Current flows through a capacitor.
4. To what is 1 F of capacitance equal?
5. What are the common units of capacitance?

19.2   FACTORS AFFECTING CAPACITANCE

The capacitance of a capacitor depends on three factors:

- The area of the plates
- The distance between the plates
- The dielectric constant of the material between the plates

The plate area affects the value of capacitance in the same manner that the size of a container affects the amount of water that it can hold. A capacitor with a large plate area can store more charges than a capacitor with a small plate area. Simply stated, the larger the plate area, the larger the capacitance.

Electrostatic lines of force are strongest when the charged particles that create them are close together. When the charged particles are moved further apart, the lines of force weaken, and the ability to store a charge decreases. (Remember from Coulomb's law that the force of attraction or repulsion is inversely proportional to the distance squared between the charges.)

When the diagram of a capacitor is magnified, it can be seen that the presence of electrical charges on the electrodes induces charges in the dielectric. These induced charges determine something called *permittivity*. Permittivity refers to how well a dielectric material can establish electrostatic lines of force (electric flux). (It is comparable to permeability and electromagnetic lines of force.) The *relative permittivity*, or *dielectric constant* (ε_r), is a measure of how good a material is for the production of electric flux.

The dielectric constant for a vacuum (and, for all practical purposes, air, as well) is 1 and is taken as a reference. For any other material, the constant is more than 1, depending on how much electric flux would be produced if the material were substituted for a vacuum as the path between the plates. The relative permittivity is unitless, because it is a ratio of the absolute permittivity, ε, of a material to the absolute permittivity, ε_o, of a vacuum. ($\varepsilon_r = \varepsilon/\varepsilon_o$.) This number varies, depending on the type of material used for the dielectric. In many instances, ε_r is also designated as K, to represent a constant. The higher the dielectric constant (K), the higher the capacitance.

Table 19.1 lists some common values of the dielectric constant for given materials.

TABLE 19.1
Selected dielectric constants

Material	Dielectric Constant, ε_r (or K), 25°C
Vacuum	1.0
Air	1.0006
Plastics, general	2 to 6
Paper	2 to 6
Mica	5.4 to 8.7
Aluminum oxide	10
Tantalum oxide	26
Barium titanate	1000 to 3000
Ceramics	12 to 400,000

19.2.1 The Capacitor Equation

The equation used to compute the value of capacitance from physical parameters is

$$C = \frac{(8.85 \times 10^{-12})\,KA}{D} \tag{19.3}$$

where C = capacitance (F)
  8.85×10^{-12} F/m, permittivity of a vacuum
  K = dielectric constant
  A = area of one plate (m^2)
  D = distance between plates (m)

EXAMPLE 19.4

Determine the capacitance of two metal plates, each 5×6 cm, that are separated by 0.5 mm. The dielectric is air.

Solution $C = \dfrac{(8.85 \times 10^{-12}\ \text{F/m})\,KA}{D}$

$\qquad = \dfrac{(8.85 \times 10^{-12}\ \text{F/m}) \times 1 \times (5.0 \times 10^{-2}\ \text{m} \times 6.0 \times 10^{-2}\ \text{m})}{0.5 \times 10^{-3}\ \text{m}}$

$\qquad = 53.1 \times 10^{-12}\ \text{F}$

$\qquad = 53.1\ \text{pF}$

Practice Problems 19.3   Answers are given at the end of the chapter.
1. Calculate the capacitance for a capacitor with the following physical characteristics: $K = 40$, $A = 5$ cm^2, and $d = 0.02$ cm.
2. Calculate the capacitance for a capacitor with the following physical characteristics: $K = 8$, plates are 5×5 cm, $d = 2$ cm.

It should be obvious by studying equation (19.3) that to get a large capacitance in a small package (the trend in modern electronics), you need to use a dielectric with a high *K*. In fact, you might say, why make commercial capacitors with any of the materials having low values of *K*? The answer generally lies with other capacitor characteristics, such as stability (to temperature variations), voltage ratings, and the like. Some low *K* values have these characteristics.

19.2.2 Dielectric Strength

In selecting or substituting a capacitor for use, consideration must be given to (1) the value of capacitance desired and (2) the amount of voltage to be applied across the capacitor. If the voltage applied across the capacitor is too great, the dielectric can break down and arcing will occur between the capacitor plates. When this happens the capacitor becomes a short circuit (a conductor), and the flow of direct current through it can cause damage to other electronic parts. Each capacitor has a voltage rating (*DC working voltage*) that should not be exceeded.

Dielectric strength is the ability of a dielectric to withstand a potential difference without arcing across the insulator. A dielectric's breakdown voltage, V_{BR}, is a function of its material and thickness. The dielectric strength of selected materials is given in volts per mil (1 mil = 0.001 in.) in Table 19.2.

TABLE 19.2
Dielectric strength of selected materials

Material	Dielectric Strength (V/mil)
Air	76
Bakelite	150–500
Mica	600–1500
Paper	1250
Ceramics	600–1250

A capacitor that may be safely charged to 120 V DC cannot be safely subjected to an alternating voltage or a pulsating direct voltage having an effective (rms) value of 120 V. An alternating voltage of 120 V AC has a peak value of 170 V. As a rule of thumb, a capacitor should be selected so that its working voltage is at least *50% greater* than the highest effective voltage to be applied to it.

SECTION 19.2 SELF-CHECK

Answers are given at the end of the chapter.

1. Define the term *permittivity*.
2. The distance separating two plates of a capacitor is reduced by 4 and the area of each plate is reduced by 4. How is the capacitance affected?
3. What should the working voltage be when selecting a capacitor for a particular application?

19.3 CAPACITOR TYPES

Capacitors may be classified as either *variable* or *fixed*. Additional classifications are usually based on the type of *dielectric* used to make the capacitor. Many capacitors also come in chip format for use in surface-mount technology.

19.3.1 Variable Capacitors

Figure 19.5 illustrates a variable capacitor that uses air as the dielectric material. In this case, the capacitance value can be changed by rotating the shaft, because this changes the effective plate area and, hence, the capacitance. The plates of this capacitor are called the *stator* and the *rotor*. Capacitors similar to this typically can change capacitance value over a 10:1 range. Such capacitors are still used as tuning elements in inexpensive portable radios. The capacitance value ranges from a few picofarads to approximately 500 pF.

FIGURE 19.5
A variable capacitor that uses air for the dielectric.

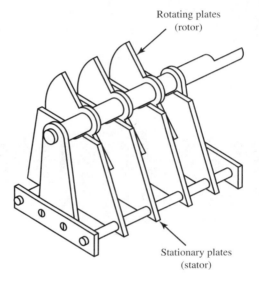

Small-value variable capacitors are also available. These are known as *trimmer* caps. A common example has two plates separated by a sheet of mica. A screw adjustment is used to vary the distance between the plates, thereby changing the capacitance. This is illustrated in Figure 19.6. Trimmer caps also use plastic and ceramic as the dielectric.

FIGURE 19.6
Trimmer capacitor.

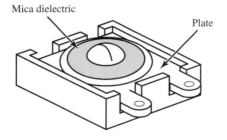

19.3.2 Fixed Capacitors (by Dielectric Type)

Figure 19.7 shows a broad selection of various types of fixed capacitors. This section gives details of their characteristics.

FIGURE 19.7
Assorted fixed capacitors.

Mica

Mica was one of the first materials used to make capacitors because of its ability to remain stable under varying parameters. It is also a natural dielectric. The two techniques used to form the capacitors are by stacking the mica in sheets through the silvered-mica process or by the use of tin-lead foil to separate the mica sheets.

Due to the inherent characteristics of the dielectric, mica capacitors are inexpensive, small, and readily available. They exhibit good stability and high reliability. They are used in circuits requiring precise frequency filtering, bypassing, and coupling.

Paper

Paper capacitors are made of alternate layers of paper and metal foil rolled together, forming a tubular structure. They usually have a band at one end to identify the lead that is connected to the outside metal foil. Grounding this lead improves noise immunity. Their disadvantage lies in their sensitivity to temperature changes. They are rarely used today.

Ceramic

Ceramic materials have a large range of dielectric characteristics and are often used to make chip capacitors as shown in Figure 19.8. Consequently, ceramics are the most versatile dielectric materials currently available. In commercial practice, the dielectric is made from finely powdered materials, the principal one being barium titanate. Disc elements

FIGURE 19.8
Ceramic disc capacitor: (a) pressed die ceramic dielectric; (b) disc ceramic with lead wires.

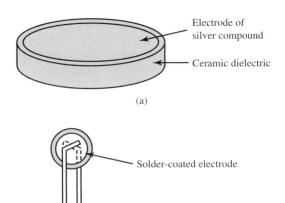

are pressed in dies and then fired at high temperatures (Figure 19.8(a)). The outer surface can easily be soldered, and wires are usually attached as seen in Figure 19.8(b). The hairpin-shaped wires are springy enough to hold the ceramic elements while the assembly is dipped in solder. The lower end of the hairpin is cut off later. These dipped discs are among the cheapest capacitors available.

A much more sophisticated design is called the *monolithic* ceramic capacitor. It offers much higher capacitance per unit volume (*C/V*). Figure 19.9 shows the capacitor in cross-sectional view and in simplified form. Only two electrodes are seen here, but 20 or 30 electrodes or more are very common in commercial practice. The additional electrodes increase the capacitance because more layers of the dielectric are used. This type of capacitor is often used for temperature compensation and thus is marked with temperature-coefficient indicators. NPO indicates a zero temperature coefficient, whereas P100 and N100 indicate positive and negative temperature coefficients of ±100 ppm/°C, respectively. (Note that a 1.0-μF capacitor contains 1 million pF. Thus, for the aforementioned temperature-compensating capacitors, there would be a change of $+100$ pF and -100 pF per degree Celsius, respectively.)

High-*K* ceramic capacitors have larger capacitance values than low-*K* types, but they are not nearly as stable with changes in temperature, voltage, and frequency.

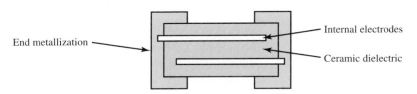

FIGURE 19.9
Monolithic ceramic element.

Plastic Film

In plastic-film capacitors, the dielectric consists of a thin film of plastic (from the family of thermoplastics). Some popular plastic dielectrics include polystyrene, polyester (mylar), polypropylene, and polycarbonate. Plastic-film capacitors are designed for use in circuit applications that require high insulation resistance and low losses over a wide temperature range.

Metallized-Film Capacitors

Metallized-film capacitors use electrodes that are vacuum-deposited on various types of plastic dielectric film instead of free-standing metal foil. Their thickness is much less than foil electrodes. They have a self-healing characteristic called *clearing*. If a breakdown by either a hole or contaminant occurs in the metal film, a tiny area of the film surrounding the breakdown point burns away. This leaves the capacitor operable, but with a slightly reduced capacitance.

19.3.3 Electrolytic Capacitors

Aluminum Oxide

Electrolytic capacitors have the largest amount of capacitance for a given physical size. The dielectric consists of a thin oxide layer (0.01 μm), which is formed by electrochemical action during the manufacturing process. Figure 19.10 illustrates a popular type of aluminum electrolytic capacitor. Sheets of aluminum foil are separated by paper that has been saturated with a conductive paste called the *electrolyte* (Figure 19.10(a)). In Figure

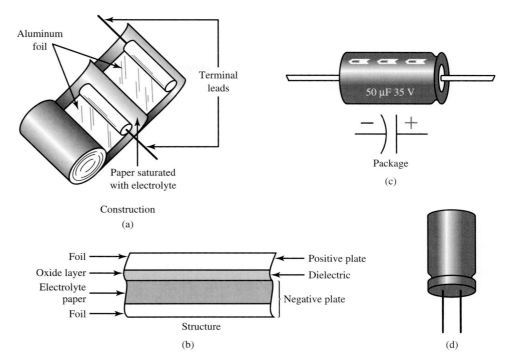

FIGURE 19.10
A polarized aluminum electrolyte capacitor: (a) construction; (b) structure; (c) axial package; (d) radial package.

19.10(b), you should note the following:

1. The foil with the aluminum oxide coating is the positive plate.
2. The electrolyte is the negative plate.
3. The dielectric is the very thin layer of aluminum oxide.

Because the aluminum oxide layer is very thin and because the area of the aluminum sheets is large, the capacitance-to-volume ratio is also large. Unfortunately, the thin oxide layer limits the maximum voltage that can be applied across the capacitor to relatively low values. In addition, the thin oxide layer results in relatively low values of leakage resistance and, consequently, large leakage currents.

Note that this capacitor is *polarized,* which means that the capacitor works properly only if the correct polarity is observed when the capacitor is put into a circuit. On the plastic sleeve that covers the body of the capacitor are value, voltage, and polarity markers. A large arrow marked (−) points to the negative terminal (this is illustrated in Figure 19.10(c), along with its symbol). Safe operating voltages are limited to about 450 V. The oxide dielectric blocks current flow in one direction but offers low resistance in the opposite direction; it is therefore limited to *DC applications.* If AC components are present, the sum of the peak AC plus the applied DC voltage should never exceed the DC rating. A voltage reversal of just a few volts causes a breakdown of the film and a destruction of the capacitor.

Note that the leads of the capacitor in Figure 19.10(c) go through the center axis of the body. Leads of this type are known as *axial* leads. The capacitor leads in Figure 19.10(d) come out of the body side by side. These are known as *radial* leads. They easily lend themselves to through-hole pc board applications, because they do not have to be bent at a 90° angle.

A nonpolarized electrolytic can be made by forming oxide coating on both plates or internally connecting two polarized capacitors in series-opposing polarity.

When aluminum oxide electrolytic capacitors are not used, their oxide layers tend to deteriorate. For this reason, such capacitors are said to have a short *shelf life*. To reuse, they should be brought up to the correct DC working voltage slowly and then left at this potential for at least 30 min. This process helps in reforming the oxide coating. (Tantalum oxide capacitors were developed to overcome this problem and other inherent disadvantages.)

Tantalum Oxide

Tantalum electrolytics have become the preferred type where high reliability and long service life are paramount considerations. Tantalum is a *valve* metal (as is aluminum), upon which one may grow very uniform and stable oxides with good dielectric properties. The dielectric constant of tantalum is very high, yielding a high capacitance per unit volume. Tantalum capacitors may utilize only 15% of the area normally required by an aluminum/paper capacitor of the same capacitance value.

Tantalum capacitors contain either liquid or solid electrolytes and come in three types. The liquid electrolyte (usually sulfuric acid) forms the wet-slug, or the tantalum-foil, capacitor. The solid electrolytic capacitor uses a dry material, manganese dioxide, as the electrolyte. The solid tantalum is the most common because of its excellent temperature characteristics.

The solid tantalum capacitors possess a unique "healing" mechanism, which results in a failure rate apparently decreasing forever. The manganese dioxide (MnO_2) provides the healing mechanism. If a fault, perhaps some impurity, produces an imperfection in the dielectric layer, a heavy current flows through that minute area when a DC potential is applied to the capacitor. The current also flows through the MnO_2 immediately adjacent to the fault. Resistance of the MnO_2 to this current flow causes localized heating. As the temperature of MnO_2 rises, this material is converted to a lower oxide of manganese, such as MnO_3, with much higher resistivity. The increase in resistance decreases the current flow and circumvents a short circuit from developing. This process is illustrated in Figure 19.11.

FIGURE 19.11
Fault in a tantalum capacitor.

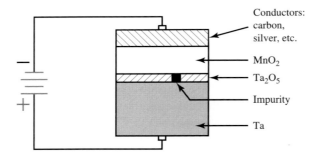

Conductors: carbon, silver, etc.

MnO_2

Ta_2O_5

Impurity

Ta

Table 19.3 summarizes various attributes of some of the major capacitor types. It may be used as a guide to the selection of the appropriate capacitor to use for a given application.

TABLE 19.3
Capacitor-selection guide

Types →	Mica	Paper	Ceramic	Plastic Film (Polyester)	Metallized Plastic Film (Polypropylene)	Aluminum Oxide	Tantalum Oxide (Solid)
Value Range	1 pF to 0.1 μF	100 pF to 100 μF	1 pF to 100,000 pF	5000 pF to 20 μF	0.001 to 4.7 μF	0.68 μF to 220,000 μF	0.001 μF to 1000 μF
Tolerance (%)	± 1, ± 10	± 10, ± 20	± 5 to ± 20	± 10, ± 20	± 5, ± 10, ± 20	-10 to $+100$	5 to 20
Voltage Rating (V)	100 to 2500	50 to 5000	3.3 to 3,000 DC	100 to 600	To 630 V DC, to 250 V AC	to 450 V	6 to 120 V DC
Temperature Range	$-55°C$ to $+150°C$	$-30°C$ to $+100°C$	Temperature compensation to 200,000 ppm/°C	$-55°C$ to $+125°C$	$-55°C$ to $+105°C$	$-55°C$ to $+85°C$	$-55°C$ to $+85°C$
Advantages	Low loss, stable	Cost, wide range of values	NPO available high C/V ratio	Small size	High stability, close tolerance	High C/V ratio	Long shelf life, high C/V ratio
Applications	Coupling, bypassing, filters, tuned circuits	Coupling, bypassing LF, filters	Temperature compensation, filters, bypassing, comes in chip format	Coupling, bypassing	DC blocking	Coupling, filtering, bypassing	Coupling, bypassing, blocking, comes in chip format

19.4 CAPACITOR CODES

Capacitors may be marked to show their value, voltage rating, tolerance, temperature stability, and other information. Because of the myriad of ways manufacturers mark capacitor values, the interpretation of these marks tends to give everyone more than a little consternation. This section attempts to simplify the interpretation of these marks. Alphanumeric and color codes are used for either the capacitance value or for the temperature-compensation value. For ceramic dielectric capacitors, alphanumeric codes are used exclusively for temperature-compensation values. Color codes have been used for capacitance values (based on the resistor color code) for some time, but manufacturers are now favoring ink-jet and laser markings (alphanumeric) for both leaded capacitors and chip capacitors.

Capacitors *always* have their value specified in either microfarads or picofarads. As a general rule, capacitors' values in whole numbers (such as 10, 47, 100, 470) are interpreted as the value in pF units. (Note that pF is not usually printed on the body of the capacitor.) Conversely, if a capacitor is labeled in a decimal fraction (such as 0.002, 0.01, 0.68), the value is interpreted as μF units (see Figure 19.13(a) on p. 357). Capacitors of this type are *uncoded*, because they are read directly. (The notation μF or MFD may or may not be printed on the body.) The exception to this rule is with aluminum electrolytic capacitors, which have a body size large enough so that direct marking (uncoded) of their value (either whole or fractional numbers), including units and voltage rating, is easily accomplished (as shown in Figure 19.10(c)).

The capacitor's voltage rating is usually specified numerically on the body of the capacitor. Some capacitors do not show the voltage rating.

19.4.1 Class I and II Dielectrics

Military contracts have utilized specifications for electronic components for a long time. However, in June 1994, the Department of Defense (DoD) issued a memorandum that stated the department must increase access to commercial technology. To do this, they are to specify *performance specifications.* If a commercial specification meets this requirement, they must use it. This is supposed to reduce the costs of military contracts.

A common military specification for fixed ceramic capacitors is the MIL-C-123 spec. A common commercial standard for the same capacitor is the Electronics Industry Association (EIA) RS-198 standard. Because of the DoD memorandum, commercial vendors are cross-referencing their specs to military specs, thereby increasing the access to the commercial standards.

The RS-198 EIA standard specifies alphanumeric codes for defining the temperature coefficients of Class I dielectrics and the allowable change in capacitance from room temperature over certain temperature ranges of Class II dielectrics. Table 19.4 shows the temperature coefficient code for Class I dielectrics. Class I dielectrics display the most stable characteristics. The most common Class I dielectric for chip capacitors is the COG designation (shown in Table 19.4). This allows 0 ppm/°C ± 30 ppm/°C and is equivalent to the NPO (zero temperature coefficient) code defined by MIL specifications.

TABLE 19.4
Temperature coefficient of capacitance
Class I dielectrics

Temperature Coefficient of Capacitance (ppm/°C)	Multiplier	Tolerance (ppm/°C)
C = 0.0	0 = −1.0	G = ±30
M = 1.0	1 = −10	H = ±60
P = 1.5	2 = −100	J = ±120
R = 2.2	3 = −1,000	K = ±250
S = 3.3	4 = −10,000	L = ±500
T = 4.7	5 = +1.0	M = ±1,000
U = 7.5	6 = +10	N = ±2,500
—	7 = +100	—
—	8 = +1,000	—
—	9 = +10,000	—

EXAMPLE 19.5

Determine the temperature coefficient of capacitance when a ceramic capacitor is marked R2G.

Solution   Using Table 19.4 as a guide, we have

$$R2G \rightarrow -220 \text{ ppm/°C} \pm 30 \text{ ppm/°C}$$

Class II dielectrics offer much higher dielectric constants than Class I dielectrics but with less stable properties to changes in temperature, voltage, and the like. They are called *general-purpose capacitors*. Table 19.5 shows the allowable capacitance variation over temperature code for Class II dielectrics. The most common mid-*K* characteristic used in ceramic chip–capacitor manufacture is X7R, and the most common high-*K* bodies are Y5V.

TABLE 19.5
Maximum change in capacitance
Class II dielectrics

Low Temperature (°C)	High Temperature (°C)	Maximum Allowable Change in Capacitance from Room Temperature to High and Low Temperatures (Stability, %)
X = −55	2 = +45	A = ±1.0
Y = −30	4 = +65	B = ±1.5
Z = +10	5 = +85	C = ±2.2
	6 = +105	D = ±3.3
	7 = +125	E = ±4.7
	8 = +150	F = ±7.5
		P = ±10
		R = ±15
		S = ±22
		T = +22/−33
		U = +22/−56
		V = +22/−82

EXAMPLE 19.6

Determine the maximum change of capacitance from 25°C for a ceramic capacitor marked Y5V.

Solution   From Table 19.5 we have Y5V → +22%, −82% maximum change from +25°C, over a −30°C to +85°C temperature range.

19.4.2 The Three-Digit Picofarad Code

As was stated earlier, with the exception of electrolytics, whole-number value markings on capacitors usually indicate a value in picofarads, and decimal fractions indicate a value in microfarads.

Another version (and quite popular) uses a *three-digit coding system*. The three-digit code is interpreted as a capacitance value in picofarads, but a picofarad value can be converted to microfarads as necessary (by moving the decimal point six places to the left). The first two digits of the code represent significant numbers. The third digit is the multiplier, or the number of zeros to add. A letter indicates the tolerance. Table 19.6 shows this code.

TABLE 19.6
The three-digit picofarad capacitor code

First and Second Significant Digits	Third Digit	Multiplier for Third Digit	Tolerance Letter Symbol	Tolerance
0 (not used for first digit)	0	1	F	±1%
1	1	10	G	±2%
2	2	100	H	±3%
3	3	1,000	J	±5%
4	4	10,000	K	±10%
5	5	100,000	M	±20%
6	—	—	Z	+80%, −20%
7	—	—	B	±0.1 pF
8	8	0.01	C	±0.25 pF
9	9	0.1	D	±0.5 pF

EXAMPLE 19.7

For the capacitor shown in Figure 19.12, determine value, voltage rating, tolerance, temperature range, and stability over this temperature range.

FIGURE 19.12
Capacitor for Example 19.7.

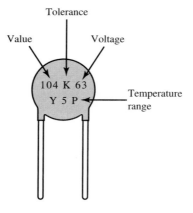

Solution $104 \rightarrow 10 \times 10{,}000$ pF
$= 100{,}000$ pF
$= 0.1\ \mu F$ (from Table 19.6)
$K \rightarrow \pm 10\%$ tolerance in value   (from Table 19.6)
$63 \rightarrow 63$ V, working voltage   (read directly)
$Y5V \rightarrow -30°$ to $+85°C$, $\pm 10\%$ stability over
this range   (from Table 19.5)

EXAMPLE 19.8

For the capacitor in Figure 19.13(a), determine the value, tolerance, temperature range, and stability.

Solution   Because the value is marked as a decimal fraction, this indicates microfarad units (the three-digit code is not valid). Table 19.5 is used to get the stability, and Table 19.6 is used to get the tolerance. Thus we have

$$0.0022 \ \mu\text{F} \quad \text{(uncoded and read directly)}$$
$$K \rightarrow \pm 10\% \quad \text{(from Table 19.6)}$$
$$\text{X7R} \rightarrow -55°\text{C to} \ +125°\text{C}, \pm 15\% \text{ stability over this range}$$
$$\text{(from Table 19.5)}$$

FIGURE 19.13
Capacitors for Examples 19.8 and 19.9.

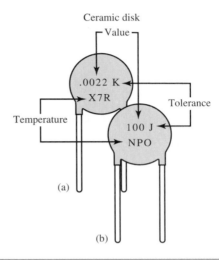

EXAMPLE 19.9

Using the capacitor in Figure 19.13(b), determine value, tolerance, and temperature coefficient.

Solution   (Note: NPO = COG.) From these, we have

$$100 \rightarrow 10 \times 1 = 10 \ \text{pF} \quad \text{(from Table 19.6)}$$
$$\text{J} \rightarrow \pm 5\% \quad \text{(from Table 19.6)}$$
$$\text{NPO} = \text{COG} \rightarrow 0 \ \text{ppm/°C} \ \pm \ 30 \ \text{ppm/°C} \quad \text{(from Table 19.4)}$$

EXAMPLE 19.10

Determine the capacitance value of a capacitor marked with the three-digit pF code of 229.

Solution   Using Table 19.6, the first two digits are significant numbers. The 9 indicates the decimal goes between the two significant numbers (multiplication by 0.1). Thus,

$$229 \rightarrow 22 \times 0.1 = 2.2 \ \text{pF}$$

(A capacitor marked 228 means that you multiply by 0.01. Thus, the value would be 0.22 pF.)

EXAMPLE 19.11

Determine the capacitance value of a capacitor marked with the three-digit pF code of 473.

Solution   473 → 47 × 1,000 = 47,000 pF   (from Table 19.6)
                              = 0.047 μF

EXAMPLE 19.12

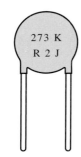

273 K
R 2 J

FIGURE 19.14
Ceramic capacitor for Example 19.12.

Determine the capacitance value of a temperature-compensating capacitor marked with the three-digit pF code of 273 at 50°C, as illustrated in Figure 19.14.

Solution
1. Using the top-row values on the capacitor and Table 19.6, we have

$$273 \rightarrow 27 \times 1000 = 27,000 \text{ pF}$$
$$= 0.027 \ \mu\text{F} \qquad \text{(at room temperature, 25°C)}$$
$$K \rightarrow \pm 10\% \text{ tolerance}$$

2. Using the bottom-row values and Table 19.4, we have

$$R \rightarrow 2.2 \text{ ppm/°C}$$
$$2 \rightarrow \text{multiplier} = -100$$
$$G \rightarrow \pm 30 \text{ ppm}$$

Thus,

$$2.2 \times (-100) = -220 \text{ ppm/°C} \ \pm \ 30 \text{ ppm}$$

3. Because the capacitor value goes down by −220 ppm (±30 ppm), we must determine how many millions are in the capacitor value.

$$\frac{0.027 \times 10^{-6} \text{ F}}{1.0 \times 10^{6}} = 2.7 \times 10^{-14} \text{ F}$$

4. Multiply this value by −220, the part change per million:

$$-220(2.7 \times 10^{-14} \text{ F}) = -5.4 \times 10^{-12} \text{ F} = -5.4 \text{ pF}$$

5. For the tolerance, multiply the value in Step 3 by ±30.

$$\pm 30(2.7 \times 10^{-14} \text{ F}) = \pm 0.81 \times 10^{-12} \text{ F} = \pm 0.81 \text{ pF}$$

6. Multiply the values by the change in the number of degrees:

$$(-5.4 \text{ pF})(25) = -135 \text{ pF}$$
$$(\pm 0.81 \text{ pF})(25) = \pm 20.3 \text{ pF}$$

7. Subtract this value from the rated value of capacitance.

$$27,000 \text{ pF} - 135 \text{ pF} \rightarrow 26,865 \text{ pF} \ \pm \ 20.3 \text{ pF}$$

(Note that if the multiplier is 5, 6, 7, 8, or 9, then the change in capacitance is positive. Therefore, in Step 7, you would add the change to the rated value of capacitance.)

Practice Problems 19.4   Answers are given at the end of the chapter.
1. Find the temperature coefficient of capacitance for a ceramic capacitor marked S7J.
2. Find the maximum change of capacitance from room temperature for a ceramic capacitor marked Z7U.
3. Determine the capacitance value of a capacitor marked with the three-digit pF code of 104.

19.4.3 Tantalum Dielectric Codes

Because of the stability of tantalum as a dielectric, and a small-volume/high-capacitance package, tantalum capacitors are very popular. Because of their high capacitance, most values are above 1 μF. In this case, whole numbers are microfarad units, not the usual picofarads as in other types. However, their value can be lower (and utilize the three-digit code) especially in nonpolarized capacitors.

Tantalum capacitance and voltage codes are not typically a problem in interpretation; however, their polarity indicators are. The polarity indicators are not standardized. This is mostly due to the availability of three types of tantalum capacitors and the various packages for these. Figure 19.15(a) shows a common way to code the tantalum capacitor, including the polarity indicator (a stripe). Note the three-digit picofarad is not used in this case. Part (b) of the figure shows another way to label tantalum capacitors. Note this capacitor uses the three-digit picofarad code and has a polarity indicator (a stripe). Part (c) shows a precision molded tantalum capacitor using the three-digit picofarad code. Note the tapered end indicates the polarity.

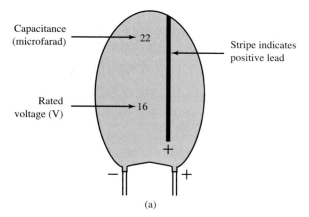

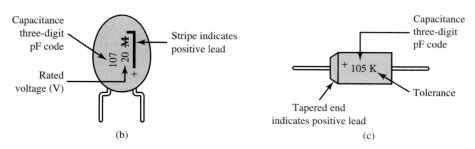

FIGURE 19.15

Tantalum capacitor codes: (a) uncoded capacitance value; (b) three-digit pF capacitance code; (c) molded tantalum utilizing three-digit pF code.

In addition to those shown in Figure 19.15, some polarized tantalum capacitors use a (+) mark to indicate the polarity, whereas others use a red dot. In most cases, the marks indicate the *positive lead;* conversely, in aluminum electrolytics the marks usually indicate the negative lead.

SECTION 19.4 SELF-CHECK

Answers are given at the end of the chapter.

1. A whole number coded on a polyester capacitor indicates a value in _____.
2. Capacitor manufacturers are now favoring _____ _____ or _____ markings for both lead and chip capacitors.

3. What characteristics do Class I dielectrics possess?
4. Class II dielectrics are called _____ _____ _____.
5. The three-digit code gives the capacitance value in _____ units.
6. Using the three-digit code, what is the third numeral when coding a value of 4.7 pF?
7. In most cases the polarity indicator used for tantalum dielectrics indicates the _____ lead.

19.5 ## CAPACITOR COMBINATIONS

19.5.1 Capacitors in Series

The overall effect of connecting capacitors in series is to move the plates of the capacitors further apart. This is shown in Figure 19.16. Notice that the junction between C_1 and C_2 has both a negative and positive charge. This causes the junction to be essentially neutral. The total capacitance of the circuit is developed between the left plate of C_1 and the right plate of C_2. Because these plates are farther apart, the total value of the capacitance in the circuit has decreased. Solving for the total capacitance (C_T) of capacitors connected in series is similar to solving for the total resistance (R_T) of resistors connected in parallel.

FIGURE 19.16
Capacitors in series.

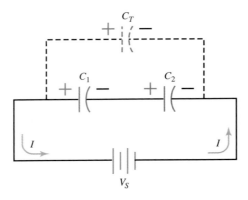

The general equation for capacitors (with all values in farads) in series is:

$$C_T = \frac{1}{1/C_1 + 1/C_2 + 1/C_3 + \cdots + 1/C_N}$$

(19.4)

EXAMPLE 19.13

Determine the total capacitance of a series circuit containing three capacitors whose values are 0.01 μF, 0.25 μF, and 50,000 pF, respectively.

Solution $C_T = \dfrac{1}{1/0.01 \times 10^{-6}\,\text{F} + 1/0.25 \times 10^{-6}\,\text{F} + 1/50,000 \times 10^{-12}\,\text{F}}$

$= 0.008\ \mu$F

19.5.2 Capacitors in Parallel

When capacitors are connected in parallel, one plate of each capacitor is connected directly to one terminal of the source, whereas the other plate of each capacitor is connected to the other terminal of the source. Figure 19.17 shows all the negative plates of the capacitors connected together and all the positive plates connected together. Therefore, C_T appears as a capacitor with a plate area equal to the sum of all the individual plate areas. As previously mentioned, capacitance is a direct function of plate area. Connecting capacitors in parallel effectively increases the plate area and thereby increases total capacitance.

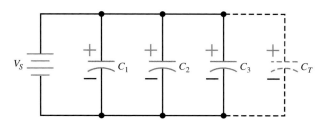

FIGURE 19.17
Capacitors in parallel.

For capacitors connected in parallel, the total capacitance is the sum of all the individual capacitors. The total capacitance of the circuit may be calculated using the equation

$$C_T = C_1 + C_2 + C_3 + \cdots + C_N \qquad \textbf{(19.5)}$$

EXAMPLE 19.14

Determine the total capacitance in a parallel capacitive circuit containing three capacitors whose values are 0.03 μF, 2.0 μF, and 0.25 μF, respectively.

Solution $C_T = 0.03\ \mu\text{F} + 2.0\ \mu\text{F} + 0.25\ \mu\text{F}$
$\qquad\qquad = 2.28\ \mu\text{F}$

Practice Problems 19.5   Answers are given at the end of the chapter.
1. Four 20-μF capacitors are connected in series. Calculate C_T.
2. Four 20-μF capacitors are connected in parallel. Calculate C_T.
3. Three capacitors of 0.033, 0.001, and 0.022 μF are connected in series. Calculate C_T.

19.5.3 Voltage Division Across Capacitors in Series

To increase the DC working voltage of capacitors, manufacturers put two capacitors in series. Thus, two 50-μF, 100-V capacitors exhibit 25 μF and 200 V when connected in series. The total working voltage is the *sum* of the individual working voltages *only* for capacitors having identical capacitance and voltage values.

For series capacitors that are not the same, the voltage division is easily determined by using the capacitance equation ($C = Q/V$). The smaller capacitance has a larger voltage drop than a larger capacitance. Because the capacitors are connected in series, they have the same charging current ($I = Q/t$, or $Q = I \times t$). Thus, in order for smaller capacitors to store the same charge as larger capacitors, they must have a larger portion of the source voltage across them. Because $V = Q/C$ and Q is the same for all series capacitors, V must be larger across the smaller capacitors. The following example illustrates this.

EXAMPLE 19.15

Two capacitors, 5 μF, 20 V and 10 μF, 30 V, are connected in series. The source is a 30-V DC supply. Find the voltage across each capacitor.

Solution   We first determine C_T and Q. (Remember, Q is the same for both capacitors, because this is a series circuit.) Using these values we can find V_{C_1} and V_{C_2}.

$$C_T = \frac{1}{1/C_1 + 1/C_2}$$

$$= \frac{1}{1/5 \ \mu\text{F} + 1/10 \ \mu\text{F}}$$

$$= 3.33 \ \mu\text{F}$$

$$Q = VC_T$$

$$= 30 \ \text{V} \times 3.33 \times 10^{-6} \ \text{F}$$

$$= 100 \ \mu\text{C}$$

$$V_{C_1} = \frac{Q}{C_1}$$

$$= \frac{100 \ \mu\text{C}}{5 \ \mu\text{F}}$$

$$= \frac{100 \times 10^{-6} \ \text{C}}{5 \times 10^{-6} \ \text{F}}$$

$$= 20 \ \text{V}$$

$$V_{C_2} = \frac{Q}{C_2}$$

$$= \frac{100 \times 10^{-6} \ \text{C}}{10 \times 10^{-6} \ \text{F}}$$

$$= 10 \ \text{V}$$

Notice that the smaller capacitor does have the larger voltage drop and that the sum of V_{C_1} and V_{C_2} equal the source (Kirchhoff's voltage law).

Practice Problems 19.6   Answers are given at the end of the chapter.
1. For the circuit in Figure 19.18, solve for C_T, Q, V_{C_1}, V_{C_2}, and V_{C_3}.

FIGURE 19.18
Circuit for Practice Problems 19.6,
Problem 1.

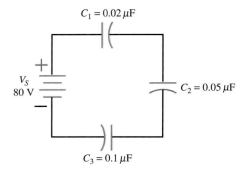

SECTION 19.5
SELF-CHECK

Answers are given at the end of the chapter.

1. When capacitors are connected in series, it is the same effect as moving the plates of a capacitor _____ _____.

2. When capacitors are connected in parallel, it is the same as making the plates of a capacitor _____.

3. The _____ capacitance in a series circuit has the _____ voltage drop.

19.6 ENERGY STORED IN A CAPACITOR

We have seen that if a capacitor is charged and the charging voltage is removed, the capacitor retains the charge. Furthermore, it has a potential difference across its terminals and has effectively become a voltage source. If a photo flashbulb is connected across the capacitor, it flashes. This shows that energy stored in the capacitor has been transferred upon discharge to a load. The capacitor is the only component besides the voltaic cell that can store electric charge.

The expression for stored energy in a capacitor is given by

$$\mathcal{E} = \tfrac{1}{2}CV^2 \qquad \textbf{(19.6)}$$

where $\mathcal{E}$ = energy stored in capacitor (J)
C = capacitance (F)
V = final voltage across the capacitor (V)

EXAMPLE 19.16

Calculate the energy stored in a 40-μF capacitor that is charged to 300 V.

Solution $\mathcal{E} = \tfrac{1}{2}CV^2$
$= \tfrac{1}{2}(40 \times 10^{-6}\,\text{F} \times (300\,\text{V})^2$
$= 1.8\,\text{J}$

Practice Problems 19.7   Answers are given at the end of the chapter.

1. Determine the number of joules stored in a 10-μF capacitor that is charged to 250 V.

SECTION 19.6 SELF-CHECK

Answers are given at the end of the chapter.

1. True or false: The capacitor retains its charge after the charging voltage is removed.

2. The energy stored in a capacitor is measured in _____.

19.7 STRAY CAPACITANCE

Stray capacitive effects occur in most components and in virtually all circuit configurations. Between any two adjacent wires is capacitance. Between two adjacent copper foils on a printed circuit board is capacitance. A coil has capacitance between the coil windings. Any component with legs mounted on a printed circuit board has capacitance between the legs. Remember that capacitance is a *physical phenomenon;* it is simply two conductors separated by an insulator. The stray capacitance associated with various components is illustrated in Figure 19.19.

Most of these stray capacitance effects are quite small and have negligible effect on circuit performance at lower operating frequencies. At lower frequencies, the stray effects of capacitance are absorbed by the "lumped" value of capacitance in the circuit. It is the lumped value of capacitance that determines low-frequency circuit performance.

There is a distributed capacitance between the windings of a coil. At higher frequencies these distributed effects and additional stray effects become more pronounced and hamper circuit performance. At microwave frequencies (1–100 GHz), the distributed

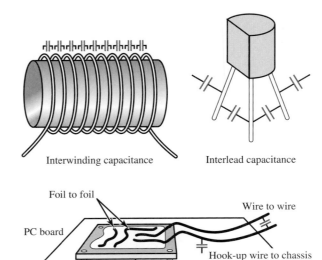

FIGURE 19.19
The effects of stray capacitance.

effects become dominant and take precedence over lumped values. Microwave transmission lines are analyzed by their distributed capacitance values.

There are things that can be done in order to keep high-frequency circuit performance from deteriorating. These include keeping lead lengths small, using larger spacings between adjacent wires, and mounting components high off the chassis. Various coil-winding techniques can be employed to reduce the effects of coil distributed capacitance.

SECTION 19.7 SELF-CHECK	Answers are given at the end of the chapter.

Answers are given at the end of the chapter.

1. What is *stray* capacitance?
2. At lower frequency operation, any stray effects of capacitance are _____ together.
3. What can be done to two adjacent wires to reduce the effects of stray capacitance?

19.8   TROUBLESHOOTING CAPACITOR FAULTS

Common failures for a capacitor include a short, an open, and a leaky condition. The short and open are checked in the same way as a resistor, using an ohmmeter. Caution must be taken to make sure the capacitor under test has been discharged. Otherwise, damage could occur to the ohmmeter. A good capacitor should indicate near 0 Ω when the test leads of the ohmmeter are first attached. Then the battery of the ohmmeter literally charges the capacitor. As this process occurs, the resistance reading on the ohmmeter moves toward infinity. An analog ohmmeter (VOM) demonstrates this process more easily in comparison to a DMM. Most DMMs have too slow a sampling rate to show the proper change in capacitor resistance.

19.8.1   Shorts and Opens

A short indicates low or zero ohms on an ohmmeter. It can be caused by a rupture of the dielectric or any scenario that causes the plates of a capacitor to touch.

An open is often caused by damage to the dielectric. Both high voltage and high temperature contribute to the damage of the dielectric. In addition, aging deteriorates the dielectric (particularly with aluminum oxide electrolytic). If open, a capacitor indicates very high or infinite ohms when the leads of the ohmmeter are attached. (The momentary low resistance and then increasing resistance as the capacitor charges are missing.)

A leaky capacitor is extremely common with aluminum oxide electrolytics. Aging usually deteriorates the dielectric causing what is effectively a partial short. Instead of infinite ohms, the resistance of the dielectric is reduced. This partial short causes some current to flow.

19.8.2 Other Types of Losses

Besides the aforementioned failures, capacitors also exhibit other forms of losses. These are more easily seen by a capacitance tester (many portable DMMs contain a capacitance tester). *Value changes* (which includes opens) often occur in capacitors and account for 25% of all capacitor failures. Better-quality testers check for leakage. *Leakage* accounts for 40% of all capacitor failures.

Dielectric absorption is a failure mode whereby a capacitor fails to discharge completely. The capacitor keeps a residual charge, acting like a battery. This affects capacitor value and accounts for 25% of all capacitor failures.

Equivalent series resistance (ESR) refers to the resistive effects of the capacitor from lead contacts, plate resistance, and the connection between the lead and plate. An aluminum electrolytic capacitor contains resistance in the electrolyte solution. ESR affects any capacitance-resistance circuit timing and accounts for 10% of all capacitor failures.

A more convenient means of defining the resistance of a capacitor is called the *dissipation factor* (DF). The dissipation factor takes into account the value of capacitance, the operating frequency, and the resistance of the capacitor (DF = $2\pi fCR$). The DF of a good capacitor is rather small and is often expressed as a percent. A high DF indicates a poor-quality capacitor.

**SECTION 19.8
SELF-CHECK**

Answers are given at the end of the chapter.

1. What are the most common failures for a capacitor?
2. What can cause a capacitor to short?
3. What failure accounts for the smallest percent of all capacitor failures?

■ SUMMARY

1. Capacitance is the property of a dielectric to store electric charge.
2. A capacitor is a device especially designed to have a certain value of capacitance.
3. A capacitor stores energy in an electrostatic field.
4. An electrostatic field can attract and repel charges, which enables a capacitor to store energy.
5. A simple capacitor consists of parallel conductors separated by a dielectric medium.
6. The source supplies the charging current to charge the capacitor, but current does not flow through the capacitor.
7. Permittivity refers to how well a dielectric material can establish electrostatic lines of force.

8. Capacitors have a voltage rating (DC working voltage) that should not be exceeded.
9. A capacitor should be selected so that its working voltage is at least 50% greater than the highest effective voltage applied to it.
10. A variable capacitor has value ranges from a few picofarads to approximately 500 pF.
11. Fixed capacitors are usually identified by the type of material that is used for the dielectric.
12. Ceramics are the most versatile dielectric materials available.
13. Electrolytic capacitors have the largest amount of capacitance for a given physical size. Their dielectrics are made of either aluminum oxide or tantalum oxide.
14. Electrolytic capacitors are polarized.

15. Capacitor values coded in whole numbers are usually interpreted as a value in picofarads, whereas decimal fractions indicate a value in microfarads.

16. An NPO (COG) temperature coefficient code on a capacitor indicates that the capacitance values do not vary with temperature changes.

17. There are a number of ways to indicate polarity on a tantalum capacitor. The marks usually indicate the positive lead.

18. The smallest capacitance in series-connected capacitors has the larger voltage across it terminals.

19. A capacitor is the only component besides a voltaic cell that can store electric charge.

20. At higher frequencies, the effects of "stray" capacitance become more noticeable.

21. Common failures for a capacitor include a short, an open, and a leaky condition.

■ IMPORTANT EQUATIONS

$$C = \frac{Q}{V} \tag{19.1}$$

$$i_c = C\left(\frac{\Delta V_C}{\Delta t}\right) \tag{19.2}$$

$$C = \frac{(8.85 \times 10^{-12})\,KA}{D} \tag{19.3}$$

$$C_T = \frac{1}{1/C_1 + 1/C_2 + 1/C_3 + \cdots + 1/C_N} \tag{19.4}$$

$$C_T = C_1 + C_2 + C_3 + \cdots + C_N \tag{19.5}$$

$$\mathscr{E} = \tfrac{1}{2}CV^2 \tag{19.6}$$

■ REVIEW QUESTIONS

1. Define the term *capacitance.*
2. What is a capacitor?
3. What are the common characteristics of electrostatic lines of force?
4. Describe the make up of a simple capacitor.
5. When a capacitor is charging, what is the source doing?
6. Define the term *farad.*
7. What are the common units of capacitance?
8. What factors affect the value of capacitance?
9. Define the term *permittivity.*
10. What does the dielectric constant indicate?
11. How is capacitance affected by plate area and the distance between plates?
12. What value should be selected for a capacitor's working voltage?
13. How are capacitors classified?
14. What is a monolithic ceramic capacitor?
15. What capacitor rating is reduced as the two plates of a capacitor are brought closer together?
16. What are some common dielectrics for fixed capacitors?

17. How is aluminum oxide formed in the electrolytic capacitor?
18. Electrolytic capacitors are limited to _____ applications.
19. What is the difference between axial and radial leads on the capacitor?
20. What is the disadvantage of aluminum oxide for use in electrolytic capacitors?
21. Why are tantalum electrolytics in favor with designers versus aluminum oxide electrolytics?
22. Capacitor values in whole numbers usually are interpreted as _____, whereas decimal fractions are interpreted as _____.
23. What are the equivalent mil spec and commercial standards for ceramic capacitors?
24. What characteristics do Class I dielectric ceramic capacitors have?
25. What characteristics do Class II dielectric ceramic capacitors have?
26. How do the polarity marks differ on tantalum versus aluminum oxide capacitors?
27. When connected in series, the smallest value of the capacitors has the _____ voltage across it.
28. Describe *stray* capacitance.
29. What are the common failures of a capacitor?
30. Describe the term *dielectric absorption.*

■ CRITICAL THINKING

1. Describe in detail the process of how a capacitor is charged, how electrostatic lines of force affect the charging process, and why current stops flowing once the capacitor is charged.
2. What is the dielectric constant and how is it quantified?
3. Describe how higher-value ceramic capacitors are manufactured.
4. Compare and contrast the various codes used for indicating capacitor value, voltage, tolerance, temperature coefficient, etc.
5. Determine the capacitance value of a temperature-compensating capacitor marked with the three-digit pF code of 472 and a temperature-compensation code of S7H at 60°C.

■ PROBLEMS

1. A capacitor stores 12,000 μF of charge with 30 V across its plates. Solve for C.
2. What voltage will be across the plates of a 0.05-μF capacitor if it stores 50 μC of charge?
3. How much charge is stored in a 10-μF capacitor with 16 V across it?
4. Determine the capacitor current when the voltage across a 12-μF capacitor changes from 2 to 20 V in 2.5 ms.
5. A constant current of 20 mA charges a 20-μF capacitor for 0.02 s. What is the voltage across the capacitor?
6. Calculate the capacitance between two metal plates having an area of 5 cm^2, $K = 2.4$, and $d = 0.04$ cm.

7. Determine the total capacitance of the capacitors shown in Figure 19.20.

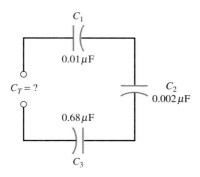

FIGURE 19.20
Circuit for Problem 7.

8. Three capacitors connected in parallel have the following values: 0.01 μF, 470 pF, and 0.0022 μF. Solve for C_T.

9. For the circuit in Figure 19.21, solve for Q, V_{C_1}, V_{C_3}, C_T, and V_S.

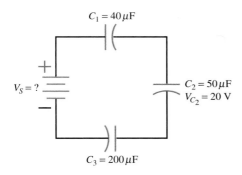

FIGURE 19.21
Circuit for Problem 9.

10. For the circuit in Figure 19.22, solve for Q, C_T, V_{C_1}, V_{C_2}, V_{C_3}, and V_{C_4}.

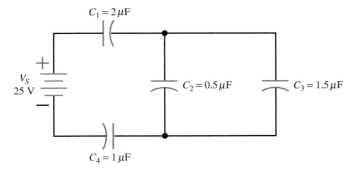

FIGURE 19.22
Circuit for Problem 10.

11. Calculate the energy stored in a 6.8-μF capacitor that is charged to 100 V.

12. Solve for the voltage across a capacitor that has 2.5 J of energy stored in a 20-μF capacitor.

■ ANSWERS TO PRACTICE PROBLEMS

PRACTICE PROBLEMS 19.1

1. $V_C = 20$ V
2. $Q = 0.4$ C
3. $C = 2.5$ μF

PRACTICE PROBLEMS 19.2

1. $i_c = 40$ mA

PRACTICE PROBLEMS 19.3

1. $C = 885$ pF
2. $C = 8.85$ pF

PRACTICE PROBLEMS 19.4

1. $+330$ ppm/°C $\pm$ 120 ppm/°C
2. $+22\%$, -56% maximum change over a -30°C to $+125$°C temperature
3. $104 \rightarrow 100{,}000$ pF $= 0.1$ μF

PRACTICE PROBLEMS 19.5

1. $C_T = 5$ μF
2. $C_T = 80$ μF
3. $C_T \approx 930$ pF

PRACTICE PROBLEMS 19.6

1. $C_T = 0.0125$ μF, $Q = 1$ μC, $V_{C_1} = 50$ V, $V_{C_2} = 20$ V, $V_{C_3} = 10$ V

PRACTICE PROBLEMS 19.7

1. $\mathscr{E} = 0.3125$ J

■ ANSWERS TO SECTION SELF-CHECKS

SECTION 19.1 SELF-CHECK

1. A parallel-plate capacitor consisting of two conductors separated by a dielectric material of uniform thickness
2. Redistributing
3. False
4. 1 F = 1 C/V
5. μF, pF

SECTION 19.2 SELF-CHECK

1. Permittivity refers to how well a dielectric material can establish electrostatic lines of force.

2. Remains the same

3. 50% greater than the highest effective voltage applied to it.

SECTION 19.3 SELF-CHECK

1. Air

2. Mica, paper, ceramic, plastic film, aluminum oxide, tantalum oxide

3. Ceramic

4. A zero temperature coefficient

5. Electrolytic

6. High reliability, long service life, high capacitance per unit volume

SECTION 19.4 SELF-CHECK

1. pF

2. Ink jet, laser

3. They exhibit temperature-compensation values and are very stable.

4. General-purpose capacitors

5. pF

6. 9

7. Positive

SECTION 19.5 SELF-CHECK

1. Farther apart

2. Larger

3. Smallest, largest

SECTION 19.6 SELF-CHECK

1. True

2. Joules

SECTION 19.7 SELF-CHECK

1. Capacitance between two adjacent wires, two adjacent copper foils, between legs of a component, etc.

2. Lumped

3. Space farther apart

SECTION 19.8 SELF-CHECK

1. Shorts, opens, and a leaky condition

2. A rupture of the dielectric

3. Equivalent series resistance (ESR)

CHAPTER 20

INDUCTORS AND INDUCTANCE

OBJECTIVES

After completing this chapter, you will be able to:

1. Describe the characteristics of an inductor.
2. Calculate inductance and self-induced voltage.
3. Describe the physical factors that affect inductance.
4. Describe the physical factors that affect mutual inductance.
5. Compare and contrast the various types of inductors via the core material used.
6. Compare and contrast inductor faults.

INTRODUCTION

In the previous chapter we saw that the capacitor is a device capable of storing energy in the form of an electrostatic field. The subjects of this chapter are the *inductor* and its DC characteristics. An inductor is a device that is also capable of storing energy and does this via an electromagnetic field. The ability to induce a voltage across itself with a change in current is known as self-inductance, or simply *inductance.* Inductance also opposes a change in current. That is, it opposes the starting, stopping, or changing of current. For this reason the inductor offers no opposition to steady DC. The symbol for inductance is *L* and the basic unit of inductance is the *henry* (H), named after the American physicist Joseph Henry.

You do not have to look very far to find a physical analogy to inductance. Anyone who has ever had to push a heavy load (wheelbarrow, car, etc.) is aware that it takes more work to start the load moving than it does to keep it moving. Once the load is moving, it is easier to keep it moving than to stop it again. This is because the load possesses the property of inertia. *Inertia* is the characteristic of mass that opposes a change in velocity. Inductance has the same effect on current in an electrical circuit that inertia has on the movement of a mechanical object. It requires more energy to start or stop current in an inductor than it does to keep it flowing.

Inductors are used in many of the same applications as capacitors, especially in tuned circuits. Inductors also come in a chip format for small-circuit-board applications, whereas larger units are found in various types of power supplies.

This chapter covers factors affecting inductance, characteristics, applications, and faults.

20.1 INDUCTOR BASICS

An inductor may take on any number of physical forms and shapes. However, it is basically nothing more than a coil of wire or an electromagnet. (Coils and electromagnets were first introduced in Chapter 15.) An inductor is sometimes referred to by such names as a *choke, impedance coil,* or *reactor.* The amount of inductance of a coil is measured by a unit called the *henry* (H) and depends on the current within the coil and the quantity of magnetic flux produced.

Basically, all inductors are made by winding a length of conductor around a core that is made either of a magnetic material (typically iron) or of an insulated material. When a magnetic core is not used, the inductor is said to have an *air core.* Both an air core and iron-core inductor and their schematic symbols are shown in Figure 20.1.

FIGURE 20.1
Inductor types and schematic symbols: (a) inductor, iron core; (b) inductor, air core.

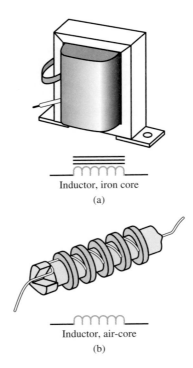

Inductor, iron core

(a)

Inductor, air-core

(b)

The physical characteristics of both the core and windings around the core affect the amount of inductance produced. A greater number of turns, a higher-permeability magnetic core, a larger core cross-sectional area, and a shorter coil length are all factors that increase inductance. These attributes are further explored in a later section.

SECTION 20.1 SELF-CHECK	Answers are given at the end of the chapter.

SECTION 20.1 SELF-CHECK

Answers are given at the end of the chapter.

1. Define the term *inductance.*
2. Describe a basic inductor.
3. Inductance is measured in _____.

20.2 SELF-INDUCTANCE

Even a perfectly straight length of conductor has some inductance. As you know, current in a conductor produces a magnetic field surrounding the conductor. If the current changes, the magnetic field changes. This causes relative motion between the magnetic field and the conductor, and an electromotive force (emf) is induced in the conductor. This emf is called a *self-induced emf* because it is induced within itself. This emf is also referred to as a *counter–electromotive force (cemf)* or *back-emf*. The polarity of the counter–electromotive force is in an opposite direction to the applied voltage of the conductor. The overall effect is to oppose a change in current magnitude. This effect is summarized by Lenz's law, which states that the induced emf in any circuit is always in a direction to oppose the effect that produced it.

If AC is the source, there is a continuous change in current, and therefore the process of a self-induced emf is continuous. This process is further described in Chapter 22. If the source is DC, there is not a continuous change in current occurring. However, it is important for you to note that there is a self-induced voltage that opposes *both* changes in current that do occur: that is, when a switch is closed and current starts to flow from the source and when the switch is opened and current ceases to flow.

If the voltage applied to the coil is removed, another point of current-flow stoppage occurs. With no movement of electrons, the magnetic field collapses. This causes a relative motion between the magnetic field and the coil conductors. The polarity of an induced voltage is such that it attempts to keep current flowing in the same direction. In other words, the induced voltage opposes the decrease of the applied voltage.

Because all circuits have conductors in them, it can be assumed that all circuits have inductance. However, inductance has its greatest effect only when there is a change in current. Inductance does *not* oppose current, only a *change* in current. Where current is constantly changing, as in an AC circuit, inductance has a more pronounced effect. If the source is DC, then the only opposition to the current occurs as the current is starting or stopping (that is, changing). A steady-state DC produces no change in current; thus, there is no change in the magnetic field and no induced voltage.

20.2.1 The Self-Induced Voltage Equation

An inductance of 1 H can be defined when a change in current of 1 A per second causes an induced voltage of 1 V. The counter-emf, or induced voltage, can be calculated by

$$V_{\text{ind}} = L\left(\frac{\Delta I}{\Delta t}\right) \tag{20.1}$$

where L = inductance (H)
  ΔI = increment of change of current (A)
  Δt = increment of change with respect to time (s)

The symbol Δ (Greek letter delta) means "change in." Equation (20.1) is just a variation of equation (15.1), $V_{\text{ind}} = N(\Delta\phi/\Delta t)$, which is an equation for induced voltage in terms of how much magnetic flux is cut by a conductor per second.

A henry is a large unit of inductance and is used with relatively large inductors (typically those with iron cores). With smaller inductors, the *millihenry* and *microhenry* are used. A millihenry (mH) is equal to 0.001 H, or 1×10^{-3} H, and a microhenry (μH) is equal to 0.000001 H, or 1×10^{-6} H.

| EXAMPLE 20.1 | How much self-induced voltage occurs across a 4-H inductance produced by a current change of 10 A/s? |

Solution $V_{\text{ind}} = L\left(\dfrac{\Delta I}{\Delta t}\right)$

$= 4\,\text{H}\left(\dfrac{10\,\text{A}}{1\,\text{s}}\right)$

$= 40\,\text{V}$

EXAMPLE 20.2

The current through a 200 mH inductor changes from 0 to 200 mA in 4 μs. How much is V_{ind}?

Solution $V_{ind} = L\left(\dfrac{\Delta I}{\Delta t}\right)$

$\qquad\qquad = 200 \times 10^{-3}\,\text{H}\left(\dfrac{200 \times 10^{-3}\,\text{A}}{4 \times 10^{-6}\,\text{s}}\right)$

$\qquad\qquad = 10{,}000\,\text{V}$

Note in this example that a relatively small inductance produced a high self-induced voltage. This occurred because of the fast change (small Δt) in current.

Practice Problems 20.1   Answers are given at the end of the chapter.
1. Calculate the inductance of a coil that induces 100 V when the current changes by 5 mA in 10 μs.

SECTION 20.2 SELF-CHECK

Answers are given at the end of the chapter.

1. What is a cemf?
2. What is the cemf for a steady-state DC current?
3. Under what conditions can a small inductance produce a large cemf?

20.3   FACTORS AFFECTING COIL INDUCTANCE

The value of inductance depends entirely upon the physical construction of the coil and can be measured directly only by special lab instruments. The physical factors that affect the inductance of a coil include the number of turns in the coil, the diameter of the coil, the coil length, the type of material used in the core, and the number of layers of winding in the coil. These are noted as follows:

1. A greater number of turns (N) increases the inductance, because more voltage can be induced due to the greater flux linkage between turns. Inductance actually increases with N^2. Doubling the number of turns increases the inductance by a factor of four. Adding more layers of windings effectively increases the number of turns.

Lab Reference: The effects of changing a circuit's inductance are demonstrated in Exercise 30.

2. The diameter of the coil affects the area (A) through which the magnetic field links. A larger-diameter coil requires more wire for a given number of turns than a small-diameter coil. Thus, with a larger diameter there are more lines of force; hence there is a greater cemf and, therefore, a greater inductance. (Note that $A = \pi r^2$.)
3. Inductance decreases as the coil length (l) increases because the magnetic field is less concentrated due to the greater distance between the turns.
4. The core material also affects the magnetic field. A high-permeability (μ_r) core allows more flux to form, thereby increasing inductance. The core also concentrates the flux in less space, yielding a greater inductance.

20.3.1   The Inductance Equation

All the physical factors affecting inductance can be placed into an equation to calculate the inductance. This is shown as follows:

$$L = \frac{\mu N^2 A}{l} \qquad\qquad\qquad \textbf{(20.2)}$$

where μ = permeability of core material; $\mu = \mu_o \mu_r$
$[\mu_0 = 1.26 \times 10^{-6}$ H/m (absolute permeability of air)$]$
$[\mu_r$ = relative permeability of core material, from Table 14.2$]$
N = number of turns
A = cross-sectional area (m²)
l = length of core (m)

EXAMPLE 20.3

Determine the inductance for the air-core coil shown in Figure 20.2.

FIGURE 20.2
A single-layer air-core inductor.

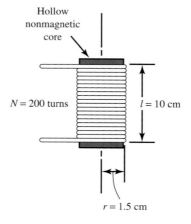

Hollow
nonmagnetic
core

$N = 200$ turns

$l = 10$ cm

$r = 1.5$ cm

Solution $\mu = \mu_o \mu_r = 1.26 \times 10^{-6} \times 1 = 1.26 \times 10^{-6}$ H/m
$A = \pi r^2 = \pi \times (1.5 \times 10^{-2})^2 = 706.9 \times 10^{-6}$ m²

$$L = \frac{\mu N^2 A}{l}$$

$$= \frac{1.26 \times 10^{-6} \times 200^2 \times 706.9 \times 10^{-6}}{10 \times 10^{-2}}$$

$$= 356.3 \times 10^{-6} \text{ H}$$

$$= 356.3 \ \mu\text{H}$$

EXAMPLE 20.4

If the core material is changed in Example 20.4 to silicon iron, what is the new value of inductance?

Solution $\mu_r = 7000$ (for silicon iron from Table 14.2)
$L = 7000 \times 356.3 \times 10^{-6}$ H
$= 2.49$ H

Practice Problems 20.2   Answers are given at the end of the chapter.
1. Calculate the inductance for a coil with an air core, $N = 200$ turns, $A = 5$ cm², and $l = 5$ cm.

**SECTION 20.3
SELF-CHECK**

Answers are given at the end of the chapter.

1. Inductance is _____ proportional to the number of turns.
2. Why does inductance decrease as the coil length increases?

20.4 MUTUAL INDUCTANCE (L_M)

Whenever two coils are located so that the flux from one coil links with the turns of the other coil, a change of flux in one coil causes an emf to be induced in the other coil. This allows the energy from one coil to be transferred or coupled to the other coil. The two coils are said to be coupled, or linked, by the property of *mutual inductance*. This situation is illustrated in Figure 20.3. Here, the varying magnetic field of L_1 cuts L_2 and induces a voltage into L_2. If a load is connected across L_2, current will flow as a result of the induced voltage.

We have already seen a common application of mutual inductance. The *transformer* of Chapter 18 utilized mutual inductance in order for transformer action to occur.

FIGURE 20.3
Mutual inductance between two coils.

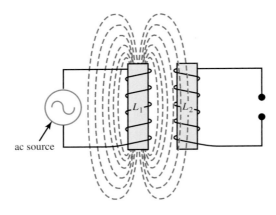

20.4.1 Factors Affecting Mutual Inductance

The amount of mutual inductance that exists between two coils depends on the *degree of coupling* between them. For example, if all the magnetic lines of force produced by one coil cut the turns of another coil, the degree of coupling between the two coils is 100%. Similarly, if none of the magnetic lines of force produced by one coil cut any turns in an adjacent coil, the degree of coupling is 0%.

The degree of coupling between two coils is indicated by the *coefficient of coupling, k*. A coefficient of coupling of 1 ($k = 1$) represents 100% coupling of the magnetic lines of force. Similarly, for degrees of coupling of 60%, 50%, and 0%, k has values of 0.6, 0.5, and 0, respectively. Clearly the possible range in values for k is $0 \leq k \leq 1$.

The term *tight coupling* refers to k-values close to 1. Tight coupling indicates that a majority of the flux produced by one coil cuts the turns of an adjacent coil. In practice, tight coupling is usually achieved by winding two coils on the same magnetic core (for example, a transformer) or by placing two air-core coils very close to each other. Conversely, the term *loose coupling* refers to k-values close to 0 and indicates that very little of the flux produced by one coil cuts the turns of an adjacent coil.

In practice, loose coupling results when two coils are very far apart or at right angles to each other, as shown in Figure 20.4(a). In Figure 20.4(b), the two coils are closer together and are positioned so that some of the flux of the first coil cuts the turns of the second coil. Consequently, a small voltage is induced in the second coil. In Figure 20.4(c), the two coils are effectively wound on the same magnetic core. The voltage induced in the second coil is maximum.

The terms *tight* and *loose* coupling are also used in later courses that deal with tuned circuits. Here, the degree of coupling affects the frequency response of the tuned circuit.

If you know the value of the coefficient of coupling (k) for two coils, you can compute the mutual inductance from the following equation:

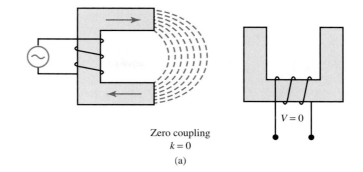

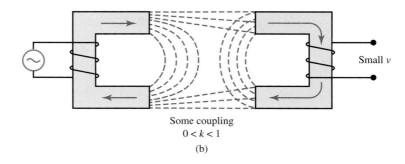

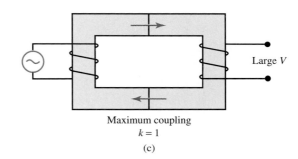

FIGURE 20.4
The concept of mutual inductance: (a) zero coupling; (b) some coupling; (c) maximum coupling.

$$L_M = k\sqrt{L_1 L_2} \tag{20.3}$$

where L_M = mutual inductance (H)
           k = coefficient of coupling
     $L_1 L_2$ = self-inductance of each coil (H)

EXAMPLE 20.5

Two coils have inductances of 8 mH and 4.7 mH, respectively. If the coefficient of coupling between them is 0.82, what is the mutual inductance?

Solution $L_M = k\sqrt{L_1 L_2}$
            $= 0.82\sqrt{8\ \text{mH} \times 4.7\ \text{mH}}$
            $= 5.03\ \text{mH}$

Practice Problems 20.3   Answers are given at the end of the chapter.
1. Find L_M when $k = 0.5$, $L_1 = 100\ \mu\text{H}$, and $L_2 = 200\ \mu\text{H}$.

20.5   INDUCTOR TYPES

Inductors are usually classified by the type of material used to make the core. The following is a brief description of the more common types of inductors available. Figure 20.5 illustrates many types of inductors and shows their schematic symbols.

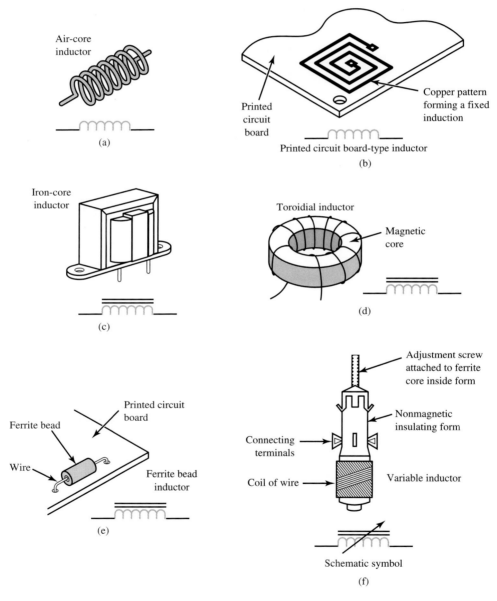

Air-core inductor

(a)

Printed circuit board

Copper pattern forming a fixed induction

Printed circuit board-type inductor

(b)

Iron-core inductor

(c)

Toroidal inductor

Magnetic core

(d)

Ferrite bead

Printed circuit board

Wire

Ferrite bead inductor

(e)

Adjustment screw attached to ferrite core inside form

Nonmagnetic insulating form

Connecting terminals

Coil of wire

Variable inductor

Schematic symbol

(f)

FIGURE 20.5
Typical coils and their schematic symbols: (a) air-core inductor; (b) printed circuit board inductor; (c) iron-core inductor; (d) toroidal inductor; (e) ferrite bead inductor; (f) variable inductor.

20.5.1 Air Core

Air-core coils (Figure 20.5(a)) are wound on either a hollow or solid nonmagnetic core. Here, the core simply provides support for the turns of wire. Where heavy, rigid wire is used to wind the coil, an actual core may not be required. Air-core coils have low values of inductance, cannot be magnetically saturated, and are used primarily in high-frequency applications.

20.5.2 Iron Core

In iron-core inductors, the core material is iron (Figure 20.5(c)) or one of its alloys. Because iron has a large relative permeability, large values of inductance are possible. Typically, iron-core coils, or chokes, have inductances ranging from less than 1 H to 40 H or more. They can become physically quite large. Hysteresis and eddy-current losses limit iron-core chokes to power-line and audio-frequency applications. Their cores are made of a laminated sheet material, just as in transformers, in order to reduce eddy currents. A soft, iron magnetic material such as silicon steel is used to reduce hysteresis losses.

20.5.3 Powdered-Iron Core

Powdered-iron cores consist of finely powdered iron or iron alloys that have been mixed with a binder material. The function of the binder is not only to hold the powdered iron particles together, but also to electrically insulate one particle from another. This construction reduces eddy-current losses in the same way that laminated sheets do in transformers or iron-core chokes. Powdered iron is a magnetic material that has an inherent distributed air gap. The distributed air gap allows the core to store higher levels of magnetic flux when compared to other magnetic materials, such as ferrites. This characteristic allows a higher DC current level to flow before the inductor saturates.

20.5.4 Ferrite Core

Ferrites (Figure 20.5(e)) are ceramic materials composed of iron and various other elements. These materials are excellent magnetic conductors and poor electrical conductors. Consequently, eddy-current losses are inherently small. The exact chemical composition of a ferrite determines its specific properties. Some ferrites are suitable for low-frequency applications, whereas others are more suitable for high-frequency applications.

One type of high-frequency choke is made by sliding a small, cylindrical-shaped ferrite bead on a wire. Due to its high permeability, the ferrite bead increases the inherent inductance of the wire. Naturally, only small inductance values may be obtained in this way, but such inductors are very useful in high-frequency applications. Because inductors oppose a change in current, these beads oppose the change in current associated with high-frequency currents and tend to block them.

20.5.5 Toroidal Core

A toroidal-core coil (Figure 20.5(d)) gets its name from the shape of the core, which resembles a donut. The core material is usually powdered iron or ferrite. Toroidal coils offer reasonably high values of inductance with comparatively little wire. In addition, virtually all the flux flows in the core, resulting in very little flux leakage loss.

20.5.6 Moveable (Variable) Core

Moveable variable coils (Figure 20.5(f)) have powdered-iron or ferrite cores that can be moved in and out of the coil via a screw-type adjustment. Consequently, the inductance value can be varied. These variable inductors are often used with capacitors to form tuned circuits in radio receivers. Variable inductors should be adjusted only with a plastic or nonmetallic alignment tool, because any metal object in the vicinity of the inductor interferes with the magnetic field and changes the inductance of the inductor.

20.5.7 Printed Circuit Board Core

Here, the inductor is simply a spiral of copper foil on a printed circuit board (Figure 20.5(b)). In the microwave frequency range (above 500 MHz), the inductors are called *strip line* or *microstrip* devices. The principal advantage of this is that the coil is formed when the printed circuit board is being made. Obviously, only small inductance values are possible with this technique, which limits its use to very high-frequency applications.

Besides indicating inductor types by core material, inductors are characterized by applications. Table 20.1 shows some selected inductor types, descriptions, and design applications.

TABLE 20.1
Selected inductor applications guide

Inductor Type	Description	Design Application
Chip (SMD) inductors	Used where increased density and lower assembly costs are required	Shorter traces, improved performance at high frequencies, very popular in cellular phone technology
Molded inductors	Axial leaded, small size	Used where core must be protected from environmental conditions; frequency range typically >50 kHz
Shielded inductors	Axial leaded, small size	Used for high-reliability applications, where magnetic coupling cannot be tolerated; vital where board is densely packed, telecommunications, such as disk drives
Dipped inductors	Axial or radial leaded, lower cost	Used in less harsh environments than molded inductors; can be designed for higher-current capability, lower R, higher Q
Chokes (high current)	Utilize ferrite or powdered iron cores, have fewer turns, less DC resistance, critical for high current applications	Available in low- and high-frequency types; high-frequency RF chokes most common, usually above 100 MHz
Wideband chokes (ferrite beads)	Used in various types of filters	Used to suppress a selected frequency or band of frequencies; used in RF circuits VHF and UHF
Toroids	Provide higher impedance and lower loss than solenoidal inductors of same material	Are self-shielding, magnetically balanced, and typically smaller in size
Air coils	Used where core saturation cannot be tolerated. Have low inductance	Used in tuners and other RF tank circuits
Variable inductors	Used in adjustable resonant circuits; tuning by moving adjustable core	Typically used at frequencies above 50 kHz in applications such as LC resonant tank circuits

20.6 INDUCTOR COMBINATIONS

20.6.1 Inductors in Series

When inductors are connected in series and are far enough apart or are shielded so that no coupling occurs between them, the total inductance is found by simply adding the inductances. Thus,

$$L_T = L_1 + L_2 + L_3 + \cdots + L_N \tag{20.4}$$

When two mutually coupled coils are connected in series, the total (equivalent) inductance is affected by their fields either series-aiding or series-opposing each other. The coil windings can be wound to produce either. The presence of mutual inductance complicates the calculation of the equivalent inductance. In equation form,

$$L_T = L_1 + L_2 \pm 2L_M \tag{20.5}$$

where
L_1 = coil 1 inductance (H)
L_2 = coil 2 inductance (H)
$\pm 2L_M \rightarrow +2L_M$ for series-aiding inductances or
$-2L_M$ for series-opposing inductances (H)

EXAMPLE 20.6

Two mutually coupled coils with 50-mH and 35-mH inductances, respectively, are connected in a series-aiding connection. If $k = 0.81$, calculate the equivalent inductance.

Solution
$$L_M = k\sqrt{L_1 L_2} \quad \text{(equation (20.3))}$$
$$= 0.81\sqrt{50 \text{ mH} \times 35 \text{ mH}}$$
$$= 33.9 \text{ mH}$$
$$L_T = L_1 + L_2 + 2L_M \quad \text{(series-aiding)}$$
$$= 50 \text{ mH} + 35 \text{ mH} + 2(33.9 \text{ mH})$$
$$= 152.8 \text{ mH}$$

Practice Problems 20.4 Answers are given at the end of the chapter.
1. Determine the equivalent inductance of two series coils (no mutual coupling) having inductances of 2 mH and 14.7 mH.
2. Determine the equivalent inductance of the two coils in Problem 1 if they are series-aiding with a coefficient of coupling of 0.89.

20.6.2 Inductors in Parallel

For inductors connected in parallel without mutual coupling, their equivalency is found the same way as with resistors in parallel.

$$L_T = \frac{1}{1/L_1 + 1/L_2 + \cdots + 1/L_N} \tag{20.6}$$

EXAMPLE 20.7

Three inductors with values of 15 mH, 20 mH, and 35 mH, respectively, are connected in parallel with no mutual induction between them. Find the equivalent inductance.

Solution $L_T = \dfrac{1}{(1/L_1 + 1/L_2 + 1/L_3)}$

$= \dfrac{1}{1/15 \text{ mH} + 1/20 \text{ mH} + 1/35 \text{ mH}}$

$= 6.89 \text{ mH}$

Practice Problems 20.5   Answers are given at the end of the chapter.
1. Determine the equivalent inductance of three coils connected in parallel (no mutual coupling) having inductances of 15 mH, 30 mH, and 50 mH.

Mutually coupled inductors connected in parallel are even more complicated, as their equivalent inductance equations show. For practical considerations, mutually coupled parallel connections are rarely used. Therefore, only the equations are shown (no examples are given).

$\dfrac{1}{L_T} = \dfrac{1}{L_1 + L_M} + \dfrac{1}{L_2 + L_M}$ (aiding fields)

$\dfrac{1}{L_T} = \dfrac{1}{L_1 - L_M} + \dfrac{1}{L_2 - L_M}$ (opposing fields)

SECTION 20.6 SELF-CHECK

Answers are given at the end of the chapter.

1. If inductors are placed far enough apart, how much mutual inductance occurs?
2. If the magnetic fields between two coils oppose each other, the equivalent inductance _____.
3. True or false: The equivalency of parallel-connected inductors without mutual coupling is determined in the same way as for resistors in parallel.

20.7   ENERGY STORED IN AN INDUCTOR

We have seen how the current supplied by a voltage source creates a magnetic field in an inductor. The energy supplied by the source is stored in the inductor's magnetic field and does the work of producing the induced voltage when the flux changes.

The amount of energy stored in the magnetic field for any value of current may be expressed in the following equation. (Note that the energy stored by the inductor is measured in joules, the same unit of measurement for energy stored in a capacitor.)

$$\mathcal{E} = \tfrac{1}{2} L I^2 \tag{20.7}$$

where $\mathcal{E}$ = energy stored in inductor (J)
L = inductance (H)
I = current (A)

EXAMPLE 20.8

Determine the energy stored in an inductor when $L = 10$ H and the current is 20 mA.

Solution $\mathcal{E} = \tfrac{1}{2} L I^2$
$= \tfrac{1}{2} \times 10 \text{ H} \times (20 \times 10^{-3} \text{ A})^2$
$= 0.002 \text{ J}$

Practice Problems 20.6   Answers are given at the end of the chapter.
1. Determine the energy stored in an inductor when $L = 500$ mH and the current is 18 mA.

20.7.1   Opening the Circuit

If an inductor has stored energy and then the circuit is opened, the magnetic field collapses and the energy is returned to the circuit in the form of an induced voltage, which tends to keep the current flowing. All the energy is available for inducing voltage, because no energy is dissipated by the magnetic field. If you take into account any circuit resistance or coil resistance, the I^2R loss that occurs from the induced current dissipates all the energy after a period of time.

SECTION 20.7
SELF-CHECK

Answers are given at the end of the chapter.

1. The energy stored in an inductor is measured in _____.
2. What happens when the circuit is opened and the magnetic field collapses?

20.8   STRAY INDUCTANCE

All conductors, whether in the form of wiring, copper foil on a printed circuit board, or component leads, possess inductance. The inductive effects of wiring or leads that are not included in the circuit inductors are considered to be *stray inductance*. The effects of any inductance are more pronounced when the current is *changing*. Just as we saw with stray capacitance, stray inductive effects are very small (less than 1 μH) and negligible at lower frequencies.

At higher frequencies (particularly at microwave frequencies), stray inductive effects become more pronounced. Lead lengths must be kept short. (Surface mounted devices minimize lead-length problems.) Carbon resistors are preferred over wirewound resistors at high frequencies. Wirewound resistors can be made noninductive by the type of winding applied. Adjacent turns are wound so that the current is in the opposite direction; thus, the magnetic fields cancel. This technique is applied to connecting leads by twisting the pair of leads to reduce the effects of inductance.

SECTION 20.8
SELF-CHECK

Answers are given at the end of the chapter.

1. True or false: At lower frequencies, the effects of stray inductance can be ignored.
2. Why are carbon resistors preferred over wirewound resistors at high frequencies?

20.9   INDUCTOR LOSSES AND FAULTS

20.9.1   Inductor Losses

Inductor losses are of the same type previously discussed with transformers. These were the core losses of hysteresis and eddy currents. We have already discussed how to minimize these losses.

An additional loss is caused by *flux-leakage loss*. This loss is brought about with the passage of magnetic flux outside the path along which it can do useful work. For mutually inductive coils, this type of loss reduces the induced voltage of the second coil.

Another loss is caused by a phenomenon known as the *skin effect*. You know that a varying current causes a counter-emf. The cemf is greater at the center of the conductor, hence, little current flows there. The current is forced to flow through the "outer skin" of the conductor. The skin effect is effectively a *copper loss* (I^2R), because the size of the conductor has been reduced as far as the current is concerned. When required, skin-effect losses are minimized by using hollow wire to make the inductor. Because no current flows in the center of the wire, there is no need for a conductor path there.

In addition to hysteresis, eddy current, and flux-leakage losses, real coils dissipate power due to the resistance of their windings. This loss is also referred to as a *copper*, or I^2R, *loss* to distinguish it from the *core losses* discussed previously. The effects of coil resistance on circuit performance are taken into account in later chapters having to do with inductors.

20.9.2 Troubleshooting Inductor Faults

Inductors fail in one of two ways. They change value, which includes an open, or they develop one or more shorted turns. Value changes in an inductor can be located by an inductance-bridge or value-only tester. Value changes account for 25% of all inductor faults.

An open inductor is quite easy to detect. An ohmmeter indicates infinity ("O.L" on a DMM) when testing an open inductor. Note that an open could be caused by a bad solder connection on a printed circuit board. Opens account for 25% of all inductor faults.

Shorts in inductors are difficult to locate even with the most expensive testers. Detecting a partial short with an ohmmeter is almost impossible, because the resistance change is so small. A test known as the *ringing test* is an industry standard for checking coils with shorts. The tester creates a magnetic field in the coil and then checks the number of "rings" produced by a collapsing field into the coil. The rings are a form of AC current caused when the field collapses, induces a current, expands the magnetic field, then collapses again, and keeps repeating. The number of rings is a function of the energy stored by the inductance versus the energy lost by coil resistance. A good coil should produce a minimum of 10 rings before the energy is dampened out. A shorted coil has less inductance, so fewer rings occur. The test results for a good coil are illustrated in Figure 20.6. Shorted turns in inductors account for 50% of all inductor faults.

FIGURE 20.6
Ring test to check for shorted coils in inductor: (a) test setup; (b) a good coil should produce a minimum of ten rings before the energy is dampened out.

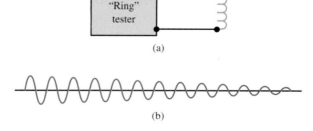

(a)

(b)

SECTION 20.9 SELF-CHECK

Answers are given at the end of the chapter.

1. What is flux-leakage loss?
2. Describe the *skin effect*.
3. An open inductor can be checked with a(n) _____.
4. The best way to check for shorts in an inductor is with a _____ tester.

■ SUMMARY

1. An inductor stores energy in an electromagnetic field.

2. The ability to induce a voltage across itself with a change in current is known as self-inductance, or simply inductance.

3. Inductance also opposes a change in current.

4. The symbol for inductance is L, and the basic unit of inductance is the henry (H).

5. An inductor is nothing more than a coil of wire or an electromagnet.

6. The self-induced emf is also known as a counter-emf (cemf).

7. Lenz's law states that the induced emf in any circuit is always in a direction to oppose the effect that produced it.

8. Inductance does not oppose current, only a change in current.

9. The physical factors that affect inductance are the number of turns in the coil, the diameter of the coil, the coil length, the type of material used in the core, and the number of layers of winding in the coil.

10. Mutual induction occurs when the energy from one coil is transferred or coupled to another coil.

11. The transformer utilizes the effects of mutual induction.

12. Inductors are usually classified by the type of material used to make the core.

13. At microwave frequencies the copper foil on the printed circuit board is known as a *microstrip* device.

14. The energy stored in an inductor is measured in joules.

15. The inductive effects of wiring or leads that are not included in the circuit inductors are considered to be stray inductance.

16. For low-frequency applications, stray inductance effects are negligible, but for high-frequency applications the effects are more pronounced.

17. Besides losses already covered that are associated with transformers, additional losses in an inductor are flux-leakage loss and skin-effect loss.

18. Skin effect is a copper loss as the size of the conductor has been reduced as far as the current is concerned.

19. Value changes account for 25% of all inductor faults.

20. Opens account for 25% of all inductor faults.

21. Shorted turns account for 50% of all inductor faults.

22. A test known as the ringing test is an industry standard for checking coils with shorts.

23. When ring-tested, an inductor should produce 10 or more rings.

■ IMPORTANT EQUATIONS

$$V_{ind} = L\left(\frac{\Delta I}{\Delta t}\right) \tag{20.1}$$

$$L = \frac{\mu N^2 A}{l} \tag{20.2}$$

$$L_M = k\sqrt{L_1 L_2} \tag{20.3}$$

$$L_T = L_1 + L_2 + L_3 + \cdots + L_N \tag{20.4}$$

$$L_T = L_1 + L_2 \pm 2L_M \tag{20.5}$$

$$L_T = \frac{1}{1/L_1 + 1/L_2 + \cdots + 1/L_N} \tag{20.6}$$

$$\mathscr{E} = \tfrac{1}{2}LI^2 \tag{20.7}$$

■ REVIEW QUESTIONS

1. What is an inductor?

2. Define the term *inductance*.

3. What is the difference between *inductance* and *self-inductance*?

4. Describe the physical makeup of an inductor.

5. What is a *self-induced emf*?

6. In a DC circuit, when does the current change?

7. True or false: Inductance has its greatest effect only when there is a change in current.

8. The unit of inductance is the _____.

9. What is the relationship between inductance and the number of turns of the coil?

10. How does a larger-diameter coil increase the value of inductance?

11. What effect does coil length have on the value of inductance?

12. Define the term *mutual inductance*.

13. Name a common application for mutual inductance.

14. Define the term "degree of coupling."

15. What is the difference between *tight* and *loose* coupling?

16. How are inductors classified?

17. What are the applications for air-core inductors?

18. What losses are associated with iron-core inductors?

19. What is the purpose of the binder used in powdered-iron cores?

20. What are the applications for ferrites?

21. How is a variable inductor made? What slug position produces the greatest inductance?

22. When inductors connected in series with no mutual coupling between them, the total inductance is found the same way as resistors in _____.

23. By what mechanism does an inductor store energy?

24. Is stored energy dissipated by the magnetic field?

25. What is stray inductance?

26. Under what circumstances does stray inductance become more pronounced?

27. Describe *skin-effect* loss.

28. What can be used to check for an open in an inductor?

29. Can an ohmmeter be used to check for a short in an inductor?

30. What accounts for the greatest number of faults in an inductor?

■ CRITICAL THINKING

1. Describe the term *inductance* in relation to inertia.

2. What are the physical factors that make up inductance and how are they related to the value of inductance?

3. Describe the various types of inductors and their applications.
4. Describe the term *stray inductance,* tell how it relates to the frequency of operation, and describe what can be done to minimize its effects.
5. Describe the *ring test* used to test for shorts in an inductor.

■ PROBLEMS

1. Calculate the inductance of a coil that induces 150 V when its current changes by 5 mA in a period of 10 μs.
2. Calculate the induced voltage when the current changes by 25 mA in 40 μs in a 100-mH inductor.
3. Calculate the inductance of a coil that induces 100 V when its current changes by 10 mA in 1.0 μs.
4. Calculate the induced voltage when the current changes from 100 mA to 500 mA in 5 ms in a 8-H inductor.
5. Calculate the inductance of a coil that induces 50 mV when its current changes at the rate of 40 mA in 5 ms.
6. Determine the inductance of a coil that has a core with a μ_r of 500, $N = 25$ turns, $A = 2$ cm^2, and $l = 2.0$ cm.
7. Calculate the inductance of a coil with an air core, $N = 200$ turns, $A = 5$ cm^2, and $l = 5$ cm.
8. A coil having 8000 turns has an inductance of 3.2 H. How much is the inductance if the number of turns is reduced by half for the same coil length?
9. Two inductors have a coefficient of coupling of 0.3 and values of 50 mH and 150 mH. Calculate L_M.
10. Two inductors have $K = 0.5$ and values of 1 H and 3 H. Calculate L_M.
11. Two 40-mH coils have a mutual inductance of 32 mH. Calculate K.
12. Two inductors without L_M are connected in series. Determine L_T if their values are 400 μH and 250 mH.
13. Three 33-mH coils are in parallel with no L_M. Calculate L_T.
14. Two coils of 8 H each are series-aiding connected with $L_M = 0.8$. Calculate L_T.
15. Two coils of 50 μH each are connected in series with $L_M = 15$ μH series-opposing. Calculate L_T and K.
16. For Problem 15, calculate L_T if L_M is series-aiding.
17. Determine the energy stored in an inductor when $L = 100$ mH and the current is 100 mA.
18. Determine the energy stored in an inductor when $L = 50$ μH and the current is 20 mA.
19. An inductor is storing 0.005 J. If the current is 100 mA, determine the inductance.
20. Determine the energy stored in an inductor when $L = 2.5$ H and the current is 250 mA.

■ ANSWERS TO PRACTICE PROBLEMS

PRACTICE PROBLEMS 20.1

1. $L = 200$ mH

PRACTICE PROBLEMS 20.2

1. $L = 504$ μH

PRACTICE PROBLEMS 20.3

1. $L_M = 70.7$ μH

PRACTICE PROBLEMS 20.4

1. $L_T = 16.7$ mH
2. $L_T = 26.4$ mH

PRACTICE PROBLEMS 20.5

1. $L_T = 8.33$ mH

PRACTICE PROBLEMS 20.6

1. $\mathscr{E} = 81$ μJ

■ ANSWERS TO SECTION SELF-CHECKS

SECTION 20.1 SELF-CHECK

1. The ability to induce a voltage across itself with a change in current is known as inductance.
2. A coil of wire
3. Henries (H)

SECTION 20.2 SELF-CHECK

1. Another name for self-induced voltage. The polarity of the induced voltage is in an opposite direction to the applied voltage, hence the name counter-emf.
2. Zero
3. When the change in time is very small

SECTION 20.3 SELF-CHECK

1. Directly
2. Inductance decreases as coil length increases because the magnetic field is less concentrated due to the greater distance between the turns.

SECTION 20.4 SELF-CHECK

1. When energy from one coil is transferred or coupled to another coil.
2. Transformer

SECTION 20.5 SELF-CHECK

1. By their core material
2. High
3. Microstrip

SECTION 20.6 SELF-CHECK

1. None
2. Decreases
3. True

SECTION 20.7 SELF-CHECK

1. Joules
2. The energy is returned to the circuit in the form of an induced voltage, which tends to keep the current flowing.

SECTION 20.8 SELF-CHECK

1. True
2. They do not contain wire resistive elements.

SECTION 20.9 SELF-CHECK

1. When magnetic flux passes outside the path along which it can do useful work
2. When current is forced to flow along the outer skin of a conductor due to cemf generated within the center of the conductor
3. Ohmmeter
4. Ring

CHAPTER 21

RC AND RL TRANSIENT RESPONSES

OBJECTIVES

After completing this chapter, you will be able to:

1. Define the term *time constant* and show its relevance to capacitor/inductor voltage and current levels.
2. Calculate time constants.
3. Compare and contrast the capacitor charge and discharge cycles.
4. Utilize the universal time constant chart and exponential equations in solving for unknown circuit values.
5. Describe how the time constant affects waveshaping of nonsinusoidal waveforms.
6. Compare and contrast the differentiator and the integrator.

INTRODUCTION

In the previous two chapters the terms *capacitance* and *inductance* were defined. We saw that both offer an opposition to a circuit parameter: the capacitor to a change in voltage and the inductor to a change in current. In those chapters we were primarily concerned with their abilities to store energy. Now we take a look at the time required for charging and discharging of current in the capacitor or for expanding or collapsing the magnetic field of the inductor.

To determine the amount of time needed to charge or discharge a capacitor, you must determine the *time constant* of the capacitive-resistive circuit. For this circuit the time constant is the product of resistance and capacitance (RC) and is denoted by τ ($\tau = RC$). Likewise, to determine the amount of time needed to expand or collapse the magnetic field in an inductor, you must determine the time constant (L/R) of the inductive-resistive circuit. A time constant is the time needed for a change of 63.2% in the voltage across a capacitor or the current through the inductor.

In addition, time constants also allow us to examine the *transient responses* in series RC and RL circuits and determine how steady-state conditions are reached. A transient response is a temporary condition involving a changing voltage or current that exists only until a steady-state value of either voltage or current is reached. Transient responses are associated with nonsinusoidal voltage and current waveforms (for example, square, rectangular, and triangular) and with a DC source being switched on and off. The effects of inductance (L) and capacitance (C) on nonsinusoidal waveforms is to produce a *waveshape change*. This effect can be analyzed by means of the time constant for inductive and capacitive circuits. Depending on the time constant, *RC* and *RL* circuits are used to provide *filtering, waveshaping,* and *timing.*

The capacitor rather than an inductor is more commonly associated with transient responses and waveshaping. This is because capacitors are smaller, are more economical, and do not have strong magnetic fields.

21.1   *RC* CHARGE AND DISCHARGE CYCLES

21.1.1   Charge Cycle

A voltage divider containing resistance and capacitance is connected in a circuit by means of a switch, as shown at the top of Figure 21.1. Such a series arrangement is called an *RC series circuit.*

In explaining the charge and discharge cycles of an *RC* series circuit, the time interval from time t_0 (time 0, when the switch is first closed) to time t_1 (time 1, when the capacitor reaches full charge or discharge potential) is used. Note that the time from t_0 to t_1 is *not* one time constant, τ, but rather, the time to fully charge or discharge the capacitor. As is shown later, the time to fully charge or discharge the capacitor is equal to five time constants.

When the switch in Figure 21.1 is moved to the charge position at t_0, the circuit voltage rises instantly to source voltage (V_S) and the charging current (i_C) is maximum. This is shown in the graphs of Figure 21.1(a) and 21.1(b), respectively. (At the first instant of time, the capacitor acts as a short, because the number of electrons on each plate is equal; thus, no potential difference exists.) Also at this time the voltage across the resistor (v_R) is maximum. This is shown in the graph of Figure 21.1(c). As time elapses toward time t_1, there is a continuous decrease in charging current flowing into the capacitor. The decreasing current flow is caused by the voltage buildup (series-opposing to the source) across the capacitor. At time t_1, the capacitor has reached full charge and has stored maximum energy in its electrostatic field. Also at this time the voltage across the resistor is zero. The graph in Figure 21.1(d) represents the voltage (v_C) across the capacitor.

You should remember that capacitance opposes a change in voltage. This is shown by comparing the graphs of Figure 21.1(a) and 21.1(d). In part (a), the voltage changes instantly from zero to source, whereas the voltage across the capacitor to part (d) took the entire time interval from time t_0 to time t_1 to reach maximum. The capacitor voltage has to wait until the charge current has stored enough energy before it can change.

21.1.2   Discharge Cycle

When the switch in Figure 21.1 is moved to the discharge position, the capacitor discharge cycle begins (this is illustrated at the top of Figure 21.2).

At the first instant, circuit voltage attempts to go from source voltage to zero, as shown in the graph of Figure 21.2(a). Remember, though, that the capacitor has stored energy in an electrostatic field during the charge cycle.

Because of the change in switch positions, the capacitor now has a path for discharge current to follow. At t_0, there is maximum discharge current (i_D) from the bottom plate of the capacitor through the resistor to the top plate of the capacitor. As time progresses toward t_1, the discharge current steadily decreases until at time t_1 it reaches zero, as shown in the graph of Figure 21.2(b).

The discharge current causes a corresponding voltage drop across the resistor, as shown in the graph of Figure 21.2(c). Note that the voltage across the resistor is decreasing and the voltage across the capacitor is also decreasing. At time t_1 the voltage (v_C) reaches zero.

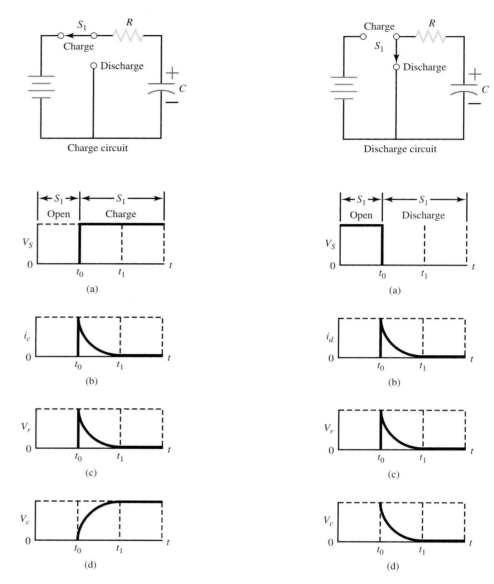

FIGURE 21.1
Charge of an RC series circuit:
(a) source voltage (V_S) rises in-
stantly at t_0; (b) charge current
(i_C) is maximum at t_0; (c) resistor
voltage (v_R) is maximum at t_0;
(d) capacitor voltage (v_C) rises
from t_0 to t_1.

FIGURE 21.2
Discharge of an RC series circuit:
(a) V_S attempts drop to zero at t_0;
(b) discharge current maximum at
t_0; (c) resistor voltage (v_R) de-
creases from t_0; (d) capacitor volt-
age (v_C) decreases from t_0.

By comparing the graphs in (a) and (d) of Figure 21.2, you can see the effect that capacitance has on a change in voltage. If the circuit had not contained a capacitor, the voltage
would have ceased at the instant the switch was moved to the discharge position. Because
the capacitor is in the circuit, voltage is applied to the circuit until the capacitor has discharged
completely at t_1. The effect of capacitance, then, has been to oppose a change in voltage.

21.1.3 High Capacitive Discharge Current

At the instant of discharge, the capacitor's discharge path is typically a smaller resistance
than its charge path. This means at the first instant of time, the capacitor can support a

large discharge current through the load, as demonstrated by the V_C/R_L ratio. The discharge current comes from the energy stored in the capacitor. The capacitor is acting as a voltage source during this time.

The high discharge current is short-lived, because after time t_0 the discharge current immediately starts decreasing toward zero. This is shown in the graph of Figure 21.2(b).

A common application for this high discharge current is to fire flashbulbs on cameras. A flashbulb needs more current to ignite than a small camera battery can supply. The battery is used to charge the capacitor, and the bulb is used as a discharge path for the capacitor. The high discharge current needs to last only long enough to ignite the bulb.

**SECTION 21.1
SELF-CHECK**

Answers are given at the end of the chapter.

1. At the first instant of time, the capacitor acts as a _____.
2. Define the term *capacitance*.
3. Name a common application for the high discharge current in a capacitor.

21.2 *RC* TIME CONSTANT

The time required to charge a capacitor to 63.2% of maximum voltage or to discharge it to 36.8% of its final voltage is known as the *time constant,* denoted by τ, of the circuit. The charge and discharge curves of a capacitor are shown in Figure 21.3.

The value of the time constant (τ) in seconds is equal to the product of the circuit resistance in ohms and the circuit capacitance in farads (RC). Mathematically, this is expressed as

$$\tau = RC \tag{21.1}$$

Note that during the second time constant, the charge on the capacitor accumulates another 63.2% of voltage on top of what it already has accumulated. This takes the charge on the capacitor up to 86.5% (63.2% $\times$ 36.8% $+$ (the initial) 63.2%). The constant of 63.2% accumulates through every time constant that passes. After *five time constants* the accumulated voltage across the capacitor is approximately 99.3%. For practical reasons, the capacitor is considered fully charged at this time, although theoretically the capacitor never achieves full charge. The charge continues at a rate of 63.2% of whatever voltage is left.

Note that τ is a unit of time, and as we will see later for every unit of this time, the charge accumulating on the capacitor is $0.632(V_S - V_C)$. Some common combinations of units for the *RC* time constant are

$$\text{M}\Omega \times \mu\text{F} = \text{s}$$
$$\text{k}\Omega \times \mu\text{F} = \text{ms}$$
$$\text{M}\Omega \times \text{pF} = \mu\text{s}$$

Note that the time in seconds is directly proportional to both R and C. Thus, if either one increases in value, the time constant increases also.

EXAMPLE 21.1

What is the time constant of a 0.01-μF capacitor in series with a 2-kΩ resistance?

Solution $\tau = RC$
$$= 2 \times 10^3 \ \Omega \times 0.01 \times 10^{-6} \ \text{F}$$
$$= 0.02 \ \text{ms}, \quad \text{or} \quad 20 \ \mu\text{s}$$

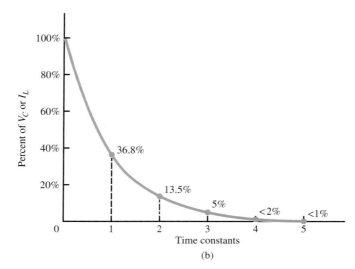

Lab Reference: Voltages are calculated and measured, resulting in the graphing of a similar chart, in Exercise 29.

FIGURE 21.3
Charge and discharge curves: (a) capacitor voltage charge (or I_L growth) curve; (b) capacitor voltage discharge (or I_L decay) curve.

EXAMPLE 21.2	What is the time constant of a 10-μF capacitor in series with a 100-kΩ resistance?

Solution $\tau = RC$
$\qquad\qquad = 100 \times 10^3\ \Omega \times 10 \times 10^{-6}\ F$
$\qquad\qquad = 1\ s$

Practice Problems 21.1   Answers are given at the end of the chapter.
1. What is τ for a 0.22-μF capacitor in series with a 4.7-kΩ resistor?
2. What series resistance is needed to cause a τ of 1.2 ms with a 4.7-μF capacitor?

The voltage waveform produced by the capacitor acquiring a charge is known as an *exponential* waveform, and the voltage across the capacitor is said to rise exponentially.

An exponential rise is also referred to as a *natural increase* and is based on the natural logarithm (base 2.71828). Later we will see how to predict the charge or discharge voltage across a capacitor by using an equation that employs the natural logarithm.

SECTION 21.2
SELF-CHECK

Answers are given at the end of the chapter.

1. Define the term *time constant*.
2. After _____ time constants the capacitor is assumed to be fully charged.
3. The voltage waveform produced by the capacitor acquiring a charge is known as an _____ waveform.

21.3   UNIVERSAL TIME CONSTANT CHART

Because the source voltage and the values of R and C (or R and L) in a circuit are usually known, a *universal time constant chart* (Figure 21.4) can be used to find the voltage or current values for any amount of time.

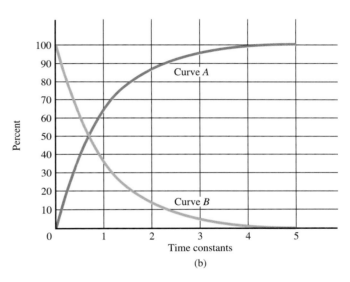

TABLE 21.1
Time constant factors

Factor (Time Constant)	Amplitude (%)
0.2	≈20
0.5	≈40
0.7	≈50
1.0	63.2
2.0	86.5
3.0	95.0
4.0	98.2
5.0	99.3

(b)

FIGURE 21.4

Universal time constant curve: Curve A: v_C charging or i_L growth; Curve B: i_C charging, v_C discharging, or i_L decay.

Curve *A* is a plot for either capacitor voltage during charge or inductor current during growth. Curve *B* is a plot for either capacitor voltage during discharge, resistor voltage during charge, or inductor current during decay.

The time scale (horizontal scale) is graduated in terms of the time constant, so that the curves may be used for any value of resistive-capacitive or resistive-inductive time constant. The voltage or current scales (vertical scales) are graduated in terms of percentage of maximum voltage or current so that the curves may be used for any value of voltage or current. If the time constant and the initial or final voltage for the circuit in question are known, the voltages across various parts of the circuit can be obtained from values on the curves for any time after a switch is closed, either on charge or discharge. The same reasoning is true for the current in the circuit.

Table 21.1 lists amplitude percentages for associated time constant factors.

EXAMPLE 21.3

A 50-kΩ resistor is connected in series with a 40-μF capacitor. With a DC source voltage of 50 V applied, what is the charge across the capacitor after a period of 6 s? (Assume v_C is starting from 0 V.)

Solution $\tau = RC$
$= 50 \times 10^3 \, \Omega \times 40 \times 10^{-6} \, F$
$= 2 \, s$
$\#\tau = \frac{6 \, s}{2 \, s}$
$= 3$
$3\tau = 95\%$ (from Curve *A*)
$v_C = 0.95 \times 50 \, V$
$= 47.5 \, V$

EXAMPLE 21.4

A 10-kΩ resistor is connected in series with a 0.01-μF capacitor with a DC voltage source of 20 V. What is the voltage drop across the resistor after a period of 200 μs? (Assume v_C is starting from 0 V.)

Solution $\tau = RC$
$= 10 \times 10^3 \, \Omega \times 0.01 \times 10^{-6} \, F$
$= 100 \, \mu s$
$\#\tau = \frac{200 \, \mu s}{100 \, \mu s}$
$= 2$
$2\tau = 13.5\%$ (from Curve *B*)
$v_R = 0.135 \times 20 \, V$
$= 2.7 \, V$

Practice Problems 21.2   Answers are given at the end of the chapter.
1. What is the voltage across a 5-μF capacitor connected in series with a 22-kΩ resistor after 330 ms? (Assume v_C is starting from 0 V with $V_S = 20$ V.)
2. For Problem 1, determine the voltage across the resistance after 440 ms.

**SECTION 21.3
SELF-CHECK**

Answers are given at the end of the chapter.
1. What does Curve *A* represent on the universal time constant chart?
2. The percent of voltage accumulated across the capacitor starting from 0 V after four time constants is _____.

21.4   *RC* EXPONENTIAL EQUATIONS

21.4.1   Capacitor-Charging Equations

The universal time constant chart provides a graphical approach to solving series *RC* and *RL* circuits. These problems are more easily solved when the number of time constants is a whole number, because interpreting the percentage from the curves is easier for whole-number multiples. For in-between values you have to interpolate. For any in-between value or to determine current or voltage values at any particular instant, the use of an *exponential equation* is far more accurate. Furthermore, a good scientific calculator makes these problems easy to do.

Capacitor Voltage

For an RC circuit in which C is charging, the following equation can be used to calculate the capacitor voltage, v_C, at any point along Curve A of Figure 21.4.

$$v_C = V_S(1 - e^{(-t/\tau)}) \tag{21.2}$$

where v_C = capacitor voltage (V)
   V_S = source voltage (V)
   e = base of natural logarithms (2ndF ln on Sharp calculator)
   t = the time interval for charging (s)
   τ = RC time constant (s)

EXAMPLE 21.5

A 22-kΩ resistor is in series with a 0.002-μF capacitor and a 12-V battery. What is the charge on the capacitor after a time of 160 μs? (Assume the capacitor voltage is starting from 0 V.)

Solution $\tau = RC$
   $= 22 \times 10^3 \ \Omega \times 0.002 \times 10^{-6} \ F$
   $= 44 \ \mu s$
   $v_C = V_S(1 - e^{(-t/\tau)})$
   $= 12(1 - e^{(-160E-06/44E-06)})$
   $= 11.68 \ V$

Optional Calculator Sequence		

The Sharp calculator sequence for this example is as follows.

Step	Keypad Entry	Top Display Response
1	12(	12(
2	1−	12(1−
3	2ndF, ln	12(1−e^
4	(	12(1−e^(
5	±160Exp−6	
6	÷	12(1−e^(−160E−06/
7	44Exp−6	
8	))	12(1−e^(−160E−06/44E−06))
9	=	11.68382423
	Answer	11.68 (bottom display)

A point to consider is that $v_C + v_R = V_S$ (KVL). If you have calculated v_C, as demonstrated in Example 21.5, and you wish to find v_R, just subtract v_C from V_S.

Practice Problems 21.3   Answers are given at the end of the chapter.

1. What is the charge on a capacitor after 82 μs for a 500-Ω resistor in series with a 0.068-μF capacitor? The source voltage is 10 V. (Assume the capacitor voltage is starting from 0 V.)

Capacitor Current

The equation for the capacitor-charging current at any particular time is given by

$$i_C = \frac{V}{R} e^{(-t/\tau)} \tag{21.3}$$

where i_C = charging current (A)

V = source voltage (V)
R = circuit resistance (Ω)
e = base of natural logarithms (2ndF ln)
t = the time interval for charging (s)
τ = RC time constant (s)

EXAMPLE 21.6

Determine the capacitor current for the problem in Example 21.5 at the instant of 160 μs.

Solution $i_C = \dfrac{V}{R}\, e^{(-t/\tau)}$

$$= \left(\frac{12\text{ V}}{22 \times 10^3\ \Omega}\right) e^{(-160\text{E}-06/44\text{E}-06)}$$

$$= 14.4\ \mu\text{A}$$

Optional Calculator Sequence

The Sharp calculator sequence for this problem is as follows.

Step	Keypad Entry	Top Display Response
1	(12	(
2	÷	(12/
3	22Exp3)	(12/22E03)
4	2ndF, ln	(12/22E03)e^
5	(±	(12/22E03)e^(−
6	160Exp−6	(12/22E03)e^(−
7	/	(12/22E03)e^(−160E−06/
8	44Exp−06	(12/22E03)e^(−160E−06/
9	)	(12/22E03)e^(−160E−06/44E−06)
10	=	14.3716259×10⁻⁰⁶
	Answer	14.4×10⁻⁶ (bottom display)

EXAMPLE 21.7

Determine the charging current for the problem in Example 21.5 at 1 μs into the charging period.

Solution $i_C = \dfrac{V}{R}\, e^{(-t/\tau)}$

$$= \left(\frac{12\text{ V}}{22 \times 10^3\ \Omega}\right) e^{(-1\text{E}-06/44\text{E}-06)}$$

$$= 533.2\ \mu\text{A}$$

Notice that the capacitor current, i_C, in Example 21.7 is approximately equal to the maximum current of V_S/R ($12/22 \times 10^3 = 545.5\ \mu$A). This is because at the first instant of time the capacitor acts as a short; thus, $i_C = V_S/R$. The time given for the capacitor charge, 1 μs, is extremely close to time zero; thus, almost all the source voltage is dropped across R and close to maximum current flows.

Practice Problems 21.4   Answers are given at the end of the chapter.
1. Determine the capacitor current for Practice Problem 21.3 at the instant of 109 μs.

Charging Voltage Time

The following equation is used to calculate the length of time required for the voltage across a capacitor to rise to a certain value. It is found by transposing equation (21.2).

$$t = -\tau \ln\left(1 - \frac{v_C}{V}\right) \tag{21.4}$$

where $\tau = RC$ time constant (s)
   ln = natural logarithm, base *e*
   v_C = capacitor voltage at any point on the charge curve
      for which time is desired
   V = initial value of the voltage

EXAMPLE 21.8

Determine the length of time for a capacitor to reach 35 V when the source is 50 V, $R = 1$ MΩ, and $C = 0.68$ μF.

Solution $\tau = RC$

$$= 1 \times 10^6 \ \Omega \times 0.68 \times 10^{-6} \ F$$

$$= 0.68 \ s$$

$$t = -\tau \ln\left(1 - \left(\frac{v_C}{V}\right)\right)$$

$$= (-0.68 \ s) \ln\left(1 - \frac{35 \ V}{50 \ V}\right)$$

$$= 819 \times 10^{-3} \ s$$

$$= 819 \ ms$$

Optional Calculator Sequence

The Sharp calculator sequence for this problem is as follows.

Step	Keypad Entry	Top Display Response
1	±.68	—
2	ln	−0.68ln
3	(1	−0.68ln(
4	−	−0.68ln(1−
5	(35	−0.68ln(1−(
6	÷	−0.68ln(1−(35/
7	50))	−0.68ln(1−(35/50))
8	=	818.7015069×10⁻³
	Answer	819×10⁻³ (bottom display)

Practice Problems 21.5   Answers are given at the end of the chapter.
1. Calculate the length of time for the capacitor in Practice Problem 21.3 to reach 7.5 V.

21.4.2 Capacitor-Discharging Equations

Capacitor Voltage

For an *RC* circuit in which *C* is discharging, the following equation can be used to calculate the capacitor voltage, v_C, at any point along curve *B* in Figure 21.4.

$$v_C = Ve^{(-t/\tau)} \tag{21.5}$$

where V = initial value of the voltage (V)
        e = base of natural logarithms (2ndF ln)
        t = time interval for discharge (s)
        τ = *RC* time constant (s)

EXAMPLE 21.9

Determine the voltage across a fully charged capacitor after a discharge period of 1.05 s, when $R = 5.6$ kΩ, $C = 100$ μF, and the voltage source is 24 V.

Solution $\tau = RC$
           $= 5.6 \times 10^3 \, \Omega \times 100 \times 10^{-6} \, F$
           $= 0.56$ s
      $v_C = Ve^{(-t/\tau)}$
         $= (24 \text{ V})e^{(-1.05/0.56)}$
         $= 3.68$ V

Practice Problems 21.6   Answers are given at the end of the chapter.
1. Determine the voltage across a fully charged capacitor after a discharge period of 31 ms when $R = 10$ kΩ and $C = 2.2$ μF; the voltage source is 20 V.

Capacitor Current

The capacitor discharge current at any instant can be found by

$$i_C = \frac{V}{R}e^{(-t/\tau)} \tag{21.6}$$

where V = initial capacitor voltage (V)
        R = circuit resistance (Ω)
        e = base of natural logarithms (2ndF ln)
        t = time interval while discharging (s)
        τ = *RC* time constant (s)

EXAMPLE 21.10

Determine the discharge capacitor current in Example 21.9 after a discharge of 2 s.

Solution $i_C = \frac{V}{R}e^{(-t/\tau)}$

          $= \left(\frac{24 \text{ V}}{5.6 \times 10^3 \, \Omega}\right)e^{(-2/0.56)}$

          $= 0.12$ mA

Practice Problems 21.7   Answers are given at the end of the chapter.
1. Determine the discharge capacitor current for Practice Problem 21.6 after a discharge of 47 ms.

Discharging Voltage Time

The length of time required for a voltage to decrease to a certain value in an *RC* circuit is determined by the following equation:

$$t = \tau \ln\left(\frac{V}{v_C}\right) \tag{21.7}$$

where $\tau = RC$ time constant (s)
 $\ln$ = natural logarithm, base e
 V = initial value of voltage
 v_C = capacitor voltage at any point on Curve *B*

EXAMPLE 21.11

A 10-μF capacitor is charged to a value of 50 V and then is discharged into a 1-kΩ resistor. Determine the length of time required for the voltage to decrease from 21 to 13 V.

Solution $\tau = RC$
$$= 1 \times 10^3 \, \Omega \times 10 \times 10^{-6} \, F$$
$$= 10 \text{ ms}$$

$$t = \tau \ln\left(\frac{V}{v_C}\right)$$
$$= (10 \times 10^{-3} \, s) \ln\left(\frac{21 \text{ V}}{13 \text{ V}}\right)$$
$$= 4.8 \text{ ms}$$

Practice Problems 21.8   Answers are given at the end of the chapter.
1. Determine the length of time required for the voltage across the capacitor in Practice Problem 21.6 to decrease from 18 to 9 V.

SECTION 21.4 SELF-CHECK

Answers are given at the end of the chapter.

1. True or false: The universal time constant chart is better suited to problems involving whole-number multiples.
2. The exponential equation relies on the use of the _____ log.

21.5   *L/R* TIME CONSTANT

The time constant for an *RL* circuit is found by the ratio *L/R*. The *L/R* time constant is a valuable tool for use in determining the time required for current in an inductor to reach a specified value. Figure 21.3 can also be used for an *RL* circuit. As shown in Figure 21.3(a), one *L/R* time constant is the time required for the current in an inductor to increase (grow) to 63.2% of the maximum current. Each time constant is equal to the time required for the current to increase by 63.2% of the difference in value between the current flowing in the inductor and the maximum current. Maximum current flows in the inductor after five *L/R* time constants are complete. Remember that an inductor opposes a change in current; thus, it takes time for the current to reach maximum. (The opposition to the change in current is manifested by the self-induced voltage (series-opposing to the source.))

When an *RL* circuit is deenergized, the circuit current decreases (decays) to zero in five time constants at the same rate that it previously increased (shown in Figure 21.3(b)).

The value of the time constant in seconds is equal to the inductance in henrys divided by the circuit resistance in ohms (L/R). The equation used to calculate one L/R time constant is

$$\tau = \frac{L}{R} \tag{21.8}$$

EXAMPLE 21.12

What is the time constant of a 10-H coil with 100 Ω of series resistance?

Solution $\tau = \dfrac{L}{R}$

$\qquad\quad = \dfrac{10\ \text{H}}{100\ \Omega}$

$\qquad\quad = 0.1\ \text{s}$

EXAMPLE 21.13

What is the time constant of a 10-mH coil in series with a 5.6-kΩ resistance?

Solution $\tau = \dfrac{L}{R}$

$\qquad\quad = \dfrac{10 \times 10^{-3}\ \text{H}}{5.6 \times 10^{3}\ \Omega}$

$\qquad\quad = 1.79\ \mu\text{s}$

Note that the time constant is directly proportional to the value of inductance and inversely proportional to circuit resistance. Thus, a greater inductance or a smaller circuit resistance increases the time constant.

Practice Problems 21.9   Answers are given at the end of the chapter.
1. What is the time constant of a 10-mH coil with 200 Ω of series resistance?
2. What is the value of inductance needed for a 3.5-ms time constant if the resistance is 500 Ω?

21.5.1   High Voltage Produced when an *RL* Circuit Is Opened

We have seen that there is a high discharge current the first instant of time when a capacitor is discharged. A somewhat similar response occurs in an inductor. Here, there is a high voltage produced when the magnetic field of the inductor collapses when the circuit is opened. The high voltage can be much higher than the applied voltage. This was demonstrated in Section 20.2.1. Many circuits that utilize inductive devices (such as relay contacts) employ diode protection devices to ensure that damage is not done by the high voltage produced from a collapsing magnetic field. Figure 21.5 shows a diode placed across the contacts of a relay. When the magnetic field collapses, producing a high voltage, the diode turns on and absorbs the energy.

SECTION 21.5 SELF-CHECK

Answers are given at the end of the chapter.

1. One *L/R* time constant is the time required for the current to increase to _____ of the maximum current.
2. What happens when the magnetic field of an inductor collapses?

FIGURE 21.5
Diode protection across relay contacts. When the magnetic field of the relay collapses, producing a high voltage, the diode turns on and absorbs the energy.

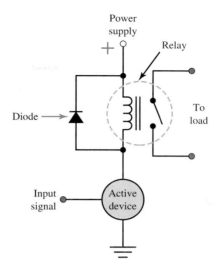

Lab Reference: The effects of the magnetic field collapsing in an inductor are demonstrated in Exercise 30.

21.6 *L/R* EXPONENTIAL EQUATIONS

21.6.1 Inductor Current Equation

The exponential equation for solving for the inductor current at any particular time is just a modified form of equation (21.2). It has been modified to solve for inductor current rather than capacitor voltage. The current buildup follows Curve *A* of Figure 21.4. The equation is

$$i_L = \frac{V_S}{R}\left(1 - e^{(-t/\tau)}\right) \tag{21.9}$$

where $\dfrac{V_S}{R}$ = steady-state value of the current (A)

e = base of natural logarithms (2ndF ln)

t = elapsed time from start to buildup (s)

$\tau = \dfrac{L}{R}$ time constant (s)

EXAMPLE 21.14

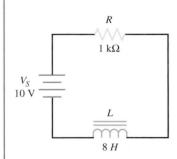

FIGURE 21.6
Circuit for Examples 21.14 and 21.15.

The circuit in Figure 21.6 has an 8-H inductor and a 1-kΩ resistor connected in series to a 10-V source. Calculate the inductor current, i_L, at $t = 6$ ms when the switch is closed.

Solution $\tau = \dfrac{L}{R}$

$= \dfrac{8\ \text{H}}{1 \times 10^3\ \Omega}$

$= 8$ ms

$i_L = \dfrac{V_S}{R}\left(1 - e^{(-t/\tau)}\right)$

$= \left(\dfrac{10\ \text{V}}{1 \times 10^3\ \Omega}\right)\left(1 - e^{(-6\text{E}-03/8\text{E}-03)}\right)$

$= 5.28$ mA

21.6.2 Inductor Voltage Equation

The voltage across the inductor follows Curve B of Figure 21.4 at the same time the current is building up. The voltage starts from maximum and then decays. The inductor voltage can be found by

$$v_L = V_S(e^{(-t/\tau)}) \tag{21.10}$$

where V_S = source voltage (V)

e = base on natural logarithms (2ndF ln)

t = time after closing switch (s)

$\tau = \dfrac{L}{R}$ time constant (s)

EXAMPLE 21.15

For the circuit in Figure 21.6, determine v_L at $t = 6$ ms.

Solution $\tau = 8$ ms   (from Example 21.14)
$$v_L = V_S(e^{(-t/\tau)})$$
$$= (10\ \text{V})(e^{(-6E-03/8E-03)})$$
$$= 4.72\ \text{V}$$

A point to consider is that $v_L + v_R = V_S$ (KVL). Thus, if you have calculated the value of v_L, all you have to do to solve for v_R is to subtract v_L from the source.

Practice Problems 21.11 Answers are given at the end of the chapter.
1. For Practice Problem 21.10, determine v_L at $t = 1.5\ \mu s$.

**SECTION 21.6
SELF-CHECK**

Answers are given at the end of the chapter.

1. Using an exponential equation for *RL* circuits, finding inductor current is equivalent to finding _____ _____ for an *RC* circuit.
2. The voltage across an inductor follows Curve _____ of Figure 21.4.
3. How is voltage in an *RL* circuit expressed by KVL?

21.7   LONG, MEDIUM, AND SHORT TIME CONSTANTS

Modifying nonsinusoidal waveforms into other useful waveshapes can be done by using *RC* and *RL* circuits with an appropriate time constant. *RC* circuits are used more frequently than *RL* circuits. Several reasons for this were given in the introduction. In addition, *RC* circuits can easily be modified to provide any time constant desired. With inductors, the internal series resistance and the distributed capacitance between windings cause additional problems.

21.7.1 Long Time Constant

A *long time constant* is defined as one in which the *RC* time constant (τ) is at least five times longer, in time, than the pulse width of the applied waveform. As a result, the capacitor of a series *RC* circuit accumulates very little charge, and v_C remains small.

21.7.2 Medium Time Constant

A *medium time constant* is defined as one in which the *RC* time constant (τ), in time, is equal to the pulse width of the applied waveform. As a result, the voltage across the capacitor of a series *RC* circuit falls between that of a long and short time constant circuit.

21.7.3 Short Time Constant

A *short time constant* is defined as one in which the time constant is no more than one-fifth the pulse width, in time, of the applied voltage. Here, the capacitor quickly charges to the applied voltage and remains there until the input drops to zero. Then the capacitor quickly discharges to zero.

**SECTION 21.7
SELF-CHECK**

Answers are given at the end of the chapter.

1. Waveshaping can be done by using a _____ waveform and an appropriate time constant circuit.
2. Define a long time constant.

21.8 *RC* INTEGRATOR

A waveshaping circuit known as an *integrator* provides an output voltage that is directly proportional to the area under the input voltage waveform. This process is accomplished by a series *RC* circuit in which the output is taken across the capacitor. (This is also known as a long-time-constant circuit.) An integrator is a simple *waveform converter.* The input to such a circuit can be DC, sinusoidal, or nonsinusoidal waveforms. For waveform conversion, the most common inputs used to an integrator are a square wave and a triangular wave. As a guide, good integration results when component values are chosen such that $RC(\tau) \geq 20T$. This means that the *RC* time constant (τ), in time, should be greater than or equal to 20 times the period of the input waveform.

Figure 21.7 illustrates the *RC* integrator circuit and various input-output combinations.

**SECTION 21.8
SELF-CHECK**

Answers are given at the end of the chapter.

1. Define the term *integrator.*
2. In an *RC* integrator the output is taken across the _____.
3. Good integration results when τ is _____.

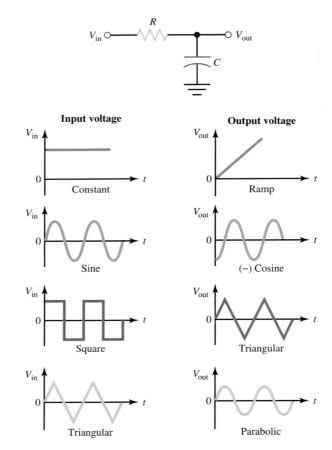

FIGURE 21.7
RC integrator input and output waveforms.

21.9 *RC* DIFFERENTIATOR

A waveshaping circuit known as a *differentiator* provides an output voltage that is directly proportional to the slope of the input voltage. This process is accomplished by a series *RC* circuit in which the output is taken across the resistor. (This is also known as a short-time-constant circuit.) Like the integrator, the differentiator is a simple waveform converter. A common resistor output is a narrow spike that is used as a trigger for pulse and digital circuits. As a guide, good differentiation is obtained if component values are chosen such that $RC(\tau) \leq T/20$. This means that the *RC* time constant (τ), in time, should be less than $\frac{1}{20}$ the period of the input wave.

Figure 21.8 illustrates the *RC* differentiator circuit and various input-output combinations. The most common of these is to convert a square wave into spikes or a triangular wave into a square wave.

**SECTION 21.9
SELF-CHECK**

Answers are given at the end of the chapter.

1. Define the term *differentiator.*
2. For differentiation in an *RC* circuit, the output is taken across the _____.

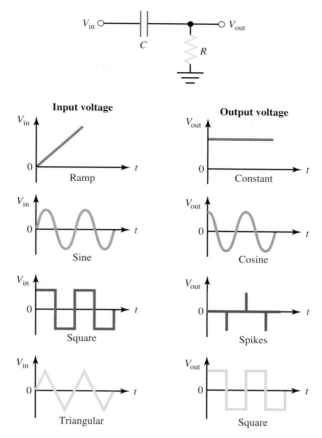

FIGURE 21.8
RC differentiator input and output waveforms.

■ SUMMARY

1. The time constant defines the amount of time needed to charge and discharge a capacitor or the amount of time needed to expand or collapse the magnetic field in an inductor.

2. For an *RC* circuit the time constant is found by multiplying $R \times C$, whereas for an *RL* circuit it is found by dividing L by R.

3. *RC* and *RL* circuits are used to provide filtering, waveshaping, and timing.

4. At the first instant of time the capacitor acts as a short.

5. As the capacitor charges, there is a decreasing charge current.

6. In the charging mode, capacitor voltage increases and resistor voltage decreases; in the discharging mode, both capacitor voltage and resistor voltage decrease.

7. The effect of capacitance is to oppose a change in voltage.

8. When a capacitor discharges, there is usually a high discharge current. An application for this is to fire a flashbulb.

9. One *RC* time constant is the time required to charge the capacitor to 63.2% of maximum. This constant accumulates through every time constant that passes.

10. The time constant is directly proportional to both R and C.

11. The voltage waveform produced by a capacitor acquiring a charge is known as an exponential waveform.

12. The universal time constant chart can be used to find the voltage or current values for any amount of time.

13. An exponential equation provides far more accuracy than a universal time constant chart.

14. An inductor opposes a change in current; thus, it takes time for its current to reach maximum.

15. When the magnetic field collapses in an inductor, there is usually a high voltage produced (that can be much higher than the applied voltage).

16. For circuit protection, inductive devices often employ the use of a diode to absorb the energy from a collapsing magnetic field.

17. By modifying nonsinusoidal waveforms, useful waveshapes can be obtained by using *RC* or *RL* circuits with an appropriate time constant.

18. The *RC* time constant is used more often than the *L/R* time constant.

19. An integrator provides an output voltage that is directly proportional to the area under the input voltage waveform.

20. A differentiator provides an output voltage that is directly proportional to the slope of the input voltage.

■ IMPORTANT EQUATIONS

$$\tau = RC \tag{21.1}$$

$$v_C = V_S(1 - e^{(-t/\tau)}) \tag{21.2}$$

$$i_C = \frac{V}{R} e^{(-t/\tau)} \tag{21.3}$$

$$t = -\tau \ln\left(1 - \frac{v_C}{V}\right) \tag{21.4}$$

$$v_C = V e^{(-t/\tau)} \tag{21.5}$$

$$i_C = \frac{V}{R} e^{(-t/\tau)} \tag{21.6}$$

$$t = \tau \ln\left(\frac{V}{v_C}\right) \tag{21.7}$$

$$\tau = \frac{L}{R} \tag{21.8}$$

$$i_L = \frac{V_S}{R}(1 - e^{(-t/\tau)}) \tag{21.9}$$

$$v_L = V_S(e^{(-t/\tau)}) \tag{21.10}$$

■ REVIEW QUESTIONS

1. Define the term *time constant*.
2. How do you calculate a time constant?
3. What is a transient response?
4. In what type of circuits are the *RC* or *L/R* time constant used?
5. Why are *RC* circuits more common than *RL* circuits for wave-shaping?
6. A capacitor opposes a change in _____.
7. At the first instant of time, the capacitor acts as a _____.
8. Why is there usually a high discharge current with capacitors?
9. Name an application for the high capacitor-discharge current.
10. After five time constants the charge on the capacitor is _____%.
11. Describe the waveform produced by the capacitor acquiring a charge.
12. On a universal time constant chart, what do the vertical and horizontal axes represent?
13. The *RC* time constant is _____ proportional to the values of *R* and *C*.
14. For what multiples is the universal time constant chart better suited?
15. Show how KVL is true for an *RC* time constant circuit.
16. When a capacitor discharges, it acts as a _____ _____.
17. Why is a high voltage produced when the magnetic field of the inductor collapses?
18. What is the purpose of the diode in Figure 21.5?
19. Define a *long* time constant.
20. Define a *medium* time constant.
21. Define a *short* time constant.

22. Describe an integrator.
23. What are the common inputs and outputs for an integrator?
24. Describe a differentiator.
25. Where is the output taken in a differentiator?

■ CRITICAL THINKING

1. Describe in detail the capacitor charge and discharge cycles.
2. Why does a capacitor act as a short at the first instant of charge?
3. Why is the time constant for an *RL* circuit *L/R* rather than *L* × *R* (like the product of *R* and *C* in an *RC* circuit)?
4. Compare and contrast the integrator and differentiator.

■ PROBLEMS

1. Calculate the *RC* time constant for a 15-kΩ series resistance and a capacitance of 0.001 μF.
2. For Problem 1, how long will it take for the capacitor to reach V_S if the source voltage is 18 V?
3. For Problem 1, what is the instantaneous value of charge current after a time of 5 μs?
4. Repeat Problem 3 for a time of 2.7 μs.
5. The time constant of an *RC* circuit is 50 μs. If *R* = 2 kΩ, determine *C*.
6. The time constant of an *RC* circuit is 1 ms. If *C* = 0.05 μF, determine *R*.
7. For the circuit in Figure 21.9, determine v_C at the following time intervals. (Assume the capacitor is fully charged.)
   (a) 0 s (b) 100 μs (c) 200 μs (d) 500 μs (e) 1000 μs

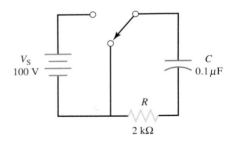

FIGURE 21.9
Circuit for Problem 7.

8. For the circuit in Figure 21.9, repeat Problem 7 but find i_C at the same time intervals.
   (a) 0 s (b) 100 μs (c) 200 μs (d) 500 μs (e) 1000 μs

For problems 9–15 solve for the unknowns listed. Each problem relates to the simple *RC* circuit shown in Figure 21.10. (All problems are for the charging cycle of the capacitor.)

	v_R	v_C	V_S	R	C	t	i_C
9.			20 V	150 kΩ	0.033 μF	1.5 ms	
10.	50 V		90 V	5 kΩ	250 μF		

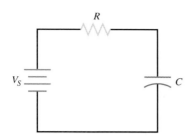

FIGURE 21.10
Circuit for Problems 9–15.

v_R	v_C	V_S	R	C	t	i_C
11. 25 V	65 V		75 kΩ	0.05 μF		
12.		2 V	100 kΩ	200 pF	35 μs	
13.	60 V	400 V		25 μF	25 s	
14. 6 V	150 V		91 kΩ	0.05 μF		
15. 60 V		210 V	390 kΩ		6 ms	

16. Determine the *L/R* time constant for a series resistance of 1.5 kΩ and inductance of 40 mH.

17. Calculate i_L for the circuit in Problem 16 at the following time intervals. (The source voltage is 60 V.)

(a) 0 s (b) 50 μs (c) 75 μs (d) 100 μs (e) 133.3 μs

18. Repeat Problem 17 by calculating v_L for the same circuit conditions.

(a) 0 s (b) 50 μs (c) 75 μs (d) 100 μs (e) 133.3 μs

■ ANSWERS TO PRACTICE PROBLEMS

PRACTICE PROBLEMS 21.1

1. $\tau = 1.03$ ms
2. $R = 255\ \Omega$

PRACTICE PROBLEMS 21.2

1. $v_C = 19$ V
2. $v_R = 0.36$ V

PRACTICE PROBLEMS 21.3

1. $v_C = 9.1$ V

PRACTICE PROBLEMS 21.4

1. $i_C = 810\ \mu$A

PRACTICE PROBLEMS 21.5

1. $t = 47.1\ \mu$s

PRACTICE PROBLEMS 21.6

1. $v_C = 4.89$ V

PRACTICE PROBLEMS 21.7

1. $i_C = 0.24$ mA

PRACTICE PROBLEMS 21.8

1. $t = 15.25$ ms

PRACTICE PROBLEMS 21.9

1. $\tau = 50\ \mu$s
2. $L = 1.75$ H

PRACTICE PROBLEMS 21.10

1. $i_L = 9.7$ mA

PRACTICE PROBLEMS 21.11

1. $v_L = 12.4$ V

■ ANSWERS TO SECTION SELF-CHECKS

SECTION 21.1 SELF-CHECK

1. Short
2. Capacitance opposes a change in voltage.
3. To fire a flashbulb

SECTION 21.2 SELF-CHECK

1. A time constant is the time needed for a change of 63.2% in voltage across the capacitor.
2. 5
3. Exponential

SECTION 21.3 SELF-CHECK

1. The voltage across the capacitor during charge, or the inductor current during growth
2. 98.2%

SECTION 21.4 SELF-CHECK

1. True
2. Natural

SECTION 21.5 SELF-CHECK

1. 63.2%
2. It induces a voltage into the coil that can be higher than the source voltage.

SECTION 21.6 SELF-CHECK

1. Capacitor voltage
2. B
3. $v_L + v_R = V_S$

SECTION 21.7 SELF-CHECK

1. Nonsinusoidal
2. A long time constant is one in which the *RC* time constant is at least five times longer than the pulse width of the applied waveform.

SECTION 21.8 SELF-CHECK

1. An integrator provides an output that is directly proportional to the area under the input voltage waveform.
2. Capacitor
3. $\geq 20T$

SECTION 21.9 SELF-CHECK

1. A differentiator provides an output voltage that is directly proportional to the slope of the input voltage.
2. Resistor

PART VI

AC REACTIVE CIRCUITS

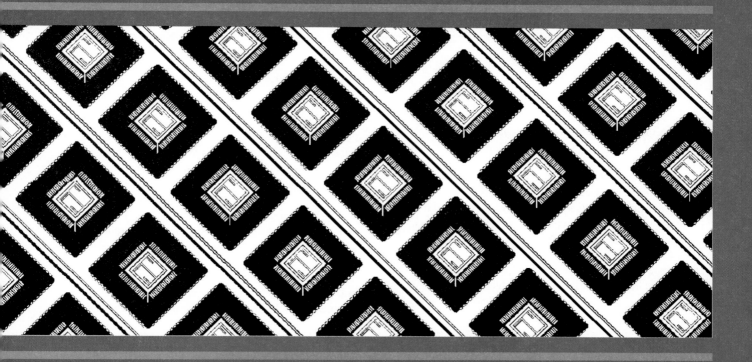

CHAPTER 22

CAPACITIVE AND INDUCTIVE REACTANCE

OBJECTIVES

After completing this chapter, you will be able to:

1. Define the terms *capacitive reactance* and *inductive reactance*.
2. Define the term *quality factor* (Q).
3. Describe the phase relationship between current and voltage in a purely capacitive AC circuit.
4. Describe the phase relationship between current and voltage in a purely inductive AC circuit.
5. Calculate unknowns such as X_C, X_L, f, C, L, and Q in AC circuits.
6. Compare and contrast conductor resistance and effective resistance as it relates to an inductor.

INTRODUCTION

We have already learned how capacitance and inductance individually behave in a direct current (DC) circuit. In addition we saw the effects of changing voltage to capacitance and varying current to inductance. In this chapter, we consider the current and voltage relationships of a capacitor and an inductor in a *sinusoidal AC circuit*. Because a sinusoidal AC voltage changes continuously, there is some rms capacitor current at all times. The ratio of capacitor voltage to current (V_C/I) is called *capacitive reactance* (X_C) and is the opposition in ohms provided by the capacitor. The reactance is inversely proportional to both the frequency and the capacitance. Thus, for direct current, the capacitor is an open circuit. This allows it to be used to block direct current and to pass alternating current.

Similarly, for an inductor the current is always changing in a sinusoidal AC circuit. There is some effective opposition to the current drawn by a coil at a given frequency. This is called *inductive reactance* (X_L) and is determined by both inductance and frequency. The reactance is directly proportional to both these terms. Because of inductive reactance, a coil has a greater opposition to alternating current than it does to direct current. This property is taken advantage of in a DC power supply filter circuit to reduce the AC ripple component. Inductors tend to block AC and pass DC.

The *quality factor* (Q) of a coil is expressed in terms of inductive reactance and is a figure of merit to show how inductive a coil is compared to its internal resistance.

Its internal resistance (effective resistance, R_e) comprises both DC and AC factors. Q is the ratio of the coil's inductive reactance to effective resistance (X_L/R_e). This may also be expressed as the ratio of energy stored to energy dissipated (as heat).

22.1 AC AND THE CAPACITOR

The four parts of Figure 22.1 show the variation of the alternating voltage and current in a *purely capacitive* circuit for each quarter of one cycle. The solid line represents the voltage across the capacitor, and the dotted line represents the current. The line running through the center is the zero, or reference, point for both the voltage and current. The bottom horizontal line marks off the time of the cycle in terms of electrical degrees. (Assume that the AC voltage has been acting on the capacitor for some time before the time represented by the starting point of the sine wave in the figure.)

FIGURE 22.1

Phase relationship of voltage and current in a capacitive circuit: (a) capacitor current is maximum at 0° and then decreases; (b) capacitor current is zero at 90° and then increases in the opposite direction; (c) capacitor current is maximum (−) at 180° and then decreases; (d) capacitor current is zero at 270° and then increases in the opposite direction.

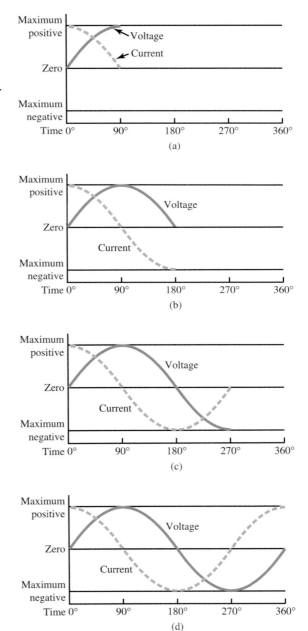

At the beginning of the first quarter-cycle (0° to 90°), the voltage has just passed through zero and is increasing in the positive direction. Because the zero point is the steepest part of the sine wave, the voltage is changing at its greatest rate. The charge of a capacitor varies directly with the voltage; therefore, the charge on the capacitor is also changing at its greatest rate at the beginning of the first quarter-cycle. In other words, the greatest number of electrons are moving off one plate and onto the other plate. Thus, the capacitor current is at its maximum value, as part (a) of the figure shows.

As the voltage proceeds toward the maximum at 90°, its rate of change becomes less and less; hence, the current must decrease toward zero. At 90° the voltage across the capacitor is maximum, the capacitor is fully charged, and there is no further movement of electrons from plate to plate. Thus, the current at 90° is zero.

At 90° the voltage stops increasing in the positive direction and starts to decrease. It is still a positive voltage, but to the capacitor the decrease in voltage means that its negative plate, which has just accumulated an excess of electrons, must lose some electrons (discharge). The current flow, therefore, must reverse direction. Part (b) of the figure shows the current curve to be below the zero reference during the second quarter-cycle (90° to 180°).

At 180° the voltage has dropped to zero (Figure 21.1(c)). This means that for a brief instant the electrons are equally distributed between the two capacitor plates; the current is maximum because the rate of change of voltage is maximum. Just after 180° the voltage has reversed polarity and starts building up to its maximum negative peak (at 270°). During this third quarter-cycle, the rate of voltage change gradually decreases as the charge builds to a maximum at 270°. At this point the capacitor is fully charged, and it carries the full impressed voltage. The current is zero at this point. The conditions are exactly the same as the end of the first quarter-cycle (90°), but the polarity is reversed.

Just after 270° the impressed voltage once again starts to decrease, and the capacitor must discharge some of its electrons from its negative plate. The discharging action continues through the last quarter-cycle (270° to 360°) until the impressed voltage has reached zero. At 360° everything starts over again.

22.1.1 The Capacitor's 90° Phase Shift

If you examine the complete voltage and current curves in Figure 22.1(d), you will see that the current always arrives at a certain point in the cycle 90° ahead of the voltage because of its charging and discharging action. You know that this time and place relationship between the current and voltage is called the *phase relationship*. The voltage-current phase relationship in a capacitive circuit is exactly opposite to that in an inductive circuit (as we show later). *The current through a capacitor leads the voltage across the capacitor by 90°.* (Stating it another way, the capacitor voltage lags the capacitor current by 90°.)

Another way to see the 90° phase shift is to examine the capacitor's charging current equation, $i_C = C(\Delta V/\Delta t)$ (equation (19.2)). Note that at the first instant of time, t_0, the denominator of equation (19.2) is zero; thus, i_C is maximum. Also, at this time each plate of the capacitor has the same number of electrons, the potential difference is zero, and the capacitor is acting as a short. If the current is maximum at the same time the voltage is zero, then this implies a 90° phase shift has occurred. For an AC sinusoidal circuit, the 90° phase shift continues as illustrated in Figure 22.1.

Because the plates of the capacitor are changing polarity at the same rate as the AC voltage, the capacitor seems to pass an alternating current. Actually, the electrons do not pass through the dielectric, but their rushing back and forth from plate to plate constitutes a current flow in the circuit. It is just convenient to say that the alternating current flows "through" the capacitor. By the same token, you may say that the capacitor does not pass a direct current (if both plates are connected to a DC source, current flows only long enough to charge the capacitor). In a circuit containing both DC and AC, only the AC is passed on to another circuit. The DC voltage remains across the capacitor (as a stored voltage). This DC voltage can be measured with a voltmeter. (An example of this is the coupling capacitor between amplifier stages. Just the amplified signal is passed on to the next stage. This is illustrated in Figure 22.2.)

FIGURE 22.2
The coupling capacitor. The DC reference of the signal from stage 1 is blocked by the capacitor. Stage 2 sees just the AC signal.

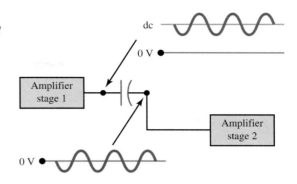

SECTION 22.1
SELF-CHECK

Answers are given at the end of the chapter.

1. At the first instant of time, the capacitor acts as a _____.
2. In a purely capacitive AC circuit, the current _____ the voltage by 90°.
3. A capacitor _____ DC and _____ AC.

22.2 CAPACITIVE REACTANCE (X_C)

So far you have been dealing with the capacitor as a device that passes AC and in which the only opposition to an alternating current has been the normal circuit resistance present in any conductor. However, capacitors themselves offer a very real opposition to current flow. This opposition arises from the fact that, at a given voltage and frequency, the number of electrons (current) that go back and forth from plate to plate is limited by the storage ability—that is, the capacitance of the capacitor. As the capacitance increases (the *RC* time constant increases), there is insufficient time for the capacitor to be charged more than a small amount before the current reverses and the capacitor discharges. Current flows very easily when the capacitor is near its discharged state, as there is very little opposition from the series-opposing voltage across the capacitor during this time. (Refer to Curve *A* versus Curve *B* in Figure 21.4. Anytime Curve *A* (v_C) is at a smaller value, the capacitor is barely charged, whereas Curve *B* (i_C) is at a higher value.

Increasing the frequency also decreases the opposition offered by a capacitor. This occurs because the allotted charge time is decreasing with increasing frequency. Thus, the capacitor remains near its discharged state. As a result, there is very little opposition from the series-opposing voltage across the capacitor during this time. The opposition a capacitor offers to AC is, therefore, inversely proportional to both frequency and capacitance. This opposition is called *capacitive reactance*. We say that capacitive reactance decreases with increasing frequency or, for a given frequency, the capacitive reactance decreases with increasing capacitance. The symbol for capacitive reactance is X_C, and the unit of measurement is the ohm (Ω).

Note that this opposition is characterized as a reactance, *not a resistance*. This is because the opposition is manifested only by an alternating current. The capacitor *reacts* to a change in voltage—hence, the term *reactance*. (Remember the capacitor blocks DC and tends to pass AC.)

The equation used to calculate capacitive reactance is

$$X_C = \frac{1}{2\pi fC} \tag{22.1}$$

where X_C = capacitive reactance (Ω)
f = frequency (Hz)
C = capacitance (F)
2π = constant, for circular motion

Lab Reference: Calculations and measurements involving X_C are demonstrated in Exercise 31.

EXAMPLE 22.1

Determine the capacitive reactance of a 10-μF capacitor when the frequency is 1 kHz.

Solution $X_C = \dfrac{1}{2\pi fC}$

$\qquad = \dfrac{1}{2\pi \times 1 \times 10^3 \text{ Hz} \times 10 \times 10^{-6} \text{ F}}$

$\qquad = 15.9 \; \Omega$

EXAMPLE 22.2

Determine the capacitive reactance of a 0.001-μF capacitor when the frequency is 4 kHz.

Solution $X_C = \dfrac{1}{2\pi fC}$

$\qquad = \dfrac{1}{2\pi \times 4 \times 10^3 \text{ Hz} \times 0.001 \times 10^{-6} \text{ F}}$

$\qquad = 39.8 \; \text{k}\Omega$

Practice Problems 22.1   Answers are given at the end of the chapter.
1. Calculate X_C when $C = 0.01 \; \mu$F and the frequency is 1 MHz.
2. Calculate X_C when $C = 2200 \; \mu$F and the frequency is 60 Hz.

22.2.1   Calculating Capacitance from X_C

In some applications, it may be necessary to find the value of capacitance required for a desired amount of X_C. By modifying equation (22.1) we have

$$C = \frac{1}{2\pi f X_C} \qquad\qquad\qquad (22.2)$$

EXAMPLE 22.3

What C is needed for an X_C of 12 kΩ at a frequency of 100 kHz?

Solution $C = \dfrac{1}{2\pi f X_C}$

$\qquad = \dfrac{1}{2\pi \times 100 \times 10^3 \text{ Hz} \times 12 \times 10^3 \; \Omega}$

$\qquad = 132.6 \; \text{pF}$

Practice Problems 22.2   Answers are given at the end of the chapter.
1. What C is needed for an X_C of 100 kΩ at a frequency of 1 kHz?
2. What C is needed for an X_C of 500 Ω at a frequency of 50 kHz?

22.2.2   Calculating Frequency from X_C

To find the frequency at which a capacitor has a specified amount of X_C, we modify equation (22.1):

$$f = \frac{1}{2\pi C X_C} \qquad\qquad\qquad (22.3)$$

At what frequency does a 0.01-μF capacitor have an X_C of 800 Ω?

Solution $f = \dfrac{1}{2\pi C X_C}$

$= \dfrac{1}{2\pi \times 0.01 \times 10^{-6}\,\text{F} \times 800\,\Omega}$

$= 19.9\ \text{kHz}$

How much current flows in the circuit of Example 22.4 if the source is 120 V AC?

Solution $I_{\text{rms}} = \dfrac{V_C}{X_C}$

$= \dfrac{120\ \text{V}}{800\ \Omega}$

$= 150\ \text{mA}$

Practice Problems 22.3   Answers are given at the end of the chapter.
1. At what frequency does a 0.022-μF capacitor have an X_C of 1000 Ω?

**SECTION 22.2
SELF-CHECK**

Answers are given at the end of the chapter.

1. Define the term *capacitive reactance*.
2. Why is X_C measured in ohms?
3. What is the proportionality of f and C to X_C?

22.3   SERIES- OR PARALLEL-CONNECTED X_C

22.3.1   Series-Connected X_C

Because capacitive reactance is an opposition expressed in ohms, series- or parallel-connected reactances are combined the same way as resistances. As shown in Figure 22.3, series capacitive reactances are added arithmetically. Thus,

$$X_{C_T} = X_{C_1} + X_{C_2} + \cdots + X_{C_N} \tag{22.4}$$

FIGURE 22.3
Finding X_{C_T} for series-connected reactances.

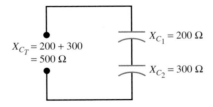

$X_{C_T} = 200 + 300$
$= 500\ \Omega$

$X_{C_1} = 200\ \Omega$

$X_{C_2} = 300\ \Omega$

22.3.2   Parallel-Connected X_C

For parallel-connected reactances, the equivalent reactance is calculated by the reciprocal formula (the same as for resistances) as shown in Figure 22.4. Thus,

$$X_{C_{EQ}} = \dfrac{1}{1/X_{C_1} + 1/X_{C_2} + \cdots + 1/X_{C_N}} \tag{22.5}$$

FIGURE 22.4
Finding $X_{C_{EQ}}$ for parallel-connected reactances.

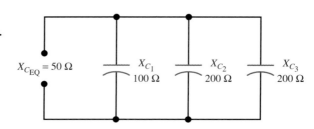

Answers are given at the end of the chapter.

1. When capacitors are connected in series, the individual reactances are _____ in order to find the total reactance.
2. When several capacitive reactances are connected in parallel, the total reactance is found the same way as _____ in parallel.

22.4 AC AND THE INDUCTOR

A circuit that has pure resistance has the alternating current through it and the voltage across it *rising and falling together.* This is illustrated in Figure 22.5(a), which shows the sine waves for current and voltage in a purely resistive circuit having an AC source. The current and voltage do not have the same amplitude, but they are *in phase.*

In the case of a circuit having *pure inductance,* the opposing force of the counter-emf is enough to keep the current from remaining in phase with the applied voltage. You learned that in a DC circuit containing pure inductance, the current took time to rise to maximum even though the full applied voltage was maximum immediately.

With an AC voltage, in the first quarter-cycle (0° to 90°), the voltage is continually increasing (Figure 22.5(b)). If there were no inductance in the circuit, the current would also increase during the first quarter-cycle, as in the purely resistive circuit of part (a). Because inductance opposes any change in current flow, no current flows during the first quarter-cycle. In the next quarter-cycle (90° to 180°), the voltage again decreases to zero; current begins to flow in the circuit and reaches a maximum value at the same instant the voltage reaches zero (Figure 22.5(c)). The applied voltage now begins to build up to the maximum in the other direction, to be followed by the resulting current. When the voltage again reaches its maximum at the end of the third cycle (270°), all values are exactly opposite those of the first half-cycle (Figure 22.5(d)). The applied voltage leads the resulting current by one quarter-cycle or 90°. To complete the full 360° cycle of the voltage, the voltage again decreases to zero and the current builds to a maximum value (Figure 22.5(e)).

22.4.1 The Inductor's 90° Phase Shift

From an examination of the voltage and current waveforms in Figure 22.5(e) we can see that *the current through the inductor always lags the voltage across the inductor by* 90°. (Stating it another way, the inductor voltage leads the inductor current by 90°.)

Another way to see the 90° phase shift is to examine the inductor's self-induced voltage equation, $V_{ind} = L\Delta i/\Delta t$ (equation (20.1)). At time t_0, the denominator of equation (20.1) is zero; thus, the inductor voltage is maximum. Because the inductor voltage is a counter-emf, it opposes the source; thus the current is zero. The inductor is acting as an open. If the voltage is maximum at the same instant the current is zero, then this implies a 90° phase shift has occurred. For an AC sinusoidal circuit, the 90° phase shift continues as illustrated in Figure 22.5.

FIGURE 22.5
Voltage and current waveforms in an inductive circuit:
(a) pure resistance circuit, V and I in phase; (b) induc-
tive circuit, voltage rises to maximum at 90° and no
current flows from 0–90°; (c) inductive circuit, voltage
decreases, current starts increasing; (d) inductive circuit,
voltage increasing to maximum (−) at 270°, current
decreasing; (e) inductive circuit, voltage decreasing to
zero, current increasing.

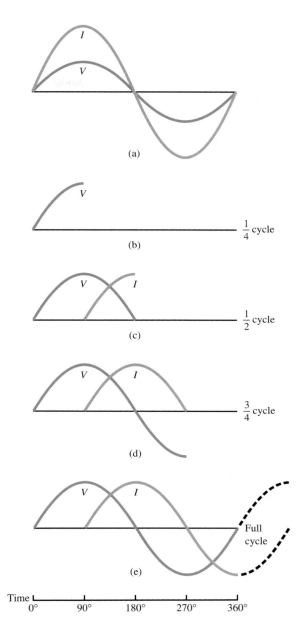

22.4.2 ELI the ICEman

Capacitors and inductors operate essentially in the same way, in that they both store energy and then return it to the circuit. However, they have completely opposite reactions to voltage and current. To help you remember the phase relationships between voltage and current for both capacitors and inductors you may wish to use the following memory aid:

ELI the ICEman

This phrase states that voltage (symbol E) leads current (I) in an inductive (L) circuit (abbreviated as ELI). In addition, current (I) leads voltage (E) in a capacitive (C) circuit (abbreviated as ICE). If the circuit is purely inductive or capacitive, the phase angle between voltage and current is always 90°.

SECTION 22.4 SELF-CHECK

Answers are given at the end of the chapter.

1. In a purely inductive circuit, _____ leads _____ by 90°.
2. At the first instant of time, the inductor acts like a(n) _____.

22.5 INDUCTIVE REACTANCE (X_L)

When the current flowing through an inductor continuously reverses itself, as in the case of an AC source, the inertia effect of the CEMF is greater than with DC. The greater the amount of inductance (L), the greater the opposition from this inertia effect. Also, the faster the reversal of current, the greater this inertial opposition. The opposing force that an inductor presents to the flow of alternating current cannot be called resistance, because it is not the result of friction within a conductor. The name given to it is *inductive reactance,* because of the reaction of the inductor to a changing alternating current. Inductive reactance is measured in ohms (Ω), and its symbol is X_L.

Lab Reference: Calculations and measurements involving X_L are demonstrated in Exercise 32.

As you know, the induced voltage in a conductor is proportional to the rate at which magnetic lines of force cut the conductor. The greater the rate (the higher the frequency), the greater the CEMF. Also, the induced voltage increases with an increase in inductance; the more ampere-turns, the greater the CEMF. Inductive reactance, then, increases with an increase of frequency and with an increase of inductance. Inductors tend to pass DC and block AC. The equation for inductive reactance is as follows:

$$X_L = 2\pi f L \tag{22.6}$$

where X_L = inductive reactance (Ω)
  f = frequency (Hz)
  L = inductance (H)
  2π = constant, for circular motion

EXAMPLE 22.6

Determine the inductive reactance of a 20-H coil when the frequency is 60 Hz.

Solution $X_L = 2\pi f L$
  $= 2\pi \times 60 \text{ Hz} \times 20 \text{ H}$
  $= 7.54 \text{ k}\Omega$

EXAMPLE 22.7

Determine the inductive reactance of a 20-mH coil when the frequency is 100 MHz.

Solution $X_L = 2\pi f L$
  $= 2\pi \times 100 \times 10^6 \text{ Hz} \times 20 \times 10^{-3} \text{ H}$
  $= 12.6 \text{ M}\Omega$

EXAMPLE 22.8

Determine the inductor current if the source is 120 V AC for the problem in Example 22.6.

Solution $I_{rms} = \dfrac{V_L}{X_L}$

  $= \dfrac{120 \text{ V}}{7.54 \times 10^3 \ \Omega}$

  $= 15.9 \text{ mA}$

Practice Problems 22.4 Answers are given at the end of the chapter.
1. Calculate X_L when $L = 100$ mH and $f = 20$ kHz.
2. Calculate X_L when $L = 8$ H and $f = 400$ Hz.
3. Calculate X_L when $L = 5\ \mu$H and $f = 100$ MHz.

22.5.1 Calculating Inductance from X_L

Very often X_L can be determined by Ohm's law. Then, if the frequency is known, L can be calculated as follows:

$$L = \frac{X_L}{2\pi f} \tag{22.7}$$

EXAMPLE 22.9

If the source voltage is 24 V AC and the inductor current is 10 mA with a frequency of 15 kHz, determine the value of L for the circuit.

Solution $X_L = \dfrac{V_L}{I_{rms}}$

$\qquad = \dfrac{24\text{ V}}{10 \times 10^{-3}\text{ A}}$

$\qquad = 2.4\text{ k}\Omega$

$\qquad L = \dfrac{X_L}{2\pi f}$

$\qquad = \dfrac{2.4 \times 10^3\ \Omega}{2\pi \times 15 \times 10^3\text{ Hz}}$

$\qquad = 25.5\text{ mH}$

Practice Problems 22.5 Answers are given at the end of the chapter.
1. Determine L when $f = 22$ kHz and X_L is 100 Ω.
2. Determine L when $f = 900$ MHz and X_L is 100 kΩ.

22.5.2 Calculating Frequency from X_L

The frequency can be solved by modifying equation (22.6) as follows:

$$f = \frac{X_L}{2\pi L} \tag{22.8}$$

EXAMPLE 22.10

At what frequency does an inductance of 1 H have a reactance of 2000 Ω?

Solution $f = \dfrac{X_L}{2\pi L}$

$\qquad = \dfrac{2000\ \Omega}{2\pi \times 1\text{ H}}$

$\qquad = 318\text{ Hz}$

Practice Problems 22.6 Answers are given at the end of the chapter.
1. At what frequency does an inductance of 500 mH have an X_L of 5 kΩ?
2. At what frequency does an inductance of 45 μH have an X_L of 20 kΩ?

22.6 SERIES- OR PARALLEL-CONNECTED X_L

22.6.1 Series-Connected X_L

Because inductive reactance is an opposition in ohms, the values of X_L are combined the same way as resistances. With series-connected inductive reactances, the total is the sum of the individual reactances, as shown in Figure 22.6. The equation for this is (assuming no mutual inductance)

$$X_{L_T} = X_{L_1} + X_{L_2} + \cdots + X_{L_N} \tag{22.9}$$

FIGURE 22.6
Combining series-connected X_L.

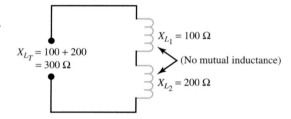

22.6.2 Parallel-Connected X_L

For the case of parallel-connected inductive reactances, the combined reactance is calculated by the reciprocal formula. As shown in Figure 22.7, the combined reactance is (assuming no mutual inductance)

$$X_{L_{EQ}} = \frac{1}{1/X_{L_1} + 1/X_{L_2} + \cdots + 1/X_{L_N}} \tag{22.10}$$

FIGURE 22.7
Combining parallel-connected X_L.

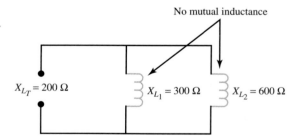

22.7 THE QUALITY FACTOR, Q

We have seen that a coil has a *figure of merit* called the *quality factor* (Q). This is an indication of the coil's ability to store energy (and later release it) rather than dissipate it as heat. It is also a function of how inductive a coil is, compared to its internal resistance. The greater the internal resistance, the less ability the coil has to produce induced voltage. More fundamentally, Q can be defined as the ratio of reactive power to real power dissipated in the coil's resistance. Thus,

$$Q = \frac{P_L}{P_{R_i}} = \frac{I^2 X_L}{I^2 R_i} = \frac{X_L}{R_i}$$

$$Q = \frac{X_L}{R_i} \tag{22.11}$$

EXAMPLE 22.11

A coil has an X_L of 800 Ω and an internal resistance of 12 Ω. Calculate the value of Q for this coil.

Solution $Q = \dfrac{X_L}{R_i}$

$\qquad = \dfrac{800 \ \Omega}{12 \ \Omega}$

$\qquad = 66.7$

As you can see, Q is a unitless quantity. In this example, Q is simply the ratio of 66.7:1. Thus, for this coil, at a particular frequency, X_L is 66.7 times greater than R_i. *As a general rule of thumb, a Q of 10 or more is considered high.* The chapter on resonance (Chapter 25) shows practical uses for varying values of Q for a coil.

Practice Problems 22.7   Answers are given at the end of the chapter.
1. Determine Q when X_L is 1 kΩ and R_i is 50 Ω.

22.7.1 Effective Resistance

At low frequencies, R_i is just the DC resistance of the conductor in the coil. As the operating frequency increases, radio-frequency (RF) coils exhibit losses greater than those from just conductor resistance. At higher frequencies, the effects of eddy currents, hysteresis, and other losses become more apparent. The losses that occur at a high frequency are known as AC resistance in order to differentiate them from DC conductor losses.

We have already seen the effects caused by eddy currents and hysteresis in AC circuits. The phenomenon known as *skin effect* is one type of loss that occurs at higher frequencies. Skin effect was first characterized in Chapter 20 as having to do with inductor losses. Skin effect reduces the conductor's cross-sectional area, thus increasing resistance. In many high-frequency applications, the transmission line (conductor) is hollow (and, as we saw in Chapter 20, the same holds for coils), because little current flows in the center anyway.

We can say that, collectively, hysteresis, eddy currents, and skin effects make up the AC resistance of the coil. When this is added to the DC resistance, we have a new value that can be called the *effective resistance, R_e*. At higher frequencies, then

$$Q = \frac{X_L}{R_e} \tag{22.12}$$

EXAMPLE 22.12

Determine the Q of a coil when X_L is 500 Ω and the effective resistance from several losses is a total of 50 Ω.

Solution $Q = \dfrac{X_L}{R_e}$

$\qquad = \dfrac{500\ \Omega}{50\ \Omega}$

$\qquad = 10$

Reducing Effective Resistance

Previously, we saw how to reduce the effects of hysteresis and eddy currents. This included the removal of the magnetic core material and its replacement with an air core. Air-core coils on a cardboard form are quite common at higher frequencies. To reduce the impact of skin effect, stranded wire can be made with separate strands insulated from each other and braided so that each strand is as much as the outer surface as all the other strands. This wire is called *Litz wire.*

As the frequency is raised from a low value, the Q tends to increase as well. This is because X_L is directly proportional to the frequency. However, at higher frequencies (RF), the Q tends to remain constant with frequency. This is because R_e increases with frequency at approximately the same rate that X_L increases, maintaining an almost constant ratio for Q.

SECTION 22.7 SELF-CHECK

Answers are given at the end of the chapter.

1. At low frequencies, the only loss in an inductor is due to _____.
2. Describe *skin effect* loss.
3. What makes up effective resistance?

■ SUMMARY

1. In a purely capacitive AC circuit, the current leads the voltage by 90°.

2. At the first instant of time, the capacitor acts as a short.

3. Electrons do not pass through the dielectric of the capacitor.

4. For an AC circuit, we can say that a capacitor passes current.

5. Between stages of an amplifier, the capacitor blocks the DC component while passing the AC amplified signal.

6. The opposition a capacitor offers to AC is called capacitive reactance (X_C) and is measured in ohms (Ω).

7. X_C is inversely proportional to both f and C.

8. Series- or parallel-connected capacitive reactances are combined the same way as resistances.

9. In a circuit with pure resistance, current and voltage are in phase.

10. Current in an inductor always lags the voltage across the inductor by 90°.

11. The current always lags the voltage in an inductor because any change in current creates a counter-emf that opposes the change in current.

12. At the first instant of time, the inductor acts as an open.

13. X_L is directly proportional to both f and L.

14. Series- or parallel-connected inductive reactances are combined the same way as resistances.

15. A coil has a figure of merit called the quality factor (Q).

16. Q can be defined as the ratio of reactive power to real power dissipated in the coil's resistance.

17. A Q greater than 10 is considered high.

18. A coil's AC resistance is a compilation of the effects of eddy currents, hysteresis, and skin effect.

19. Skin effect is a loss that effectively reduces a conductor's cross-sectional area and thereby increases its resistance.

20. At higher frequencies the Q tends to remain constant with frequency.

21. Litz wire is often used to make coils that have reduced skin effects.

■ IMPORTANT EQUATIONS

$$X_C = \frac{1}{2\pi f C} \tag{22.1}$$

$$C = \frac{1}{2\pi f X_C} \tag{22.2}$$

$$f = \frac{1}{2\pi C X_C} \tag{22.3}$$

$$X_{C_T} = X_{C_1} + X_{C_2} + \cdots + X_{C_N} \tag{22.4}$$

$$X_{C_{EQ}} = \frac{1}{1/X_{C_1} + 1/X_{C_2} + \cdots + 1/X_{C_N}} \tag{22.5}$$

$$X_L = 2\pi f L \tag{22.6}$$

$$L = \frac{X_L}{2\pi f} \tag{22.7}$$

$$f = \frac{X_L}{2\pi L} \tag{22.8}$$

$$X_{L_T} = X_{L_1} + X_{L_2} + \cdots + X_{L_N} \tag{22.9}$$

$$X_{L_{EQ}} = \frac{1}{1/X_{L_1} + 1/X_{L_2} + \cdots + 1/X_{L_N}} \tag{22.10}$$

$$Q = \frac{X_L}{R_i} \tag{22.11}$$

$$Q = \frac{X_L}{R_e} \tag{22.12}$$

■ REVIEW QUESTIONS

1. Define the term *capacitive reactance*.
2. Why is X_C measured in ohms?
3. Describe the phase shift in a purely capacitive circuit.
4. Does a capacitor actually pass current flow through it? Why or why not?
5. At the first instant of time a charging capacitor acts as a _____.
6. A capacitor passes _____ and blocks _____.
7. Name an application where an AC signal has a DC component that would need to be blocked.
8. Why is X_C denoted as a reactance rather than a resistance, because it is measured in ohms?
9. What affects the value of X_C?
10. How are capacitive reactances combined when they are series- or parallel-connected?
11. The current and voltage in a purely resistive circuit are _____ _____.
12. Define the term *inductive reactance*.
13. Describe the phase shift in a purely inductive circuit.
14. At the first instant of time the inductor acts as a(n) _____.
15. Inductors tend to pass _____ and block _____.
16. What affects the value of X_L?
17. How are inductive reactances combined when they are series- or parallel-connected?

18. What is the quality factor (Q)?
19. What parameters affect the value of Q?
20. Describe the term *effective resistance* as it relates to an inductor.
21. What is *skin effect*?
22. What can be done to reduce skin effects?
23. Why does Q remain constant with an increase in frequency?

■ CRITICAL THINKING

1. Describe in detail the phase shift that occurs in a capacitor with an AC voltage applied.
2. Why does a charging capacitor act as a short at the first instant of time?
3. Describe in detail the phase shift that occurs in an inductor with an AC voltage applied.
4. Why does an inductor act as an open at the first instant of time?
5. Describe the makeup of the term AC resistance as it relates to an inductor.

■ PROBLEMS

1. Calculate X_C for a 50-pF capacitor at the following frequencies:
   (a) 1 kHz (b) 10 kHz (c) 100 kHz (d) 500 kHz (e) 1 MHz
2. Calculate X_C for a 0.02-μF capacitor at the following frequencies:
   (a) 1 kHz (b) 10 kHz (c) 100 kHz (d) 500 kHz (e) 1 MHz
3. Calculate X_C for a 4.7-μF capacitor at the following frequencies:
   (a) 1 kHz (b) 10 kHz (c) 100 kHz (d) 500 kHz (e) 1 MHz
4. Calculate X_C for a 1-pF capacitor at the following frequencies:
   (a) 1 MHz (b) 500 MHz (c) 1 GHz (d) 5 GHz (e) 10 GHz
5. What is the capacitive reactance of any capacitor at 0 Hz (DC)?
6. For the circuit in Figure 22.8, solve for the unknowns listed.

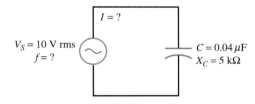

FIGURE 22.8
Circuit for Problem 6.

7. For the circuit in Figure 22.9, solve for the unknowns listed.

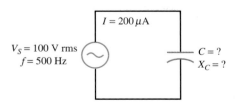

FIGURE 22.9
Circuit for Problem 7.

8. For the circuit in Figure 22.10, solve for the unknowns listed.

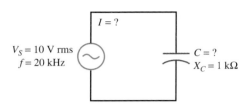

FIGURE 22.10
Circuit for Problem 8.

9. Calculate X_L for a 100-mH coil at the following frequencies.
   (a) 2 kHz (b) 30 kHz (c) 90 kHz (d) 250 kHz (e) 500 kHz

10. Calculate the X_L of a 2.5-H coil at the following frequencies.
    (a) 2 kHz (b) 30 kHz (c) 100 kHz (d) 250 kHz (e) 500 kHz

11. Calculate X_L for a 33-μH coil at the following frequencies.
    (a) 200 Hz (b) 2 kHz (c) 500 kHz (d) 100 MHz (e) 1 GHz

12. What is the inductive reactance of any coil at a frequency of 0 Hz (DC)?

13. For the circuit in Figure 22.11, solve for the unknowns listed.

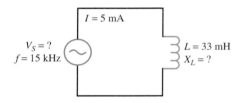

FIGURE 22.11
Circuit for Problem 13.

14. For the circuit in Figure 22.12, solve for the unknowns listed.

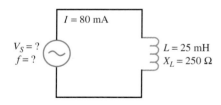

FIGURE 22.12
Circuit for Problem 14.

15. For the circuit in Figure 22.13, solve for the unknowns listed.

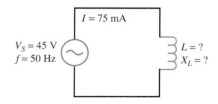

FIGURE 22.13
Circuit for Problem 15.

16. Determine Q when $X_L = 200\ \Omega$ and $R_i = 8\ \Omega$.
17. Determine X_L when $Q = 50$ and $R_i = 15\ \Omega$.
18. Determine R_i when $Q = 33$ and $X_L = 1.5$ kΩ.
19. Determine Q when $X_L = 1.8$ kΩ and $R_e = 15.7\ \Omega$.
20. Determine Q when R_e is made up of cumulative losses of 5 Ω, 3.3 Ω, and 12 Ω and $X_L = 2.2$ kΩ.

■ ANSWERS TO PRACTICE PROBLEMS

PRACTICE PROBLEMS 22.1

1. $X_C = 15.9\ \Omega$
2. $X_C = 1.2\ \Omega$

PRACTICE PROBLEMS 22.2

1. $C = 1.59$ nF, or 0.0016 μF
2. $C = 6.37$ nF, or 0.0064 μF

PRACTICE PROBLEMS 22.3

1. $f = 7.23$ kHz

PRACTICE PROBLEMS 22.4

1. $X_L = 12.57$ kΩ
2. $X_L = 20.1$ kΩ
3. $X_L = 3.14$ kΩ

PRACTICE PROBLEMS 22.5

1. $L = 723\ \mu$H
2. $L = 17.68\ \mu$H

PRACTICE PROBLEMS 22.6

1. $f = 1.59$ kHz
2. $f = 70.7$ MHz

PRACTICE PROBLEMS 22.7

1. $Q = 20$

■ ANSWERS TO SECTION SELF-CHECKS

SECTION 22.1 SELF-CHECK

1. Short
2. Leads
3. Blocks, passes

SECTION 22.2 SELF-CHECK

1. The opposition to an AC current offered by a capacitor is called capacitive reactance.
2. Because it limits current flow just as resistance does
3. X_C is inversely proportional to both f and C.

SECTION 22.3 SELF-CHECK

1. Added
2. Resistors

SECTION 22.4 SELF-CHECK

1. Voltage, current
2. Open

SECTION 22.5 SELF-CHECK

1. Directly
2. True
3. Ohms (Ω)

SECTION 22.6 SELF-CHECK

1. True
2. Reciprocal

SECTION 22.7 SELF-CHECK

1. The DC resistance of the conductor
2. It is the reduction of a conductor's cross-sectional area due to the fact that electrons are forced to flow on the outer skin of the conductor.
3. Any DC resistance plus losses due to eddy current, hysteresis, and skin effect

CHAPTER 23

COMPLEX NUMBERS FOR AC ANALYSIS

OBJECTIVES

After completing this chapter, you will be able to:

1. Describe complex numbers and show their use in mathematics and in electronics.
2. Know the rules for mathematical operations with complex numbers.
3. Differentiate between rectangular and polar forms of complex numbers.
4. Define the term *impedance* and calculate its value from either rectangular or polar forms.
5. Use the Pythagorean theorem to solve for unknowns in right triangles.
6. Convert complex numbers from rectangular form to polar form and vice versa.

INTRODUCTION

In DC circuits we were concerned only with the magnitude of various electrical quantities (volts, amps, ohms, and the like). In fact, in DC circuits, electrical quantities have *only* magnitude. There is no associated phase angle or phase shift with their quantities. A quantity that can be expressed with just a magnitude is known as a *scalar quantity.*

Complex numbers make up a numerical system that can be used to describe the magnitude and a phase angle for various quantities. This makes complex numbers very useful in the analysis of AC circuits containing capacitance or inductance along with resistance. *Impedance* is the total opposition to current flow in an AC circuit. It can be purely resistive, purely reactive, or a combination of both. Because there is an associated phase shift between the current and voltage of reactive quantities, their electrical quantities need to be expressed with both a magnitude and a phase angle.

We used the term phasor to describe various AC quantities in Chapter 16. A *phasor* (also called a vector) is a quantity that has both a magnitude and a direction (phase angle) made with respect to a reference. It is technically more correct to describe an electrical quantity as a phasor. This is because a phasor has both magnitude and direction, and it also represents a quantity that varies in magnitude with *time* (such as alternating voltage or current). A vector also has both magnitude and direction, but it is more properly used to represent physical quantities, such as force or velocity. Because impedance, resistance, and reactance do not vary with respect to time, they are often represented by vector diagrams. In either case, the *length* of an arrow represents its magnitude, whereas the *direction* of the arrow (with respect to a reference) represents its phase angle.

Our study of AC circuits utilizes complex numbers in two forms. The *polar form* is just a phasor in which the magnitude and phase angle of a complex number are expressed in angular form with polar coordinates. An example of the polar form is $7.07\angle45°$. The other form is known as the *rectangular* form and is based on the Gaussian plane, where a real number is expressed on the x-axis and an imaginary number is expressed on the y-axis. (This is also called a Cartesian coordinate system.) Rather than use *i* for an imaginary number, in electronics circuit analysis the letter *j* (known as the *j operator*) is used instead, in order to avoid confusion with *i* being used as the symbol for current. The rectangular form equivalent of the polar quantity $7.07\angle45°$ is $5 + j5$. For most applications, we generally work with the polar form. Both forms are shown graphically in Figure 23.1.

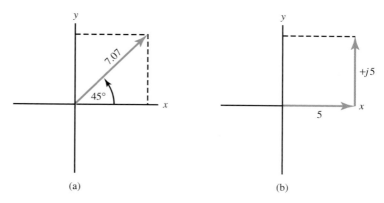

FIGURE 23.1
Complex number representation: (a) polar form; (b) rectangular form.

Further details about working with complex numbers in both rectangular and polar form are given in this chapter. In addition, their association to electrical quantities is also described.

23.1   THE REAL-NUMBER LINE

The real-number line, which shows positive and negative numbers (positive to the right of zero and negative to the left of zero) indicates more than the relative magnitude of a number. It also shows the phase relationships between positive and negative numbers. Positive numbers are represented to the right along a line. The associated phase angle is 0°. On the other hand, negative numbers are represented to the left along a line. The associated phase angle is 180°. This is illustrated in Figure 23.2. Note that 0° and 180° are diametrical opposites.

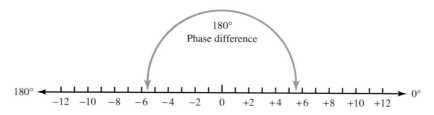

FIGURE 23.2
The real-number line showing positive and negative numbers.

Answers are given at the end of the chapter.

1. Positive numbers are represented to the _____ on the real-number line.
2. Negative numbers are represented to the _____ on the real-number line.
3. Positive and negative numbers on the real-number line are _____° apart.

23.2   THE *J* OPERATOR

The *j operator* represents an *imaginary unit* and is defined as $j = \sqrt{-1}$. It is important to understand that j is just a symbol for $\sqrt{-1}$. We write j instead of $\sqrt{-1}$ because it is easier to work with.

In mathematics, the square root of a negative number is called an *imaginary* number. The j operator allows us to simplify imaginary numbers by separating j: $\sqrt{-4} = \sqrt{-1}\sqrt{4} = j\sqrt{4} = j2$. Note that $-j$ is not the same as j. They are as different as -1 and 1.

In order to multiply and divide imaginary numbers, we must first separate the j operator. The rules for multiplication and division of radicals do not apply here. The following example shows how to work with the j operator.

Simplify $(\sqrt{-9})(\sqrt{-16})$:

First separate the j operator:

$$(\sqrt{-9})(\sqrt{-16}) = (j\sqrt{9})(j\sqrt{16})$$

Then multiply the j's and simplify the radicals:

$$(j\sqrt{9})(j\sqrt{16}) = j^2(3)(4)$$

Replace j^2 with -1:

$$j^2(3)(4) = (-1)(3)(4) = -12$$

The final number is a real number and has no j operator. Note that if you try to use the usual rules for radicals, before separating out the j's, you would get $\sqrt{(-9)(-16)} = \sqrt{144} = 12$, which is not the correct answer.

23.2.1   Multiplication by the *j* Operator

When you multiply repeatedly by the j operator, a cycle occurs that repeats after four multiplications. The pattern is

$$j^0 = +1 \rightarrow 0°$$
$$j^1 = j \rightarrow 90°$$
$$j^2 = -1 \rightarrow 180°$$
$$j^3 = j^2(j) = -j \rightarrow 270° \quad (\text{or } -90°)$$
$$j^4 = (j^2)(j^2) = (-1)(-1) = +1 \rightarrow 0°$$
$$j^5 = (j^4)(j) = +1(j) = j \rightarrow 90°$$

The j operator performs the operation of a *90° rotation* each time you multiply by it. The specific rule is that each time you multiply by j, it is the same as rotating counterclockwise by 90° on the Gaussian plane. This can be shown graphically: the horizontal axis is the real-number axis containing the numbers 1 and -1, whereas the vertical axis is the imaginary-number axis containing j and $-j$. This is illustrated in Figure 23.3.

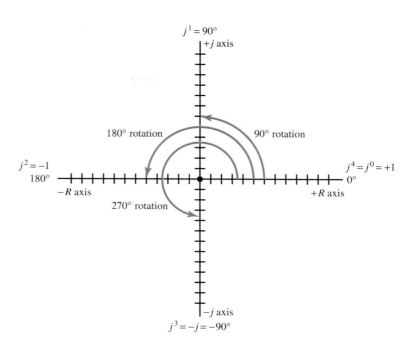

FIGURE 23.3
The j operator: a 90° ccw rotation occurs each time j is multiplied by itself.

23.2.2 Using the *j* Operator

To summarize, in mathematics the *j* operator is used to simplify imaginary numbers, whereas in electronics circuit analysis the *j* operator is used to indicate a direction for complex numbers. This is important when working with phasors in AC circuits. (Remember that both inductance and capacitance have 90° phase shifts associated with their voltages and currents.) The real numbers 4 and −4 indicate a magnitude and direction on the horizontal axis, whereas $j5$ and $-j5$ indicate a magnitude and direction on the vertical axis. Thus, $j5$ means 5 units with a leading phase angle of 90°, and $-j5$ means 5 units with a lagging phase angle of −90°.

**SECTION 23.2
SELF-CHECK**

Answers are given at the end of the chapter.

1. What does the *j* operator represent?
2. The *j* operator performs the operation of a _____° rotation each time you multiply by it.
3. How is the *j* operator used for electronics circuit analysis?

23.3 MATHEMATICAL OPERATIONS WITH COMPLEX NUMBERS

Mathematical operations can be done with complex numbers whether they are expressed in rectangular or polar form. However, real numbers and *j* terms cannot be combined directly, because they are 90° out of phase. For addition or subtraction in rectangular form, add or subtract the real and the *j* terms separately. Multiplication in rectangular form involves squaring the *j* term. Division in rectangular form involves the use of the conjugate of the denominator. The following examples demonstrate some traditional arithmetic operations for the rectangular form.

Optional Calculator Feature

Note that the examples shown represent arithmetic operations performed by the use of a conventional scientific calculator. The Sharp calculator has a complex-number mode (Mode 1). When in this mode, complex numbers in *either* form, rectangular or polar, may be added, subtracted, multiplied, or divided directly. These features are demonstrated in the next chapter.

23.3.1 Rectangular Form

Addition and Subtraction

The procedure is to combine like terms and then add or subtract the real and j terms separately:

$$(8 + j5) + (5 + j4) = 8 + 5 + j5 + j4 = 13 + j9$$
$$(8 - j5) + (5 - j4) = 8 + 5 - j5 - j4 = 13 - j9$$
$$(8 + j5) + (5 - j4) = 8 + 5 + j5 - j4 = 13 + j1$$
$$(8 + j5) + (5 - j8) = 8 + 5 + j5 - j8 = 13 - j3$$

Multiplication

Multiply by using the distributive property.

$$(8 + j5) \times (5 + j4) = 40 + j32 + j25 + j^2 20$$
$$= 40 - 20 + j32 + j25$$
$$= 20 + j57$$

Division

Division requires the use of the complex conjugate of the divisor. The complex conjugate of $5 + j4$ is $5 - j4$. Note that the use of the complex conjugate leaves the denominator with real terms only, the sum of the squares of the coefficients.

$$\frac{8 + j5}{5 + j4} = \frac{8 + j5}{5 + j4} \times \frac{5 - j4}{5 - j4}$$
$$= \frac{40 - j32 + j25 - j^2 20}{25 - j20 + j20 - j^2 16}$$
$$= \frac{40 - j7 + 20}{25 + 16}$$
$$= \frac{60 - j7}{41}$$
$$= 1.46 - j0.17$$

Note that multiplication and division require a more complete understanding of the j operator in arithmetic applications. Thus, in most cases, only addition and subtraction are done manually in rectangular form. To multiply or divide numbers in rectangular form, they are first converted to polar form, the arithmetic operation is completed, and then they are converted back to rectangular form.

23.3.2 Polar Form

The following examples demonstrate the traditional arithmetic operations in polar form on a conventional scientific calculator.

Addition and Subtraction

Addition and subtraction cannot be done directly in polar form unless the angles involved are the same. To add and subtract when the angles are not the same, the polar form must be converted to rectangular form, the addition or subtraction must be done, and then the result must be converted to polar form. (This conversion is assumed to be done on a conventional scientific calculator. Examples of direct polar to rectangular ($P \rightarrow R$) conversion are demonstrated in a later section.) Therefore, only multiplication and division are done manually in polar form.

Multiplication

The rules of multiplication and division of polar terms follow the rules for numbers with exponents (such as powers of ten). To find a product, multiply the magnitudes and add the angles algebraically.

$$16\angle 25° \times 8\angle 51° = 128\angle 76°$$
$$16\angle 25° \times 8\angle -51° = 128\angle -26°$$
$$16\angle -25° \times 8\angle -51° = 128\angle -76°$$

Division

To divide polar terms, divide the magnitudes and subtract the angles algebraically.

$$\frac{16\angle 25°}{8\angle 51°} = 2\angle -26°$$

$$\frac{16\angle 25°}{8\angle -51°} = 2\angle 76°$$

$$\frac{16\angle -25°}{8\angle -51°} = 2\angle 26°$$

**SECTION 23.3
SELF-CHECK**

Answers are given at the end of the chapter.

1. With which complex-number form is it easier to add and subtract numbers?
2. With which complex-number form is it easier to multiply and divide numbers?

23.4 IMPEDANCE

Impedance is defined as the total opposition to current flow in an AC circuit. Impedance might be just pure resistance, pure reactance, or a combination of both. The value depends on what components are in the AC circuit and the frequency of the source. Impedance usually refers to a combination of resistance and reactance. The symbol for impedance is the letter *Z,* and the unit of measurement is the ohm (Ω).

By modifying Ohm's law we have $Z = V_s/I_T$. Because the magnitude of impedance is the ratio between voltage and current, the impedance has a phase angle associated with it that is the difference between the voltage and current phase angles. In most AC circuit analyses, V_s is used as the *reference* at an angle of 0°. (Remember, a phasor's angle is always taken with respect to a reference.) Thus, impedance has a negative or positive phase angle, depending on whether the circuit is dominated by capacitance or inductance reactances.

Vector diagrams containing *R, X,* and *Z* are referred to as impedance triangles and can prove to be very useful in problem-solving AC circuits. Two examples of impedance triangles are illustrated in Figure 23.4. The diagram in (a) represents an inductance and resistance connected in series, whereas the impedance diagram in (b) represents capacitance and resistance connected in series. If a circuit contains both inductance and capacitance, the two reactances are *subtracted* from one another (yielding a net reactance), because they are both located on the vertical axis but in opposite directions.

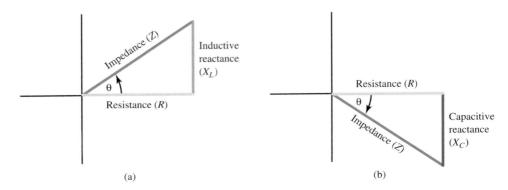

FIGURE 23.4
Impedance vector diagram: (a) resistance and inductive reactance in series; (b) resistance and capacitive reactance in series.

The natural form for impedance is in the rectangular form designated as $Z = R \pm jX_{Net}$. This is the universal form of the equation. For series inductive and resistive circuits the rectangular form for impedance is $Z = R + jX_L$, whereas for series capacitive and resistive circuits, it is $Z = R - jX_C$. In polar form, impedance is designated as $Z \angle \theta$ and is found by the conversion from rectangular to polar. The conversion from rectangular to polar ($R \rightarrow P$) is demonstrated in the sections that follow.

Note that in series-connected circuits, the total impedance may be found as $Z_T = Z_1 + Z_2 + Z_3 + \cdots + Z_N$, whereas in parallel-connected circuits the total impedance may be found as $Z_T = 1/(1/Z_1 + 1/Z_2 + 1/Z_3 + \cdots + 1/Z_N)$. This is the same process as for individual X_C and X_L series- or parallel-connected reactances. The use of Ohm's law may also be used as necessary.

**SECTION 23.4
SELF-CHECK**

Answers are given at the end of the chapter.

1. Define the term *impedance*.
2. What is used as the reference when determining the phase angle associated with impedance?
3. When a circuit contains both X_C and X_L (series-connected), the net reactance is the _____ of the two.
4. State the natural form for impedance.

23.5   PYTHAGOREAN THEOREM

The impedance triangle of a series *RL* or *RC* circuit (Figure 23.4) can be solved geometrically by using the Pythagorean theorem. The Greek Pythagoras formally utilized this theorem around 520 B.C. The theorem states that in a right triangle, the sum of the squares of the sides equals the square of the hypotenuse. Mathematically this is stated as $a^2 + b^2 = c^2$. This equation may be applied to the impedance triangle, because it also is a right triangle. Thus, in electronic circuit analysis, side *a* (adjacent side) represents *R* (resistance), side *b* (opposite side) represents *X* (either inductive or capacitive), and the hypotenuse, c, represents impedance (Z). This is illustrated in Figure 23.5. The impedance, Z, of a series circuit can be found using:

$$Z = \sqrt{R^2 + X^2} \qquad\qquad (23.1)$$

where Z = impedance of a series circuit (Ω)
   R = resistance (Ω)
   X = net reactance (Ω)

Examples utilizing this equation are shown in a later section. (Note that this equation can be modified for parallel RC or RL circuits where the focus is on currents. Then $I_T = \sqrt{I_R^2 + I_X^2}$.)

FIGURE 23.5
Right triangle and Pythagorean theorem.

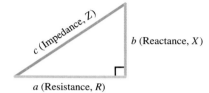

SECTION 23.5
SELF-CHECK

Answers are given at the end of the chapter.

1. State the Pythagorean theorem.
2. In electronics, what do right-triangle sides a, b, and c represent?

23.6   CALCULATING THE IMPEDANCE PHASE ANGLE

Note that in Figure 23.4, angle theta (θ) represents the phase angle for the impedance of a series circuit in polar form. This angle is a displacement from the positive x-axis and may be found using trigonometry, because the impedance triangle is a right triangle. The phase angle is found by

$$\theta = \tan^{-1}\left(\frac{X}{R}\right) \qquad\qquad (23.2)$$

Lab Reference: The measurement of phase using the oscilloscope is demonstrated in Exercise 33.

where θ = series circuit impedance phase angle (degrees)
   $\tan^{-1}$ = inverse tangent (or arctangent) function
   X = net reactance (Ω)
   R = resistance (Ω)

Note that for a series RL circuit, this angle is positive, whereas for a series RC circuit, this angle is negative. Examples utilizing this equation are shown in the next section. (Alternatively, the phase angle can also be found by $\theta = \tan^{-1}(V_X/V_R)$.)

SECTION 23.6
SELF-CHECK

Answers are given at the end of the chapter.

1. What does angle theta (θ) represent?
2. For a net RL circuit, θ is _____.
3. For a net RC circuit, θ is _____.

| 23.7 | **CONVERSION BETWEEN RECTANGULAR AND POLAR COORDINATES** |

23.7.1 Converting Rectangular to Polar ($R \rightarrow P$)

In polar coordinates, the distance out from the center is the magnitude of the phasor, Z. Its displacement above or below the positive x-axis is represented by the phase angle, θ. We have already seen how to convert from rectangular to polar form. (Equations (23.1) and (23.2) are used to do this.)

EXAMPLE 23.1

Convert the rectangular form $3 + j4$ to its polar-coordinate equivalent.

Solution
$$Z = \sqrt{R^2 + X^2}$$
$$= \sqrt{3^2 + 4^2}$$
$$= \sqrt{25}$$
$$= 5$$

Notice that the answer for Z (5) is larger than either term (R or X) but smaller than their sum. This is always the case and is a good check to verify that the problem is done correctly.

$$\theta = \tan^{-1}\left(\frac{X}{R}\right)$$
$$= \tan^{-1}\left(\frac{4}{3}\right)$$
$$= 53.1°$$

The polar form is $5\angle 53.1°$. Note that θ is greater than 45°, because X is greater than R. If R is greater than X, then θ is less than 45°.

Optional Calculator Sequence

The Sharp calculator sequence for Z in this problem is as follows.

Step	Keypad Entry	Top Display Response
1	√, (	√(
2	$3x^2$	√(3^2
3	+	√(3^2+
4	$4x^2$	√(3^2+4^2
5	)	√(3^2+4^2)
6	=	√(3^2+4^2)=
	Answer	5.00 (bottom display)

The Sharp calculator sequence for θ in this problem is as follows.

Step	Keypad Entry	Top Display Response
1	2ndF tan	$\tan^{-1}$
2	(	$\tan^{-1}$(
3	4/	$\tan^{-1}$(4/
4	3)	$\tan^{-1}$(4/3)
5	=	$\tan^{-1}$(4/3)=
	Answer	53.13 (bottom display)

EXAMPLE 23.2

Convert $8 - j6$ to its polar-coordinate equivalent.

Solution $Z = \sqrt{8^2 + (-6)^2}$
$$= \sqrt{64 + 36}$$
$$= \sqrt{100}$$
$$= 10$$
$$\theta = \tan^{-1}\left(\frac{-6}{8}\right)$$
$$= -36.9°$$

So, the polar form is $10\angle -36.9°$.

Practice Problems 23.1 Answers are given at the end of the chapter.
1. Convert $10 + j15$ to its polar-coordinate equivalent.
2. Convert $75 - j50$ to its polar-coordinate equivalent.
3. Convert $45 - j45$ to its polar-coordinate equivalent.

23.7.2 Converting Polar to Rectangular $(P \rightarrow R)$

The impedance, $Z\angle\theta$, is the hypotenuse of a right triangle with sides formed by the real term (R) and the j term (X) in rectangular coordinates. The sides R and X (horizontal and vertical) of the right triangle can be determined by trigonometry as follows:

$$R = Z \cos \theta \qquad\qquad (23.3)$$

$$X = Z \sin \theta \qquad\qquad (23.4)$$

EXAMPLE 23.3

Convert $5\angle 53.1°$ to rectangular coordinates. (*Note:* Make sure your calculator is in the **deg** mode.)

Solution $R = Z \cos \theta$
$$= 5 \cos 53.1°$$
$$= 3$$
$$X = Z \sin \theta$$
$$= 5 \sin 53.1°$$
$$= 4$$

Thus, $5\angle 53.1° = 3 + j4$.

Optional Calculator Sequence

On the Sharp calculator the value for R is found as follows. (X is solved using the same process.)

Step	Keypad Entry	Top Display Response
1	5cos	5cos
2	53.1	5cos
3	=	5cos53.1=
	Answer	3.00 (bottom display)

EXAMPLE 23.4

Convert $10\angle -36.9°$ to rectangular coordinates.

Solution $R = Z \cos \theta$
$$= 10 \cos (-36.9°)$$
$$= 8$$
$$X = Z \sin \theta$$
$$= 10 \sin (-36.9°)$$
$$= -6$$

Thus, $10\angle -36.9° = 8 - j6$.

Practice Problems 23.2   Answers are given at the end of the chapter.
1. Convert $50\angle -45°$ to rectangular coordinates.
2. Convert $200\angle 23°$ to rectangular coordinates.

**SECTION 23.7
SELF-CHECK**

Answers are given at the end of the chapter.

1. A phasor's distance out from the center represents its _____.
2. A phasor's displacement above or below the positive *x*-axis represents its _____.
3. True or false: When using the Pythagorean theorem to solve for *Z*, *Z* is always larger than either *R* or *X* but is smaller than their sum.

23.8   **CALCULATOR SHORTCUTS: AN EASIER WAY**

Many scientific calculators contain conversion keys that make going from rectangular to polar (and vice versa) extremely fast and convenient. On some calculators the conversion keys are labeled $R \rightarrow P$ and $P \rightarrow R$. (Check your calculator manual for the exact procedure.)

> ### Optional Calculator Sequence
>
> On the Sharp calculator the easiest method for converting from $R \rightarrow P$ or $P \rightarrow R$ is to simply put the calculator into the complex-number mode. This is done by pressing the **MODE** key and then entering **1** for the complex-number mode. When this is done the calculator defaults to the rectangular mode, as evidenced by the *xy* in the upper-left corner of the top display.

23.8.1 $R \rightarrow P$

To convert from rectangular to polar, the process is as follows.

EXAMPLE 23.5

> ### Optional Calculator Sequence
>
> Convert $3 + j4$ into its polar coordinate equivalent.
>
> *(continued)*

Solution From Mode 1, press **2ndF rθ,** (rθ is above the numeral 8 in yellow), which puts the calculator into the mode to which we are converting (the polar form). Now enter **3, +, i, 4,** and **=.** The value for *Z* is shown as 5. Note the → in the upper-right corner of the display. This reminds you to press the large white → key to see the phase angle displayed. The phase angle is displayed as ∠53.1. By repeatedly pressing the ← and → keys, you can toggle back and forth between the values for *Z* and ∠θ.

Practice Problems 23.3 Answers are given at the end of the chapter.
1. Using the calculator shortcut, convert 5 + *j*15 to its polar coordinate equivalent.
2. Repeat Problem 1 for 45 − *j*15.

23.8.2   P → R

To convert from polar to rectangular, the process is as follows.

EXAMPLE 23.6

Optional Calculator Sequence

Convert 5∠53.1° to its rectangular form.

Solution Make sure the calculator is in the complex-number mode. (Remember, it defaults to the rectangular form, as shown by the *xy* in the upper-left corner of the display when **Mode = 1** is selected.) Leave the calculator in the *xy* mode, because this is the form to which we are converting. Enter **5, ∠, 53.1,** and **=.** The rectangular form is displayed in two parts. The real number, *R*, is shown as 3 with → in the upper-right corner to remind you to press this key to see the reactive value, *X*, as 4. Note that the reactive part has an *i* on the right side to remind you that this is the reactive part.

Practice Problems 23.4 Answers are given at the end of the chapter.
1. Convert 150∠74° to its rectangular-coordinate equivalent.
2. Convert 30∠ −64° to its rectangular-coordinate equivalent.

SECTION 23.8 SELF-CHECK

Answers are given at the end of the chapter.
1. On the Sharp calculator, the easiest method for conversion between rectangular and polar is to put the calculator in the _____ mode.
2. On the Sharp calculator, the complex number mode is mode _____.

■ SUMMARY

1. A quantity that can be expressed with magnitude only is known as a scalar quantity.
2. Complex numbers refer to a numerical system that includes the magnitude and a phase angle for various quantities.
3. Impedance is the total opposition to current flow in an AC circuit.
4. Impedance can be purely reactive, purely resistive, or a combination of both.
5. A phasor is a quantity that has both a magnitude and a direction.

6. It is technically more correct to describe an electrical quantity as a phasor.

7. Complex numbers can be expressed in two forms.

8. The polar form of a complex number is just a phasor in which the magnitude and phase angle of a complex number are expressed in an angular form with polar coordinates.

9. The rectangular form of a complex number is based on the Gaussian plane, where a real number is shown on the *x*-axis and an imaginary number is shown on the *y*-axis.

10. In electronics, we use the letter *j* to represent an imaginary number.

11. The *j* operator represents an imaginary unit and is equal to $\sqrt{-1}$.

12. Addition and subtraction are more easily done in rectangular form.

13. Multiplication and division are more easily done in polar form.

14. The *j* operator performs the operation of a 90° rotation each time you multiply by it.

15. In electronics we use the *j* operator to indicate a direction for complex numbers.

16. Vector diagrams containing *R*, *X*, and *Z* are referred to as impedance triangles and are very useful in problem solving.

17. If a circuit contains both *L* and *C*, their reactive quantities subtract from each other, because they are both on the *y*-axis but in opposite directions.

18. The natural form for impedance is the rectangular form, where it is designated as $Z = R \pm jX_{net}$.

19. The impedance triangle of a series *RC* or *RL* circuit can be solved geometrically by the use of the Pythagorean theorem.

20. In electronics, for an impedance triangle side *a* represents *R*, side *b* represents *X*, and side *c* represents *Z*.

■ IMPORTANT EQUATIONS

$$Z = \sqrt{R^2 + X^2} \tag{23.1}$$

$$\theta = \tan^{-1}\left(\frac{X}{R}\right) \tag{23.2}$$

$$R = Z\cos\theta \tag{23.3}$$

$$X = Z\sin\theta \tag{23.4}$$

■ REVIEW QUESTIONS

1. Define the term *scalar quantity*.
2. Define the term *complex number*.
3. Define the term *impedance*.
4. What makes up impedance?
5. Define the term *phasor*.
6. What is the difference between a phasor and a vector?
7. What is the polar form of a complex number?
8. What is the rectangular form of a complex number?

9. In rectangular form, where are real and imaginary numbers represented?

10. With which form of complex numbers does a technician generally work?

11. Besides indicating a magnitude, what else does the real-number line represent?

12. Describe the *j* operator.

13. How is the *j* operator used in mathematics and in electronics?

14. Describe an imaginary number.

15. What arithmetic operations are more easily done in rectangular form?

16. What arithmetic operations are more easily done in polar form?

17. Describe the vector diagram for impedance.

18. What is the natural form for impedance?

19. State the Pythagorean theorem.

20. How is the Pythagorean theorem utilized for an impedance triangle?

■ CRITICAL THINKING

1. Compare and contrast the terms *scalar, phasor,* and *vector*.

2. Describe a complex number and in what forms it may be represented.

3. Describe in detail how the *j* operator performs a 90° rotation each time you multiply by it.

4. Why is it easier to add and subtract in rectangular form and multiply and divide in polar form?

5. Describe how the Pythagorean theorem is used with an impedance triangle.

■ PROBLEMS

Complete the following table for the unknowns listed.

Converting Rectangular to Polar Form

	$R\,(\Omega)$	$X_C\,(\Omega)$	$X_L\,(\Omega)$	$Z\,(\Omega)$	$\theta\,(°)$
1.	2	4	—		
2.	15	—	45		
3.	6	8	—		
4.	5	10	20		
5.	24	6	—		
6.	50	100	50		
7.	100	—	40		
8.	35	85	30		
9.	4	8	12		
10.	36	—	82		

Complete the following table for the unknowns listed.

Converting Polar to Rectangular Form

	$R\,(\Omega)$	$X_C\,(\Omega)$	$X_L\,(\Omega)$	$Z\,(\Omega)$	$\theta\,(°)$
11.				10	36.9°
12.				7.07	45°
13.				7.07	−45°
14.				100	−23.5°
15.				24	67°
16.				20	53.1°
17.				64	−15°
18.				52	67°
19.				212	−40°
20.				36	10°

■ ANSWERS TO PRACTICE PROBLEMS

PRACTICE PROBLEMS 23.1

1. $18\angle 56.3°$
2. $90.1\angle -33.7°$
3. $63.6\angle -45°$

PRACTICE PROBLEMS 23.2

1. $35.4 - j35.4$
2. $184.1 + j78.1$

PRACTICE PROBLEMS 23.3

1. $15.8\angle 71.6°$
2. $47.4\angle -18.4°$

PRACTICE PROBLEMS 23.4

1. $41.3 + j144.2$
2. $13.2 - j27$

■ ANSWERS TO SECTION SELF-CHECKS

SECTION 23.1 SELF-CHECK

1. Right
2. Left
3. 180

SECTION 23.2 SELF-CHECK

1. An imaginary unit ($j = \sqrt{-1}$)
2. 90
3. To indicate a direction for complex numbers

SECTION 23.3 SELF-CHECK

1. Rectangular
2. Polar

SECTION 23.4 SELF-CHECK

1. Impedance is the total opposition to current flow in an AC circuit.
2. The source voltage at $\angle 0°$
3. Difference
4. $Z = R \pm jX_{net}$

SECTION 23.5 SELF-CHECK

1. $a^2 + b^2 = c^2$
2. $a = R, b = X, c = Z$

SECTION 23.6 SELF-CHECK

1. θ represents the phase angle for the impedance of a series circuit in polar form.
2. Positive
3. Negative

SECTION 23.7 SELF-CHECK

1. Magnitude
2. Direction
3. True

SECTION 23.8 SELF-CHECK

1. Complex number
2. 1

CHAPTER 24

RCL AC CIRCUIT ANALYSIS

OBJECTIVES

After completing this chapter, you will be able to:

1. Demonstrate proficiency when working with complex numbers for circuit analysis.
2. Calculate unknowns for series AC *RC*, *RL*, and *RCL* circuits.
3. Calculate unknowns for parallel AC *RC*, *RL*, and *RCL* circuits.
4. Calculate unknowns for AC complex branch impedances.
5. Define the terms *susceptance* and *admittance*.
6. Compare and contrast true power, reactive power, and apparent power.
7. Define the power factor and describe its meaning to AC circuits.

INTRODUCTION

Generally speaking, the currents and voltages in DC circuits are constant. In *RC* and *RL* DC circuits, abrupt changes due to the signal source, or a switch, produce responses that exponentially rise or decay from an initial constant value to a final constant value. This transient condition lasts for a relatively short time, equal to approximately five time constants. Recall that the effects of capacitance and inductance were significant *only* during the relatively brief transient period—when the responses were changing.

An AC source is one that periodically reverses polarity. Thus, current flows first in one direction and then the other. Because the responses are continually changing, capacitance and inductance oppose the associated changes in AC voltage or the flow of AC current. This opposition by a capacitor or an inductor is called *reactance.*

In circuits containing both resistance and reactance, the total opposition to the flow of AC current is called *impedance.* Because resistance is represented by a real number and reactance, by an imaginary number, impedance is represented by a *complex number.* Also, because resistance, reactance, and impedance all limit the flow of AC current, each quantity is expressed in ohms (Ω). By using complex numbers, the techniques of DC circuit analysis can be extended to include the analysis of AC circuits.

The primary emphasis in this chapter is on AC circuit analysis, similar to what was done in DC circuits. Practical applications of *RCL* circuits are covered in Chapter 26. In addition to analyzing an AC circuit's ohmic, voltage, or current values, we look at AC power. We examine the concepts of real power, apparent power, and power factor.

24.1 SERIES AC CIRCUITS

Series AC circuits have the same current flowing through each component, and the voltage divides across each component. The focus of series AC circuits, then, is on the voltages, because the voltage source is equal to the phasor sum of the individual component voltages. As will be shown, Kirchhoff's voltage law (KVL) can be applied to AC circuits.

24.1.1 Series *RC* Circuits

When a capacitance and a resistance are connected in series (Figure 24.1(a)) to an AC source, the sine-wave voltage across the capacitance, V_C, is out of phase with the sine-wave voltage across the resistance, V_R. The stored charge in the capacitor tends to oppose the applied voltage, and V_C lags V_R by 90°.

FIGURE 24.1
Series RC circuit; (a) circuit; (b) impedance diagram.

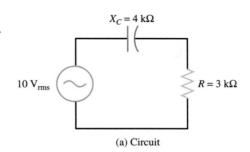

(a) Circuit

Lab Reference: Measurements and calculations for a series RC circuit are demonstrated in Exercise 34.

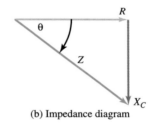

(b) Impedance diagram

In a series circuit, the current is the same throughout and is in phase with the voltage across the resistance. Therefore, the current also leads V_C by 90°. The voltage source phasor is referenced to $\angle0°$. The current and voltage phasor diagram for the series *RC* circuit of Figure 24.1 is shown in Figure 24.2.

FIGURE 24.2
Voltage and current phasor diagram for circuit in Figure 24.1(a).

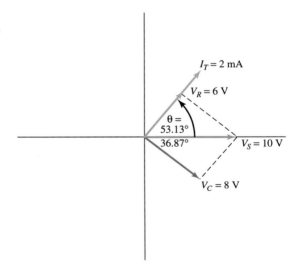

The traditional way to determine Z and θ (illustrated in the impedance diagram of Figure 24.1(b)) is through the use of the Pythagorean theorem and the trigonometric tangent function, as demonstrated in the previous chapter.

Optional Calculator Sequence

The Sharp calculator can solve AC problems using the same techniques that were learned in DC circuits by using the complex mode (Mode 1) and entering the terms in polar form. Thus, we have

$$Z = R\angle 0° + X_C \angle -90° \tag{24.1}$$

(The calculator displays θ by using the $\rightarrow$ key after Z is calculated. θ has a negative sign in a series RC circuit as it is associated with Z.)

$$I_T = \frac{V_S \angle 0°}{Z\angle \theta} \tag{24.2}$$

$$V_R = I_T \angle \theta \times R\angle 0° \tag{24.3}$$

$$V_C = I_T \angle \theta \times X_C \angle -90° \tag{24.4}$$

Note that X_C is determined from the value of capacitance and the source frequency: $X_C = 1/(2\pi f C)$. For our purposes in this chapter, X_C has already been calculated and given in ohms. From Kirchhoff's voltage law (KVL), in a series RC circuit we have

$$V_S = V_C + V_R \tag{24.5}$$

The following example illustrates these points.

EXAMPLE 24.1

Find the circuit values shown in Figure 24.1 (Z, θ) and in Figure 24.2 (I_T, V_S, V_C, V_R).

Solution With the Sharp calculator in the complex mode (Mode 1) and in the $\rightarrow r\theta$ (polar) mode:

$$\begin{aligned} Z &= R\angle 0° + X_C \angle -90° \\ &= 3\,\text{k}\Omega\angle 0° + 4\,\text{k}\Omega\angle -90° \\ &= 5\,\text{k}\Omega\angle -53.1° \end{aligned}$$

The Sharp calculator sequence for Z in this problem is as follows.

Step	Keypad Entry	Top Display Response
1	3, Exp, 3, $\angle$, 0	3E03$\angle$
2	+	3E03$\angle$0+
3	4, Exp, 3, $\angle$, −90	3E03$\angle$0+4E03$\angle$−
4	=	3E03$\angle$0+4E03$\angle$−90=$\rightarrow$
5	Answer	5.00E03 (bottom display)
6	$\rightarrow$	$\angle$−53.1 (bottom display)

The other unknowns are just as easily determined with the Sharp calculator in the complex-number mode, as follows:

$$\begin{aligned} I_T &= \frac{V_S \angle 0°}{Z\angle \theta} \\ &= \frac{10\,\text{V}\angle 0°}{5\,\text{k}\Omega\angle -53.1°} \\ &= 2\,\text{mA}\angle 53.1° \end{aligned}$$

$$V_R = I_T \angle \theta \times R \angle 0°$$
$$= 2\,\text{mA} \angle 53.1° \times 3\,\text{k}\Omega \angle 0°$$
$$= 6\,\text{V} \angle 53.1°$$
$$V_C = I_T \angle \theta \times X_C \angle -90°$$
$$= 2\,\text{mA} \angle 53.1° \times 4\,\text{k}\Omega \angle -90°$$
$$= 8\,\text{V} \angle -36.9°$$
$$V_S = V_R + V_C$$
$$= 6\,\text{V} \angle 53.1° + 8\,\text{V} \angle -36.9°$$
$$= 10\,\text{V} \angle 0°$$

(Note that using traditional computations, Kirchhoff's voltage law could be shown only by first converting the polar-form voltages into rectangular form, adding, and then converting them back to polar form or by the use of the Pythagorean theorem.)

For Example 24.1 you should observe the following:

1. The voltage across the resistor is in phase with the current flowing through it. In all circuits, resistor voltage and current are always in phase with each other.
2. The current flowing through the capacitor leads the voltage across the capacitor by 90° (ELI the *ICEman*). Because the resistor voltage is in phase with the current, the voltage across the resistor also leads the capacitor voltage by 90°.
3. The *phasor sum* of the resistor and capacitor voltages equals the source voltage. This is illustrated graphically in Figure 24.2 and mathematically by Kirchhoff's voltage law in the solution to Example 24.1.
4. An angle exists between the applied voltage, V_S, and the current, I_T. This angle is called the *circuit phase angle, θ*. This angle is associated with impedance in series *RC* or *RL* circuits. As you will see later, the circuit phase angle is important for determining how much power the source supplies to the circuit.
5. Because this is an *RC* circuit, θ is negative, and because X_C is larger than R, θ is greater than $\angle -45°$.

Practice Problems 24.1   Answers are given at the end of the chapter.
1. Determine Z, θ, I_T, V_C, and V_R for the circuit in Figure 24.3.

FIGURE 24.3
Circuit for Practice Problems 24.1.

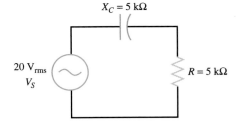

$X_C = 5\,\text{k}\Omega$

20 V$_{\text{rms}}$
V_S

$R = 5\,\text{k}\Omega$

24.1.2   Series *RL* Circuits

Replacing X_C with X_L in the equations for a series *RC* circuit modifies them for use in series *RL* circuits.

Optional Calculator Sequence

The Sharp calculator procedure for the equations is as follows:

$$Z = R \angle 0° + X_L \angle 90° \tag{24.6}$$

(The calculator will display θ if you use the $\rightarrow$ key after Z is calculated. θ has a positive sign in a series RL circuit as it is associated with Z.)

$$I_T = \frac{V_S \angle 0°}{Z \angle \theta} \tag{24.7}$$

$$V_R = I_T \angle \theta \times R \angle 0° \tag{24.8}$$

$$V_L = I_T \angle \theta \times X_L \angle 90° \tag{24.9}$$

$$V_S = V_R + V_L \tag{24.10}$$

EXAMPLE 24.2

Find Z, θ, I_T, V_R, and V_L for the circuit in Figure 24.4. Prove that KVL is true for this circuit.

FIGURE 24.4
Series RL circuit for Example 24.2.

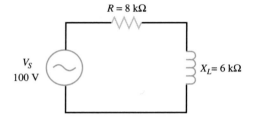

Lab Reference: Measurements and calculations for a series *RL* circuit are demonstrated in Exercise 35.

Solution

$$\begin{aligned}
Z &= R\angle 0° + X_L \angle 90° \\
&= 8 \text{ k}\Omega\angle 0° + 6 \text{ k}\Omega\angle 90° \\
&= 10 \text{ k}\Omega\angle 36.9° \\
I_T &= \frac{V_S \angle 0°}{Z \angle \theta} \\
&= \frac{100 \text{ V}\angle 0°}{10 \text{ k}\Omega\angle 36.9°} \\
&= 10 \text{ mA}\angle -36.9° \\
V_R &= I_T \angle \theta \times R \angle 0° \\
&= 10 \text{ mA}\angle -36.9° \times 8 \text{ k}\Omega\angle 0° \\
&= 80 \text{ V}\angle -36.9° \\
V_L &= I_T \angle \theta \times X_L \angle 90° \\
&= 10 \text{ mA}\angle -36.9° \times 6 \text{ k}\Omega\angle 90° \\
&= 60 \text{ V}\angle 53.1° \\
V_S &= V_R + V_L \\
&= 80 \text{ V}\angle -36.9° + 60 \text{ V}\angle 53.1° \\
&= 100 \text{ V}\angle 0°
\end{aligned}$$

For the problem in Example 24.2, you should observe the following:

1. The voltage across the resistor is in phase with the current flowing through it. In all circuits, resistor voltage and current are always in phase with each other.
2. The current flowing through the inductor lags the voltage across the inductor by 90° (*ELI* the ICE*man*). Because the resistor voltage is in phase with the current, the voltage across the resistor also lags the inductor voltage by 90°.

3. The *phasor sum* of the resistor and inductor voltages equals the source voltage. This is demonstrated mathematically by Kirchhoff's voltage law in the solution to Example 24.2.

4. An angle exists between the applied voltage, V_S, and the current, I_T. This angle is called the *circuit phase angle, θ*. As you will see later, the circuit phase angle is important for determining how much power the source supplies to the circuit.

5. Because this is an *RL* circuit, θ is positive, and because R is larger than X_L, θ is less than 45°.

Practice Problems 24.2   Answers are given at the end of the chapter.

1. Determine Z, θ, I_T, V_L, and V_R for the circuit in Figure 24.5.

FIGURE 24.5
Circuit for Practice Problems 24.2.

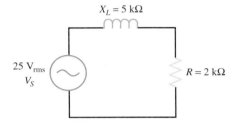

24.1.3   Series *RCL* Circuits

When an inductance and a capacitance are connected in series in an AC circuit, the reactances oppose each other. This is because inductive reactance has a phase angle of 90°, whereas capacitive reactance has a phase angle of −90°. The phasors have opposite directions and subtract from each other. (This is also true for the voltage phasors of an *RLC* AC circuit.) Thus, the circuit becomes a *net RC* or *RL* circuit. The traditional solution to the total impedance of the *RLC* series circuit in rectangular form is

$$Z = R + j(X_L - X_C) = R \pm jX_{net}$$

As we saw in the previous chapter, starting from the rectangular form, the Pythagorean theorem and the inverse tangent function are used to solve for the polar form of Z and θ, respectively.

Lab Reference: Measurements and calculations for a series *RCL* circuit are shown in Exercise 36.

Optional Calculator Sequence

Using the Sharp calculator in the complex-number mode, the solution to the total impedance of the *RLC* series circuit can be solved in polar form as:

$$Z = R\angle 0° + X_L\angle 90° + X_C\angle -90° \tag{24.11}$$

(θ is displayed by using the → key after Z is calculated and is positive or negative, depending on whether X_L or X_C is the larger quantity.)

Note that this form of the equation for impedance is modified from equations (24.1) and (24.6) and is comparable to the total resistance equation for series DC circuits. I_T, V_R, V_L, and V_C are found as before (from Ohm's law). For KVL, equation (24.5) and equation (24.10) are combined to form

$$V_S = V_R + V_C + V_L \tag{24.12}$$

EXAMPLE 24.3

Find Z, θ, I_T, V_R, V_C, and V_L for the circuit in Figure 24.6. Prove that KVL is true for the circuit. Draw the voltage phasor diagram for the circuit.

FIGURE 24.6
Series RCL circuit for Example 24.3.

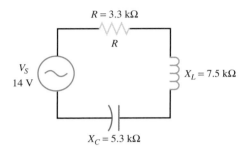

Solution

$$Z = R\angle 0° + X_L\angle 90° + X_C\angle -90°$$
$$= 3.3\ \text{k}\Omega\angle 0° + 7.5\ \text{k}\Omega\angle 90° + 5.3\ \text{k}\Omega\angle -90°$$
$$= 3.97\ \text{k}\Omega\angle 33.7°$$

$$I_T = \frac{V_S\angle 0°}{Z\angle \theta}$$
$$= \frac{14\ \text{V}\angle 0°}{3.97\ \text{k}\Omega\angle 33.7°}$$
$$= 3.52\ \text{mA}\angle -33.7°$$

$$V_R = I_T\angle \theta \times R\angle 0°$$
$$= 3.52\ \text{mA}\angle -33.7° \times 3.3\ \text{k}\Omega\angle 0°$$
$$= 11.6\ \text{V}\angle -33.7°$$

$$V_C = I_T\angle \theta \times X_C\angle -90°$$
$$= 3.52\ \text{mA}\angle -33.7° \times 5.3\ \text{k}\Omega\angle -90°$$
$$= 18.7\ \text{V}\angle -123.7°$$

$$V_L = I_T\angle \theta \times X_L\angle 90°$$
$$= 3.52\ \text{mA}\angle -33.7° \times 7.5\ \text{k}\Omega\angle 90°$$
$$= 26.4\ \text{V}\angle 56.3°$$

$$V_S = V_R + V_C + V_L$$
$$= 11.6\ \text{V}\angle -33.7° + 18.7\ \text{V}\angle -123.7° + 26.4\ \text{V}\angle 56.3°$$
$$\approx 14\ \text{V}\angle 0°$$

See Figure 24.7.

For the problem in Example 24.3, you should note the following:

1. In the voltage phasor diagram, V_C is shown with an angle greater than $-90°$. This does not present a problem, because V_C and V_L are opposites and are subtracted from each other. Thus, the net circuit has phasors that are less than 90° and that fit into a graphical right-triangle solution.
2. Both V_C and V_L are greater than V_S. Depending on the *RLC* component values, either V_C or V_L or both can be greater than V_S. (In a DC circuit, this is not possible.) Remember that V_C and V_L are electrical opposites, and the net reactive voltage is *less* than V_S. V_S is the phasor sum of V_R and V_{net}.
3. The reason for the high reactive voltages is the fact that because both reactive components are present in the series *RLC* circuit, they are subtracted from each other. This,

FIGURE 24.7
Voltage phasor diagram for Example 24.3.

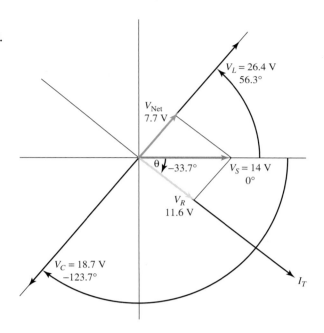

in turn, reduces the impedance of the circuit and causes a high current to flow. A special case (occurring at one frequency only) is where the reactances totally cancel and cause a maximum current. This is known as *series resonance*. Resonance is the subject of the next chapter.

Practice Problems 24.3 Answers are given at the end of the chapter.
1. Determine Z, θ, I_T, V_C, V_L, and V_R for the circuit in Figure 24.8.

FIGURE 24.8
Circuit for Practice Problems 24.3.

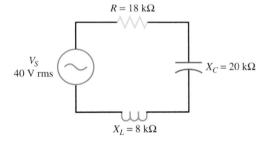

**SECTION 24.1
SELF-CHECK**

Answers are given at the end of the chapter.

1. For a series AC *RC* circuit, θ is _____.
2. For a series AC *RC* circuit, the current _____ the voltage across the capacitor by _____°.
3. What is used as the reference for AC circuit analysis?

24.2 PARALLEL AC CIRCUITS

Parallel AC circuits, like their DC counterparts, are the opposites of series AC circuits. In a parallel AC circuit, the voltage is the same across each branch, and the current divides through each component. The analysis of parallel AC circuits, therefore, focuses on the currents in the same way as the analysis of series circuits focused on the voltages.

24.2.1 Parallel *RC* Circuits

Figure 24.9(a) shows an *RC* parallel circuit containing a resistor and a capacitor in parallel. The voltage is the same across the resistance and the capacitance. The voltage source, V_S, has a reference phase angle of 0°. The current through the resistance, I_R, is in phase with V_S and, therefore, also has a phase angle of 0°. The current through the capacitance, I_C, leads V_C by 90° (ELI the ICEman).

FIGURE 24.9
Parallel RC circuit: (a) circuit; (b) current phasor diagram.

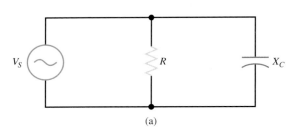

(a)

Lab Reference: Measured and calculated values for a parallel *RC* circuit are demonstrated in Exercise 34.

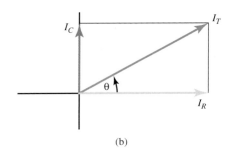

(b)

I_R and I_C are found by Ohm's law:

$$I_R = \frac{V_S \angle 0°}{R \angle 0°}$$

(24.13)

$$I_C = \frac{V_S \angle 0°}{X_C \angle -90°}$$

(24.14)

Note in Figure 24.9(b) that I_C is shown at a positive 90° angle ($I_C = V_C \angle 0°/X_C \angle -90°$). The total current, I_T, is the phasor sum of I_R and I_C, also shown in the figure. Traditionally, I_T is solved with the Pythagorean theorem, whereas the phase angle, θ, is solved with the inverse tangent function, as was shown in the previous chapter.

Optional Calculator Sequence

By using the Sharp calculator, I_T is solved as the phasor sum of I_R and I_C:

$$I_T = I_R \angle 0° + I_C \angle 90°$$

(24.15)

(θ is displayed by using the → key after I_T is calculated. θ has a positive sign for a parallel *RC* circuit.)

Note that the phase angle, θ, is associated with I_T in parallel circuits instead of impedance, as was the case in series circuits. This is because in parallel circuits the right triangle is drawn with the phasors that are additive (for example, the current phasors). Equation (24.15) is comparable to the calculation for total current in a DC parallel circuit. The equivalent impedance, Z_{EQ}, is determined by Ohm's law:

$$Z_{EQ} = \frac{V_S \angle 0°}{I_T \angle \theta}$$

(24.16)

EXAMPLE 24.4

Find I_R, I_C, I_T, θ, and Z_{EQ} for the circuit in Figure 24.10. Draw the current phasor diagram for this circuit.

FIGURE 24.10
Circuit for Example 24.4.

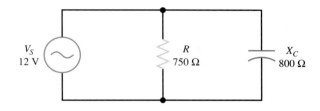

V_S 12 V R 750 Ω X_C 800 Ω

Solution

$$I_R = \frac{V_S \angle 0°}{R \angle 0°}$$

$$= \frac{12\ \text{V} \angle 0°}{750\ \Omega \angle 0°}$$

$$= 16\ \text{mA} \angle 0°$$

$$I_C = \frac{V_S \angle 0°}{X_C \angle -90°}$$

$$= \frac{12\ \text{V} \angle 0°}{800\ \Omega \angle -90°}$$

$$= 15\ \text{mA} \angle 90°$$

$$I_T = I_R \angle 0° + I_C \angle 90°$$

$$= 16\ \text{mA} \angle 0° + 15\ \text{mA} \angle 90°$$

$$= 21.9\ \text{mA} \angle 43.2°$$

$$Z_{EQ} = \frac{V_S \angle 0°}{I_T \angle \theta}$$

$$= \frac{12\ \text{V} \angle 0°}{21.9\ \text{mA} \angle 43.2°}$$

$$= 547.9\ \Omega \angle -43.2°$$

See Figure 24.11.

Practice Problems 24.4   Answers are given at the end of the chapter.
1. Solve for I_R, I_C, I_T, θ, and Z_{EQ} for the circuit shown in Figure 24.12.

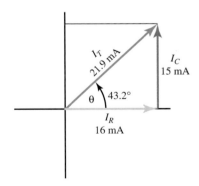

FIGURE 24.11
Current phasor diagram for Example 24.4.

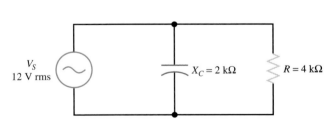

V_S 12 V rms $X_C = 2\ \text{k}\Omega$ $R = 4\ \text{k}\Omega$

FIGURE 24.12
Circuit for Practice Problems 24.4.

24.2.2   Parallel *RL* Circuits

Figure 24.13(a) shows an *RL* parallel circuit containing a resistance and an inductance. The voltage across the resistance and the voltage across the inductance are the same. The phase angle of the voltage source, V_S, is the reference of $\angle 0°$.

FIGURE 24.13
Parallel RL circuit: (a) circuit; (b) current phasor diagram.

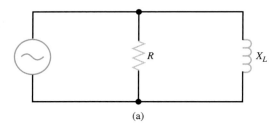

(a)

Lab Reference: Measured and calculated values for a parallel *RL* circuit are demonstrated in Exercise 35.

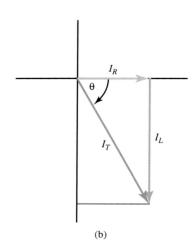

(b)

The current through the resistance I_R is in phase with V_S and, therefore, has the same phase angle (0°). The current through the inductance, I_L, is out of phase with the voltage and lags V_S by 90°. Therefore, I_L also lags I_R by 90° and has a phase angle of $-90°$ (ELI the ICEman). I_L is found by modifying equation 24.14:

$$I_L = \frac{V_S \angle 0°}{X_L \angle 90°}$$

(24.17)

I_R and the impedance are found as before by using Ohm's law.

Optional Calculator Sequence

The total current is the phasor sum of I_R and I_L, as shown in Figure 24.13(b).

$$I_T = I_R \angle 0° + I_L \angle -90°$$

(24.18)

(θ is displayed by using the $\rightarrow$ key after I_T is calculated and has a negative sign for a parallel *RL* circuit.)

EXAMPLE 24.5

Given the *RL* circuit in Figure 24.14, determine I_R, I_L, I_T, θ, and Z_{EQ}. Draw the current phasor diagram.

FIGURE 24.14
Parallel RL circuit for Example 24.5.

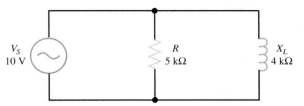

Solution

$$I_R = \frac{V_S\angle 0°}{R\angle 0°}$$

$$= \frac{10\ \text{V}\angle 0°}{5.0\ \text{k}\Omega\angle 0°}$$

$$= 2.0\ \text{mA}\angle 0°$$

$$I_L = \frac{V_S\angle 0°}{X_L\angle 90°}$$

$$= \frac{10\ \text{V}\angle 0°}{4.0\ \text{k}\Omega\angle 90°}$$

$$= 2.5\ \text{mA}\angle -90°$$

$$I_T = I_R\angle 0° + I_L\angle -90°$$

$$= 2.0\ \text{mA}\angle 0° + 2.5\ \text{mA}\angle -90°$$

$$= 3.2\ \text{mA}\angle -51.3°$$

$$Z_{EQ} = \frac{V_S\angle 0°}{I_T\angle \theta}$$

$$= \frac{10\ \text{V}\angle 0°}{3.2\ \text{mA}\angle -51.3°}$$

$$= 3.1\ \text{k}\Omega\angle 51.3°$$

See Figure 24.15.

FIGURE 24.15
Current phasor diagram for Example 24.5.

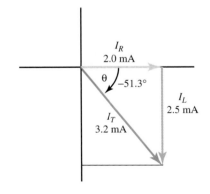

Practice Problems 24.5 Answers are given at the end of the chapter.
1. Determine I_R, I_L, I_T, θ, and Z_{EQ} for the circuit in Figure 24.16.

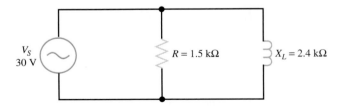

FIGURE 24.16
Circuit for Practice Problems 24.5.

24.2.3   Parallel *RCL* Circuits

When an inductance and a capacitance are connected in parallel, the inductive current opposes the capacitive current. The currents oppose each other because the phase angle of
I_L is $-90°$, whereas the phase angle of I_C is $90°$. The phasors have opposite directions
and subtract from each other. Consider the *RLC* parallel circuit in Figure 24.17(a). The
current phasor diagram is shown in Figure 24.17(b). Note that the reactive currents are
subtracted, yielding a net I_X current. The total current of the *RLC* circuit in rectangular
form is

$$I_T = I_R + j(I_C - I_L) = I_R \pm jI_{X_{net}}$$

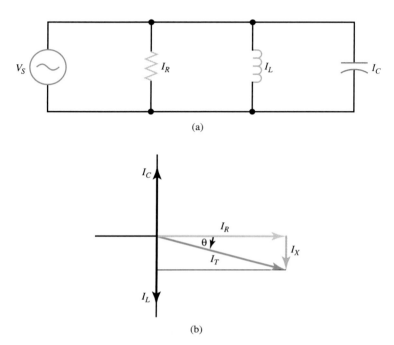

(a)

(b)

FIGURE 24.17
Parallel RLC circuit: (a) circuit; (b) current phasor diagram.

Using traditional mathematics, I_T is solved with the Pythagorean theorem, whereas
θ is found by the inverse tangent function.

Optional Calculator Sequence

Using the Sharp calculator in the complex-number mode, I_T is found by combining
equations (24.15) and (24.18):

$$I_T = I_R\angle0° + I_C\angle90° + I_L\angle-90° \tag{24.19}$$

(θ is displayed by using the $\rightarrow$ key after I_T is calculated and is positive or negative,
depending on whether I_L or I_C is the larger quantity.)

$I_R, I_C, I_L,$ and Z_{EQ} are found by using equations (24.13), (24.14), (24.17), and (24.16),
respectively. θ is displayed as before after I_T is calculated. As a check, Z_{EQ} can be calculated by the reciprocal equation, the same as for resistance in a parallel DC circuit.

Optional Calculator Sequence

The Sharp calculator does this in one step:

$$Z_{EQ} = \cfrac{1}{\cfrac{1}{R\angle 0°} + \cfrac{1}{X_C\angle -90°} + \cfrac{1}{X_L\angle 90°}} \qquad \textbf{(24.20)}$$

EXAMPLE 24.6

Using Figure 24.18, determine I_R, I_L, I_C, I_T, θ, and Z_{EQ}. Use the reciprocal equation (equation (24.20)) as a check for Z_{EQ}.

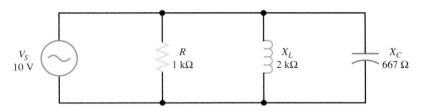

FIGURE 24.18
Circuit for Example 24.6.

Solution

$$I_R = \frac{V_S\angle 0°}{R\angle 0°}$$

$$= \frac{10\text{ V}\angle 0°}{1\text{ k}\Omega\angle 0°}$$

$$= 10\text{ mA}\angle 0°$$

$$I_L = \frac{V_S\angle 0°}{X_L\angle 90°}$$

$$= \frac{10\text{ V}\angle 0°}{2\text{ k}\Omega\angle 90°}$$

$$= 5\text{ mA}\angle -90°$$

$$I_C = \frac{V_S\angle 0°}{X_C\angle -90°}$$

$$= \frac{10\text{ V}\angle 0°}{667\ \Omega\angle -90°}$$

$$= 15\text{ mA}\angle 90°$$

$$I_T = I_R\angle 0° + I_C\angle 90° + I_L\angle -90°$$

$$= 10\text{ mA}\angle 0° + 15\text{ mA}\angle 90° + 5\text{ mA}\angle -90°$$

$$= 14.1\text{ mA}\angle 45°$$

$$Z_{EQ} = \frac{V_S\angle 0°}{I_T\angle \theta}$$

$$= \frac{10\text{ V}\angle 0°}{14.1\text{ mA}\angle 45°}$$

$$= 709\ \Omega\angle -45°$$

As a check,

$$Z_{EQ} = \cfrac{1}{\cfrac{1}{1 \text{ k}\Omega \angle 0°} + \cfrac{1}{2 \text{ k}\Omega \angle 90°} + \cfrac{1}{667 \text{ } \Omega \angle -90°}}$$

$$= 707 \text{ } \Omega \angle -45°$$

For Example 24.6 you should note the following:

1. I_L was larger than I_T. Depending on component values, in a parallel *RLC* circuit either I_L or I_C or both can be larger than I_T. (In a DC circuit this is not possible.)
2. I_L and I_C are electrical opposites and are subtracted from each other. Thus, the net reactive current (I_X) is less than I_T. I_T is the phasor sum of I_R and I_X.
3. The phase angle, θ, is associated with I_T rather than Z, as is the case in series AC circuits.

Practice Problems 24.6   Answers are given at the end of the chapter.
1. Solve for I_R, I_L, I_C, I_T, θ, and Z_{EQ} for the circuit shown in Figure 24.19.

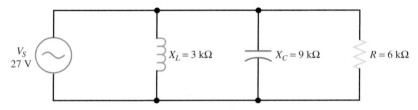

FIGURE 24.19
Circuit for Practice Problems 24.6.

24.2.4 Susceptance and Admittance

In parallel DC circuits, the concept of conductance, G, is useful to help analyze the circuit. Conductance is defined as the reciprocal of resistance ($G = 1/R$) and is measured in siemens (S). Conductance is additive in parallel DC circuits.

Similarly, in parallel AC circuits the concepts of *susceptance, B,* and *admittance, Y,* are used to help analyze an AC circuit. They are phasors, defined as the reciprocals of reactance and impedance, respectively, and are also measured in siemens. They are calculated as follows:

$$B = \frac{1}{X}$$

$$Y = \frac{1}{Z}$$

Because of the reciprocal relationship, B has a negative phase angle for an inductive reactance and a positive phase angle for a capacitive reactance. This is the same as the phase angles for inductive and capacitive currents.

For a resistance, R, in parallel with a net reactance, the total admittance, Y, can be solved as:

$$Y = \frac{1}{R} \pm j\frac{1}{X} = G \pm jB$$

This rectangular form of admittance can be converted to the polar form using methods previously shown with impedance. These methods include the use of right-triangle solutions and a calculator.

Thus, in parallel AC circuits, it may be easier to add the reciprocal functions to find impedance rather than solve for branch currents. We could make up a plethora of example problems utilizing the reciprocal functions for you to practice, but more important is that you are familiar with the terms susceptance and admittance. They are commonly used in a high-frequency (microwave) communications course that employs a Smith chart. Therefore, if you take a course in microwave communications, you will undoubtedly see the terms susceptance and admittance again.

**SECTION 24.2
SELF-CHECK**

Answers are given at the end of the chapter.

1. In parallel AC circuits, θ is associated with _____.
2. In a parallel AC circuit containing R, L, and C, the reactive currents _____.
3. For a parallel AC circuit containing R and L, θ is _____.
4. The reciprocal of impedance is _____.

24.3 COMBINING COMPLEX BRANCH IMPEDANCES

By combining *RLC* components in various series and parallel combinations, we can create rather complex-looking AC circuits. These can be analyzed using techniques that were employed in series-parallel DC circuits. The net result is to determine the series equivalent circuit for the complex circuit.

Similar to two resistances in parallel in a DC circuit, the equivalent impedance of two impedances in parallel is equal to their product divided by their sum:

$$Z_{EQ} = \frac{Z_1 Z_2}{Z_1 + Z_2}$$

Note that a traditional solution to this equation requires the conversion of Z_1 and Z_2 to rectangular form so that the denominator can be added and then a conversion back to polar form to complete the division.

In addition, the equivalent impedance for two or more parallel impedances can also be expressed in terms of reciprocals:

$$Z_{EQ} = \frac{1}{\dfrac{1}{Z_1} + \dfrac{1}{Z_2} + \dfrac{1}{Z_3}} \tag{24.21}$$

Note that a traditional solution to this equation also requires the conversion of the denominator to rectangular form for addition and then back to polar form to complete the division. As will be shown via an example problem, this conversion is not necessary with the Sharp calculator.

Besides utilizing the reciprocal equation, parallel branch currents can be solved directly and added to find the total current. Then using Ohm's law, the equivalent (total) impedance can be found by dividing V_S by I_T. The branch currents are found by first determining the branch impedances from the addition of component values of each branch. Then using Ohm's law, the branch currents are found by dividing V_S by the branch impedance.

Optional Calculator Sequence

The following example illustrates these points.

EXAMPLE 24.7

Determine the branch impedances, branch currents, I_T, θ, and Z_{EQ} for the circuit in Figure 24.20.

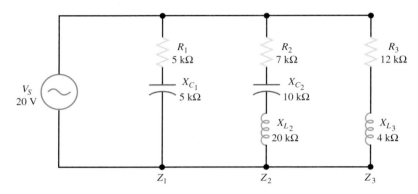

FIGURE 24.20
Circuit for Example 24.7.

Solution Using the Sharp calculator in the complex polar-number mode, solve for the branch impedances by adding the component values of each branch. (Note that there is a series circuit within each branch.)

$$Z_1 = R\angle 0° + X_C\angle -90°$$
$$= 5\text{ k}\Omega\angle 0° + 5\text{ k}\Omega\angle -90°$$
$$= 7.07\text{ k}\Omega\angle -45°$$
$$Z_2 = R\angle 0° + X_C\angle -90° + X_L\angle 90°$$
$$= 7\text{ k}\Omega\angle 0° + 10\text{ k}\Omega\angle -90° + 20\text{ k}\Omega\angle 90°$$
$$= 12.2\text{ k}\Omega\angle 55°$$
$$Z_3 = R\angle 0° + X_L\angle 90°$$
$$= 12\text{ k}\Omega\angle 0° + 4\text{ k}\Omega\angle 90°$$
$$= 12.65\text{ k}\Omega\angle 18.4°$$

Solve for the branch currents by using Ohm's law.

$$I_1 = \frac{V_S\angle 0°}{Z_1}$$
$$= \frac{20\text{ V}\angle 0°}{7.07\text{ k}\Omega\angle -45°}$$
$$= 2.83\text{ mA}\angle 45°$$
$$I_2 = \frac{V_S\angle 0°}{Z_2}$$
$$= \frac{20\text{ V}\angle 0°}{12.2\text{ k}\Omega\angle 55°}$$
$$= 1.64\text{ mA}\angle -55°$$
$$I_3 = \frac{V_S\angle 0°}{Z_3}$$
$$= \frac{20\text{ V}\angle 0°}{12.65\text{ k}\Omega\angle 18.4°}$$
$$= 1.58\text{ mA}\angle -18.4°$$

Solve for the total current by adding the branch currents.

$$I_T = I_1 + I_2 + I_3$$
$$= 2.83 \text{ mA}\angle 45° + 1.64 \text{ mA}\angle -55° + 1.58 \text{ mA}\angle -18.4°$$
$$= 4.44 \text{ mA}\angle 2.05°$$

Solve for the equivalent impedance using Ohm's law.

$$Z_{EQ} = \frac{V_S \angle 0°}{I_T \angle \theta}$$

$$= \frac{20 \text{ V} \angle 0°}{4.44 \text{ mA}\angle 2.05°}$$

$$= 4.5 \text{ k}\Omega\angle -2.05°$$

As a check for Z_{EQ}, use the reciprocal equation (equation (24.21)) to solve for the equivalent impedance. This is done directly on the Sharp calculator in the complex-number mode.

$$Z_{EQ} = \frac{1}{\dfrac{1}{Z_1} + \dfrac{1}{Z_2} + \dfrac{1}{Z_3}}$$

$$= \frac{1}{\dfrac{1}{7.07 \text{ k}\Omega\angle -45°} + \dfrac{1}{12.2 \text{ k}\Omega\angle 55°} + \dfrac{1}{12.65 \text{ k}\Omega\angle 18.4°}}$$

$$= 4.5 \text{ k}\Omega\angle -2.04°$$

Practice Problems 24.7 Answers are given at the end of the chapter.
1. Determine Z_1, Z_2, Z_3, I_1, I_2, I_3, I_T, θ, and Z_{EQ} for the circuit shown in Figure 24.21.

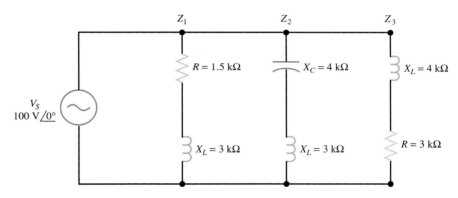

FIGURE 24.21
Circuit for Practice Problems 24.7.

Answers are given at the end of the chapter.

1. The net result for an AC circuit with complex branch impedances is to determine the _____ equivalent circuit.
2. In an AC circuit with complex branch impedances, Z_{EQ} can be found by _____ law or the _____ equation.

24.4 AC POWER

In AC circuits, both current and voltage are continually changing in a sinusoidal manner. The instantaneous AC power is the product of the current and voltage at each instant in time. Figure 24.22 illustrates how the instantaneous AC power changes with time in purely resistive, capacitive, and inductive circuits. In each case, the instantaneous power varies sinusoidally at a frequency twice that of the current and voltage. In a purely resistive circuit, the current and voltage are in phase with each other.

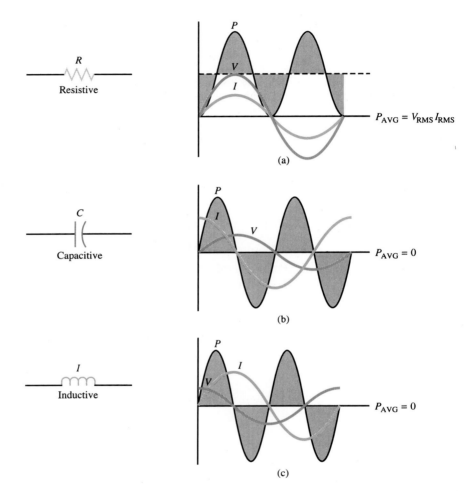

FIGURE 24.22
Power dissipation in purely resistive, capacitive, and inductive circuits: (a) resistive; (b) capacitive; (c) inductive.

To obtain the instantaneous power curve in Figure 24.22(a), simply multiply the instantaneous current and voltage values and plot the resulting products. When a positive quantity (current) is multiplied by another positive quantity (voltage), the product is positive. Also, whenever two negative values are multiplied, the product is positive. For these reasons, the instantaneous power curve in a purely resistive circuit consists of all positive values.

The average power is also positive, as shown in Figure 24.22(a). The average power equals the product of the rms values of current and voltage in the circuit. In purely capacitive and inductive circuits, the current and voltage are 90° out of phase with each other. At certain instants, the current is positive and the voltage is negative, and vice versa. At other instants, both the current and voltage are positive or both are negative.

For these reasons, the instantaneous power curves in purely capacitive and inductive circuits consist of both positive and negative values, as shown in Figure 24.22(b) and (c), respectively. Note that the positive and negative portions in the instantaneous power curves in a purely capacitive and inductive circuit are equal. Thus, the average power dissipated by an ideal capacitor or inductor is *zero*. *Ideal reactive components store energy and then return it to the circuit at a later time.* The energy they store is called *reactive power* and is measured in volts-amperes-reactive, abbreviated *VAR*.

Thus, when an AC circuit contains resistance and reactance, the power indicated by the product of the voltage source and the total current gives what is known as the *apparent power,* which is measured in volt-amperes (V·A). Apparent power (AP) comprises reactive power and true power. *True power* (TP), or real power, is the power actually dissipated by circuit resistance and is measured in watts (W). True power is always less than apparent power. The point to remember here is that *only* resistance dissipates power.

Figure 24.23 illustrates the complex power triangle. Note that apparent power (AP) is the vector sum of true power (TP) and reactive power.

FIGURE 24.23
Complex power triangle.

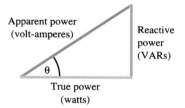

Apparent power
(volt-amperes)

Reactive
power
(VARs)

θ

True power
(watts)

24.4.1 Calculating True Power

The basic equations for computing power in DC circuits can be used for true power in AC circuits if the values of current and voltages are rms values. These are $P = I^2R = V^2/R = IV_R$.

EXAMPLE 24.8

Determine the true power for a series *RL* circuit when $R = 5\,\text{k}\Omega$, $X_L = 5\,\text{k}\Omega$, $Z = 7.07\,\text{k}\Omega$, $\theta = 45°$, and $V_S = 20\,\text{V}$.

Solution
$$I_T = \frac{V_S}{Z}$$
$$= \frac{20\,\text{V}}{7.07\,\text{k}\Omega}$$
$$= 2.82\,\text{mA}$$
$$TP = I^2R$$
$$= (2.82\,\text{mA})^2 \times 5\,\text{k}\Omega$$
$$= 0.0398\,\text{W} \quad (39.8\,\text{mW})$$

Practice Problems 24.8 Answers are given at the end of the chapter.
1. Solve for the true power in a series *RC* circuit when $R = 10\,\text{k}\Omega$, $X_C = 10\,\text{k}\Omega$, and V_S is 30 V.

24.4.2 Calculating Reactive Power

The reactive power is the power returned to the source by the reactive components of the circuit. Any of the DC power equations can be modified to solve for reactive power. These become I^2X, V_X^2/X, and IV_X.

EXAMPLE 24.9

Using the circuit values in Example 24.8, determine the reactive power.

Solution   reactive power $= I^2X$
$$= (2.82 \text{ mA})^2 \times 5 \text{ k}\Omega$$
$$= 0.04 \text{ VAR} \qquad (40 \text{ mVAR})$$

Practice Problems 24.9   Answers are given at the end of the chapter.
1. Solve for the reactive power for the circuit in Practice Problems 24.8.

24.4.3   Calculating Apparent Power

Apparent power is the power that appears to the source because of the circuit impedance. Because the impedance is the total opposition to AC, the apparent power is the power the voltage source sees. Apparent power is the combination of true power and reactive power. Apparent power is not found by simply adding true power and reactive power, just as impedance is not found by the simple addition of resistance and reactance.

To calculate apparent power, you may use any of the DC equations for total power, modified to the following:

$$AP = I_T^2 Z = \frac{V_S^2}{Z} = I_T V_S$$

EXAMPLE 24.10

Using the circuit values from Example 24.8, determine the apparent power.

Solution $AP = I_T V_S$
$$= 2.82 \text{ mA} \times 20 \text{ V}$$
$$= 0.056 \text{ V·A} \ (56.4 \text{ mV·A})$$

Practice Problems 24.10   Answers are given at the end of the chapter.
1. Solve for apparent power for the circuit in Practice Problems 24.8.

**SECTION 24.4
SELF-CHECK**

Answers are given at the end of the chapter.

1. The instantaneous power curve in a purely resistive AC circuit contains all _____ values.
2. The average power dissipated by an ideal capacitor or inductor is _____.
3. The unit of measure for reactive power is the _____.

24.5   POWER FACTOR

A *power factor* is a number (represented as a decimal fraction or percentage) that represents that portion of the apparent power dissipated as true power in a circuit. Thus, the power factor is equal to:

$$PF = \frac{TP}{AP} \tag{24.22}$$

Going a step further, another equation for the power factor can be developed by substituting the equations for true power and apparent power in the equation for power factor. This yields

$$PF = \frac{I_R^2 R}{I_T^2 Z}$$

In a series circuit, the current is the same, and I_R equals I_T. Therefore, this power factor equation (valid for series circuits only) becomes

$$PF = \frac{R}{Z} \tag{24.23}$$

For a parallel circuit, equation (24.23) is modified as:

$$PF = \frac{I_R}{I_Z} \tag{24.24}$$

(Note that $I_Z = I_T$.)

You should recognize the R/Z ratio in equation (24.23) as two sides of a right triangle that correspond to the cosine function. This was shown in Figure 24.16, with true power (TP) the adjacent side of the right triangle and corresponding to resistive power (R). Also, apparent power (AP) is the hypotenuse, representing power from circuit impedance (Z). Thus, the power factor equals

$$PF = \cos \theta \tag{24.25}$$

Now you can see how the circuit phase angle (θ) affects the power distribution in AC circuits. The cosine of θ represents the resistive portion of the impedance and that which dissipates the power as heat. A smaller angle, meaning the circuit is more resistive, has more power dissipated in the form of heat.

EXAMPLE 24.11

Using the circuit values from Example 24.8, determine the power factor.

Solution $PF = \dfrac{R}{Z}$

$$= \frac{5 \text{ k}\Omega}{7.07 \text{ k}\Omega}$$

$$= 0.707, \quad \text{or} \quad 70.7\%$$

Combining equations (24.23), (24.24), and (24.25) gives an equation for true power that is universal and can be used in any type of circuit:

$$TP = I_T V_S \cos \theta \tag{24.26}$$

EXAMPLE 24.12

Using equation (24.26), calculate the TP using the values from Example 24.8 to see how the calculations compare.

Solution TP = 2.82 mA × 20 V × cos 45°
= 0.0399 W   (39.9 mW)

Practice Problems 24.11 Answers are given at the end of the chapter.
1. Solve for the power factor for the circuit in Practice Problems 24.8.

24.5.1 Power Factor Correction

The apparent power in an AC circuit has been described as the power the source sees. As far as the source is concerned, the apparent power is the power that must be provided to the circuit. The source does this by feeding a line current to the load, across which is the source voltage. You also know that the true power is the power actually dissipated as heat in the circuit. The difference between apparent power and true power is wasted, because, in reality only true power is consumed. The ideal situation is to have the power factor of

1 (unity), or 100%. If the energy supplied by the electrical utility is being stored by reactance, then that energy is not available to contribute to the work being done in a factory.

The expression *correcting the power factor* refers to reducing the reactance (and therefore the reactive power) in a circuit. Because most manufacturers employ inductive loads (motors, etc.), some form of power factor correction needs to be done in order to utilize the power supplied more fully and to get better rates from the electrical utility. The correction is performed by the user or the electrical utility by placing a capacitor in parallel to the load. The capacitor tends to cancel the inductance of the load. This, in turn, tends to reduce the line current supplied and reduces the manufacturer's cost of electrical power.

SECTION 24.5
SELF-CHECK

Answers are given at the end of the chapter.

1. Define the term *power factor.*
2. What is the unit of measure for the power factor?
3. In a typical power factor correction, a _____ is placed across the load.

■ SUMMARY

1. The opposition to an AC voltage or current by a capacitor or inductor is called reactance.
2. The total opposition to the flow of AC current is called impedance.
3. Resistance is represented by a real number, reactance by an imaginary number, and impedance by a complex number.
4. By using complex numbers, the techniques of DC circuit analysis can be extended to include the analysis of AC circuits.
5. In RCL circuits, V_S is used as the reference phasor at $\angle 0°$.
6. The current through a capacitor leads the voltage across the capacitor by 90°.
7. The phasor sum of the resistor, capacitor, and inductor voltages equals the source voltage.
8. The angle between the applied voltage and the current in a series RCL circuit is called the circuit phase angle.
9. The circuit phase angle is important for determining how much power the source supplies to the circuit.
10. The current flowing through an inductor lags the voltage across the inductor by 90°.
11. Depending on the series RLC component values, either V_C or V_L or both can be greater than V_S.
12. V_C and V_L are electrical opposites in a series RCL circuit.

13. Series resonance is a special case where X_C and X_L totally cancel in a series RLC circuit.
14. In a series RCL circuit, Z is always greater than the net reactance or resistance but smaller than their sum.
15. In a parallel RCL circuit, θ is associated with I_T.
16. In a parallel RCL circuit, I_L and I_C are electrical opposites.
17. Susceptance is the reciprocal of reactance, whereas admittance is the reciprocal of impedance.
18. The reciprocal functions are commonly used in high-frequency communications circuit analysis.
19. The average power dissipated by an ideal capacitor or inductor is zero.
20. Apparent power is the product of V_S and I_T and comprises reactive power and true power.
21. True power is the power actually dissipated by circuit resistance.
22. Reactive power is the power stored and then returned to the source by the reactive components of the circuit.
23. A power factor is a number that represents that portion of the apparent power dissipated as true power in a circuit.
24. The expression *correcting the power factor* refers to reducing the reactive power in a circuit and is typically corrected by placing a capacitor across the inductive loads used in manufacturing.

▪ IMPORTANT EQUATIONS

$$Z = R\angle 0° + X_C\angle -90° \tag{24.1}$$

$$I_T = \frac{V_S\angle 0°}{Z\angle \theta} \tag{24.2}$$

$$V_R = I_T\angle \theta \times R\angle 0° \tag{24.3}$$

$$V_C = I_T\angle \theta \times X_C\angle -90° \tag{24.4}$$

$$V_S = V_C + V_R \tag{24.5}$$

$$Z = R\angle 0° + X_L\angle 90° \tag{24.6}$$

$$I_T = \frac{V_S\angle 0°}{Z\angle \theta} \tag{24.7}$$

$$V_R = I_T\angle \theta \times R\angle 0° \tag{24.8}$$

$$V_L = I_T\angle \theta \times X_L\angle 90° \tag{24.9}$$

$$V_S = V_R + V_L \tag{24.10}$$

$$Z = R\angle 0° + X_L\angle 90° + X_C\angle -90° \tag{24.11}$$

$$V_S = V_R + V_C + V_L \tag{24.12}$$

$$I_R = \frac{V_S\angle 0°}{R\angle 0°} \tag{24.13}$$

$$I_C = \frac{V_S\angle 0°}{X_C\angle -90°} \tag{24.14}$$

$$I_T = I_R\angle 0° + I_C\angle 90° \tag{24.15}$$

$$Z_{EQ} = \frac{V_S\angle 0°}{I_T\angle \theta} \tag{24.16}$$

$$I_L = \frac{V_S\angle 0°}{X_L\angle 90°} \tag{24.17}$$

$$I_T = I_R\angle 0° + I_L\angle -90° \tag{24.18}$$

$$I_T = I_R\angle 0° + I_C\angle 90° + I_L\angle -90° \tag{24.19}$$

$$Z_{EQ} = \frac{1}{\dfrac{1}{R\angle 0°} + \dfrac{1}{X_C\angle -90°} + \dfrac{1}{X_L\angle 90°}} \tag{24.20}$$

$$Z_{EQ} = \frac{1}{\dfrac{1}{Z_1} + \dfrac{1}{Z_2} + \dfrac{1}{Z_3}} \tag{24.21}$$

$$PF = \frac{TP}{AP} \tag{24.22}$$

$$PF = \frac{R}{Z} \tag{24.23}$$

$$PF = \frac{I_R}{I_Z} \tag{24.24}$$

$$PF = \cos \theta \tag{24.25}$$

$$TP = I_T V_S \cos \theta \tag{24.26}$$

▪ REVIEW QUESTIONS

1. Define the term *impedance*.
2. What are complex numbers and how are they used?
3. Is KVL true for a series *RCL* AC circuit?
4. If X_C is larger than R in a series *RC* AC circuit, θ is _____ and _____ than 45°.
5. If R is larger than X_C in a series *RC* AC circuit, θ is _____ and _____ than 45°.
6. What is the circuit phase angle and what does it indicate?
7. If X_L is larger than R in a series *RL* AC circuit, θ is _____ and _____ than 45°.
8. If R is larger than X_L in a series *RL* AC circuit, θ is _____ and _____ than 45°.
9. Why do X_C and X_L cancel in a series *RCL* AC circuit?
10. Why are V_C and V_L electrical opposites in a series *RCL* AC circuit?
11. What is used as the reference phasor in a series AC circuit?
12. θ is associated with _____ in a parallel AC circuit.
13. I_L and I_C are electrical opposites in a(n) _____ *RCL* AC circuit.
14. Define the term *susceptance*.
15. Define the term *admittance*.
16. Where are susceptance and admittance used?
17. The instantaneous power curve in a purely resistive circuit consists of all _____ values.
18. Why is the average power dissipated in an ideal capacitor or inductor equal to zero?
19. Define the term *apparent power*.
20. Define the term *true power*.
21. Define the term *reactive power*.
22. Define the term *power factor*.

▪ CRITICAL THINKING

1. Show how complex numbers are utilized in *RCL* AC circuits.
2. How is it possible for V_L, V_C, or both to be larger than V_S in a series *RCL* AC circuit?
3. Show how the average resistive power is a positive value, whereas the reactive power is zero.
4. What is meant by *power factor correction* and how is it done?

■ PROBLEMS

Solve for each unknown listed in the following table for a series *RCL* AC circuit.

	R	X_C	X_L	Z	θ	I_T	V_R	V_C	V_L	V_S
1.	3 kΩ	4 kΩ	—						—	10 V
2.	12 kΩ	—	8 kΩ					—		20 V
3.		—	8 kΩ	10 kΩ				—		12 V
4.	5 kΩ	5 kΩ	—						—	15 V
5.	4 kΩ	4 kΩ	8 kΩ							24 V
6.	20 kΩ	5 kΩ	17 kΩ							12 V
7.	3 kΩ		8 kΩ	5 kΩ						20 V
8.	34 kΩ	56 kΩ	12 kΩ							18 V
9.	23 kΩ	17 kΩ	4 kΩ							15 V
10.	4.7 kΩ	8.2 kΩ	12 kΩ							10 V

Solve for each unknown listed in the following table for a parallel *RCL* AC circuit.

	R	X_C	X_L	I_R	I_C	I_L	I_T	θ	Z_{EQ}	V_S
11.	3 kΩ	—	4 kΩ		—					10 V
12.	5 kΩ	5 kΩ	—			—				12 V
13.	10 kΩ	—	10 kΩ		—					50 V
14.	12 kΩ	24 kΩ	36 kΩ							15 V
15.	24 kΩ	6 kΩ	2 kΩ							10 V
16.	4.7 kΩ	10 kΩ	5 kΩ							15 V
17.	6.8 kΩ	12 kΩ	20 kΩ							18 V
18.	1 kΩ	3 kΩ	2 kΩ							5 V
19.	500 Ω	3 kΩ	4 kΩ							1 V
20.	50 kΩ	18 kΩ	24 kΩ							24 V

21. Solve for Z_1, Z_2, Z_3, I_1, I_2, I_3, I_T, θ, and Z_{EQ} for the circuit in Figure 24.24.

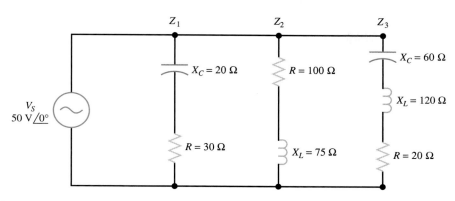

FIGURE 24.24
Circuit for Problem 21.

22. Solve for Z_1, Z_2, Z_3, I_1, I_2, I_3, I_T, θ, and Z_{EQ} for the circuit in Figure 24.25.

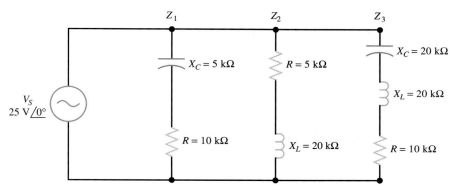

FIGURE 24.25
Circuit for Problem 22.

23. Solve for Z_1, Z_2, Z_3, I_1, I_2, I_3, I_T, θ, and Z_{EQ} for the circuit in Figure 24.26.

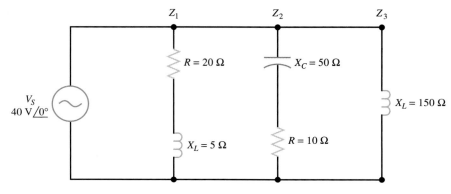

FIGURE 24.26
Circuit for Problem 23.

24. Solve for AP, TP, reactive power, and the power factor for each series circuit shown in Problems 1–5.

25. Solve for AP, TP, reactive power, and the power factor for each parallel circuit shown in Problems 11–15.

■ ANSWERS TO PRACTICE PROBLEMS

PRACTICE PROBLEMS 24.1

1. $Z = 7.07 \text{ k}\Omega\angle-45°$, $I_T = 2.83 \text{ mA}\angle45°$, $V_C = 14.1 \text{ V}\angle-45°$, $V_R = 14.1 \text{ V}\angle45°$

PRACTICE PROBLEMS 24.2

1. $Z = 5.39 \text{ k}\Omega\angle68.2°$, $I_T = 4.64 \text{ mA}\angle-68.2°$, $V_L = 23.2 \text{ V}\angle21.8°$, $V_R = 9.2 \text{ V}\angle-68.2°$

PRACTICE PROBLEMS 24.3

1. $Z = 21.6 \text{ k}\Omega\angle-33.7°$, $I_T = 1.85 \text{ mA}\angle33.7°$, $V_C = 37 \text{ V}\angle-56.3°$, $V_L = 14.8 \text{ V}\angle123.7°$, $V_R = 33.3 \text{ V}\angle33.7°$

PRACTICE PROBLEMS 24.4

1. $I_R = 3 \text{ mA}\angle0°$, $I_C = 6 \text{ mA}\angle90°$, $I_T = 6.71 \text{ mA}\angle63.4°$, $Z = 1.79 \text{ k}\Omega\angle-63.4°$

PRACTICE PROBLEMS 24.5

1. $I_R = 20 \text{ mA}\angle0°$, $I_L = 12.5 \text{ mA}\angle-90°$, $I_T = 23.6 \text{ mA}\angle-32°$, $Z_{EQ} = 1.27 \text{ k}\Omega\angle32°$

PRACTICE PROBLEMS 24.6

1. $I_R = 4.5 \text{ mA}\angle0°$, $I_L = 9 \text{ mA}\angle-90°$, $I_C = 3 \text{ mA}\angle90°$, $I_T = 7.5 \text{ mA}\angle-53.1°$, $Z_{EQ} = 3.6 \text{ k}\Omega\angle53.1°$

PRACTICE PROBLEMS 24.7

1. $Z_1 = 3.35 \text{ k}\Omega\angle63.4°$, $Z_2 = 1 \text{ k}\Omega\angle-90°$, $Z_3 = 5 \text{ k}\Omega\angle53.1°$, $I_1 = 29.9 \text{ mA}\angle-63.4°$, $I_2 = 100 \text{ mA}\angle90°$, $I_3 = 20 \text{ mA}\angle-53.1°$, $I_T = 62.6 \text{ mA}\angle66°$, $Z_{EQ} = 1.6 \text{ k}\Omega\angle-66°$

PRACTICE PROBLEMS 24.8

1. TP = 44.9 mW

PRACTICE PROBLEMS 24.9

1. Reactive power = 44.9 mVAR

PRACTICE PROBLEMS 24.10

1. AP = 63.6 mV·A

PRACTICE PROBLEMS 24.11

1. PF = 0.71

■ ANSWERS TO SECTION SELF-CHECKS

SECTION 24.1 SELF-CHECK

1. Negative
2. Leads, 90
3. $V_S\angle0°$

SECTION 24.2 SELF-CHECK

1. I_T
2. Cancel (subtract)
3. Negative
4. Admittance

SECTION 24.3 SELF-CHECK

1. Series
2. Ohm's, reciprocal

SECTION 24.4 SELF-CHECK

1. Positive
2. Zero
3. VAR

SECTION 24.5 SELF-CHECK

1. A power factor is a number that represents that portion of the apparent power dissipated as true power in a circuit.
2. The power factor is unitless.
3. Capacitor

CHAPTER 25

RESONANCE

OBJECTIVES

After completing this chapter, you will be able to:

1. Define the term *resonance*.
2. State the conditions under which resonance occurs and the circuit responds at resonance.
3. Differentiate the characteristics of series resonance from those of parallel resonance.
4. Solve for various unknowns associated with series or parallel resonant circuits.
5. Compare and contrast the circuit parameters of Q and BW.
6. Compare and contrast an ideal tank circuit and a practical tank circuit.

INTRODUCTION

Because capacitive and inductive reactance vary with frequency, AC circuits with capacitors and inductors exhibit a number of characteristics not found in DC or purely resistive AC circuits. One of these characteristics is the phenomenon of *resonance*. This occurs at *one particular frequency only,* when X_L is equal to X_C. At resonance, the only opposition to current flow is from any circuit resistance, which is usually, in part, the effective resistance of the inductor.

The main application of resonance is in RF (communications) tuned circuits. A *tuned circuit* allows for a signal of a desired frequency or a band of frequencies to be selected from many available. With resonant tuned circuits we can tune radios, television receivers, transmitters, and electronics equipment in general. The resonant circuit can select one band of frequencies for processing and send it to the output. This occurs even though many different frequencies are present at the input. *Selectivity* is a general term used to describe a circuit that desires to accept a signal for processing at one band of frequencies from many available. Resonance can occur with both series and parallel LC circuits, with different characteristics for each.

25.1 THE RESONANT EFFECT

Figure 25.1(a) illustrates a series *RLC* circuit. At any frequency, the impedance of the circuit is the vector sum of the resistance and reactances. The relationship between R, X_L, and X_C for the series *RLC* circuit is indicated by the vector diagram in Figure 25.1(b). Similarly, Figure 25.1(c) illustrates how the magnitude of X_L, X_C, and Z vary with frequency. In this figure note that X_L increases with frequency, whereas X_C decreases. *Resonance* is defined as the frequency at which the circuit phase angle, θ, is $0°$ (note that resonance occurs at *one frequency only* for a specific value of L and C). For the series *RLC* circuit, this condition occurs when the magnitude of the inductive and capacitive reactances are equal ($X_L = X_C$). Note that the impedance curve has a minimum value that is equal to R_S at the resonant frequency, f_O.

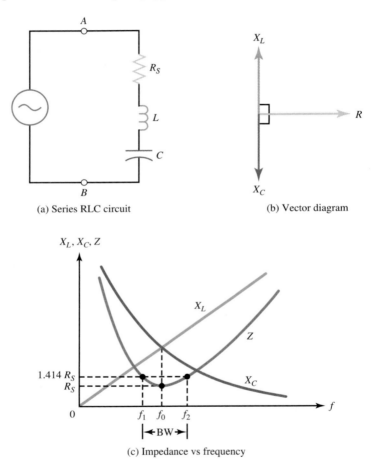

(a) Series RLC circuit

(b) Vector diagram

(c) Impedance vs frequency

FIGURE 25.1
The series RLC circuit: (a) series RLC circuit. (b) vector diagram. (c) impedance versus frequency.

Notice the peculiar responses for X_L and X_C in Figure 25.1. The curve for X_L is a straight line, whereas that for X_C is curved. Both reactances are calculated from linear equations, but the equation for X_C is an inverse function ($1/2\pi fC$). The inverse function results in a curved response for X_C.

25.1.1 Resonant Frequency Equation

By equating $2\pi fL$ to $1/2\pi fC$ (the equations for X_L and X_C, respectively) and solving for f yields:

$$2\pi fL = \frac{1}{2\pi fC}$$

$$1 = 4\pi^2 f^2 LC$$

$$f^2 = \frac{1}{4\pi^2 LC}$$

$$f_O = \frac{1}{2\pi \sqrt{LC}}$$

(25.1)

The subscript *o* simply indicates this is the resonant frequency. The *o* represents *output* frequency, *operating* frequency, or even *oscillating* frequency, depending on one's interpretation. (Some electronics sources use f_R to indicate the resonant frequency.)

EXAMPLE 25.1

Determine the resonant frequency for a series *RLC* circuit when $L = 10$ mH and $C = 0.1$ μF.

Solution $\quad f_O = \dfrac{1}{2\pi \sqrt{LC}}$

$$= \frac{1}{2\pi \sqrt{10 \times 10^{-3} \text{ H} \times 0.1 \times 10^{-6} \text{ F}}}$$

$$= 5.03 \times 10^3 \text{ Hz}$$

$$= 5.03 \text{ kHz}$$

Optional Calculator Sequence

The Sharp calculator sequence for this problem is as follows.

Step	Keypad Entry	Top Display Response
1	1/	1/
2	2π	$1/2\pi$
3	$\sqrt{}$	$1/2\pi\sqrt{}$
4	(10Exp−03	$1/2\pi\sqrt{}($
5	x	$1/2\pi\sqrt{}(10\text{E}{-}03*$
6	0.1Exp−06)	$1/2\pi\sqrt{}(10\text{E}{-}03*0.1\text{E}{-}06)$
7	=	$1/2\pi\sqrt{}(10\text{E}{-}03*0.1\text{E}{-}06)=$
8	Answer	5.03×10^{03}
		(bottom display)

EXAMPLE 25.2

Determine the resonant frequency for a series *RLC* circuit when $L = 1.0$ μH and $C = 4.7$ μF.

Solution $\quad f_O = \dfrac{1}{2\pi \sqrt{LC}}$

$$= \frac{1}{2\pi \sqrt{1.0 \times 10^{-6} \text{ H} \times 4.7 \times 10^{-6} \text{ F}}}$$

$$= 73.4 \times 10^3 \text{ Hz}$$

$$= 73.4 \text{ kHz}$$

EXAMPLE 25.3

Determine the resonant frequency for a series RLC circuit when $L = 5.0\,\mu\text{H}$ and $C = 2.7\text{ pF}$.

Solution $f_O = \dfrac{1}{2\pi\sqrt{LC}}$

$$= \dfrac{1}{2\pi\sqrt{(5.0 \times 10^{-6}\text{ H} \times 2.7 \times 10^{-12}\text{ F})}}$$

$$= 43.3 \times 10^6\text{ Hz}$$

$$= 43.3\text{ MHz}$$

Practice Problems 25.1   Answers are given at the end of the chapter.
1. Solve for the resonant frequency when $L = 250\,\mu\text{H}$ and $C = 5\text{ pF}$.
2. Solve for the resonant frequency when $L = 1\text{ mH}$ and $C = 10\,\mu\text{F}$.

Note that the individual values of L and C (or their product) are inversely proportional to the resonant frequency. Thus, either one or both can be varied in order to change the resonant frequency. In most tuned circuits, either L or C is a variable component (variable capacitors are more common). Thus, the resonant frequency can be adjusted as necessary.

Equation (25.1) can be rearranged to indicate the value of capacitance required to resonate with a particular value of inductance, and vice versa. Thus,

$$C = \dfrac{1}{4\pi^2 f_O^2 L} \tag{25.2}$$

$$L = \dfrac{1}{4\pi^2 f_O^2 C} \tag{25.3}$$

EXAMPLE 25.4

Determine the value of capacitance needed in order to resonate a circuit at 100 kHz when $L = 10\text{ mH}$.

Solution $C = \dfrac{1}{4\pi^2 f_O^2 L}$

$$= \dfrac{1}{4\pi^2 \times (100 \times 10^3\text{ Hz})^2 \times 10 \times 10^{-3}\text{ H}}$$

$$= 253.3 \times 10^{-12}\text{ F}$$

$$= 253.3\text{ pF}$$

Optional Calculator Sequence

The Sharp calculator sequence for this problem is as follows.

Step	Keypad Entry	Top Display Response
1	1/	1/
2	$(4\pi,x^2$	$1/(4\pi^2$
3	x(	$1/(4\pi^2($
4	100Exp3)	$1/(4\pi^2(100\text{E}03)$
5	x^2	$1/(4\pi^2(100\text{E}03)^2$
6	×10Exp−3)	$1/(4\pi^2(100\text{E}03)^2*10\text{E}{-}03)$
7	=	$1/(4\pi^2(100\text{E}03)^2*10\text{E}{-}03)=$
8	Answer	253.3×10^{-12} (bottom display)

| EXAMPLE 25.5 | Determine the value of inductance needed in order to resonate a circuit at 25 kHz when $C = 0.022 \ \mu F$. |

Solution

$$L = \frac{1}{4\pi^2 f_O^2 C}$$

$$= \frac{1}{4\pi^2 \times (25 \times 10^3 \text{ Hz})^2 \times 0.022 \times 10^{-6} \text{ F}}$$

$$= 1.84 \times 10^{-3} \text{ H}$$

$$= 1.84 \text{ mH}$$

Practice Problems 25.2 Answers are given at the end of the chapter.

1. Determine the value of capacitance needed in order to resonate a circuit at 1 MHz if $L = 33 \ \mu H$.
2. Determine the value of inductance needed in order to resonate a circuit at 500 kHz if $C = 40$ pF.

SECTION 25.1 SELF-CHECK

Answers are given at the end of the chapter.

1. Define the term *resonance*.
2. Impedance is _____ for a series resonant circuit.
3. The values of L and C are _____ proportional to the resonant frequency.

25.2   SERIES RESONANCE

When the conditions for resonance occur in a series *RLC* circuit ($\theta = 0°$), the circuit's condition is known as *series resonance*. Figure 25.2 contains a great deal of information about a series resonant circuit. By carefully examining it, you can deduce the following:

1. For frequencies less than f_O, $X_C > X_L$. Thus, the net reactance is capacitive. For this reason, the circuit appears to the source as a series *RC* circuit, as illustrated in Figure 25.2(b). This condition is described by saying the circuit exhibits a *capacitive complexion*.
2. For frequencies greater than f_O, $X_L > X_C$, which results in a net inductive reactance, as illustrated in Figure 25.2(d). This condition is described by saying the circuit exhibits an *inductive complexion*.
3. When the source frequency equals f_O, $X_L = X_C$. Consequently, the net reactance is zero, making the impedance purely resistive and equal to R_S, as shown in Figure 25.2(c).
4. Because the impedance is reduced to R_S, a series resonant circuit produces a very high current.

25.2.1   The Quality Factor, Q, and Bandwidth, BW

An important figure of merit that determines many characteristics of a resonant circuit is the *quality factor, Q*. It was previously defined as the ratio of energy stored (and later returned) by the inductive reactance to energy dissipated by circuit resistance. It is calculated as follows:

$$Q = \frac{X_L}{R_S} \tag{25.4}$$

Lab Reference: Series resonant
measured and calculated values are
demonstrated in Exercise 37.

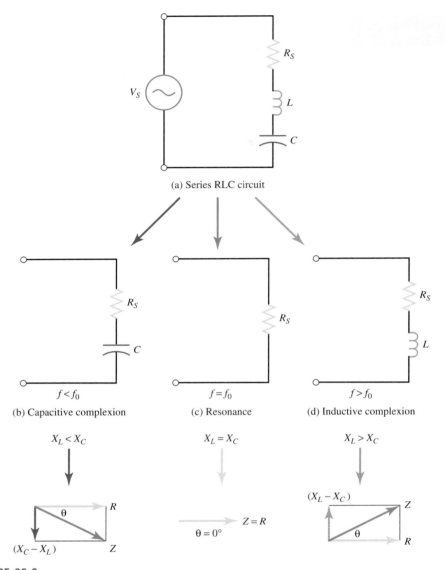

FIGURE 25.2
*Equivalent circuits for the series RLC circuit for frequencies below, at, and above the
resonant frequency: (a) series RLC circuit; (b) capacitive complexion; (c) resonance;
(d) inductive complexion.*

EXAMPLE 25.6

Given a series circuit in which $X_L = 1\ k\Omega$ and $R_S = 50\ \Omega$, determine the Q of the coil.

Solution $\quad Q = \dfrac{X_L}{R_S}$

$\qquad = \dfrac{1\ k\Omega}{50\ \Omega}$

$\qquad = 20$

Practice Problems 25.3   Answers are given at the end of the chapter.
1. Given a series circuit in which $X_L = 2.5\ k\Omega$ and $R_S = 100\ \Omega$, determine the Q of
the coil.

The value for Q is dimensionless. It is simply a numerical factor without any units. For RLC circuits, a value of Q less than 10 is considered low. Values between 10 and 300 are considered as a high Q. A value greater than 300 is considered a very high Q.

The Q has the same value if it is calculated with X_C instead of X_L, because they are equal at resonance. However, the Q of a circuit is generally associated with X_L, because usually the coil contains the series resistance of the circuit or at least a part of it.

Note in Figure 25.1(c) that for frequencies on either side of f_o, the impedance increases rapidly. Thus, the circuit responds well only to the narrowband of frequencies between what is identified as f_1, the lower cutoff frequency, and f_2, the upper cutoff frequency. At either f_1 or f_2, the impedance has a magnitude of $1.414R_S$ (shown in Figure 25.1). This indicates that the current flowing in the circuit at f_1 or f_2 has a magnitude of $V_S/1.414R_S$, or 70.7% (0.707) of the maximum value of V_S/R_S. In addition, the power supplied to the circuit at f_1 or f_2 is 50%. Hence, f_1 and f_2 are also referred to as the *half-power* points. As will be shown later, the half-power points are referred to as the -3-dB points.

The band of frequencies between f_1 and f_2 is called the circuit *bandwidth* (BW). Frequencies between these two points produce a current in the series resonant RLC circuit that is 70.7% of I_{max} or greater. Because the current levels for those frequencies within the band are higher, they are passed on, whereas frequencies outside this band are rejected. This process allows a tuned circuit to be selective and accept only those frequencies that have 70.7% or more of the maximum current. An interesting relationship exists between the bandwidth, resonant frequency, and quality factor. Specifically,

$$\text{BW} = \frac{f_o}{Q} \tag{25.5}$$

Thus, for a given resonant frequency, the higher Q is, the narrower the BW, and vice versa. High-Q circuits produce a "sharper" resonance effect. This is because the value of Q determines the *shape factor* of the impedance curve. High-Q circuits produce taller, narrow-bandwidth curves, whereas lower-Q circuits produce lower, wider-bandwidth curves. By observation you can see that to shape the curve to any desired bandwidth at a particular resonant frequency, one has only to adjust the value of R_S or X_L, as necessary. A higher Q can also be obtained by increasing X_L through the increase of L and the decrease of C. In this way, the resonant frequency remains the same, but the Q varies.

An example of where a very high Q is needed is in the telecommunications field. In order for an analog cellular phone company to be economically sound, as many subscribers as possible are allocated within the available bandwidth of the telco. To keep the channels separated so that cross talk does not occur between adjacent channels, a band-pass filter (Chapter 26) with a very high Q is employed. The tall, narrow response of the filter provides the necessary channel separation for optimum circuit performance.

The equations for f_1 and f_2 are rather complicated. In most circuits, the following approximate equations are adequate when Q is greater than 2 ($Q > 2$):

$$f_1 \approx f_o - \frac{\text{BW}}{2} \tag{25.6}$$

$$f_2 \approx f_o + \frac{\text{BW}}{2} \tag{25.7}$$

EXAMPLE 25.7

Figure 25.3 illustrates current-versus-frequency curves for two RLC series circuits. Note that each curve has the same resonant frequency (1 kHz) and current value (10 A) at resonance. Calculate the bandwidth for the circuits represented by each curve. Also, estimate the cutoff frequencies, f_1 and f_2, for each.

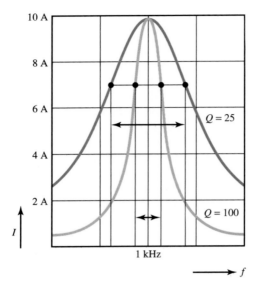

FIGURE 25.3
Bandwidth decreases as Q increases.

Solution For $Q = 25$:

$$\text{BW} = \frac{f_O}{Q}$$

$$= \frac{1 \text{ kHz}}{25}$$

$$= 40 \text{ Hz}$$

$$f_1 \approx f_O - \frac{\text{BW}}{2}$$

$$\approx 1 \text{ kHz} - \frac{40 \text{ Hz}}{2}$$

$$\approx 980 \text{ Hz}$$

$$f_2 \approx f_O + \frac{\text{BW}}{2}$$

$$\approx 1 \text{ kHz} + \frac{40 \text{ Hz}}{2}$$

$$\approx 1020 \text{ Hz}$$

For $Q = 100$:

$$\text{BW} = \frac{f_O}{Q}$$

$$= \frac{1 \text{ kHz}}{100}$$

$$= 10 \text{ Hz}$$

$$f_1 \approx f_O - \frac{\text{BW}}{2}$$

$$\approx 1 \text{ kHz} - \frac{10 \text{ Hz}}{2}$$

$$\approx 995 \text{ Hz}$$

$$f_2 \approx f_O + \frac{\text{BW}}{2}$$

$$\approx 1\text{ kHz} + \frac{10\text{ Hz}}{2}$$

$$\approx 1005\text{ Hz}$$

Practice Problems 25.4   Answers are given at the end of the chapter.
1. Calculate the values of f_1 and f_2 for a resonant circuit having $Q = 50$ and $f_O = 790$ kHz.

25.2.2   The Magnification of V_s

In a series resonant circuit, Q also indicates the factor by which the source voltage is multiplied in order to obtain the magnitude of the inductor and capacitor voltages at resonance. This is shown as

$$V_L = V_C = QV_S \tag{25.8}$$

EXAMPLE 25.8

Given $V_S = 10$ V, $X_L = 100\ \Omega$, and $R_S = 10\ \Omega$, determine V_L for a series RLC resonant circuit.

Solution $Q = \dfrac{X_L}{R_S}$

$$= \frac{100\ \Omega}{10\ \Omega}$$

$$= 10$$

$$V_L = QV_S$$

$$= 10 \times 10\text{ V}$$

$$= 100\text{ V}$$

Practice Problems 25.5   Answers are given at the end of the chapter.
1. Given $V_S = 25$ V, $X_L = 2.5$ kΩ, and $R_S = 100$, determine V_L for a series RLC resonant circuit.

Remember, the reason that V_L and V_C are greater than the source voltage is simply that when X_L and X_C cancel, a large circuit current is produced. The large current flowing through both reactances produces a rather substantial voltage drop.

25.2.3   Summary of Series Resonant Circuits

The following example and statements summarize the series RLC resonant circuit.

EXAMPLE 25.9

Calculate the value of C required for resonance in Figure 25.4(a). Also calculate the current and voltages across the various components.

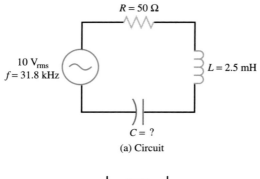

(a) Circuit

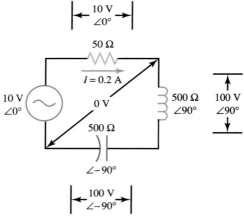

(b) Frequency domain responses

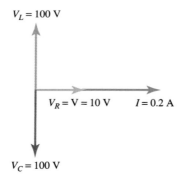

(c) Phasor diagram

FIGURE 25.4
Circuit and frequency responses for Example 25.9: (a) circuit; (b) frequency domain responses; (c) phasor diagram.

Solution

Given $L = 2.5$ mH and $f_O = 31.8$ kHz

$$C = \frac{1}{4\pi^2 f_O^2 L}$$

$$= \frac{1}{4\pi^2 \times (31.8 \times 10^3 \text{ Hz})^2 \times 2.5 \times 10^{-3} \text{ H}}$$

$$= 10 \times 10^{-9} \text{ F}$$

$$= 0.01 \ \mu\text{F}$$

Because the circuit is resonant, $Z = R_S = 50 \ \Omega$. Thus,

$$I_T = \frac{V_S}{R_S}$$

$$= \frac{10 \text{ V}}{50 \ \Omega}$$

$$= 0.2 \text{ A}$$

$$X_L = 2\pi f L$$

$$= 2\pi \times (31.8 \times 10^3 \text{ Hz}) \times (2.5 \times 10^{-3} \text{ H})$$

$$= 500 \ \Omega$$

In a series circuit, X_L has a vector at an angle of 90°. At resonance the magnitudes of X_L and X_C are equal, but they have opposite phase shifts. Thus, X_C must equal $500 \ \Omega\angle-90°$ for the circuit in Figure 25.4.

$$V_R = 0.2 \text{ A}\angle 0° \times 50 \ \Omega\angle 0°$$

$$= 10 \text{ V}\angle 0°$$

$$V_L = 0.2 \text{ A}\angle 0° \times 500 \ \Omega\angle 90°$$

$$= 100 \text{ V}\angle 90°$$

$$V_C = 0.2 \text{ A}\angle 0° \times 500 \ \Omega\angle-90°$$

$$= 100 \text{ V}\angle-90°$$

If you prefer, V_L and V_C may be calculated as follows:

$$Q = \frac{X_L}{R_S}$$

$$= \frac{500 \ \Omega}{50 \ \Omega}$$

$$= 10$$

$$V_L = V_C = QV_S$$

$$= 10 \times 10 \text{ V}$$

$$= 100 \text{ V}$$

Naturally, V_L leads I_T by 90°, and V_C lags I_T by 90° (ELI the ICEman). For this reason, the voltage across the series combination of L and C is zero. Figure 25.4(b) and (c) summarizes the various responses.

The series resonant circuit has the following important characteristics:

1. The impedance is purely resistive, minimum, and equals the series resistance, R_S. At resonance, the source voltage and current are in phase ($\theta = 0°$).
2. The current is maximum and equals V_S/R_S.
3. The source voltage is dropped across the series resistance, R_S.
4. The power factor of the circuit at resonance is 1 (100%).
5. V_L and V_C are equal in magnitude and 180° out of phase with each other.
6. The magnitude of the inductor and capacitor voltage are greater than the source voltage by a factor of Q at resonance.

SECTION 25.2
SELF-CHECK

Answers are given at the end of the chapter.

1. State the criteria needed for series resonance.
2. Above f_o, the series resonant circuit acts _____.
3. For what values is Q considered to be a *high* Q?
4. Define the term *bandwidth*.
5. What does the value of Q determine?
6. For a series resonant circuit, the values of V_L and V_C are _____ times V_S.

25.3   PARALLEL RESONANCE

25.3.1   The Ideal Tank Circuit

The parallel combination of a capacitor and an inductor is called a *tank circuit*. This is because at resonance the circuit acts as a *storage device*. If the circuit resistance is negligible, the energy from the source is exchanged between the inductance and the capacitance at a rate corresponding the frequency of the applied signal. This exchange continues even if the signal source is removed. This is due to the constant charge and discharge of the capacitor and the expansion and collapsing of the magnetic field in the inductor. The oscillations of charge and discharge that occur at the natural *LC* resonant frequency produce what is known as a *flywheel effect.* The oscillations continue (flywheel) until such time as their energy is expended by any circuit resistance.

An ideal tank circuit is illustrated in Figure 25.5(a). This circuit is considered ideal because it does not contain resistance. As before, resonance occurs when $X_L = X_C$. Thus, the formula for the resonant frequency, f_o, is the same as it was for the series *RLC* circuit ($f_O = 1/2\pi\sqrt{LC}$).

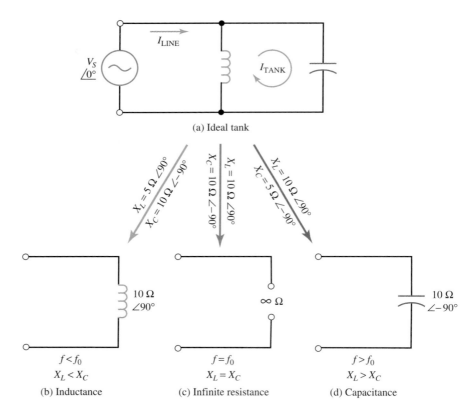

FIGURE 25.5
Characteristics of an ideal tank circuit: (a) ideal tank; (b) inductance; (c) infinite resistance; (d) capacitance.

At any frequency, the impedance of the tank circuit is

$$Z = \frac{X_L \times X_C}{X_L + X_C} \tag{25.9}$$

For frequencies less than f_O, $X_C > X_L$. Consequently, the circuit exhibits an *inductive complexion,* which is just the opposite of the series circuit. To understand why this happens, assume that at some frequency less than f_O, $X_C = 10\ \Omega\angle-90°$ and $X_L = 5\ \Omega\angle90°$. Thus,

$$Z = \frac{10\ \Omega\angle{-90°}\ \times\ 5\ \Omega\angle{90°}}{10\ \Omega\angle{-90°}\ +\ 5\ \Omega\angle{90°}}$$

$$= 10\ \Omega\angle{90°}$$

As you can see, the circuit appears to the source to be an inductive reactance of $10\ \Omega\angle{90°}$. Another way to see this is to realize that because X_L is less than X_C, then I_L is greater than I_C. From the point of view of current, the circuit has an inductive complexion. For frequencies greater than f_O, $X_L > X_C$. Similarly, it can be shown that the circuit exhibits a capacitive complexion.

As the source frequency approaches the resonant frequency of the tank, X_L approaches the value of X_C. Thus, the denominator of equation (25.6) approaches zero. This indicates that the impedance of the ideal tank is *infinite* at f_O. As far as the source is concerned, it sees an open circuit. For this reason, the line current in Figure 25.5(a) is zero at f_O. This does not mean that the inductor and capacitor currents are zero, because the source voltage is still connected across each component. At resonance then (for an ideal circuit),

$$Z_{\text{tank}} = \infty\ \Omega$$
$$I_{\text{line}} = 0\ \text{A}$$
$$X_L = X_C$$
$$I_L = \frac{V_S\angle{0°}}{X_L\angle{90°}}$$
$$I_C = \frac{V_S\angle{0°}}{X_C\angle{-90°}}$$

Because I_L lags V_S by 90° and I_C leads V_S by 90°, I_L and I_C are 180° out of phase with each other. This means that when the current is flowing in one direction through L, an equal current is flowing in the opposite direction through C. Thus, the phasor sum of I_L and I_C is zero at f_O—which, of course, equals the line current. Incidentally, the current flowing through L and C is called the *tank current*, I_{tank}. Figure 25.5(b), (c), and (d) summarizes the action of the ideal tank circuit.

25.3.2 A Real Tank Circuit

In a real tank circuit, the reactive branch currents, I_L and I_C, are not equal. This is because the inductor contains some effective resistance; thus, its current is lower than I_C. The Q of the inductor at parallel resonance is

$$Q = \frac{X_L}{R_S}$$

Notice that this is the same equation used in series resonant circuits (equation (25.4)). This value is also the Q of the parallel resonant circuit with no shunt (parallel) damping resistance.

For parallel resonance, the main criteria are

1. Zero phase angle,
2. Maximum Z and minimum I_{line},
3. $X_L = X_C$.

When the Q is less than 10 (low Q), these three criteria do not occur at the same frequency. For high-Q ($Q \geq 10$) circuits, we consider the inductor's effective resistance to be negligible, and the three effects are acceptably achievable. Parallel resonant circuits are designed so that they have as small a loss as possible, which means they exhibit a preferred higher-Q value.

The Q of a real tank circuit is found by examining the tank current and the line current. The tank current (I_L or I_C) equals the line current magnified by the factor of Q (an opposite effect to that which happens in series resonant circuits with V_L or V_C and V_S). Thus, Q is equal to

$$Q = \frac{I_{\text{tank}}}{I_{\text{line}}} \tag{25.10}$$

$$I_{\text{tank}} = I_L = I_C = Q\,(I_{\text{line}}) \tag{25.11}$$

Another form of Q is

$$Q = \frac{Z}{X_L} \tag{25.12}$$

$$Z = QX_L \tag{25.13}$$

This shows that the impedance of a real-tank circuit is Q times the inductive reactance at resonance. The problem with equation (25.12) (and real tank calculations) is that the impedance is difficult to determine without a direct measurement. The problem with an ideal tank (and ideal tank calculations) is that the impedance is infinite. In either case (ideal or real), the impedance value is difficult to work with. To solve this problem, a practical tank uses a shunt- (parallel-) connected *damping resistor.* It is called a damping resistor because it effectively loads the tank and lowers its impedance. This resistor simplifies the calculations for circuit analysis, as shown in the next section.

25.3.3 Damping of Parallel Resonant Circuits

A practical tank circuit is derived by first converting a series inductor with associated series resistance to a parallel inductor with associated parallel resistance. The associated capacitance needed to complete the resonant circuit is the same for either series or parallel resonance. The inductor conversion is shown in Figure 25.6, along with the conversion calculations. (Note that these calculations are for a $Q < 10$ (low Q).)

Lab Reference: The conversion of a series resonant circuit to a practical parallel resonant circuit is demonstrated in Exercise 38.

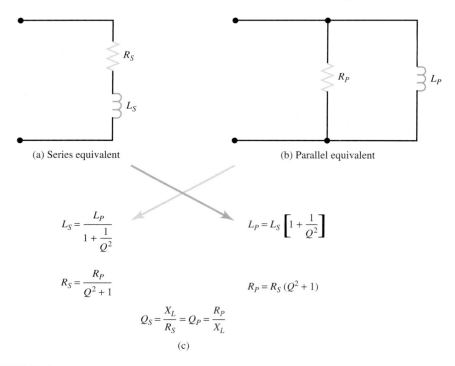

(a) Series equivalent   (b) Parallel equivalent

$$L_S = \frac{L_P}{1 + \dfrac{1}{Q^2}} \qquad\qquad L_P = L_S\left[1 + \frac{1}{Q^2}\right]$$

$$R_S = \frac{R_P}{Q^2 + 1} \qquad\qquad R_P = R_S\,(Q^2 + 1)$$

$$Q_S = \frac{X_L}{R_S} = Q_P = \frac{R_P}{X_L}$$

(c)

FIGURE 25.6

Conversion of series inductance and resistance to parallel inductance and resistance (low-Q circuit): (a) series equivalent; (b) parallel equivalent; (c) conversion equations.

Many practical tank circuits try to achieve a $Q > 10$; thus for values of $Q \geq 10$, the equations in Figure 25.6 are modified to the following approximate equations:

$$R_P = Q^2 R_S \tag{25.14}$$

$$R_S = \frac{R_P}{Q^2} \tag{25.15}$$

$$L_S = L_P \tag{25.16}$$

The Q and impedance of the parallel resonant circuit are

$$Q = \frac{R_P}{X_L} \tag{25.17}$$

$$Z = R_P \tag{25.18}$$

EXAMPLE 25.10

A 50.64-mH inductor has a resistance of 157 Ω. The inductor is placed in parallel with a 0.005-μF capacitor to form a tank circuit that is resonant at 10 kHz. Calculate the Q of the inductor, determine the tank impedance at resonance, and show the equivalent circuit with an R_P.

Solution   At the 10-kHz resonant frequency,

$$X_L = 2\pi f L$$
$$= 2\pi \times (10 \times 10^3 \text{ Hz}) \times (50.64 \times 10^{-3} \text{ H})$$
$$= 3.18 \text{ k}\Omega$$

$$X_C = \frac{1}{2\pi f C}$$

$$= \frac{1}{2\pi \times (10 \times 10^3 \text{ Hz}) \times (0.005 \times 10^{-6} \text{ F})}$$

$$= 3.18 \text{ k}\Omega$$

The Q of the inductor is

$$Q = \frac{X_L}{R_S}$$

$$= \frac{3.18 \times 10^3 \ \Omega}{157 \ \Omega}$$

$$= 20.3$$

Because $Q \geq 10$, we can use the approximate equations to determine the parallel equivalent circuit.

$$Z = R_P = Q^2 R_S$$
$$= (20.3)^2 \times 157 \ \Omega$$
$$= 64.7 \text{ k}\Omega$$
$$L_P = L_S = 50.64 \text{ mH}$$

See Figure 25.7.

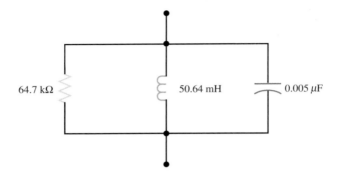

FIGURE 25.7
Parallel equivalent circuit for Example 25.10.

EXAMPLE 25.11

Figure 25.8 illustrates a parallel resonant circuit. Calculate the current through each component, Q, and BW, and estimate the upper and lower cutoff frequencies.

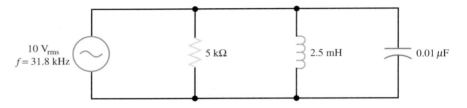

FIGURE 25.8
Circuit for Example 25.11.

Solution   Because the circuit is resonant, $Z = R_P = 5$ kΩ. Thus,

$$I_{\text{line}} = \frac{V_S}{R_P}$$

$$= \frac{10 \text{ V}\angle 0°}{5 \text{ k}\Omega\angle 0°}$$

$$= 2 \text{ mA}\angle 0°$$

$$X_L = 2\pi fL$$

$$= 2\pi \times (31.8 \times 10^3 \text{ Hz}) \times (2.5 \times 10^{-3} \text{ H})$$

$$= 500 \text{ }\Omega\angle 90°$$

At resonance $X_L = X_C$. Thus, $X_C = 500 \text{ }\Omega\angle -90°$. Because $V_S = 10$ V, we have

$$I_L = \frac{10 \text{ V}\angle 0°}{500 \text{ }\Omega\angle 90°}$$

$$= 20 \text{ mA}\angle -90°$$

$$I_C = \frac{10 \text{ V}\angle 0°}{500 \text{ }\Omega\angle -90°}$$

$$= 20 \text{ mA}\angle 90°$$

$$Q = \frac{R_P}{X_L}$$

$$= \frac{5 \text{ k}\Omega}{500 \text{ }\Omega}$$

$$= 10$$

If you prefer, I_L and I_C may be calculated as follows:

$$I_{tank} = I_C = I_L = Q\,(I_{line}) = 10 \times 2\text{ mA} = 20\text{ mA}$$

Naturally, I_L lags V_S by 90° and I_C leads V_S by 90°.

$$\text{BW} = \frac{f_O}{Q}$$

$$= \frac{31.8\text{ kHz}}{10}$$

$$= 3.18\text{ kHz}$$

$$f_1 \approx f_O - \left(\frac{\text{BW}}{2}\right)$$

$$\approx 31.8\text{ kHz} - \frac{3.18\text{ kHz}}{2}$$

$$\approx 31.8\text{ kHz} - 1.59\text{ kHz}$$

$$\approx 30.21\text{ kHz}$$

$$f_2 \approx f_O + \left(\frac{\text{BW}}{2}\right)$$

$$\approx 31.8\text{ kHz} + \frac{3.18\text{ kHz}}{2}$$

$$\approx 31.8\text{ kHz} + 1.59\text{ kHz}$$

$$\approx 33.39\text{ kHz}$$

Practice Problems 25.6   Answers are given at the end of the chapter.

1. For the circuit in Figure 25.9, find the value of R_P in order to convert the circuit to a practical tank circuit. Also determine f_O, X_L, and the inductor Q.

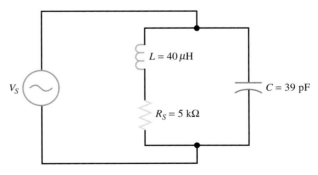

FIGURE 25.9
Circuit for Practice Problems 25.6, Problem 1.

2. For the circuit in Figure 25.10, determine f_O, X_L, X_C, I_L, I_C, Q, BW, f_1, and f_2.

FIGURE 25.10
Circuit for Practice Problems 25.6, Problem 2.

25.3.4 Summary of a Practical Parallel Resonant Circuit

The following statements summarize a practical parallel resonant circuit.

1. The impedance is maximum and equals the shunt resistance, R_P.
2. Line current is minimum and equals V_S/R_P.
3. The entire source current flows through R_P.
4. I_L and I_C are equal in magnitude and 180° out of phase with each other.
5. The magnitude of the inductor and capacitor currents are greater than the line current by the factor of Q at resonance.
6. $Q = R_P/X_L$.

SECTION 25.3 SELF-CHECK

Answers are given at the end of the chapter.

1. Why is a parallel resonant circuit called a *tank* circuit?
2. Above parallel resonance, the circuit acts _____.
3. What is the impedance of an ideal tank circuit?
4. For a practical tank circuit, the impedance is equal to _____.

■ SUMMARY

1. X_L is equal to X_C at one particular frequency, known as the resonant frequency.
2. Resonance is defined as the frequency at which the circuit phase angle is 0°.
3. Individual values of L or C or their product are inversely proportional to the resonant frequency.
4. At series resonance, for frequencies less than f_O the circuit acts capacitively.
5. At series resonance, the impedance is equal to R_S.
6. The quality factor, Q, is the ratio of energy stored by the inductive reactance to energy dissipated by circuit resistance.
7. A $Q < 10$ is considered low, values between 10 and 300 are considered high, and a value greater than 300 is considered very high.
8. The cutoff frequencies are denoted by f_1 (lower cutoff frequency) and f_2 (upper cutoff frequency).
9. The cutoff frequencies occur at the frequency where the current is 70.7% of its maximum.
10. The difference between f_1 and f_2 is called the bandwidth, BW.
11. The value of Q determines the *shape factor* of the circuit and, thus, the value of BW.
12. At series resonance, $Z = R_S$ and is minimum, whereas the current is maximum.
13. A parallel resonant circuit is known as a tank circuit.
14. An ideal parallel resonant circuit has an infinite impedance.
15. In order to simplify parallel resonant circuit analysis, a shunt-connected damping resistor is connected across the tank.
16. For a practical tank circuit, the impedance is maximum and equal to R_P.
17. For a practical tank circuit, the line current is minimum, whereas the tank current is maximum.

■ IMPORTANT EQUATIONS

$$f_O = \frac{1}{2\pi\sqrt{LC}} \tag{25.1}$$

$$C = \frac{1}{4\pi^2 f_O^2 L} \tag{25.2}$$

$$L = \frac{1}{4\pi^2 f_O^2 C} \tag{25.3}$$

$$Q = \frac{X_L}{R_S} \tag{25.4}$$

$$\text{BW} = \frac{f_O}{Q} \tag{25.5}$$

$$f_1 \approx f_O - \frac{\text{BW}}{2} \tag{25.6}$$

$$f_2 \approx f_O + \frac{\text{BW}}{2} \tag{25.7}$$

$$V_L = V_C = QV_S \tag{25.8}$$

$$Z = \frac{X_L \times X_C}{X_L + X_C} \tag{25.9}$$

$$Q = \frac{I_{\text{tank}}}{I_{\text{line}}} \tag{25.10}$$

$$I_{\text{tank}} = I_L = I_C = Q\,(I_{\text{line}}) \tag{25.11}$$

$$Q = \frac{Z}{X_L} \tag{25.12}$$

$$Z = QX_L \tag{25.13}$$

$$R_P = Q^2 R_S \tag{25.14}$$

$$R_S = \frac{R_P}{Q^2} \tag{25.15}$$

$$L_S = L_P \qquad (25.16)$$

$$Q = \frac{R_P}{X_L} \qquad (25.17)$$

$$Z = R_P \qquad (25.18)$$

■ REVIEW QUESTIONS

1. Define the term *resonance*.
2. Define the term *selectivity*.
3. State the conditions for series resonance to occur.
4. When the frequency is less than f_O, the series *RLC* circuit acts _____.
5. When the frequency is greater than f_O, the series *RLC* circuit acts _____.
6. Define the term *quality factor*.
7. For what values of Q is the Q considered low, high, and very high?
8. What are the lower and upper cutoff frequencies?
9. Define the term *bandwidth*.
10. How does the value of Q relate to the shape factor of the impedance curve?
11. At series resonance what are the values of V_L and V_C?
12. At series resonance the impedance is _____ and equal to _____.
13. At series resonance the current is _____.
14. How can the value of V_L and V_C be greater than V_S at series resonance?
15. Define the term *tank circuit*.
16. The impedance of an ideal tank circuit is _____.
17. The line current of an ideal tank circuit is _____.
18. Why is the line current different in a real tank circuit compared to an ideal tank?
19. What is the purpose of a *damping* resistor for a parallel resonant circuit?
20. At parallel resonance, the impedance is _____ and equal to _____.
21. At parallel resonance, the line current is _____.
22. I_L and I_C are greater than I_{line} by the factor of _____.

■ CRITICAL THINKING

1. Show how Q affects the *shape factor* of the impedance curve and its effect on the BW.
2. What are the major differences between series resonance and parallel resonance?
3. Show how the value of series current establishes the cutoff frequencies, f_1 and f_2, and how this relates to the factor of 0.707.
4. Compare and contrast an ideal tank circuit with a real tank circuit with respect to how the impedances, currents, and Q factors are determined in each case.
5. Show how to convert a real tank circuit into a practical tank circuit.

■ PROBLEMS

1. Determine the resonant frequency for the following values of L and C: $L = 10$ mH, $C = 0.01$ μF; $L = 25$ μH, $C = 100$ pF; $L = 2.2$ H, $C = 0.0022$ μF; $L = 250$ mH, $C = 5$ μF.
2. Determine the value of capacitance needed in order to resonate a circuit for the given inductances at the following frequencies.

   $C =$ _____ $L = 100$ mH $f_O = 25$ kHz
   $C =$ _____ $L = 25$ mH $f_O = 100$ kHz
   $C =$ _____ $L = 10$ μH $f_O = 1$ MHz

3. Determine the value of inductance needed in order to resonate a circuit for the given capacitances at the following frequencies.

   $L =$ _____ $C = 0.068$ μF $f_O = 455$ kHz
   $L =$ _____ $C = 100$ pF $f_O = 2.5$ MHz
   $L =$ _____ $C = 0.003$ μF $f_O = 790$ kHz

Complete the following table for all unknowns listed for a series resonant circuit.

	C	L	f_O	$R_S(\Omega)$	Q	f_1	f_2	BW
4.	0.01 μF	10 mH		20				
5.	10 pF	100 μH		50				
6.		22 mH	100 kHz		25			
7.	0.005 μF	200 mH		150				
8.	0.68 μF		500 kHz	33				
9.	2.2 pF	10 μH		100				
10.	0.05 μF	500 μH		40				

11. Convert the circuit in Figure 25.11 into a practical tank circuit and then determine f_O, Q, f_1, f_2, and BW.

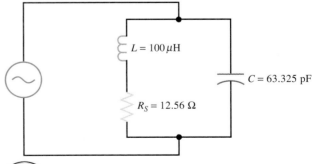

FIGURE 25.11
Circuit for Problem 11.

12. Convert the circuit in Figure 25.12 into a practical tank circuit and then determine f_O, Q, f_1, f_2, and BW.

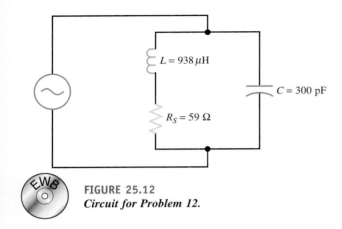

FIGURE 25.12
Circuit for Problem 12.

13. A tank circuit has an $X_L = 5\ k\Omega$ with negligible internal resistance and a shunt R_P of 200 kΩ. Calculate Q and, for a resonant frequency of 500 kHz, determine f_1, f_2, and BW.

14. Find the highest and lowest values of C needed to tune an AM radio broadcast band of 535–1600 kHz using an inductor of 250 μH.

15. Find the highest and low values of C needed to tune a FM radio broadcast band of 88–108 MHz using an inductor of 5 nH.

■ ANSWERS TO PRACTICE PROBLEMS

PRACTICE PROBLEMS 25.1

1. $f_O = 4.5$ MHz
2. $f_O = 1.59$ kHz

PRACTICE PROBLEMS 25.2

1. $C = 768$ pF
2. $L = 2.53$ mH

PRACTICE PROBLEMS 25.3

1. $Q = 25$

PRACTICE PROBLEMS 25.4

1. $f_1 = 782.1$ kHz, $f_2 = 797.7$ kHz

PRACTICE PROBLEMS 25.5

1. $V_L = 625$ V

PRACTICE PROBLEMS 25.6

1. $f_O = 4.03$ MHz, $X_L = 1$ kΩ, $Q = 200$, $R_P = 200$ kΩ
2. $f_O = 6.50$ kHz, $X_L = 1.23$ kΩ, $X_C = 1.22$ kΩ, $I_L = 40.98$ mA, $I_C = 40.98$ mA, $Q = 16.4$, BW = 396 Hz, $f_1 = 6.302$ kHz, $f_2 = 6.698$ kHz

■ ANSWERS TO SECTION SELF-CHECKS

SECTION 25.1 SELF-CHECK

1. Resonance is a condition that occurs at one frequency only, when $X_L = X_C$.
2. Minimum
3. Inversely

SECTION 25.2 SELF-CHECK

1. For series resonance, the phase shift must equal 0°.
2. Inductive
3. Values between 10 and 300 are considered as high Q.
4. The band of frequencies that produce a current of 0.707 or higher of maximum current is denoted as the bandwidth. They occur between the cutoff frequencies of f_1 and f_2.
5. Q determines the shape factor of the impedance curve.
6. Q

SECTION 25.3 SELF-CHECK

1. Because the parallel resonant circuit acts as a storage device
2. Capacitively
3. Infinite
4. R_P

PART VII

AC CIRCUIT APPLICATIONS

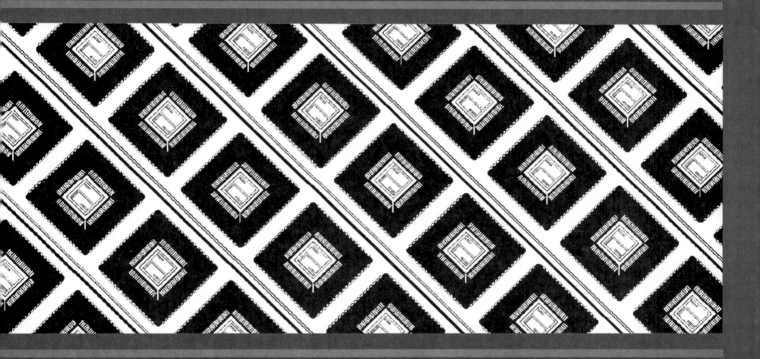

CHAPTER 26

COUPLING AND FILTER CIRCUITS

OBJECTIVES

After completing this chapter, you will be able to:

1. Demonstrate the use of logarithms and the application of decibels to electronic circuit analysis.
2. Solve for unknowns utilizing decibels (both power and voltage ratios).
3. Compare and contrast direct and indirect coupling methods.
4. Define the term *filter* and evaluate various types of filters for circuit performance.
5. Describe the Bode plot and show its application to filter responses.
6. Describe filter applications to electronic circuits.

INTRODUCTION

We saw in the previous chapter how a tuned circuit can select a signal of a desired band of frequencies from many available. When a signal is transferred from one circuit to another, the process is known as *coupling*. A *filter*, as the name implies, is a device that removes, or filters, unwanted signals. Thus, a coupling circuit and a filter are very similar in their electrical characteristics. Often the terms are used interchangeably. The devices that are employed to couple, or filter, circuits include capacitors, inductors, resistors, and transformers.

Before we begin an analysis of how these devices can be used as couplers and filters we need to consider the *decibel*. A decibel is a common magnitude term used when signals of different frequencies and signal levels occur in circuits. These include communication circuits from audio (AF) to extremely high frequencies (EHF). Filters, in particular, use decibels to show how the amplitude varies with respect to frequency.

The *bel* is based on the logarithm of the ratio of two voltages, currents, or power levels. Named after Alexander Graham Bell, this unit of measure was originally used to provide comparisons between the power-loss properties of various telephone parameters. In most electronic circuits, the bel is too large a unit for practical applications. For this reason, the *decibel*, or dB, which is one-tenth of a bel, is used.

26.1 DECIBELS

Decibels are based on the common log (*base* 10). Natural logs are derived from the *base e*, 2.718. Among other things, natural logs are used to describe the exponential charge and discharge of the capacitor, as we saw in previous chapters.

26.1.1 Common Logarithms

Logarithms are used to simplify numerical computations involving exponents and powers of 10. Applying logarithms to decibels is a valuable skill to learn, because decibels are used in many forms of electronic circuits.

The *logarithm* of a number is defined as the exponent that indicates the power to which a base must be raised in order to produce that number. The whole-number part of the logarithm is called the *characteristic,* and the fractional part is called the *mantissa.*

Examples of the common log are as follows.

$$
\begin{array}{rcl}
10^2 = 100 & \rightarrow & \log 100 = 2 \\
10^3 = 1000 & \rightarrow & \log 1000 = 3 \\
10^{4.25} = 17{,}782 & \rightarrow & \log 17{,}782 = 4.25 \\
10^{-1} = 0.1 & \rightarrow & \log 0.1 = -1.0 \\
10^{-2.5} = 0.003162 & \rightarrow & \log 0.003162 = -2.5
\end{array}
$$

Logarithms are commonly used in many scientific and technical calculations. For instance, amplifier gain, signal attenuation, and coupling factor are often expressed as the logarithm of a power ratio. Thus, anyone involved with technical applications benefits by being able to perform logarithmic computations.

The advent of the electronic calculator has made the calculation of a logarithm simple. To find the logarithm of a number, all that is necessary is to enter the number into the calculator and press the **log** button. (*Note:* There are two log buttons on a calculator; one is labeled **log** and the other is labeled **ln.** *Log* refers to the common log, base 10, and *ln* refers to the natural log, base $e = 2.718$. Also note that some D.A.L. (direct algebraic logic) calculators require that you press **log** first, then enter the number, and then press equals (=). (The Sharp calculator is an example of such a calculator.)

EXAMPLE 26.1

Find the log of 40. (This example uses the Sharp calculator.)

Solution
1. Press **log.**
2. Enter **40.**
3. Press =.
4. Read the display.

$$\log 40 = 1.60206$$

(*Note:* The log of a negative number is undefined.)

Practice Problems 26.1   Answers are given at the end of the chapter.
1. Find the logs of the following numbers: 1.0, 0.75, 500, 2000.

26.1.2 Antilogarithms

If you are given the logarithm of a number and asked to find the number it represents, reverse the procedure used to find the log.

EXAMPLE 26.2

Find the antilog of 2.301303. (This example uses the Sharp calculator.)

Solution
1. Press **2nd F.**
2. Press **log.**
3. Enter **2.301303.**
4. Press **=.**
5. Read the display.

   antilog 2.301303 = 200

Practice Problems 26.2   Answers are given at the end of the chapter.
1. Find the antilogs of the following numbers: 1.0, 2.487, −0.512, and 6.0789.

26.1.3   Decibels and Power Ratios

The decibel is the common unit of measure for power amplification or attenuation. Many items of test equipment are calibrated in decibels (dB), and yet many otherwise experienced technical personnel do not understand the term. Actually, the concept of the decibel is simple and the method of calculating decibels is not difficult. The decibel is not linear; it is a logarithmic unit for expressing a *power ratio*. Note that a decibel represents a power ratio, not an actual power. By using logs and decibels, the computations for finding gain or loss are greatly simplified.

The usefulness of all this becomes apparent when we think about how an ear perceives loudness. An ear is very sensitive. The softest audible sound has a power of about 0.000000000001 W/m^2 (1 pW/m^2), whereas the threshold of pain is around 1 W/m^2. This gives a total range of about 120 dB. Our judgment of relative levels of loudness is somewhat logarithmic. If a sound has approximately 10 times the power of a reference (10 dB), we hear it as twice as loud. If we merely double the power (3 dB), the difference is just noticeable.

Gains, or amplification of power, are represented by *positive* values. Losses, or attenuation of power, are represented by *negative* values. When the output power and the input power are equal, the power ratio is 1, and the decibel level is 0 dB. Decibels may be added, subtracted, or combined algebraically to indicate the composite effect of more than one device. The power ratio in decibels is as follows:

$$\text{power ratio (dB)} = 10 \log\left(\frac{\text{power out}}{\text{power in}}\right)$$

This equation may be found in various forms. Common examples are

$$\text{dB} = 10 \log\left(\frac{P_{\text{out}}}{P_{\text{in}}}\right) \tag{26.1}$$

$$\text{dB} = 10 \log\left(\frac{P_2}{P_1}\right) \tag{26.2}$$

EXAMPLE 26.3

An amplifier has an input of 5 W, and the output is 10 W. What is the gain in decibels?

$$\textit{Solution} \quad dB = 10 \log\left(\frac{P_{out}}{P_{in}}\right)$$

$$= 10 \log\left(\frac{10\ W}{5\ W}\right)$$

$$= 10 \log 2$$

$$= 3.01$$

EXAMPLE 26.4

A filter has an input of 1.0 W; the output is 0.5 W. How many decibels is this?

$$\textit{Solution} \quad dB = 10 \log\left(\frac{P_{out}}{P_{in}}\right)$$

$$= 10 \log\left(\frac{0.5\ W}{1\ W}\right)$$

$$= -3.01$$

Practice Problems 26.3 Answers are given at the end of the chapter.
1. A circuit has an input of 20 mW, and the output is 900 mW. How many decibels is this?
2. A circuit has an input of 50 mW, and the output is 35 mW. How many decibels is this?

In Chapter 25, the half-power points (cutoff frequencies) were referred to as the -3-dB points. This is because at the cutoff frequencies, the output power is reduced by 50% of the input power $(0.5(\text{power}) = -3\ dB)$. Thus, the cutoff frequencies are also referred to as the -3-dB frequencies.

To find the overall gain of several stages, either multiply the power ratios together or convert them to decibels and add.

EXAMPLE 26.5

Assume that an amplifier circuit has four stages, each of which amplifies the input signal by two. What is the total gain of this circuit in dB?

$$\textit{Solution} \quad dB = 10 \log\left(\frac{P_{out}}{P_{in}}\right)$$

$$= 10 \log 2$$

$$= 3$$

$$\text{total gain} = 3\ dB + 3\ dB + 3\ dB + 3\ dB = 12\ dB$$

Alternatively, total gain $= 2 \times 2 \times 2 \times 2 = 16$

$$dB = 10 \log\left(\frac{P_{out}}{P_{in}}\right)$$

$$= 10 \log 16$$

$$= 12$$

Practice Problems 26.4 Answers are given at the end of the chapter.
1. A multistage amplifier circuit has the following gain per stage: 4 dB, 6.7 dB, 5 dB, and 3.2 dB. What is the overall gain of the circuit?

Figure 26.1 shows a typical communications block diagram. Amplifier stages usually have gain, whereas the transmission cable has losses associated with its length and frequency of use. Notice that the output is 10 dB higher than the input (10 dB − 20 dB + 20 dB = 10 dB).

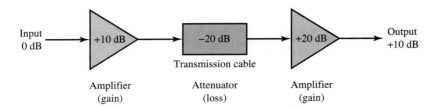

FIGURE 26.1
Decibel notation in a typical communications circuit.

Table 26.1 lists some common gains and losses (in dB) and their equivalent power ratios.

TABLE 26.1
Selected dB gains and losses versus equivalent power ratios

Gain (dB)	Equivalent Power Ratio
1.0	1.25
3.0	2.00
10.0	10.0

Loss (dB)	Equivalent Power Ratio
−1.0	0.8
−3.0	0.5
−10.0	0.1

26.1.4   Decibels and Voltage Ratios

In some circuits you might be examining voltage ratios instead of power ratios. To work these in terms of their logarithmic values requires that the voltage ratio (V_{out}/V_{in}) calculation is done using equal impedances. If the impedances are not equal, an adjustment to the formula must be made. (We assume equal impedances in our problems.)

Because V^2 is proportional to power, the voltage ratio is multiplied by 2 (20 log instead of 10 log). The new equation is

$$dB = 20 \log \left(\frac{V_{out}}{V_{in}} \right) \tag{26.3}$$

EXAMPLE 26.6

The input voltage to an amplifier is 2.5 mV. The output voltage is 3600 mV. What is the gain of the amplifier in decibels?

Solution $dB = 20 \log \left(\dfrac{V_{out}}{V_{in}} \right)$

$= 20 \log \left(\dfrac{3600 \text{ mV}}{2.5 \text{ mV}} \right)$

$= 20 \log 1440$

$= 63.2$

EXAMPLE 26.7

The output voltage of a filter is 0.707 V, and the input is 1.0 V. What is this in decibels?

Solution $dB = 20 \log \left(\dfrac{V_{out}}{V_{in}} \right)$

$= 20 \log \left(\dfrac{0.707 \text{ V}}{1 \text{ V}} \right)$

$= -3.01$

Practice Problems 26.5   Answers are given at the end of the chapter.
1. What is the gain in decibels of a circuit that has a 10-V output and a 100-mV input?
2. What is the loss in decibels of a circuit that has a 3.5-V output and a 7.2-V input?

Table 26.2 lists some common gains and losses (in decibels) and their equivalent voltage ratios.

TABLE 26.2
Selected gains and losses versus equivalent voltage ratios

Gain (dB)	Equivalent Voltage Ratio
1.0	1.118
3.0	1.414
6.0	2.0
20.0	10.0

Loss (dB)	Equivalent Voltage Ratio
−1.0	0.894
−3.0	0.707
−6.0	0.5
−20	0.1

26.1.5   Decibel Reference Levels

Decibels are derived from the ratios of two quantities. They do not represent an actual power but are the ratio of one power to another. To represent an actual power, not a power ratio, decibels are referenced to a *relative power level*. The relative power level becomes a *reference* and replaces the input power value. The concept of relative power comes into play by asking the question, Relative to what? This question has no single answer.

The standard reference level is 0 dB (ratio = 1). It was chosen to be some convenient value for an application in question. Acousticians deal with positive values and call their measurements *dB SPL* (sound pressure level). The reference value for sound pressure is 20 μPa; for sound power, it's 1 pW. For electrical engineers and electronic technicians, there are several references available. The abbreviation dB is often followed by a letter or letters to indicate the reference that is intended for a particular application. One example is the *dBm*. A dBm represents the power level relative to 1 mW. There are reference levels for voltage as well. Table 26.3 lists some common references for decibels.

As you advance in your study of electronics (particularly for communication circuits), you will find many additional examples of decibel problems utilizing the references in Table 26.3.

Table 26.4 lists sound pressure levels in decibels for various sources of sound. This may help answer the question, How loud is loud?

TABLE 26.3
Common references for decibels

Standard	Reference Value
dBW	1 W
dbm	1 mW (at 600 Ω)
dBmV	1 mV
dBV	1 V

TABLE 26.4
Typical sound pressure levels

dB	Sound Source
5	Softest audible sound
40	Library
60	Conversation at 1 m
67	Executive airport minimum for "loud" jet
70	Average street noise
90	Subway train
97	Very loud executive airport jet
110	Rock band
120	Nearby thunder
140	Threshold of pain
200	Saturn moon rocket at close range

SECTION 26.1 SELF-CHECK

Answers are given at the end of the chapter.

1. Define the term *logarithm*.
2. A power gain is represented by a _____ value in decibels.
3. A −3-dB voltage value indicates that the voltage ratio is _____.
4. The cutoff frequencies are noted as the _____ dB points.

26.2 COUPLING CIRCUITS

26.2.1 Direct Coupling

Direct coupling is one type of coupling from one circuit to another. The other type is *indirect coupling*. Circuits that employ direct coupling have the current of the input circuit flowing through a common impedance. A voltage drop occurs across the common impedance, which represents the output voltage of the circuit. Direct coupling is characterized by wideband frequency response, because there is essentially no circuit tuning taking place. Figure 26.2 shows three typical direct-coupled circuits. Note that in resistive coupling (Figure 26.2(a)) the output voltage does not vary with a change in frequency. For inductive coupling (Figure 26.2(b)) there is an increase in output voltage as the frequency increases. For capacitive coupling (Figure 26.2(c)) there is a decrease in output voltage as the frequency increases.

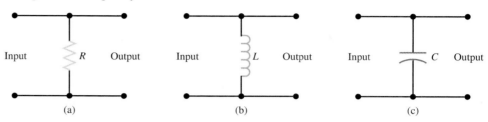

FIGURE 26.2
Examples of direct-coupled circuits: (a) resistive coupled; (b) inductive coupled; (c) capacitive coupled.

Direct-coupled circuits can be used to transfer all frequencies from one circuit to another. When it is necessary to process very low-frequency signals (including DC), direct coupling is used. Direct coupling permits both the alternating current of the signal and any direct current components of the signal to be transferred. In situations where only the alternating current must be passed, indirect coupling is necessary.

26.2.2 Indirect Coupling

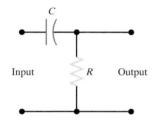

FIGURE 26.3
Indirect coupling utilizing capacitor and resistor.

In addition to being used in direct-coupled circuits, the capacitor can also be employed to couple circuits indirectly. Figure 26.3 shows an example of indirect coupling using a capacitor and a resistor. We know from basic capacitor theory that capacitors block DC while passing AC. If the reactance of the coupling capacitor is small (typically $0.1R$ with respect to the lowest frequency to be applied) compared with the resistance, practically all the AC signal appears across the resistance. In addition, the associated phase shift is kept small.

The disadvantage of capacitive coupling is the difficulty in making a good impedance match between circuits. Consequently, power transfer can be hindered. In addition, besides the DC component, low-frequency signals tend to be blocked by the capacitor as well.

Transformer coupling is considered to be another form of indirect coupling. This is because there is no physical connection between the input and output. In this type of coupling, the signal is transferred via the mutual inductance between the primary and secondary coils. Figure 26.4 shows this form of coupling. Note that the DC component associated with the signal is not transferred to the output.

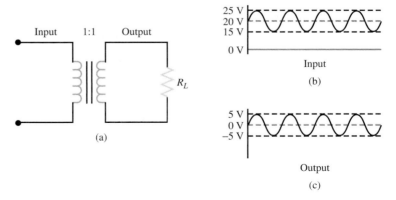

FIGURE 26.4
Indirect coupling utilizing a transformer: (a) circuit; (b) input signal with DC component; (c) output signal without DC component.

Transformer coupling is popular in radio frequency (RF) circuits (AM and FM radios), because it provides good impedance matching and allows for tuning. In many applications, a fixed capacitor is placed across the transformer coils. This provides a mechanism for tuning the frequency response of the circuit, because the transformer inductance can be varied via an adjustable core.

26.3   FILTER CIRCUITS

Filters are used to pass or block a specific range of frequencies. They are either passive or active. *Passive filters* use resistors, capacitors, and/or inductors and do not contain amplifying devices to increase signal strength. Passive filters have an output level (either power or voltage) that is always less than the input. The reason for this is that passive filters attenuate the input signal due to the fact that they are inserted in the line between the input and output. Circuit reactance stores energy, whereas circuit resistance absorbs energy and dissipates it as heat. In either case this act causes an *insertion loss*. The loss represents the difference in the output-signal level before the insertion of the filter and the signal level after insertion. Passive filters are the subject of this chapter.

Active filters employ an amplifying device to compensate for the insertion loss. They are typically included in a text on semiconductor devices.

Filters can be classified into four main types, depending on which frequency components of the input signal are passed to the output. The four types of filters are the low-pass, the high-pass, the band-pass, and the band-stop (reject), or notch.

26.3.1   Low-Pass Filters

A *low-pass filter* employs a circuit that has a relatively constant output from 0 Hz (DC) up to a cutoff, or critical, frequency (f_c). Frequencies above f_c are attenuated. The attenuation is caused by an effective shunt to ground when using an *RC* low-pass circuit (Figure 26.5(a)). In an *RL* low-pass circuit (Figure 26.5(b)), the frequencies above f_c are blocked. In either case, the degree of attenuation increases with frequency. Frequencies below f_c are said to be in the *passband* and provide an output level nearly equal to the input. Low-pass filters are designed using a combination of resistors, inductors, and capacitors.

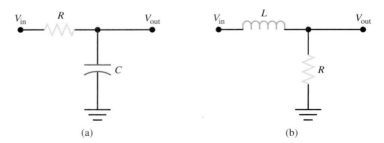

(a)   (b)

FIGURE 26.5
Low-pass filters: (a) RC low-pass filter; (b) RL low-pass filter.

Notice in the *RC* circuit (the more common of the two) that the output is taken across the capacitor (the capacitor is in parallel to the load) and that X_C is larger than R for low frequencies; thus, most of the input signal is dropped across the capacitor. The cutoff frequency occurs when $X_C = R$ or $X_L = R$. At this frequency, $V_{out} = 0.707\ V_{in}$ and $P_{out} = 0.5P_{in}$. The cutoff frequency is also known as the half-power point. In either case (voltage or power) the equivalent decibel value is -3 dB. The use of both a series inductor and a parallel capacitor improves the filtering action by providing a sharper cutoff. The response curve for a low-pass filter is shown in Figure 26.6.

FIGURE 26.6
Response curve for a low-pass filter.

Lab Reference: The attributes of a low-pass *RC* filter are demonstrated in Exercise 39.

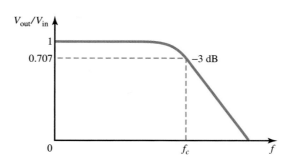

By equating X_C to R or X_L to R and solving for the frequency, we have

$$f_c = \frac{1}{2\pi RC} \tag{26.4}$$

$$f_c = \frac{R}{2\pi L} \tag{26.5}$$

EXAMPLE 26.8

Determine the cutoff frequency of a low-pass *RC* filter when $R = 2.2\ \text{k}\Omega$ and $C = 0.01\ \mu\text{F}$.

Solution $f_c = \dfrac{1}{2\pi RC}$

$$= \frac{1}{2\pi \times 2.2 \times 10^3\ \Omega \times 0.01 \times 10^{-6}\ \text{F}}$$

$$= 7.23\ \text{kHz}$$

EXAMPLE 26.9

Determine the cutoff frequency of a low-pass *RL* filter when $R = 33\ \text{k}\Omega$ and $L = 100\ \text{mH}$.

Solution $f_c = \dfrac{R}{2\pi L}$

$$= \frac{33 \times 10^3\ \Omega}{2\pi \times 100 \times 10^{-3}\ \text{H}}$$

$$= 52.5\ \text{kHz}$$

Practice Problems 26.6   Answers are given at the end of the chapter.
1. Determine the cutoff frequency of a low-pass *RC* filter when $R = 15\ \text{k}\Omega$ and $C = 0.0022\ \mu\text{F}$.
2. Determine the cutoff frequency of a low-pass *RL* filter when $R = 500\ \Omega$ and $L = 22\ \text{mH}$.

The *RC* low-pass filter is known as a lag network, because the output voltage lags the input voltage by a specified angle. The phase shift between the output voltage and the input voltage at the cutoff frequency for an *RC* low-pass filter is $\theta = -45°$. (Note that the phase shift can be anywhere between 0° and $-90°$, depending on the frequency.) For very low frequencies, X_C is large, θ is nearly zero, and $V_{out} = V_{in}$. For any phase shift, θ is found by

$$\theta = -90° + \tan^{-1}\left(\frac{X_C}{R}\right) \tag{26.6}$$

EXAMPLE 26.10

Find θ for an *RC* filter at the cutoff frequency when $R = 1\ k\Omega$ and $C = 1\ k\Omega$.

Solution $\theta = -90° + \tan^{-1}\left(\frac{X_C}{R}\right)$

$\qquad\quad = -90° + \tan^{-1}\left(\frac{1 \times 10^3\ \Omega}{1 \times 10^3\ \Omega}\right)$

$\qquad\quad = -90° + 45°$

$\qquad\quad = -45°$

Practice Problems 26.7 Answers are given at the end of the chapter.
1. Find the phase shift between V_{out} and V_{in} for an *RC* low-pass filter with $R = 2\ k\Omega$ and $X_C = 4\ k\Omega$.

The output voltage of an *RC* low-pass filter may be found by using the voltage-divider rule between X_C and R and the input voltage. This is shown as

$$V_{out} = \frac{X_C}{\sqrt{X_C^2 + R^2}} \times V_{in} \tag{26.7}$$

EXAMPLE 26.11

Find the output voltage of an *RC* low-pass filter at the cutoff frequency when $V_{in} = 10\ V$, $X_C = 1\ k\Omega$, and $R = 1\ k\Omega$.

Solution $V_{out} = \frac{X_C}{\sqrt{X_C^2 + R^2}} \times V_{in}$

$\qquad\qquad = \frac{1 \times 10^3\ \Omega}{\sqrt{(1 \times 10^3\ \Omega)^2 + (1 \times 10^3\ \Omega)^2}} \times 10\ V$

$\qquad\qquad = 7.07\ V$

Practice Problems 26.8 Answers are given at the end of the chapter.
1. Find V_{out} when $V_{in} = 20\ V$, $X_C = 10\ k\Omega$, and $R = 2\ k\Omega$ for an *RC* low-pass filter.

26.3.2 High-Pass Filters

A *high-pass filter* allows frequencies above the cutoff frequency to transfer to the output and attenuates all those frequencies below the cutoff frequency. By interchanging the components of an *RC* or *RL* low-pass filter, we obtain the equivalent high-pass filter. These are shown in Figure 26.7. For an *RC* high-pass filter, X_C is high for low frequencies and effectively attenuates them. For higher frequencies, X_C is low and passes them on to the output. The capacitor acts as a coupling capacitor and is in series to the load. The cutoff frequency is calculated using the same equations shown for the low-pass filter.

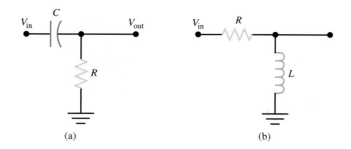

FIGURE 26.7
High-pass filters: (a) RC high-pass filter; (b) RL high-pass filter.

Figure 26.8 shows the response curve for a high-pass filter circuit. The cutoff frequency on the high-pass response curve is called the half-power point. The use of both a series capacitor and a parallel inductor improves the filtering action by providing a sharper cutoff.

FIGURE 26.8
Response curve for a high-pass filter.

Lab Reference: The attributes of a high-pass *RC* filter are demonstrated in Exercise 40.

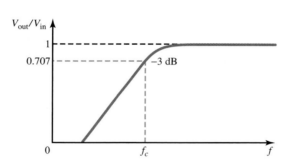

The *RC* high-pass filter is a lead network. V_{out} leads V_{in} by a specified angle. The phase angle is found by

$$\theta = \tan^{-1}\left(\frac{X_C}{R}\right) \tag{26.8}$$

The output voltage from the *RC* high-pass filter may be found by using the voltage-divider rule between R and X_C and the input voltage. It is found by

$$V_{out} = \frac{R}{\sqrt{R^2 + X_C^2}} \times V_{in} \tag{26.9}$$

EXAMPLE 26.12

Find the phase shift and V_{out} for an *RC* high-pass filter when $V_{in} = 10$ V, $R = 100$ Ω, and $X_C = 5$ kΩ.

Solution $\theta = \tan^{-1}\left(\dfrac{X_C}{R}\right)$

$\qquad\quad = \tan^{-1}\left(\dfrac{5 \times 10^3 \ \Omega}{100 \ \Omega}\right)$

$\qquad\quad = 88.9°$

$$V_{out} = \frac{R}{\sqrt{R^2 + X_C^2}} \times V_{in}$$

$$= \frac{100\ \Omega}{\sqrt{(100\ \Omega)^2 + (5 \times 10^3\ \Omega)^2}} \times 10\ V$$

$$= 0.2\ V$$

Practice Problems 26.9   Answers are given at the end of the chapter.
1. Find the phase shift and V_{out} from an *RC* high-pass filter when $V_{in} = 10$ V, $R = 5$ kΩ, and $X_C = 200$ Ω.

26.3.3 Band-Pass Filters

A *band-pass* filter is a circuit that allows a certain range or band of frequencies to pass through to the output relatively unattenuated. A band-pass filter can be made by combining a low-pass filter and a high-pass filter or employing resonant circuits. Band-pass filters have a frequency response dictated by the application. This may be anywhere from a very narrow, defined frequency response to a wide frequency response. For resonant filters, the value of *Q* (a function of inductance, frequency, and circuit resistance) helps determine the specific frequency response. One may "shape" the frequency response of a band-pass filter as required.

Figure 26.9 shows the response curve of a band-pass filter. The half-power point, f_C (−3 dB), on the low side becomes the lower frequency limit, called f_1. The low side is a high-pass filter response. The half-power point, f_C, on the high side becomes the upper frequency limit, called f_2. The high side is a low-pass filter response. The *passband,* or *bandwidth* (*BW*), is the difference between f_2 and f_1. (Note that all electronic devices have a bandwidth. Amplifiers, *RF* circuits, and even conductors and transmission cables used for communication have a bandwidth.)

FIGURE 26.9
Response curve for a band-pass filter. Bandwidth is equal to $f_2 - f_1$.

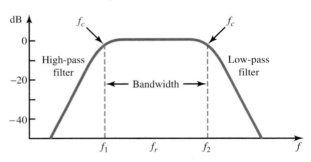

Lab Reference: The attributes of a series resonant, band-pass filter are demonstrated in Exercise 41.

In resonant band-pass filters, the *Q* factor shapes the response curve and greatly affects the values of f_1, f_2, and the bandwidth. This was shown in Chapter 25. An example of a series resonant, band-pass filter is shown in Figure 26.10. At resonance, the circuit impedance is minimum and equal to *R*. Thus, at resonance, those frequencies close to the resonant frequency are passed. All others are attenuated, because the circuit impedance increases above or below the resonant frequency.

FIGURE 26.10
Series resonant band-pass filter.

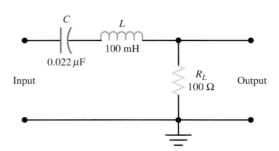

| EXAMPLE 26.13 | For the circuit shown in Figure 26.10, calculate the resonant frequency and the bandwidth. |

Solution

$$f_O = \frac{1}{2\pi\sqrt{LC}} \tag{25.1}$$

$$= \frac{1}{2\pi\sqrt{100 \times 10^{-3}\,\text{H} \times 0.022 \times 10^{-6}\,\text{F}}}$$

$$= 3.39 \text{ kHz}$$

$$X_L = 2\pi fL \tag{22.6}$$

$$= 2\pi \times 3.39 \times 10^3\,\text{Hz} \times 100 \times 10^{-3}\,\text{H}$$

$$= 2.13 \text{ k}\Omega$$

$$Q = \frac{X_L}{R} \tag{25.4}$$

$$= \frac{2.13 \times 10^3\,\Omega}{100\,\Omega}$$

$$= 21.3$$

$$\text{BW} = \frac{f_O}{Q} \tag{25.5}$$

$$= \frac{3.39 \times 10^3\,\text{Hz}}{21.3}$$

$$= 159 \text{ Hz}$$

Practice Problems 26.10   Answers are given at the end of the chapter.
1. Determine the resonant frequency, Q, and the BW for the circuit in Figure 26.10 if the values are changed to $C = 0.0015\ \mu\text{F}$, $L = 50$ mH, and $R_L = 100\ \Omega$.

Figure 26.11 shows a parallel resonant circuit connected as a band-pass filter. At parallel resonance, the high impedance of the tank causes the current to be shunted into the load resistance. Thus, there is a high voltage across both the tank and R_L at and near the resonant frequency. Off resonance, the magnitude of the tank impedance decreases sharply, and the voltage across both the tank and R_L decreases.

FIGURE 26.11
Parallel resonant band-pass filter.

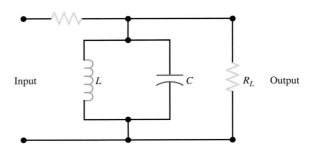

26.3.4   Band-Stop Filters

A *band-stop* filter rejects or blocks frequencies within a specified band and passes signals outside this band. A band-stop filter is also known as a band-reject, notch, wave-trap, or band-suppression filter. Like the band-pass filter, the band-stop filter can be made by combining a low-pass and a high-pass filter or by resonant circuits. The frequency response curve of the band-stop filter made by combining a low- and a high-pass filter is shown in Figure 26.12. Note that f_c of the low-pass filter is designed to be below f_c of the high-pass filter.

FIGURE 26.12
Response curve for a band-stop filter.

Lab Reference: The attributes of a band-stop filter are demonstrated in Exercise 42.

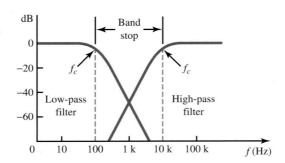

Figure 26.13 shows a band-stop filter made with series and parallel resonant circuits. Notice that the resonant circuit is connected to the load (output) opposite to the way in which it was connected for band-pass filters.

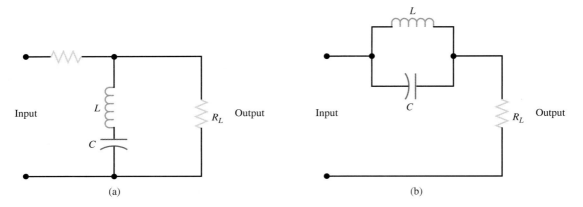

FIGURE 26.13
Resonant band-stop filters: (a) series resonant; (b) parallel resonant.

SECTION 26.3
SELF-CHECK

Answers are given at the end of the chapter.

1. Define the term *filter*.
2. Why do passive filters have an output that is always less than the input?
3. What are the four types of filters?
4. How does one shape the frequency response of a band-pass filter?
5. What is a passband, or bandwidth?

26.4 BODE PLOTS

A *Bode* (pronounced Bō-dē) *plot* is a straight-line approximation of the response of signal magnitude to frequency. It is drawn to aid in the analysis of filters, tuned circuits, amplifiers, etc. These approximations of actual curves are named after Hendrik W. Bode.

Bode plots are normally plotted on a semilog graph paper and expressed in decibels. Semilog graph paper has scales with linear graduations along the vertical axis and logarithmic graduations on the horizontal axis. For our purposes, the linear scale is labeled in decibels, and the log axis is labeled in units of frequency. The actual frequency response and the superimposed Bode plot of a low-pass filter are shown on the graph in Figure 26.14. To draw a Bode plot like that shown in Figure 26.14, a straight line is drawn along the 100-Hz, 0-dB point to the 10-kHz cutoff (*corner*) frequency. Then a straight line is drawn at 45° from the 0-dB, 10-kHz point to the −20-dB, 100-kHz point. The slope of the line is 20 dB/decade, or 6 dB/octave. The straight-line approximation, also called the *asymptotic curve,* is close to the exact curve at the −0.18-dB and −6.99-dB points, with a maximum of −3 dB at the corner frequency.

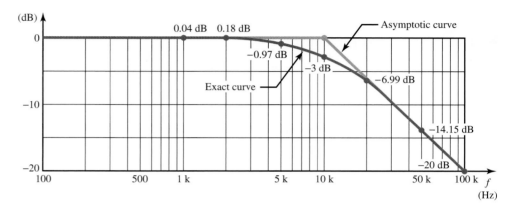

FIGURE 26.14
Bode plot for a low-pass filter.

A *decade* is a tenfold change in frequency, and an *octave* is a twofold change in frequency. Therefore, in describing the slope of the line in the Bode plot of Figure 26.14, notice that a 6-dB change takes place between 10 kHz and 20 kHz. Likewise, a 6-dB change takes place between 20 kHz (−6 dB) and 40 kHz (−12 dB), and so forth to 80 kHz. Then a −2-dB change occurs between 80 kHz and 100 kHz. Hence, (−6 dB) + (−6 dB) + (−6 dB) + (−2 dB) gives a −20-dB change. This −20-dB change occurs over a frequency range of 10 kHz to 100 kHz (a decade).

There is a natural *attenuation rate* of −20 dB/decade from either an *RC* or *RL* combination in a filter. The attenuation rate signifies how decisive the filter action is with respect to frequency response. The attenuation rate is also called the *roll-off.* One *RC* or *RL* combination within a filter is called a *pole.* Two *RC* or *RL* combinations in a filter are called a *two-pole* filter with a roll-off rate of −40 dB/decade (−12 dB/octave). Each additional pole causes an additional −20 dB (−6 dB/octave) attenuation. Active filters often employ multiple-pole filters and are commonplace for high-frequency communications circuits. Filters on a chip (an active circuit-amplifying device) are also commonplace.

Table 26.5 lists the roll-off rates for multiple-pole filters.

TABLE 26.5
Roll-off rates for multiple-pole filters

Number of Poles	Roll-off Rate (dB) (Decade/Octave)
1	−20/−6
2	−40/−12
3	−60/−18
4	−80/−24

Answers are given at the end of the chapter.

1. What is a Bode plot?
2. How are the axes of semilog paper labeled?
3. What is a decade? An octave?
4. What is the roll-off of an *RC* or *RL* combination?
5. Define the term *pole*.

26.5 FILTER APPLICATIONS

Low-pass filters are often used in conjunction with rectifiers in power supplies to smooth out the ripple voltage of the pulsating DC output. Band-pass filters are found in audio circuits to equalize sound levels and as part of a speaker crossover network. In communication circuits, a band-pass filter is used to select a narrow band of radio frequencies at the antenna from many available.

Figure 26.15 shows a passive, three-way *crossover network* for an audio system. In this system, the low-pass filter passes signals below 630 Hz to the large bass speaker, or woofer. The band-pass filter passes signals between 630 Hz and 8 kHz to the midrange speaker. A high-pass filter allows the high-frequency signals (above 8 kHz) to be passed on to the tweeter. Multiple crossover networks are called *graphic equalizers*. Because of the losses associated with passive filters, active filters are often employed in equalizers.

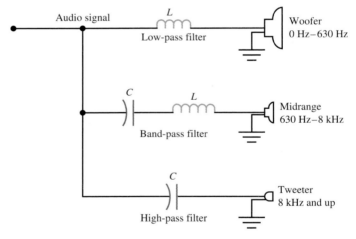

FIGURE 26.15
Passive three-way crossover network for an audio system.

Figure 26.16 shows a high-pass filter connected between two amplifier stages. The filter is designed to allow the signal to pass through the coupling capacitor unimpeded, whereas the DC voltage from the first amplifier is blocked by the capacitor and prevented from being applied to the second amplifier. The additional DC voltage from the first amplifier stage would unduly influence the operation of the second amplifier. The key is to use a capacitor value such that its X_C is one-tenth the value of R at the lowest frequency applied.

Another filter application is a band-stop (wave-trap) filter connected between an antenna and a television (TV). A common example of the wave-trap is one tuned to approximately 27 MHz (citizens band radio frequency) in order to help in the reception of Channel 2 (54 MHz) of the TV band. Note that 54 MHz is a second harmonic of the CB band. If a CB user's radio is misadjusted, the 54-MHz harmonic can cause interference in the reception of Channel 2 of the TV band.

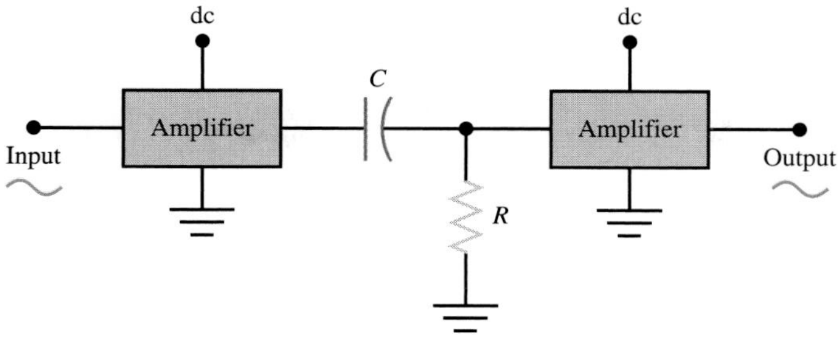

FIGURE 26.16
Example of a high-pass filter connected between amplifier stages. The capacitor blocks DC and passes the AC signal.

SECTION 26.5 SELF-CHECK

Answers are given at the end of the chapter.

1. Which type of filter is used in conjunction with a power supply?
2. What characterizes a passive crossover network versus an active crossover network?
3. What can be done to filter out an unwanted signal picked up by an antenna?

■ SUMMARY

1. Coupling is the process of transferring a signal from one circuit to another.
2. A filter is a device that removes, or filters, unwanted signals.
3. A decibel is a unit of common magnitude used when signals of different frequencies and signal levels occur in circuits.
4. The bel is based on the logarithm of the ratio of two voltages, currents, or power levels.
5. For most electronic circuitry, the decibel (dB), which is one-tenth of a bel, is used.
6. Logarithms are used to simplify numerical calculations involving exponents and powers of 10.
7. The logarithm of a number is the exponent that indicates the power to which a base must be raised in order to produce that number.
8. The decibel is not linear.
9. The decibel is a logarithmic unit for expressing a power ratio, not an actual power.
10. Gains, or amplification of power, are represented by positive decibels.
11. Losses, or attenuation of power, are represented by negative decibels.
12. To find the overall gain of several stages, either multiply the power ratios together or convert them to decibels and add.
13. Because V^2 is proportional to power, the voltage ratio is multiplied by 2 (20 log).
14. To represent an actual power, not a power ratio, decibels are referenced to a relative power level.
15. Direct coupling is one type of coupling from one circuit to another.
16. A direct coupling has the current of the input circuit flowing through a common impedance.
17. Direct-coupled circuits can be used to transfer all frequencies from one circuit to another.
18. Indirect coupling is used in a situation where only the alternating current must be passed from a composite signal containing both AC and DC components.
19. Passive filters use resistors, capacitors, or inductors and do not contain amplifying devices to increase signal strength.
20. The four types of filters are the low-pass, the high-pass, the band-pass, and the band-stop.
21. A low-pass filter has a relatively constant output from 0 Hz up to a cutoff, or critical, frequency and is known as a *lag* circuit.
22. At the critical frequency, $V_{out} = 0.707\ V_{in}$, or -3 dB.
23. A high-pass filter allows frequencies above the cutoff frequency to transfer to the output, attenuates all those frequencies below the cutoff frequency, and is known as a *lead* circuit.
24. A band-pass filter allows a certain range, or band, of frequencies to pass through to the output relatively unattenuated.
25. For a band-pass filter, the bandwidth is the difference between f_1 (the lower half-power point) and f_2 (the upper half-power point).
26. All electronic devices have a bandwidth.
27. A band-stop filter rejects, or blocks, frequencies within a specified band and passes signals outside this band.
28. A Bode plot is a straight-line approximation of the response of signal magnitude to frequency.
29. A decade is a tenfold change in frequency and an octave is a twofold change in frequency.
30. There is a natural attenuation rate of -20 dB/decade or -6 dB/octave for one *RC* or *RL* combination within a filter.
31. Filters are used in a variety of electronic applications, such as power supplies, equalizers, crossover networks, and wave traps.

■ IMPORTANT EQUATIONS

$$dB = 10 \log\left(\frac{P_{out}}{P_{in}}\right) \qquad (26.1)$$

$$dB = 10 \log\left(\frac{P_2}{P_1}\right) \qquad (26.2)$$

$$dB = 20 \log\left(\frac{V_{out}}{V_{in}}\right) \qquad (26.3)$$

$$f_c = \frac{1}{2\pi RC} \qquad (26.4)$$

$$f_c = \frac{R}{2\pi L} \qquad (26.5)$$

$$\theta = -90° + \tan^{-1}\left(\frac{X_C}{R}\right) \qquad (26.6)$$

$$V_{out} = \frac{X_C}{\sqrt{X_C^2 + R^2}} \times V_{in} \qquad (26.7)$$

$$\theta = \tan^{-1}\left(\frac{X_C}{R}\right) \qquad (26.8)$$

$$V_{out} = \frac{R}{\sqrt{R^2 + X_C^2}} \times V_{in} \qquad (26.9)$$

■ REVIEW QUESTIONS

1. Define the terms *coupling* and *filter*.
2. Define the term *decibel*.
3. For what type of electronic circuits is the decibel commonly used?
4. What is the difference between the bel and the decibel?
5. Decibels are based on _____ logs.
6. Natural logs are derived from the base _____.
7. Where are natural logs used?
8. What is a logarithm and how are they used?
9. What are the whole and fractional parts of the logarithm called?
10. On a scientific calculator, you press the _____ button in order to work problems associated with common logs.
11. The decibel is a logarithmic unit for expressing a power _____, not an actual power.
12. Gains, or amplifications, are indicated by _____ decibels.
13. Losses, or attenuation of power, are represented by _____ decibels.
14. What can be done with decibels arithmetically?
15. What are the half-power points?
16. Because V^2 is proportional to power, the voltage ratio is multiplied by 2 (_____ log instead of 10 log).
17. What do decibels represent?
18. The standard reference for decibel is _____ dB.
19. What are some common reference levels used for both power and voltage?
20. Compare and contrast direct coupling and indirect coupling.

21. Why is transformer coupling considered another form of indirect coupling?
22. Where is transformer coupling employed?
23. Describe passive filters and their use.
24. What is an active filter?
25. What are the four main types of filters?
26. Show the correlation with output voltage and the frequencies associated with the four main types of filters.
27. What is the passband, or bandwidth, of a band-pass filter?
28. What is a Bode plot and how is it used?
29. Compare the attenuation of an *RC* or *RL* combination in a filter to the decade and the octave.
30. State examples of how filters are used in electronic circuitry.

■ CRITICAL THINKING

1. Show how the logarithm provides a natural way to work with decibels.
2. Explain how the decibel represents a power ratio, not an actual power.
3. Compare and contrast direct coupling and indirect coupling and how they are used in electronic circuits.
4. Compare and contrast the four main types of filters.
5. Show how a Bode plot is made and its usefulness in the analysis of filters, tuned circuits, and the like.
6. Develop the bandwidth equations for low- and high-pass filters based on the fact that the bandwidth of a band-pass filter is $f_2 - f_1$.

■ PROBLEMS

Complete the following table for the unknowns listed.

	Number	Log	Antilog
1.	1000		—
2.	0.25		—
3.	100		—
4.	2345		—
5.	6.25		—
6.		—	3.69897004
7.		—	−0.1505805862
8.		—	2.698970004
9.		—	−0.3010299957
10.		—	6.698970004

Complete the following table for the unknowns listed.

	P_{out}	P_{in}	dB
11.	1 W	1 mW	
12.	1 W		20
13.	25 mW	500 mW	
14.	100 μW	700 mW	
15.	400 mW	2.5 W	
16.		25 mW	24
17.		10 mW	-32
18.	500 mW		40
19.	100 μW	200 mW	
20.		5 mW	100

Complete the following table for the unknowns listed.

	V_{out}	V_{in}	dB
21.	1 V	1 V	
22.	1 V	0.002 V	
23.	2.5 V		-10
24.	100 μV	2 μV	
25.	3.33 V		6
26.		50 mV	-3
27.		100 mV	20
28.	20 V		-20
29.	500 mV	20 mV	
30.		0.25 V	100

31. Determine f_c for the *RC* low-pass filter shown in Figure 26.17.

Complete the following table to determine the unknowns listed for the circuit in Figure 26.17.

		500 Hz	1 kHz	5 kHz	10 kHz
32.	X_C				
33.	θ				
34.	V_{out}				

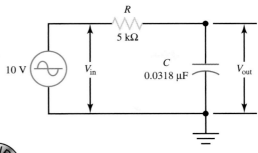

FIGURE 26.17
Circuit for Problem 31.

35. Determine f_c for the *RC* high-pass filter shown in Figure 26.18.

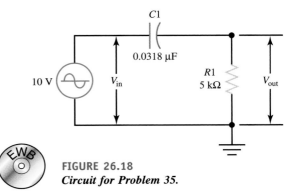

FIGURE 26.18
Circuit for Problem 35.

Complete the following table to determine the unknowns listed for the circuit in Figure 26.18.

		500 Hz	1 kHz	5 kHz	10 kHz
36.	X_C				
37.	θ				
38.	V_{out}				

39. For each circuit in Figure 26.19, identify and label the type of filter represented. (*Hint:* Pay attention to the placement of the capacitor and inductor combinations with respect to the output.)

■ ANSWERS TO PRACTICE PROBLEMS

PRACTICE PROBLEMS 26.1

1. 0, -0.125, 2.699, 3.301

PRACTICE PROBLEMS 26.2

1. 10, 306.9, 0.308, 1,199,223

PRACTICE PROBLEMS 26.3

1. 16.5 dB
2. -1.55 dB

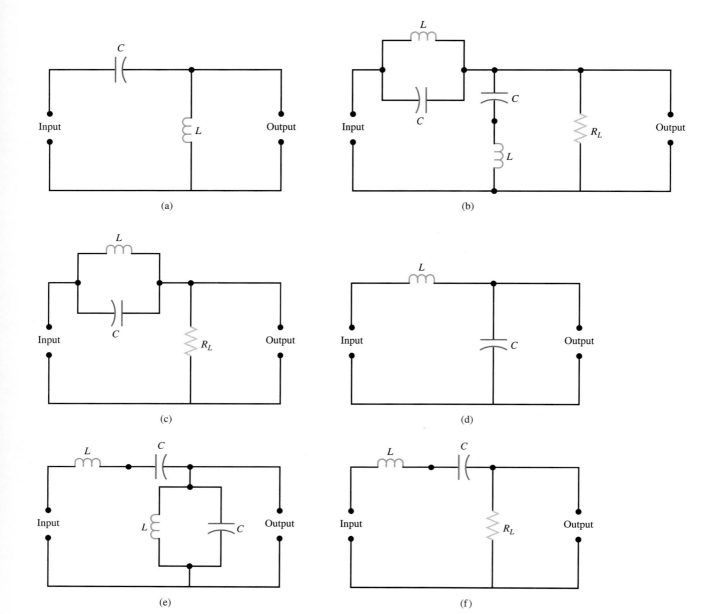

FIGURE 26.19
Circuits for Problem 39.

PRACTICE PROBLEMS 26.4

1. 18.9 dB

PRACTICE PROBLEMS 26.5

1. 40 dB
2. −6.27 dB

PRACTICE PROBLEMS 26.6

1. f_c = 4.82 kHz
2. f_c = 3.62 kHz

PRACTICE PROBLEMS 26.7

1. $\theta = -26.6°$

PRACTICE PROBLEMS 26.8

1. V_{out} = 19.6 V

PRACTICE PROBLEMS 26.9

1. $\theta = 2.29°$, V_{out} = 9.99 V

PRACTICE PROBLEMS 26.10

1. $f_O = 18.38$ kHz, $Q = 57.7$, BW $= 318.5$ Hz

■ ANSWERS TO SECTION SELF-CHECKS

SECTION 26.1 SELF-CHECK

1. The logarithm of a number is the exponent indicating the power to which a base must be raised in order to produce that number.
2. Positive $(+)$
3. 0.707
4. -3

SECTION 26.2 SELF-CHECK

1. Direct-coupling circuits have the current of the input circuit flowing through a common impedance.
2. All frequencies
3. DC
4. Transformer coupling is popular in RF circuits (communication circuits).

SECTION 26.3 SELF-CHECK

1. Filters are used to pass or block a specific range of frequencies.
2. Passive filters attenuate the input signal due to the fact that they are inserted in the line between the input and output. The cir-cuit reactances store energy and the circuit resistance dissipates energy.
3. Low-pass, high-pass, band-pass, band-stop
4. By varying the value of Q
5. BW is the difference in frequency between f_2 and f_1.

SECTION 26.4 SELF-CHECK

1. A Bode plot is a straight-line approximation of the response of signal magnitude to frequency.
2. For filters, the vertical axis (y) is linear (usually in dB), and the horizontal axis (x) is logarithmic (usually in frequency).
3. A decade is a tenfold change in frequency, whereas an octave is a twofold change in frequency.
4. -20 dB/decade or -6 dB/octave
5. One RC or RL combination within a filter is called a pole.

SECTION 26.5 SELF-CHECK

1. Low-pass
2. Passive crossover networks do not have gain or amplify the signal.
3. Insert a band-stop (wave-trap) filter.

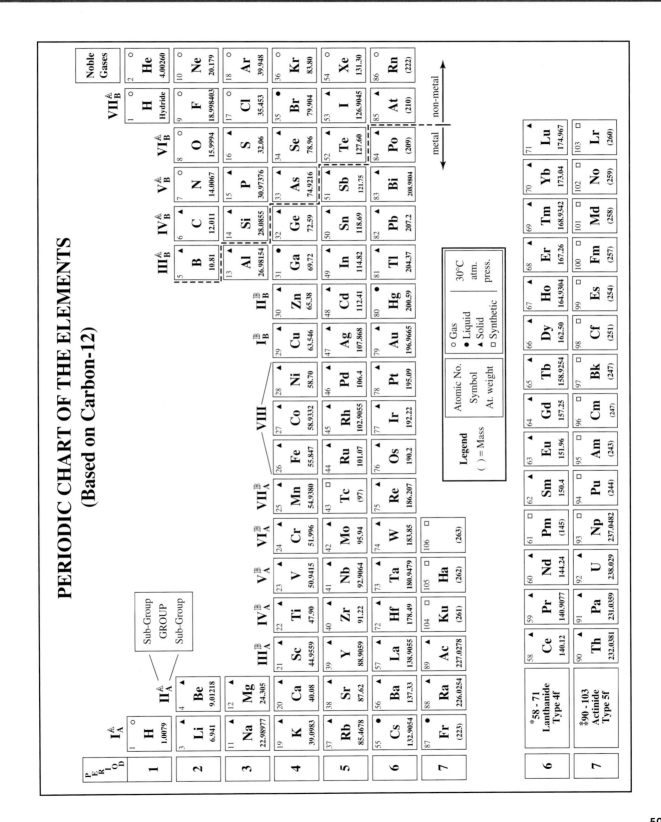

PERIODIC CHART OF THE ELEMENTS
(Based on Carbon-12)

SELECTED SCHEMATIC SYMBOLS FOR DC/AC CIRCUITS

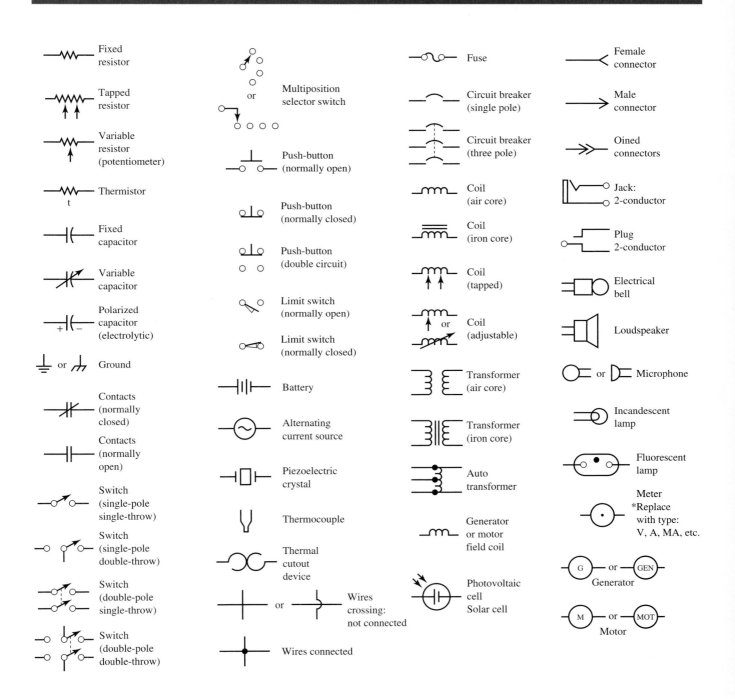

GLOSSARY OF SELECTED DC/AC TERMS

active component A component that provides gain or amplification, such as a transistor or integrated circuit.

air gap The air space between poles of a magnet.

alternating current (AC) A source of electric current whose magnitude varies with time and changes direction.

ammeter An instrument for measuring the value of current in amperes, milliamperes, or microamperes.

amp (ampere) The unit of electrical current. One ampere corresponds to the flow of 6.25×10^{18} electrons per second.

amplitude The highest value reached by voltage, current, or power during a complete cycle.

analog A system in which a signal continuously varies in amplitude with respect to time.

apparent power The product of voltage and current when they are out of phase.

audio A signal that falls within the audio range and thus can be heard.

audio frequency (AF) A frequency range generally from 20 Hz to 20 kHz.

average value For a sine-wave AC voltage or current, equal to 0.637 of the peak value.

band-pass A filter that allows a band of frequencies to be coupled to the output.

band-stop A filter that prevents a band of frequencies from being coupled to the output.

bandwidth The range of frequencies between the critical frequencies $(f_2 - f_1)$ of filter circuits.

battery A group of cells connected to increase the output voltage.

bleeder current The current from the source used to provide a stable current to the voltage-divider resistances; usually 10–20% of the load current.

branch A part of a parallel circuit.

bridge Usually a resistive circuit where voltages can be balanced for a net effect of zero between two points on the bridge.

bypass capacitor A capacitor that provides a path of low impedance to AC signals.

capacitance (C) The property that allows a capacitor to store electric charge when a potential difference exists between its terminals, measured in farads.

capacitor A device used to store electric charge.

cathode-ray oscilloscope An instrument with a cathode-ray tube (CRT) that provides a visual indication of the waveform at any point in a circuit.

cell A single source of voltage produced from chemical action.

ceramic capacitor A capacitor using ceramics as the dielectric; a common form is the multilayer capacitor.

chip Another name for an integrated circuit, or the piece of silicon on which semiconductors are created.

choke An inductor designed to present a high impedance to alternating current.

circuit breaker A protective device that opens when excessive current flows in a circuit. Can be reset.

coil A conductor wound in a series of turns.

complex number A value that has both magnitude and a direction; can be expressed in rectangular form or polar form and is used in AC circuit-analysis problems.

condenser Obsolete term for a capacitor.

conductance (G) The ability to conduct current, measured in siemens.

conventional current flow A type of current flow stemming from the days of Franklin that assumes a charge flowing from a positive to negative potential.

coulomb (C) Unit of electric charge; 1 C = 6.25×10^{18} electrons.

counter-electromotive force (cemf) Also called back-emf, the emf produced within a conductor by a changing current; the cemf opposes the source.

critical frequency Usually associated with filters, the frequency where the amplitude of the output voltage decreases by 0.707 or the power decreases by 0.5.

current divider A parallel circuit to divide the total current.

dB The abbreviation for decibel.

decade A 10:1 range of values.

decibel A logarithmic expression comparing two power or voltage levels.

dielectric An insulator used to store electric charge in a capacitor.

digital electronics The branch of electronics dealing in information transfer utilizing binary and other forms of number systems.

direct current (DC) Current flow that travels in one direction only throughout a circuit.

discrete A term used to describe a circuit made up of individual components.

DMM A digital multimeter; it does not use a moving-coil meter movement as an indicator.

earphone (earpiece) A device for converting electrical energy into sound waves that fits over or within the ear.

earth Ground; a reference or a direct connection to the earth itself representing a point of zero potential.

eddy current A circulating current within the core of an inductor or transformer caused by flux variations.

effective value Also known as root-mean-square (rms), the equivalent heating value of an AC that a DC source would produce; for sine-wave AC values, it is equal to 0.707 of the peak value.

electrolytic A type of capacitor that has a liquid or paste between the plates to increase its capacitance via the formation of an oxide coating.

electron A basic particle of an atom having a negative charge.

electron flow A current flow done via electrons traveling from negative to positive within a circuit.

emf Electromotive force, also known as electrical pressure, potential difference, or voltage; it represents the property of electrical energy that causes a current to flow.

farad (F) The unit of capacitance; because it is a very large value, it is more common to use the microfarad (μF) or the picofarad (pF).

ferrite Magnetic material that is not an electrical conductor.

filter A circuit that can separate different frequencies.

flux (Φ) Magnetic lines of force.

frequency (f) The number of complete cycles per second for any periodic waveform; measured in hertz—formally, cycles per second (cps).

full-scale deflection (FSD) The maximum value of displacement of the scale of a moving-coil meter.

function generator A piece of test equipment that produces a sine, square, or triangular waveform.

fuse A component that will easily burn out when excessive current flows.

giga (G) Metric prefix for 10^9.

ground A common or return to earth for AC power lines; the reference for DC voltage measurements.

ground plane The earth or negative rail of a circuit.

harmonic A sinusoidal component of a waveform that is a whole multiple of the fundamental frequency.

henry (H) The electromagnetic unit of inductance; 1 H occurs when a cemf of 1 V is produced when the current changes at the rate of 1 A/s.

hertz (Hz) The unit of frequency in cycles per second.

hysteresis In magnetics, the lagging behind of the magnetic induction versus the magnetizing force.

imaginary number In mathematics $\sqrt{-1}$ or a multiple of $\sqrt{-1}$; used in electronics to indicate a 90° shift on the y-axis.

impedance (Z) The opposition to current flow in a circuit containing AC current as the source, measured in ohms (Ω).

impedance matching The process of matching source impedance to load impedance for maximum power transfer.

inductance (L) The property that opposes a change in current; 1 H of inductance produces a cemf of 1 V when the current is changing at the rate of 1 A/s.

input The part of a circuit that accepts a signal for processing.

insulator Any material that resists the flow of current.

integrated circuit (IC) A circuit component consisting of a piece of semiconductor material containing both passive and active devices. A circuit comprising components within a monolithic form rather than from discrete components; also known as a chip.

IR drop The voltage drop across a resistance.

j operator Used in electronics as part of a rectangular complex number to indicate a $\pm 90°$ phase shift.

joule (J) The practical unit of work or energy: 1 J = 1 W/s.

KCL Kirchhoff's current law.

kilo (k) Metric prefix for 10^3.

kilowatt-hour (kWh) A unit of energy where 1 kW of power is expended for 1 h.

KVL Kirchhoff's voltage law.

leakage The passage of electric current that is unintended.

load That part of a circuit that takes source current and converts it into a more useful form of energy.

loading effect The decrease of voltage as load current increases.

loudspeaker A device that converts electrical energy into sound waves.

magnetism The effects of attraction and repulsion between iron and similar materials without the need for an external force.

magnitude The value of a quantity regardless of phase.

maxwell (Mx) Unit of magnetic flux, equal to one line of force in a magnetic field.

mega (M) Metric prefix for 10^6.

megohm (MΩ) One million ohms.

mesh current An assumed path of current in a closed path, used in DC circuit-analysis problems.

micro (μ) Metric prefix for 10^{-6}.

microphone A device for converting sound waves into electrical energy.

milli (m) Metric prefix for 10^{-3}.

motor A device that produces mechanical motion from electrical energy.

nano (n) Metric prefix for 10^{-9}.

ni-cad (NiCd) Nickel-cadmium rechargeable cell.

node A common connection for two or more branch currents.

octave A 2:1 range of values.

ohm (Ω) The unit of resistance.

Ohm's law Written as $I = V/R$, where I = current, V = voltage, and R = resistance.

open circuit A circuit with infinite resistance, resulting in no current.

oscilloscope A piece of test equipment used to view and measure a variety of waveforms.

output The part of a circuit where the processed signal is available.

parallel Components connected so that there are two or more branch currents.

passive component A component that does not provide amplification or gain, such as resistor, capacitor, or inductor.

PCB Printed circuit board.

peak The maximum value of a waveform in either polarity.

peak-to-peak (p-p) The amplitude between opposite peaks.

permeability The ability to concentrate magnetic flux lines.

phase angle The difference between two phasors.

phasor A line representing a magnitude and direction with respect to time.

photocell A device that changes resistance with light-energy (photons) variations.

pico (p) Metric prefix for 10^{-12}.

polar form A notation method for complex numbers giving a magnitude and a direction (phase angle), as in $Z\angle\theta$.

polarized A component or plug where polarity must be observed or be fitted in a certain way.

polyester A common type of dielectric for capacitors.

potentiometer (pot) A variable resistance with three terminals (two fixed, one adjustable), usually connected as a voltage divider.

power (P) The rate of doing work, measured in watts.

power factor The amount of energy dissipated as heat to the available energy supplied by the source.

primary The input side of a transformer; a nonrechargeable cell.

Q-factor The quality factor of an inductor; a ratio of energy stored versus energy dissipated as heat.

radio frequency (RF) That part of the spectrum from about 50 kHz to gigahertz (GHz).

RC A resistive-capacitor combination, as in RC time constant or RC coupling.

rectangular form Notation for complex numbers utilizing the j operator, as in $R \pm jX$.

relay An electromechanical device containing a coil and a set of contacts. The contacts close (electromagnetically) as the coil is activated by current flow.

resistance (R) The opposition to the flow of current.

resistor Passive component with a known resistance marked with a color code or direct alphanumeric characters to indicate the value.

rms Also known as effective voltage, the equivalent DC heating value of an AC sine-wave voltage; calculated as $0.707 V_P$.

schematic Another name for a circuit diagram utilizing symbols for the components.

secondary The output side of a transformer; a rechargeable cell.

sensitivity With respect to a moving-coil meter, the amount of current it takes to cause FSD.

series Components connected in such a way so that there is only one path for current.

series resonant circuit A series RCL circuit exhibiting minimum impedance and maximum current at one particular frequency only.

short-circuit Also known as a short, an unintended path of low resistance.

shunt Generally meaning a parallel connection.

signal generator A circuit that can produce variable amplitude and frequency signals. The signals are usually a sine wave and a square wave.

sine wave A waveform whose amplitude varies in proportion to the sine function of an angle.

single-pole, single-throw switch (SPST) A switch in which only one circuit is controlled.

SMD A surface-mounted device using surface-mount technology (SMT), where the leads are directly soldered to a PCB (does not require through-hole technology).

solar cell A cell that produces current using sunlight.

solenoid A coil that, when current flows through it, produces a magnetic flux that pushes or pulls a rod.

square wave A periodic wave that alternately assumes one of two fixed values (high and low), with the transition time between these two levels being negligible; has a 50% duty cycle.

static electricity Charges that are not in motion.

superconductivity Having low resistance at very low temperatures.

switch A device for connecting and disconnecting power to a circuit.

tank circuit A parallel resonant LC circuit.

taper The amount of resistance variation as a function of shaft rotation for a potentiometer.

temperature coefficient For resistance, how resistance varies with a change in temperature.

thermistor Resistor that varies in value according to its temperature.

time constant Time required to charge a capacitor to 63.2% of the source voltage.

tolerance The allowable percentage variation of any component from that specified.

transducer A device that receives energy in one form and supplies an output in another form.

transformer A device with two or more coil windings used to step up or down AC voltage or current.

trimmer Trimmer capacitor (trim cap), a small variable capacitor used in parallel across a larger capacitor to adjust the total capacitance a small amount. Also a trimmer potentiometer (trim pot), a small variable resistance.

turns ratio The ratio of primary to secondary turns.

VAR Unit for reactive power in volt-amperes.

vector A line representing magnitude and direction.

volt The unit of voltage.

volt-ampere (VA) Unit of apparent power.

voltage divider The voltage produced at the intersection of two components; usually made up of a series of resistors.

voltmeter A meter for reading voltage; it is one of the ranges in a multimeter.

voltmeter loading The reduction of voltage when the voltmeter (acting as a load) is placed across a device in order to measure voltage.

watt (W) The unit of power; 1 W is the product of 1 V and 1 A.

wavelength (λ) The distance between two points of corresponding phase in consecutive cycles.

Wheatstone bridge Balanced resistive circuit used for precise resistance measurements.

X_C Capacitive reactance in ohms.

X_L Inductive reactance in ohms.

ANSWERS TO ODD-NUMBERED PROBLEMS

CHAPTER 2

1. 3.6×10^3 **3.** 3.49×10^1 **5.** 3.67×10^0
7. 1.0×10^6 **9.** 5.7×10^{-5} **11.** 150×10^0
13. 12×10^0 **15.** 3.0×10^{-3} **17.** 3.3×10^{-3}
19. 13.25×10^3 **21.** Mega, M   **23.** Giga, G
25. Tera, T   **27.** Pico, p   **29.** 50 mA
31. 2.7 kΩ **33.** 120,000 mV   **35.** 2.4 kHz
37. 1.5 MΩ **39.** 0.02 mA

CHAPTER 3

1. 1.35 N

CHAPTER 4

1. 32 V   **3.** 300 J   **5.** 0.5 A   **7.** 500 μC   **9.** 20 kΩ

CHAPTER 5

1. 22 Ω,   47 kΩ,   100 kΩ,
   2.2 Ω,   1 kΩ,   15 kΩ,
   1 MΩ,   390 Ω,   2.7 kΩ
3. 22.2 kΩ $\pm$ 2%,   1.1 kΩ $\pm$ 2%,
   47.5 Ω $\pm$ 1%,   17.4 Ω $\pm$ 1%,
   75 kΩ $\pm$ 2%,   5.11 kΩ $\pm$ 1%,
   33.3 kΩ $\pm$ 2%,   5.1 kΩ $\pm$ 2%

CHAPTER 6

1. 1.6 kΩ **3.** 66.67 μA   **5.** 100 V
7. 100 Ω **9.** 99.2 V   **11.** 5.45 mA
13. 40.5 V   **15.** 40 MΩ **17.** 12 mA
19. 1 A   **21.** 24 W   **23.** 225 mW
25. 4.9 V   **27.** 1.58 A   **29.** 11.18 V
31. $P = 3.52$ kW, $R = 55$ Ω **33.** $R = 37.5$ Ω, $V = 750$ V
35. $I = 10$ mA, $V = 12$ V   **37.** $P = 80$ mW, $I = 6.67$ mA
39. $R = 1.25$ kΩ, $I = 40$ mA   **41.** 70.5%
43. 77.7%   **45.** $8.64

CHAPTER 7

1. $R_T = 70$ kΩ, $I_T = 500$ μA, $V_1 = 7.5$ V, $V_2 = 11$ V,
   $V_3 = 16.5$ V, $P_T = 17.5$ mW, $P_1 = 3.75$ mW, $P_2 = 5.5$ mW,
   $P_3 = 8.25$ mW
3. $R_T = 2.5$ kΩ, $I_T = 4$ mA, $V_1 = 1.56$ V, $V_2 = 3.64$ V,
   $V_3 = 4.8$ V, $P_T = 40$ mW, $P_1 = 6.24$ mW, $P_2 = 14.56$ mW,
   $P_3 = 19.2$ mW
5. $R_T = 75$ kΩ, $I_T = 160$ μA, $V_1 = 4.32$ V, $V_2 = 5.28$ V,
   $V_3 = 2.4$ V, $P_T = 1.92$ mW, $P_1 = 691$ μW, $P_2 = 845$ μW,
   $P_3 = 384$ μW
7. $R_T = 5$ kΩ, $I_T = 6$ mA, $V_1 = 13.2$ V, $V_2 = 10.8$ V,
   $R_2 = 1.8$ kΩ, $P_T = 180$ mW, $P_1 = 79.3$ mW,
   $P_2 = 64.8$ mW, $P_3 = 36$ mW

9. $R_T = 9$ kΩ, $I_T = 2$ mA, $R_2 = 2.2$ kΩ, $V_1 = 2.4$ V,
   $V_2 = 4.4$ V, $P_T = 36$ mW, $P_1 = 4.8$ mW, $P_2 = 8.8$ mW,
   $P_3 = 22.4$ mW
11. $R_T = 80$ kΩ, $R_3 = 10$ kΩ, $V_S = 32$ V, $V_1 = 6$ V,
   $V_2 = 8.8$ V, $V_4 = 13.2$ V, $V_{AG} = -13.2$ V, $V_{BG} = 4$ V,
   $V_{CG} = 12.8$ V, $V_{DG} = 18.8$ V, $P_T = 12.8$ mW,
   $P_1 = 2.4$ mW, $P_2 = 3.52$ mW, $P_3 = 1.6$ mW,
   $P_4 = 5.28$ mW
13. $V_S = 24$ V, $R_T = 1.2$ kΩ, $V_1 = 4$ V, $V_2 = 12$ V,
   $V_3 = 5$ V, $V_4 = 3$ V, $R_2 = 600$ Ω, $R_4 = 150$ Ω,
   $V_{AG} = 8$ V, $V_{BG} = 5$ V, $V_{CG} = -12$ V,
   $V_{AB} = 3$ V, $V_{DC} = -4$ V, $P_1 = 80$ mW, $P_2 = 240$ mW,
   $P_3 = 100$ mW, $P_4 = 60$ mW
15. R_3 short   **17.** R_2 short   **19.** 15 V

CHAPTER 8

1. $R_T = 687.5$ Ω, $I_T = 16$ mA, $I_1 = 11$ mA, $I_2 = 5$ mA,
   $P_1 = 121$ mW, $P_2 = 55$ mW, $P_T = 176$ mW
3. $R_T = 40$ Ω, $I_T = 1.25$ A, $I_1 = 0.625$ A, $I_2 = 0.5$ A,
   $I_3 = 0.125$ A, $P_1 = 31.25$ W, $P_2 = 25$ W, $P_3 = 6.25$ W,
   $P_T = 62.5$ W
5. $R_T = 750$ Ω, $I_T = 60$ mA, $I_1 = 30$ mA, $I_2 = 22.5$ mA,
   $I_3 = 7.5$ mA, $P_1 = 1.35$ W, $P_2 = 1$ W, $P_3 = 0.35$ W,
   $P_T = 2.7$ W
7. 6 kΩ **9.** $I_1 = 1.25$ A, $I_2 = 1$ A, $I_3 = 0.25$ A
11. $R_T = 3$ Ω, $I_T = 4$ A, $I_1 = 0.55$ A, $I_2 = 1.45$ A,
   $I_3 = 2$ A, $P_1 = 6.6$ W, $P_2 = 17.4$ W, $P_3 = 24$ W
13. $I_T = 600$ mA, $I_1 = 200$ mA, $I_2 = 200$ mA, $R_1 = 270$ Ω,
   $R_2 = 270$ Ω, $R_3 = 270$ Ω, $P_T = 32.4$ W,
   $P_1 = P_2 = P_3 = 10.8$ W
15. $R_{EQ} = 1.83$ kΩ, $I_T = 57.5$ mA, $I_2 = 7.5$ mA, $I_3 = 20$ mA,
   $I_4 = 25$ mA, $R_1 = 21$ kΩ, $R_3 = 5.25$ kΩ, $P_T = 6$ W,
   $P_1 = 0.5$ W, $P_2 = 0.8$ W, $P_3 = 2.1$ W, $P_4 = 2.6$ W
17. R_2 open   **19.** R_4 open

CHAPTER 9

1. 1.6 Ω **3.** 10 kΩ **5.** 9 kΩ **7.** 1.2 kΩ **9.** 12 kΩ
11. $R_{EQ} = 7.5$ kΩ, $I_T = 2$ mA, $I_1 = 1.5$ mA, $I_2 = 5$ mA,
   $I_3 = I_4 = 0.25$ mA, $I_5 = 0.5$ mA, $V_1 = 15$ V, $V_2 = 5$ V,
   $V_3 = V_4 = 2.5$ V, $V_5 = 7.5$ V
13. $R_{EQ} = 2.1$ kΩ, $I_T = 40$ mA, $I_1 = I_3 = 40$ mA,
   $I_2 = I_4 = I_7 = 20$ mA, $I_5 = I_6 = 10$ mA, $V_1 = 10.8$ V,
   $V_2 = 66$ V, $V_3 = 7.2$ V, $V_4 = 36$ V,
   $V_5 = V_6 = 10$ V, $V_7 = 20$ V
15. $R_{EQ} = 5$ kΩ, $I_T = 8$ mA, $I_1 = 8$ mA, $I_2 = 0.8$ mA,
   $I_3 = 1.2$ mA, $I_4 = 8$ mA, $I_5 = 6$ mA, $I_6 = 3.6$ mA,
   $I_7 = 2.4$ mA, $I_8 = 6$ mA
17. $R_T = 12$ kΩ, $I_T = 2$ mA, $V_S = 24$ V, $I_2 = 1.25$ mA,
   $I_5 = 0.75$ mA, $V_1 = 3.6$ V, $V_3 = 5.4$ V, $V_4 = 0.9$ V,
   $V_5 = 5.1$ V, $V_6 = 9$ V

19. $R_{EQ} = 8$ kΩ, $I_T = I_1 = 15$ mA, $V_S = 120$ V, $I_2 = 9$ mA, $I_3 = 4.5$ mA, $I_4 = 4.5$ mA, $I_5 = 6$ mA, $V_1 = 15$ V, $V_2 = 35.1$ V, $V_3 = V_4 = 36.9$ V, $V_6 = 33$ V

CHAPTER 10

1. 50 Ω, 12.5 Ω, 5.5 Ω, 2.6 Ω, 1 Ω, 0.5 Ω
3. 556 Ω, 128 Ω, 25 Ω, 5 Ω
5. 47.5 kΩ, 75 kΩ, 375 kΩ, 750 kΩ, 3.75 MΩ
7. 50 kΩ/V **9.** 10 kΩ **11.** 20 μA **13.** 4 V **15.** 3 V

CHAPTER 11

1. 20 V **3.** 2 V **5.** 9 V
7. $V_1 = 8.2$ V, $V_2 = 9.8$ V, $V_3 = 12$ V
9. $V_1 = 23.5$ V, $V_2 = 28$ V, $V_3 = 13.5$ V
11. $I_1 = 18$ A →, $I_2 = 8$ A↑, $I_4 = 2$ A↑,
$I_5 = 8$ A →, $I_6 = 8$ A↑, $I_9 = 1$ A ←,
$I_{10} = 11$ A↑, $I_{11} = 12$ A ←, $I_{13} = 6$ A↑
13. $I_1 = 12$ mA, $I_2 = 8$ mA
15. $I_1 = 15$ mA, $I_2 = 7.5$ mA, $I_3 = 7.5$ mA
17. $R_1 = 2.22$ kΩ, $R_2 = 138$ Ω, $R_3 = 232$ Ω
19. $P_1 = 45$ mW, $\frac{1}{8}$-W resistor, $P_2 = 29$ mW, $\frac{1}{8}$-W resistor, $P_3 = 276$ mW, $\frac{1}{2}$-W resistor

CHAPTER 12

1. $R_i = 1$ Ω, $R_L = 5$ Ω **3.** $V_S = 6$ V, $R_i = 12$ kΩ
5. 0 V **7.** $V_{R_1} = 9.35$ V, $V_{R_2} = 10.65$ V, $V_{R_3} = 1.35$ V
9. 5.6 V **11.** 25 mA
13. For $R_L = 10$ Ω, $I_L = 91$ mA, $V_L = 0.91$ V, $P_L = 82.8$ mW,
  $P_T = 910$ mW, eff. = 9%;
  for $R_L = 25$ Ω, $I_L = 80$ mA,
    $V_L = 2$ V, $P_L = 160$ mA, $P_T = 800$ mW, eff. = 20%;
  for $R_L = 50$ Ω, $I_L = 66.7$ mA, $V_L = 3.3$ V, $P_L = 222$ mW,
    $P_T = 667$ mW, eff. = 33.3%;
  for $R_L = 100$ Ω, $I_L = 50$ mA, $V_L = 5$ V, $P_L = 250$ mW,
    $P_T = 500$ mW, eff. = 50%;
  for $R_L = 200$ Ω, $I_L = 33.3$ mA, $V_L = 6.67$ V,
    $P_L = 222$ mW, $P_T = 333$ mW, eff. = 67%;
  for $R_L = 500$ Ω, $I_L = 16.7$ mA, $V_L = 8.3$ V, $P_L = 139$ mW,
    $P_T = 166.9$ mW, eff. = 83%;
  for $R_L = 1000$ Ω, $I_L = 9.1$ mA, $V_L = 9.1$ V, $P_L = 82.8$ mW,
    $P_T = 91$ mW, eff. = 91%
15. 16.7 V **17.** 12 V **19.** 20 V **21.** 0.8 V
23. 3 V **25.** 3.86 V

CHAPTER 13

1. 5 Ω **3.** 10 Ω **5.** 33.4 Ω **7.** 24.8 Ω **9.** 91.8 Ω

CHAPTER 14

1. 40 A·t **3.** 0.4 A **5.** 16,000 T
7. 2.5 μWb **9.** 5×10^6 A·t/Wb **11.** 1.67×10^{-3} T
13. 100 T **15.** 12 mm **17.** 36
19. 2×10^{-3} T/(A·t/m)

CHAPTER 15

1. 3 Wb/s **3.** 0 V **5.** 10,000 turns

CHAPTER 16

1. (a) 14.14 V (b) 0.707 V (c) 55.5 V (d) 10.6 V
3. (a) 28.3 V (b) 157 mA (c) 500 V (d) 565.6 V
5. (a) 0 V (b) 147 V (c) 170 V
  (d) 85 V (e) 0 V (f) −85 V
  (g) −120 V (h) −120 V (i) −44 V
7. 169.7 V **9.** 28.1 V
11. (a) 50 Hz (b) 120 Hz (c) 200 kHz (d) 25 kHz (e) 100 MHz
13. (a) 8.5 m (b) 1.55 m (c) 0.85 m (d) 0.41 m
15. 333 ps **17.** 75.2% **19.** 0 V **21.** 25 kHz
23. 800 Hz **25.** 1 kHz

CHAPTER 18

1. 25 turns **3.** 80 V **5.** 1000 V **7.** 5 A
9. $R_L = 2.6$ kΩ **11.** $N_P:N_S = 3.16:1$

CHAPTER 19

1. 400 μF **3.** 160 μF **5.** $Q = 400$ μC, $V = 20$ V
7. 0.00167 μF **9.** 50 V **11.** 0.034 J

CHAPTER 20

1. 0.3 H **3.** 10 mH **5.** 6.25 mH
7. 504 μH **9.** 26 mH **11.** 0.8
13. 11 mH **15.** $L_T = 70$ μH, $k = 0.3$
17. 500×10^{-6} J **19.** 1 H

CHAPTER 21

1. 15 μs **3.** 860 μA **5.** 0.025 μF
7. (a) 100 V (b) 60.7 V (c) 36.8 V (d) 8.2 V (e) 674 mV
9. 98 μA **11.** 333 μA **13.** 55.3 μA **15.** 154 μA
17. (a) 0 A (b) 33.86 mA (c) 37.6 mA (d) 39 mA (e) 39.73 mA

CHAPTER 22

1. (a) 3.18 MΩ (b) 318.3 kΩ (c) 31.83 kΩ (d) 6.37 kΩ
  (e) 3.18 kΩ
3. (a) 33.86 Ω (b) 3.386 Ω (c) 0.337 Ω
  (d) 0.068 Ω (e) 0.0337 Ω
5. ∞ Ω
7. $X_C = 500$ kΩ, $C = 636.6$ pF
9. (a) 1.26 kΩ (b) 18.85 kΩ (c) 56.5 kΩ (d) 157 kΩ
  (e) 314.2 kΩ
11. (a) 0.042 Ω (b) 0.42 Ω (c) 103.7 Ω (d) 20.73 Ω (e) 207.3 Ω
13. $X_L = 3.11$ kΩ, $V_S = 15.6$ V
15. $X_L = 600$ Ω, $L = 1.91$ H
17. 750 Ω **19.** 114.6

CHAPTER 23

1. $4.47\ \Omega\angle-63.4°$ **3.** $10\ \Omega\angle-53.1°$ **5.** $24.7\ \Omega\angle-14°$

7. $107.7\ \Omega\angle21.8°$ **9.** $5.67\ \Omega\angle45°$ **11.** $8\ \Omega+j6\ \Omega$

13. $5\ \Omega-j5\ \Omega$ **15.** $9.4\ \Omega+j22\ \Omega$

17. $61.8\ \Omega-j16.6\ \Omega$ **19.** $162.4\ \Omega-j136.2\ \Omega$

CHAPTER 24

1. $Z=5\ \text{k}\Omega\angle-53.1°$, $\theta=-53.1°$, $I_T=2\ \text{mA}\angle53.1°$,
$V_R=6\ \text{V}\angle53.1°$, $V_C=8\ \text{V}\angle-36.9°$

3. $\theta=53.1°$, $I_T=1.2\ \text{mA}\angle-53.1°$,
$V_R=7.2\ \text{V}\angle-53.1°$, $V_C=9.6\ \text{V}\angle36.9°$

5. $Z=5.66\ \text{k}\Omega\angle45°$, $\theta=45°$,
$I_T=4.24\ \text{mA}\angle-45°$, $V_R=9\ \text{V}\angle-45°$,
$V_C=9\ \text{V}\angle-135°$, $V_L=18\ \text{V}\angle45°$

7. $X_C=4\ \text{k}\Omega$, $Z=5\ \text{k}\Omega\angle53.1°$,
$\theta=53.1°$, $I_T=4\ \text{mA}\angle-53.1°$,
$V_R=12\ \text{V}\angle-53.1°$, $V_C=16\ \text{V}\angle-143.5°$,
$V_L=32\ \text{V}\angle36.9°$

9. $Z=26.4\ \text{k}\Omega\angle-29.5°$, $\theta=-29.5°$,
$I_T=568\ \mu\text{A}\angle29.5°$, $V_R=13\ \text{V}\angle29.5°$,
$V_C=9.7\ \text{V}\angle-60.5°$, $V_L=2.3\ \text{V}\angle119.5°$

11. $I_R=3.33\ \text{mA}\angle0°$, $I_L=2.5\ \text{mA}\angle-90°$,
$I_T=4.16\ \text{mA}\angle-36.9°$, $\theta=-36.9°$,
$Z=2.4\ \text{k}\Omega\angle36.9°$

13. $I_R=5\ \text{mA}\angle0°$, $I_L=5\ \text{mA}\angle-90°$,
$I_T=7.1\ \text{mA}\angle-45°$, $\theta=-45°$,
$Z=7.1\ \text{k}\Omega\angle45°$

15. $I_R=417\ \mu\text{A}\angle0°$, $I_C=1.67\ \text{mA}\angle90°$,
$I_L=5\ \text{mA}\angle-90°$, $I_T=3.36\ \text{mA}\angle-83°$,
$\theta=-83°$, $Z=3\ \text{k}\Omega\angle83°$

17. $I_R=2.65\ \text{mA}\angle0°$, $I_C=1.5\ \text{mA}\angle90°$,
$I_L=0.9\ \text{mA}\angle-90°$, $I_T=2.72\ \text{mA}\angle12.8°$,
$\theta=12.8°$, $Z=6.62\ \text{k}\Omega\angle-12.8°$

19. $I_R=2\ \text{mA}\angle0°$, $I_C=0.33\ \text{mA}\angle90°$,
$I_L=0.25\ \text{mA}\angle-90°$, $I_T=2\ \text{mA}\angle2.3°$,
$\theta=2.3°$, $Z=500\ \Omega\angle-2.3°$

21. $Z_1=36.1\ \Omega\angle-33.7°$, $Z_2=125\ \Omega\angle36.9°$,
$Z_3=63.2\ \Omega\angle71.6°$, $I_1=1.4\ \text{A}\angle33.7°$,
$I_2=0.4\ \text{A}\angle-36.9°$, $I_3=0.79\ \text{A}\angle-71.6°$,
$I_T=1.74\ \text{A}\angle-7.2°$, $\theta=-7.2°$,
$Z_{EQ}=28.7\ \Omega\angle7.2°$

23. $Z_1=20.6\ \Omega\angle14°$, $Z_2=51\ \Omega\angle-78.7°$,
$Z_3=150\ \Omega\angle90°$, $I_1=1.94\ \text{A}\angle-14°$,
$I_2=0.78\ \text{A}\angle78.7°$, $I_3=0.27\ \text{A}\angle-90°$,
$I_T=2\ \text{A}\angle1°$, $\theta=1°$,
$Z_{EQ}=20\ \Omega\angle-1°$

25. (11) $AP=41.6\ \text{mVA}$, $TP=33.3\ \text{mW}$,
$P_X=25\ \text{mVAR}$, $PF=0.8$;
(12) $AP=40.8\ \text{mVA}$, $TP=28.8\ \text{mW}$,
$P_X=28.8\ \text{mVAR}$, $PF=0.71$;
(13) $AP=0.35\ \text{VA}$, $TP=0.25\ \text{W}$,
$P_X=0.25\ \text{VAR}$, $PF=0.71$;
(14) $AP=19.1\ \text{mVA}$, $TP=18.8\ \text{mW}$,
$P_X=3.4\ \text{mVAR}$, $PF=0.98$;
(15) $AP=33.6\ \text{mVA}$, $TP=4.23\ \text{mW}$,
$P_X=33.3\ \text{mVAR}$, $PF=0.13$

CHAPTER 25

1. $f_1=15.92\ \text{kHz}$, $f_2=2.29\ \text{kHz}$,
$f_3=3.18\ \text{MHz}$, $f_4=142.4\ \text{Hz}$

3. $L_1=1.8\ \mu\text{H}$, $L_2=40.5\ \mu\text{H}$,
$L_3=13.53\ \mu\text{H}$

5. $f_O=5.03\ \text{MHz}$, $X_L=3.16\ \text{k}\Omega$,
$Q=63.2$, $BW=79.6\ \text{kHz}$,
$f_1=4.99\ \text{MHz}$, $f_2=5.07\ \text{MHz}$

7. $f_O=5.03\ \text{kHz}$, $X_L=6.32\ \text{k}\Omega$,
$Q=42.1$, $BW=119.5\ \text{Hz}$,
$f_1=4.97\ \text{kHz}$, $f_2=5.09\ \text{kHz}$

9. $f_O=33.93\ \text{MHz}$, $X_L=2.13\ \text{k}\Omega$,
$Q=21.3$, $BW=1.59\ \text{MHz}$,
$f_1=33.13\ \text{MHz}$, $f_2=34.73\ \text{MHz}$

11. $f_O=2\ \text{MHz}$, $X_L=1.26\ \text{k}\Omega$,
$Q=100$, $BW=20\ \text{kHz}$,
$f_1=1.99\ \text{MHz}$, $f_2=2.01\ \text{MHz}$

13. $Q=40$, $BW=1.25\ \text{kHz}$,
$f_1=499.375\ \text{kHz}$, $f_2=500.625\ \text{kHz}$

15. $C_1=654.2\ \text{pF}$, $C_2=434.3\ \text{pF}$

CHAPTER 26

1. 3   **3.** 2   **5.** 0.796   **7.** 0.707

9. 0.5   **11.** 30 dB   **13.** -13 dB   **15.** -7.96 dB

17. $6.31\ \mu\text{W}$ **19.** -33 dB   **21.** 0 dB   **23.** 7.91 V

25. 1.67 V   **27.** 1 V   **29.** 28 dB   **31.** 1 kHz

33. $\theta_1=-26.6°$, $\theta_2=-45°$, $\theta_3=-78.7°$, $\theta_4=-84.3°$

35. 1 kHz   **37.** $\theta_1=63.4°$, $\theta_2=45°$, $\theta_3=11.3°$, $\theta_4=5.7°$

39. (a) High-pass (b) band-stop (c) band-stop
(d) low-pass (e) band-pass (f) band-pass

INDEX